Mineral Resource Estimation

Mario E. Rossi • Clayton V. Deutsch

Mineral Resource Estimation

Mario E. Rossi
Geosystems International, Inc.
Delray Beach
Florida
USA

Clayton V. Deutsch
University of Alberta
Edmonton
Alberta
Canada

Additional material can be downloaded from http://extras.springer.com

ISBN 978-94-017-7650-9 ISBN 978-1-4020-5717-5 (eBook)
DOI 10.1007/978-1-4020-5717-5
Springer Dordrecht Heidelberg New York London

Printed on acid-free paper

Springer is part of Springer Science+Business Media (www.springer.com)

Preface

This book is about resource modeling. It explains important issues; it describes geological and statistical tools used in resource modeling; and presents case studies for illustration. The main intent is to avoid strict theoretical presentations, and focus on practical adaptations that result in good resource estimation practice. An understanding of the intrinsic limitations and weaknesses of the techniques and resulting models used is critical to success in resource modeling.

This book fills a knowledge gap in the mining industry. There are many books available that describe geostatistical methods for resource estimation, but they tend to emphasize theory, and provide few or no guidelines for the necessary adaptations in practical applications. Those books generally dwell in geostatistical theory with more detail than this one does. On the other hand, there are a few "practical" resource modeling books, but they are either not comprehensive enough or do not contain enough theory to support or justify the methodology and procedures recommended. We attempt to balance both aspects.

Our target audience is geologists and engineers; either students in advanced undergraduate or graduate studies, or professionals just starting out in resource estimation. These are the professionals that are most in need of learning from other people's experience.

We have attempted to collect and reflect in this book good resource modeling practices; not only from our own experiences, but also from those that we have worked with through the years. In the global mining industry, this includes mentors and colleagues from different parts of the world. It is thus also a reflection of working relationships and friendships forged through the years.

We have a debt of gratitude with many colleagues, too many to be mentioned here. Specially, those with whom we have shared many hours over light tables, computer screens, and in healthy discussions about modeling and other things. But in particular we would like to acknowledge BHP Billiton, and in particular Rick Preece, Global Practice Leader, Geology and Ore Reserves, BHP Billiton Base Metals Operations, for facilitating BHP's financial support of this project, and constant encouragement.

Contents

Introduction

1

Abstract

The estimation of mineral resources is an important task for geoscientists and mining engineers. The approaches to this challenge have evolved over the last 40 years. This book presents an overview of established current practice. The book is intended for advanced undergraduate students or professionals just starting out in resource estimation.

1.1 Objectives and Approach

Our objective is to explain important issues, describe commonly used geological and statistical tools for resource modeling, present case studies that illustrate important concepts, and summarize good resource estimation practice. Wherever possible a common thread will be maintained through the sections including details of theory and references to appendices and other authors, relevant examples, software tools available, required documentation trail for better practice, extensions to handling multiple variables, modeling of other less common variables such as metallurgical properties, and limitations and weaknesses of the assumptions and models used.

There are a wide variety of minerals of interest including industrial minerals such as gravel and potash, base metals such as copper and nickel, and precious metals such as gold and platinum. There are other spatially distributed geological variables such as coal, diamonds, and variables used to characterize petroleum reservoirs. Often, the constituent of interest has variable concentration within the subsurface. A *resource* is the tonnage and grade of the subsurface material of interest. The resource is in-situ and may not be economic to extract. A *reserve* is that fraction of a resource that is demonstrated to be technically and economically recoverable. Estimation of resources and reserves requires the construction of long-term models (life of asset) for the entire deposit, which are updated every 1–3 years of operation. Medium-term models may be built for planning one to 6 months into the future. Short-term models are built for weekly or day-to-day decisions related to grade control or detailed planning.

Constructing numerical models for long, medium or short-term resource assessment includes four major areas of work:

1. Data collection and management;
2. Geologic interpretation and modeling;
3. Grades assignment; and,
4. Assessing and managing geologic and grade uncertainty.

Data collection and management involves a large number of steps and issues. There are books on drilling and sampling theory, such as Peters (1978) and Gy (1982). The richness and complexity of these subjects cannot be covered in detail; nevertheless, it is important that the resource estimator consider subjects that affect the quality of the ultimate estimates. Some background information is provided.

Geologic interpretation and modeling requires that site specific geologic concepts and models are integrated with actual data to construct a three dimensional model of geological domains. This geologic model is a representation of those variables that control the mineralization the most and forms the basis for all subsequent estimation. Often, the geological model is the most important factor in the estimation of mineralized tonnage.

The concentrations of different elements or minerals (grades) are assigned within geological domains. The grades within the different domains may be reasonably homogeneous; however, there is always some variability within the domains. The grades are predicted at a scale relevant for the anticipated mining method. The recoverable resources are calculated considering a set of economic and technical criteria. There are a wide variety of methods available and many implementation aspects must be considered. The chosen method will

M. E. Rossi, C. V. Deutsch, *Mineral Resource Estimation,* DOI 10.1007/978-1-4020-5717-5_1,

depend on the study objectives, the available data and the professional time available to complete the study.

Resource estimates should be complemented with a measure of uncertainty. All numerical models have multiple significant sources of uncertainty including the data, the geologic interpretation, and the grade modeling. A statement quantifying the uncertainty in the predicted variables is required for good and best practices.

These four main subjects are covered in 14 chapters. Each chapter concludes with an exercise that summarizes the key points and helps interested readers test their understanding of the material presented. No solutions to the exercises are provided.

1.2 Scope of Resource Modeling

The collection, gathering, and initial analysis of data are the first steps in mineral resource modeling. Sufficient quality controls and safeguards are required to achieve an adequate degree of confidence in the data. The overall process of Quality Assurance and Quality Control (QA/QC) should encompass field practices, sampling, assaying, and data management. This is necessary to ensure confidence in the resource model.

The data are subset within different geological domains. These domains may be based on a variety of geological controls such as structure, mineralogy, alteration and lithology. Categorical variable models are constructed to subdivide the data and focus analysis in different regions of the subsurface. Domains are commonly assigned to a gridded block model. The block model must have sufficient resolution to represent the geological variations and provide the required resolution for engineering design. Of course, the number of blocks must not be too large. At the time of writing this book, it is common to use 1 to 30 million blocks. Larger models are possible, but they require more computer resources and managing multiple realizations of many variables becomes time consuming.

Statistical analyses of the available data are required before decisions can be made about geological domains. Mineralization controls interact to control the spatial distribution of grades. Compositing the original data values is common practice. This is done partly to homogenize the support of the data used in estimation, but also to reduce the variability of the dataset. Further statistical analyses are performed to understand and visualize the data distributions and to define the most appropriate form of estimation.

After defining the block model geometry and geological domains, it is necessary to assign grades. The choice of an estimation method and the formulation of plans for grade interpolation are described in later chapters. Special considerations required for simulation are also discussed.

Each step in mineral resource estimation requires assumptions and decisions that should be explicitly stated. Perceived limitations and risk areas should be documented. The process of model validation and reconciliation is iterative. The calibration of a recoverable resource model against production, if available, is particularly important to ensure future predictions are as accurate as possible. Proper and detailed documentation is required for each step. An audit trail must be created during the entire resource estimation process to allow a third party to review the modeling work. Transparency and the ability to allow for peer-reviews are essential components of the work.

1.3 Critical Aspects

The estimation of resources and reserves requires detailed consideration of a number of critical issues. Like a chain, they are linked such that the quality of the overall resource estimate will be equal to the quality of the weakest link; any one of them failing will result in an unacceptable resource estimate. Resource estimators must deal with these issues on a daily basis.

The quality of the mineral resource estimate depends firstly on the available data and the geological complexity of the deposit; however, the resource estimate is also strongly dependent on the overall technical skills and experience of the mine staff, how the problems encountered are solved, the level of attention to detail at every stage, the open disclosure of basic assumptions along with their justifications, and the quality of the documentation for each step.

The emphasis on documenting every aspect of the work is stressed throughout this book because it is the final and, possibly, the most important link in the chain. Justification and documentation of every important decision serves as quality control of the work, because it forces detailed internal reviews. In addition, it also facilitates third-party reviews and audits, which are a common requirement in industry. Some basic issues to be dealt with in resource estimation are briefly discussed next.

1.3.1 Data Assembly and Data Quality

The quality of the resource estimate is directly dependent on the quality of the data gathering and handling procedures. Many different technical issues affect the overall quality of the data. Some important ones are mentioned here.

The concept of data quality is used in a pragmatic way. The concept is that data (samples) from a certain volume will be collected and used to predict tonnages and grades of the elements of interest. Decisions are made based on geological knowledge and statistical analyses applied in conjunction with other technical information. Therefore, the numerical basis for the analyses has to be of good quality to provide for sound decision-making. This is particularly important because a very small fraction of the mineral deposit is sampled.

A second key concept is that the samples should be representative of the volume (or material) being sampled, both in a spatial sense and at the location where the sample is being taken from. *Representative* means that the sampling and analyzing process used to obtain a sample results in a value that is statistically similar to any other that we could have taken from the same volume. Therefore, the sample values are considered to be a fair representation of the true value of the sampled volume of rock. Representation in a spatial sense implies that the samples have been taken in an approximately regular or quasi-regular sampling grid, such that each sample represents a similar volume or area within the orebody of interest. This is often not the case and some correction will be required. If the samples are not representative, then an error will be introduced that will bias the final resource estimate.

In the context of data quality, the technical issues related to sample collection can be divided into those related to field work, and those related to processing of the information. Some of the most important issues in the field include (1) the location of drill holes, trenches, and pits; (2) the type of drill holes used such as open-hole percussion, reverse circulation, or diamond drill holes; (3) the drilling equipment used; (4) the sampling conditions such as the presence of highly fractured rock or groundwater; and (5) sample collection procedures. Core recovery or the sample weight should be recorded. Geologic logging of the geologic characteristics of the samples should be performed. Sample preparation and assaying procedures are critical. The related quality assurance and quality control program is a fundamental element in the process.

Deposit- and mineral-specific sample preparation and assaying protocols must be derived and adhered to throughout the sampling campaign. Heterogeneity tests (Pitard 1993; François-Bongarçon and Gy 2001) are necessary to understand sampling variances and minimize errors.

The construction and maintenance of the sampling database requires a continuous quality control program, including periodic manual and automatic checks. These checks should be performed over all the variables in the database, including grades, geologic codes, collar location and surveys, and density data. Relational databases offer the possibility of easier data handling and improved quality control. But they do not provide quality control by themselves, nor do they replace the need for periodic manual audits.

1.3.2 Geologic Model and Definition of Estimation Domains

Much geologic information is gathered during the investigations performed at different stages of a mining project. The information is used to understand the genesis of the mineral deposit, the distribution of mineralized rock, and to develop exploration criteria for increasing resources.

The level of detail in the geologic description of a deposit steadily increases as the project advances through its different stages. Economic factors are the most important ones affecting the decision of whether or not to proceed with further geologic investigations; therefore, most geologic work is orientated towards finding more mineral resources, and to some extent to more detailed general exploration.

Not all geologic information is relevant to resource estimation. Geologic investigations for resource development should concentrate on defining mineralization controls. Certain geologic details and descriptions are more useful for exploration in that they do not describe a specific mineralization control, but rather provide guidelines for mineral occurrences.

The process of defining estimation domains amounts to modeling the geological variables that represent mineralization controls. The estimation domains are sometimes based on combinations of two or more geologic variables, for which a relationship with grade can be demonstrated. For example, in the case of an epithermal gold deposit, an estimation domain can be defined as a combination of structural, oxidation, and alteration controls. In the case of a diamondiferous kimberlitic pipe, in addition to the geometry of the pipe (lithology), internal waste relics are common, such as granitic xenoliths. The frequency and volume of these within the pipe may condition the definition of estimation domains.

The determination of the estimation domains to use is based on geologic knowledge and should be supported by extensive statistical analysis (exploratory data analysis, or EDA), including variography. The procedure can take a significant amount of time, particularly when all possible combinations of the available geologic variables are studied, but it is typically worth the effort. Estimates are improved when carefully constrained by geological variables.

The definition of estimation domains is referred to as the definition of stationary zones within the deposit. An important part of stationarity is a decision of how to pool information within a specific zone within the deposit, within certain boundaries, or the deposit as a whole. Decisions are based on oxidation zones, lithologies, alterations, or structural boundaries. The stationary domains cannot be too small; otherwise, there are too few data for reliable statistical description and inference. The stationary domains cannot be too big; otherwise, the data could likely be subset into more geologically homogeneous subdivisions.

Defining the estimation domains in resource evaluation is often equivalent to defining the mineralized tonnage available in the deposit. Some units will be mostly mineralized (with the potential of becoming ore), while others will be mostly un-mineralized (almost certainly non-recoverable low-grade resources or waste). The mixing of different types of mineralization should be kept to a minimum to avoid smearing grades across geologic boundaries.

Adequate definition of the estimation domains is an important task for resource evaluation. Mixing of populations within the deposit will generally produce a sub-standard resource estimate that underestimates or overestimates grades and tonnages. It is very rare that any geostatistical technique will compensate for a poor definition of stationarity. A good definition of estimation domains means that only relevant samples are used to estimate each location.

1.3.3 Quantifying Spatial Variability

The grade values observed within a mineral deposit are not independent from each other. Spatial dependency is a consequence of the genesis of the deposit, that is, all of the geological processes that contributed to its formation. The reader is referred to Isaaks and Srivastava (1989) for an accessible discussion on the subject, as well as David (1977), Journel and Huijbregts (1978), and Goovaerts (1997) for more details.

A clear description of the spatial variability (or continuity) of the variables being modeled is desirable. Knowledge of the spatial correlation between different points in the deposit will lead to a better estimation of the mineral grade at an unknown location. The spatial variability is modeled using the variogram and related measures of spatial variability/correlation.

A spatial variability model improves the estimation of each point or block in the deposit. Parameters of the model are important. Attention should be paid to the definition of the nugget effect (the amount of randomness); the number of structures; the behavior of the variogram model near the origin; and the specification of anisotropic features. Although the spatial variability model will change depending on the estimator and available data, it should be compatible with accepted geologic knowledge. For example, the modeled anisotropies should be consistent with the spatial distribution of known geologic controls, and the variances and ranges of the models should be consistent with the overall variability observed in the data.

Geologic variables have some degree of spatial correlation. The challenges often encountered when quantifying the spatial correlation lie with the inadequacy of the data being used, inadequate definition of estimation domains, or use of estimators that are less robust with respect to skewed data. These challenges are discussed in detail in later chapters.

1.3.4 Geologic and Mining Dilution

In-situ and *recoverable* resources must be differentiated. The precise definition of recoverable varies in different parts of the world. In general, the term refers to mineralization that can be recovered and processed by mining. Any resource evaluation, in order for it to become the basis for an economic evaluation, has to be recoverable, and therefore include some dilution and ore loss. After applying constraints derived from the ability to economically mine the deposit, as well as all relevant types of dilution, the resource may become a reserve.

Some resource estimators advocate the estimation of purely geological in-situ resources, that is, an estimate of the resources that are to be found if a snapshot of the deposit at the same scale and level of detail as provided by the drill hole data and other geologic information could be taken. Thus, it would be a description of its true geologic nature, as it occurs at our scale of observation. This point of view assigns to the mining engineer and economic evaluator the task of converting the purely geologic resource into a minable reserve. This is required to realistically describe the economic potential of the deposit. In general, however, the geologist and geostatistician (resource evaluators) are better equipped to incorporate geologic dilution; otherwise, it may go uncharacterized or poorly modeled.

Mining is a large scale industrial operation; selection of large volumes is taking place over short times. Some mixing of waste with ore and ore with waste is inevitable. The failure to understand and properly estimate geologic dilution and lost ore explains most of the failures of resource estimates. Although some degree of error or uncertainty is expected, ignoring or mistreating knowledge of anticipated dilution is an invitation for disaster. An interesting discussion in layman terms about this issue can be found in Noble (1993). In the context of using a block model to estimate resources, the basic types of dilution often encountered can be summarized as:

1. Internal dilution, related to the use of small size composites to estimate large blocks, also called the volume-variance effect. The more mixing of high and low grades within the block, the more important this effect will be, as is common for example with gold mineralization.
2. The geologic (or in-situ) contact dilution, related to the mixtures of different estimation domains within blocks. One reason for grade profile changes is the existence of different geologic and mineralization domains. Mixing of grades will occur when mining near to or at contacts.
3. The operational mining dilution that occurs at the time of mining. The blasting of the rock is an important factor, since material shifts position. The loading operation is also a source of dilution and ore loss since the loader is never able to precisely dig to the exact ore limits.

An understanding of the information effect is also required. The long-term block model is not used for final selection of ore and waste. Rather, a different model is used to select ore from waste that uses much more closely-spaced data available at the time of mining. In an open pit mine the mineral boundaries and the quality are predicted using closely spaced data. The information at the time of resource estimation is quite different than at the time of mining, for which estimates will be much better.

1.3.5 Recoverable Resources: Estimation

The importance of calculating recoverable resources and reserves was recognized early on in geostatistics (Matheron 1962, 1963), but it was M. David's early work (1977) that demonstrated the practical significance of estimating recoverable reserves, while Journel and Huijbregts (1978) provided the theoretical and practical foundations for the most common methods used to estimate at different volumes.

Block model resources estimated from exploration or development drill holes (long-term models) and mine production predictions (short-term models) may show significant discrepancies. The discrepancies are even larger when compared to actual production figures which may or may not be reliable. It is desirable to minimize these discrepancies for evaluation and planning purposes. It has been shown that incorrect accounting for the volume of prediction (the volume-variance effect) is a major contributor to the discrepancies usually encountered.

The resource model contains blocks with dimensions that should relate to the spacing of the data, hopefully determined based on the quantity of information available to predict grades. Block sizes may be larger than the selective mining unit (SMU) of the operation. The smoothing effect of kriging will generally result in a grade distribution that does not match the distribution of grade of the SMUs. In addition, in-pit selection is not perfect. The grade-tonnage predictions based on blast holes may need to be corrected for unplanned dilution and other errors of estimation in the short-term model.

An integrated approach to predicting reserves and mine performance is required for more accurate predictions. Specifically, the volume-variance relationship, the selectivity of the mining operation, planned dilution and ore loss must be accounted for. Additionally, incorporating an allowance for unplanned dilution at the time of mining is reasonable.

The traditional estimation techniques provide limited flexibility to account for these factors. The estimation of recoverable resources is based on limited information about the SMU distribution of grades. There are a number of methods and techniques that help estimate point distributions, but relatively little research has been done to develop robust methods for estimating block distributions. It is a difficult task, since little is known a-priori about the SMU distribution. An important option available is the use of conditional simulation models to resolve the issues related to recoverable resources.

1.3.6 Recoverable Resources: Simulation

The traditional approach to block modeling is to estimate a single value in each block of the model, obtaining the best possible prediction in some statistical sense. This estimation can be done using non-geostatistical methods, or more commonly, some form of kriging. Although there is a need for a single estimate in each block, there are some important shortcomings in attaching only the estimated value to each block.

An alternative approach to resource evaluation is the use of conditional simulation that provides a set of possible values for each block, which represent a measure of uncertainty. The idea is to obtain a number of simulated realizations that reproduce the histogram and the variogram of the original drill hole information. The realizations are built on a fine grid. Reproducing or honoring the histogram means that the realizations will correctly represent the proportion of high and low values, the spatial complexity of the orebody, the connectivity of high and low values, and the overall grade continuity in three dimensions. These characteristics of the mineralization are important aspects that play a significant role in designing, planning, and scheduling a mining operation.

A number of issues have to be adequately resolved for the realizations to be representative of the grades of the deposit. These include, among others, choosing among several simulation techniques available, such as Sequential Gaussian (Isaaks 1990), Sequential Indicator (Alabert 1987), or others. Also, decisions about grid size, conditioning data, search neighborhoods, and treatment of high grade values must be made. It is a similar process to developing a kriging block model. Some discussions about practical implementations can be found in Deutsch and Journel (1997) and Goovaerts (1997), among others.

When a number of these realizations have been created and checked, then, for each node defined in the grid, there will be a corresponding number of different grades available. This set of multiple grades is a model of uncertainty for that node. These simulated points can be re-blocked to any block size desired such as the Selective Mining Unit SMU size of the operation. These results are used further by mining engineers.

Important parameters can be obtained from the distributions of local uncertainty such as the mean, median, and probability of exceeding of exceeding a specified cutoff grade. Therefore, the information provided by a simulation model is significantly more complete than the single estimate provided by an estimated block model. The simulation models can provide recoverable resources for any selectivity by re-blocking the simulated grades to the chosen SMU block size. It is likely that, in due time, simulation models will replace estimated block models, since they not only provide a single estimate, but also a full range of possible values.

1.3.7 Validation and Reconciliation

Checking resource models involves several steps and requires a significant amount of time and effort. There are two basic types of checks to be done: graphical and statistical.

Graphical checks involve 3-D visualization and plotting the estimated values on sections and plans. Every estimated block grade should be explained by the data surrounding it and the modeling parameters and method used. Although these graphical checks can be performed on computer screens, it is often worthwhile to have a hardcopy set of maps because of the level of detail required and the important record-keeping and audit trails. Unfortunately, this practice is disappearing, as some operations do not take the time to produce sets of geological sections and plans views on paper.

Statistical checks are both global (large scale or deposit-wide) and local (block-wise or by smaller volumes, such as monthly production volumes). The checking, validation, and reconciliation procedures should ensure the internal consistency of the model, as well as reproduction of past production if available. Some of the more basic checks are:

- The global average of the model should match the average of the declustered data distribution. This check needs to be performed for each estimation domain.
- The smoothing of the distribution of the block model grades: the comparison with respect to the predicted (SMU) grade distributions should be reasonable. If the predicted SMU and block model grade-tonnage curves are very different, it is likely that the block model has incorporated too much or too little dilution.
- The spatial and statistical relationships between the modeled variables must correspond to the relationships observed in the original data set.
- A resource model should be constructed using an alternative method. The results and differences should be as expected, given the characteristics of each method.
- The estimates should be compared to previous estimates. This should be done cautiously and considering the differences in data quantity and quality, as well as the methodology used for the different resource estimate.
- The estimates should be compared to all available historical production data. Ideally, resource models should predict past production. This provides some indication that the block model may also predict future mining.

Reconciliation against past production should be done based on pre-defined volumes of interest and according to specified error acceptance criteria. Additionally, production can provide an initial indication of the expected uncertainty of the resource model. This expected uncertainty should be expressed in the classical form of *within x% confidence limit p% of the time.*

Production information should be used with great care. Oftentimes, tonnages and grades reported by the processing plant do not adequately represent true mill feed (head) tonnages and grades, that is, the material delivered by the mine. Rather, they may be influenced by plant performance parameters, which will bias the comparisons with the head grades and tonnages reported by the mine. The implication is that reliable head tonnages and grade information are best obtained from direct sampling of the material delivered at the entrance of the plant. In some cases these comparisons may not be possible due to the characteristics of the operation such as extensive stockpiling or lack of reliable mill feed information. Often, only very general statements can be made about the quality of the reconciliation data.

1.3.8 Resource Classification

The purpose of classifying resources is to provide a global confidence assessment to the project's stakeholders including mining partners, stockholders, and financial institutions investing in the project. There are several resource and reserve classification systems used by different government agencies around the world. Most of them share in their main characteristics and objectives.

The assessment of confidence is critical for project development since sufficient resources and reserves must be known with enough confidence to be considered assets. For operating mines, continued confidence in future long-term production is also important in providing shareholder value and supporting long-term planning.

The terminology used in most guidelines for classification is purposefully vague. They must be applicable to many different types of deposits, locations and mining methods. The guidelines do not prescribe specific methodology for quantifying uncertainty or risk. Rather, there is increased reliance on the judgment of the resource estimator, formalized through the concept of a competent or qualified person. A common basis for comparison is therefore difficult to achieve, since the wording may have different meaning under different circumstances, and depends on the individuals involved. A possible solution is to attempt to describe confidence in traditional statistical terms, and as a function of production units. There is an industry trend towards using a statistical description of uncertainty to supplement traditional classification criteria.

The confidence assessment required by the shareholders of a mining project is generally global, and mostly concerned with long-term performance. This is different from the shorter-term mining risk assessment that engineers need in the day-to-day operation of the mine. Unfortunately, a global confidence assessment is frequently also used as a local measure of uncertainty, which often leads to unreasonable expectations in the resource model. Current practice for resource classification includes different methods that have conceptual similarities. Some common ones are:

- Using the number of drill holes and samples near each block is geometric in nature and easy to explain, although it frequently tends to be simplistic in its implementation.
- The kriging variance provides an index of data configuration (Chap. 8), that is, a measure of how well each block in the model is informed at the time of estimation.
- Using different search radii to estimate blocks in a stepwise process, while keeping track of when the blocks get an estimated value. The more information is used to obtain an estimate, the more certain it will be.
- Deciding according to geologic criteria what drill hole grid spacing is required for the resource to belong to a category (measured, indicated, or inferred), and then searching throughout the deposit for that nominal grid spacing, thus classifying the different areas of the deposit.

Purely geometric criteria could be supplemented with conventional statistical criteria, that is, defining the expected grade and a corresponding range of possible grades around it. For example, measured resources may be defined as those predicted to be known ±15 %, 90 % of the time for a volume equivalent to 3 months production. The model (numerical or subjective) used to come up with such a statement is most important to the effectiveness of the classification scheme.

There are shortcomings and pitfalls in the practice of resource classification. Many of these can be resolved with a defendable model of uncertainty based on geostatistical simulation. Inevitably, the process of classifying resources depends on the circumstances and conditions of the mining project being assessed in addition to purely geologic conditions and technical issues. Nevertheless, in all cases, the classification must be defendable by the professional that signs off on the resource model.

1.3.9 Optimal Drill Hole Spacing

Drill hole spacing should be optimal for a given cost-benefit analysis, which is dependent on the project development stage. New drill holes must reduce the uncertainty of the resources to a tolerable, pre-defined level, as required for project advancement.

A cost-benefit analysis of potential new drill holes requires assessing the benefit of decreasing the uncertainty of the resource model by a given amount. This amounts to quantifying the value of new information. If the consequences of errors in the resource estimates can be defined and quantified, then it is feasible to use simulated realizations to determine the economic consequences of uncertainty. This can be further refined by applying existing mine plans to the simulation models, such that, for a specific mine plan, an evaluation of the impact of new drilling on recovered reserves can be made.

In practice, this type of analysis is based on production volumes, such as metal sold in a month. If the parameters that describe metallurgical plant performance are known, then the uncertainty of the tonnages and grades fed to the mill can be directly linked to the risk of not achieving the expected production plan.

The typical question asked by the project development manager is "how many drill holes do I need?" The answer to this question requires a definition of the objectives of the new drilling in terms of uncertainty. Then, the applicable optimality criteria can be developed and the value of new drilling can be assessed. This could be expressed in dollar values, in terms of uncertainty and risk reduction, or in terms of reduction of cash flow and net present value (NPV) risk.

1.3.10 Medium- and Short-term Models

Medium- and short-term models are auxiliary models used to improve the local estimation of the long-term resources model. These are reserve models that are used in an operating mine for production purposes. Medium- and short-term models are used to improve the estimation of relatively small volumes of the deposit. This is useful because mine operations plan on smaller, shorter-term volumes. The definition of what is long-, medium-, and short-term varies from one operation to another; however, common use of the terms suggest that long-term refers to production periods of a year or longer, while medium-term refers to three to 6 months production, and short-term implying 1 month production or less. The periods chosen will be related to the budget and forecast cycles of the operation.

At most medium to large mining operations there is a yearly budget that updates the material movement and corresponding expected cash flows of the original long-term mine plan. It provides a cash flow prediction for the following year. Additionally, this budget is itself updated by a short-term forecast, usually done on a semi-annual, quarterly, or monthly basis, depending on the characteristics of the operation.

The update of the existing long-term model is accomplished by incorporating infill drilling and production information. Since this work is to be performed within a production environment, the procedures and methods used in updating the resource model are constrained by time and human resources. The definition of the most appropriate and practical methodology to update the geological and grade models can become a significant challenge.

1.3.11 Grade Control

Grade control is an important task performed at the mine on a daily basis. It is a basic, economic decision that selects the destination of each parcel of material mined. Mistakes at this stage are costly, irreversible, and can be measured in terms of cash flow losses and increased operational costs.

Grade control models are based on a large number of samples. In underground mines, production data is usually a series of tightly drilled holes, channel samples, or short holes to test production stopes. In an open pit environment, blast holes samples are obtained on closely spaced grids, according to blasting requirements. Less frequently, grade control drilling is performed separate from blast hole drilling, for example using dedicated reverse circulation (RC) drilling. In some geologic settings, surface tranches and channel samples are used as well.

Production samples are used to select ore from waste, and are affected by several sampling issues. Often, blast hole samples are not as reliable as samples obtained from exploration or RC drill holes. This is explained by a combination of drilling and field sampling methods. Sometimes, the large quantity of samples available will tend to minimize the impact of the error of a single blast hole sample.

Geologic variables are mapped in the pit or stopes, but are not always used in production control. Procedures for extracting some benefit from the local geology mapped should be implemented. The goal is to find practical ways of mapping and quickly processing geological information. The typical turnaround time for a grade control model in an open pit is 24–48 h.

Conventional grade control methods include defining grade outlines and using inverse distance, polygonal estimation, or more commonly kriging of blast hole grades. These methods do not account for the uncertainty in prediction. Alternatively, simulation of multiple realizations provides the basis for different optimization algorithms, such as the minimum-loss/maximum profit method.

In general, improvements from the simulation-based methods are evident in more erratic grade distributions and in more marginal mixed ore-type zones. More complicated grade control scenarios, such as those including multiple processing options and stockpiling, will also lend themselves to optimization through simulation based methods.

1.4 Historical Perspective

Hand-calculated sectional estimates continue to have a place in resource and reserve estimation. They have the advantages of directly accounting for expert geological interpretation and providing a first order approximation; however, they also tend to be optimistic with respect to continuity of the mineralization and the grade that can be achieved. Inverse distance and nearest-neighbor methods became popular in the early days of computer-aided mapping. The computer was used to mimic what was done by hand calculations, but hopefully faster. The implementation aspects of these techniques evolved as more sophisticated computer tools became available.

Mineral resource modeling evolved further with advances in drilling and assaying techniques, and with greater awareness of the possible pitfalls related to sample preparation and analysis. Methods used for geologic interpretation and modeling also evolved, mostly through the section-by-section interpretation and into three-dimensional modeling (wireframes and solids modeling for visualization). The occasional use of three-dimensional hand-made models was made common with the availability of computers.

Grade estimation techniques have evolved through the years, beginning with early geostatistics (Sichel 1952; Krige 1951; Matheron 1962, 1963) that attempt to predict single values into blocks. Advanced versions of these techniques are pervading industry practice and are the most commonly used methods.

The estimation of probability functions developed next, although using the same basic linear regression tools. Assumptions about statistical properties and variable transformations led to the development of probabilistic estimation of a distribution of possible values for any given block.

In more recent years the use of simulation for modeling uncertainty has become important. Geological processes have important patterns and structure, but also have uncertainty due to the chaotic nature of the processes. Characterizing the natural heterogeneity and the uncertainty that results from incomplete sampling is an important goal of mineral resource estimation.

References

Alabert FG (1987) Stochastic imaging of spatial distributions using hard and soft information. MSc Thesis, Stanford University, p 197

David M (1977) Geostatistical ore reserve estimation. Elsevier, Amsterdam

Deutsch CV, Journel AG (1997) GSLIB: geostatistical software library and user's guide, 2nd edn. Oxford University Press, New York, p 369

François-Bongarçon D, Gy P (2001) The most common error in applying 'Gy's Formula' in the theory of mineral sampling, and the history of the liberation factor. In: Edwards AC (ed) Mineral resource and ore reserve estimation—the AusIMM guide to good practice. The Australasian Institute of Mining and Metallurgy, Melbourne, p 67–72

Goovaerts P (1997) Geostatistics for natural resources evaluation. Oxford University Press, New York, p 483

Gy P (1982) Sampling of particulate materials, theory and practice, 2nd edn. Elsevier, Amsterdam

Isaaks EH (1990) The application of Monte Carlo methods to the analysis of spatially correlated data. PhD Thesis, Stanford University, p 213

Isaaks EH, Srivastava RM (1989) An introduction to applied geostatistics. Oxford University Press, New York, p 561

Journel AG, Huijbregts ChJ (1978) Mining geostatistics. Academic Press, New York

Krige DG (1951) A statistical approach to some basic mine valuation problems on the Witwatersrand. J Chem Metall Min Soc South Africa 52:119–139

Matheron G (1962, 1963) Traité de Géostatistique Appliquée, Tome I; Tome II: Le Kriegeage. I: Mémoires du Bureau de Recherches Géologiques et Minières, No. 14 (1962), Editions Technip, Paris; II: Mémoires du Bureau de Recherches Géologiques et Minières, No. 24 (1963), Editions B.R.G.M., Paris

Noble AC (1993) Geologic resources vs. ore reserves, Mining Eng February issue, pp 173–176

Peters WC (1978) Exploration and mining geology, 2nd edn. Wiley, New York

Pitard F (1993) Pierre Gy's sampling theory and sampling practice, 2nd edn. CRC Press, Boca Raton

Sichel HS (1952) New Methods in the statistical evaluation of mine sampling data. Trans Inst Min Metall Lond 61:261

Statistical Tools and Concepts

2

Abstract

Mineral resource estimation requires extensive use of statistics. In our context, *statistics* are mathematical methods for collecting, organizing, and interpreting data, as well as drawing conclusions and making reasonable decisions based on such analysis. This chapter presents essential concepts and tools required throughout the book.

2.1 Basic Concepts

A conventional presentation of statistics includes the notion of a *population* that is the virtually infinite collection of values that make up the mineral deposit. A *sample* is a representative subset selected from the population. A good sample must reflect the essential features of the population from which it is drawn. A r*andom sample* is a sample where each member of a population had an equal chance of being included in the sample. The *sample space* is the set of all possible outcomes of a chance experiment, for example a drilling campaign. The *event of a sample space* is a group of outcomes of the sample space whose members have some common characteristic. *Statistically independent events* are such that the occurrence of one event does not depend on the occurrence of other events. Sampling mineral deposits rarely fits nicely in the framework of representative samples from a statistical population; nevertheless, many concepts and tools from conventional statistics are used routinely.

Inductive statistics or *statistical inference* is attempted if the sample is considered representative. In this case, conclusions about the population can often be inferred. Since such inference cannot be absolutely certain, the language of probability is used for stating conclusions. *Descriptive statistics* is a phase of statistics that describes or analyses a given sample without inference about the population. Although our goal in mineral resource estimation is almost always inference, we use many descriptive statistics for viewing, understanding, and evaluating data.

An essential concept in statistics is *stationarity*, that is, our choice of data to pool together for common analysis. Chapter 6 describes stationarity more formally, but the concept is that data must be grouped together before any statistical calculations are attempted. Ideally, a decision of how to group the data can be made on the basis of clear geological controls, as discussed in Chap. 4. Some of the statistical tools presented in this chapter are useful to help make a choice of stationarity, but most assume that the decision has already been made and the data have been assembled into reasonable groups.

In most cases we consider continuous variables that are mass or volume fractions, that can take any value between a minimum (0%) and maximum (100%). We sometimes consider categorical or discrete variables that can take specific values from a closed set. A typical categorical variable would be lithology or mineralization type.

Statistical tools are used for several reasons, including (1) an improved understanding of the data and the mineral deposit, (2) to ensure data quality, (3) to condense information, and (4) to make inferences and predictions. In general, we are not interested in the statistics of the samples. Our goal is to go beyond the limited sample to predict the underlying population. Additionally, creative visualization of data is an important component of mineral resource estimation, partly because of its usefulness as a tool to understand data, but also to help validate spatially distributed models.

There are many good references on basic statistics. One accessible reference is Lapin (1983). This book uses a few notation conventions. Lowercase letters (x, y, z,...) denote actual values such as a measured value or a specified threshold. Uppercase letters (X, Y, Z,...) denote a random variable (RV) that is unknown. We characterize the uncertainty in a random variable with a probability distribution. A random

M. E. Rossi, C. V. Deutsch, *Mineral Resource Estimation,* DOI 10.1007/978-1-4020-5717-5_2,

variable could be the grade at an unsampled location denoted $Z(\mathbf{u})$ where **u** represents a location coordinates vector. A set of random variables is called a random function (RF). The set of grades over a stationary geologic population $\mathcal{A}$ is a random function $\{Z(\mathbf{u}), \mathbf{u} \in \mathcal{A}\}$.

2.2 Probability Distributions

Probabilities are closely associated to proportions. A probability of 0.8 or 80% assigned to an event means that the proportion of times it will occur, in similar circumstances, is 0.8 or 8/10 or 80%. The *similar circumstances* relates to our decision of stationarity. In some cases we calculate the probabilities directly through proportions. For example, the probability for a mineral grade within a particular geologic unit to be less than a particular threshold could be calculated by counting the number of samples below the threshold and dividing by the total number of data.

There are many cases, however, when probabilities cannot be calculated from proportions. This is particularly true for conditional probabilities, that is, probability values given certain a set of data events. Consider the probability that a mineral grade be less than a particular threshold given one measurement 50 m away that is twice the threshold and another measurement 75 m away that is just below the threshold. In such cases, we do not have multiple replications to calculate an experimental proportion. We must rely on probabilistic models and well established probability laws.

Probability distributions are characterized as parametric or non-parametric. A parametric distribution model has a closed analytical expression for the probability, and is completely determined by a finite number of parameters, as for example the Gaussian distribution model with parameters mean (m) and standard deviation (s) that control the center and spread of the distribution, respectively.

It is common to consider probability distributions that relate to one continuous or categorical variable at a time. Such distributions are called univariate distributions. Two examples: (1) the probability for a continuous variable to be less than a particular threshold, or (2) the probability for a particular lithology to prevail at a certain location. When we consider probability distributions of more than one variable at a time, then we call them multivariate distributions. The distribution of two variables is a bivariate distribution. For example, the probability of one grade being less than one threshold and a second grade being less than another threshold is a bivariate probability.

There are a large number of references for probability and basic statistics. Some general statistical ones and also some related to spatial data include Borradaile (2003); Davis (1986); Koch and Link (1986); Ripley (1987); and Rohatgi and Ehsanes Saleh (2000).

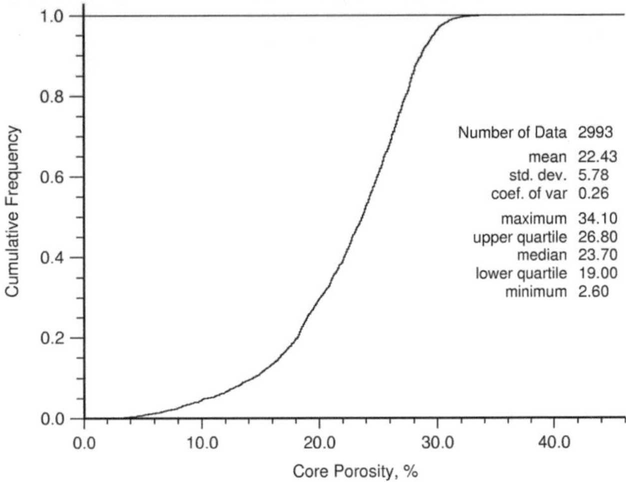

Fig. 2.1 Cumulative distribution of 2,993 data values. The cumulative frequency or probability is the probability to be less than the threshold value

2.2.1 Univariate Distributions

The cumulative distribution function (CDF) is the universal way to express a state of incomplete knowledge for a continuous variable. Consider an RV denoted by *Z*. The CDF *F(z)* is defined as:

$$F(z) = Prob\{Z \le z\} \quad \in [0,1]$$

The lowercase *z* denotes a threshold. $Prob\{\cdot\}$ denotes a probability or proportion. An example CDF is shown on Fig. 2.1; the z-variable is between 2 and 35 and is most probably between 20 and 30.

A cumulative histogram is an experimental CDF based on the data. It is useful to see all of the data values on one plot and sometimes can be used to isolate statistical populations. Cumulative histograms do not depend on a bin width, and can be created at the resolution of the data.

An important challenge is to determine how representative each sample is of the actual mineralization. This issue is discussed in more detail in Chap. 5. It is also important to determine whether the distribution of all samples adequately represents the actual grade distribution in the deposit, or whether certain weighting should be applied.

The interval probability of *Z* occurring in an interval from *a* to *b* (where $b > a$) is the difference in the CDF values evaluated at values *b* and *a*:

$$Prob\{Z \in [a,b]\} = F(b) - F(a)$$

The probability density function (PDF) is the derivative of the CDF, if it is differentiable. Applying the fundamental theorem of calculus, the CDF can be obtained by integrating the PDF:

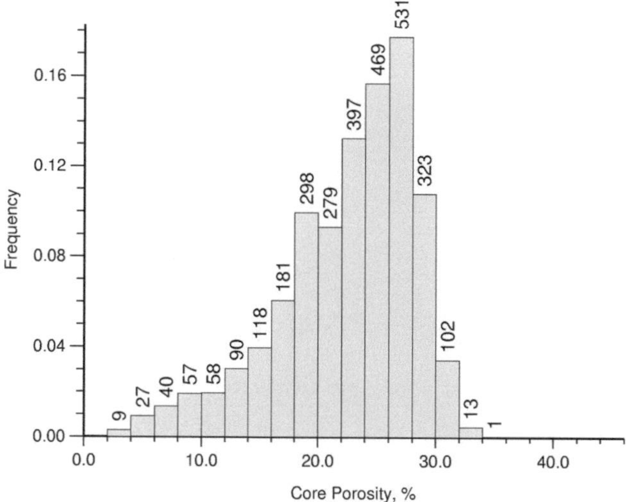

Fig. 2.2 Histogram of 2,993 data values. The common representation of the histogram has constant bin widths; the number of data in each bin is labeled on this histogram

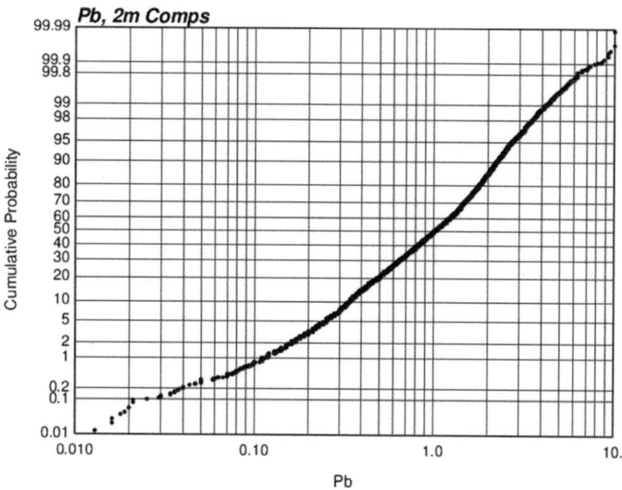

Fig. 2.3 An example of a probability plot. Data is lead concentration, on 2 m composites, on a logarithmic scale

$$f(z) = F'(z) = \lim_{dz \to 0} \frac{F(z+dz) - F(z)}{dz}$$

$$F(z) = \int_{-\infty}^{z} f(z)dz$$

The most basic statistical tool used in the analysis of data is the histogram, see Fig. 2.2. Three decisions must be made: (1) arithmetic or logarithmic scaling—arithmetic is appropriate because grades average arithmetically, but logarithmic scaling more clearly reveals features of highly skewed data distributions; (2) the range of data values to show—the minimum is often zero and the maximum is near the maximum in the data; and (3) the number of bins to show on the histogram, which depends on the number of data. The number of bins must be reduced with sparse data and it can be increased when there are more data. The important tradeoff is reduced noise (less bins) while better showing features (more bins).

The mean or average value is sensitive to extreme values (or outliers), while the median is sensitive to gaps or missing data in the middle of a distribution. The distribution can be located and characterized by selected quantiles. The spread is measured by the variance or standard deviation. The coefficient of variation (CV) is the standard deviation divided by the mean; it is a standardized, unit-less measure of variability, and can be used to compare very different types of distributions. When the CV is high, say greater than 2.5, the distribution must be combining high and low values together and most professionals would investigate whether the pool of data could be subset based on some clear geological criteria.

Sample histograms tend to be erratic with few data. Saw-tooth-like fluctuations are usually not representative of the underlying population and they disappear as the sample size increases. There are techniques available for smoothing the distribution, which not only removes such fluctuations, but also allows increasing the class resolution and extending the distributions beyond the sample minimum and maximum values. Smoothing is only a consideration when the original set of data is small, and artifacts in the histogram have been observed or are suspected. In practice, sufficient data are pooled to permit reliable histogram determination from the available data.

The graph of a CDF is also called a probability plot. This is a plot of the cumulative probability (on the Y axis) to be less than the data value (on the X axis). A cumulative probability plot is useful because all of the data values are shown on one plot. A common application of this plot is to look at changes in slope and interpret them as different statistical populations. This interpretation should be supported by the physics or geology of the variable being observed. It is common on a probability plot to distort the probability axis such that the CDF of normally distributed data would fall on a straight line. The extreme probabilities are exaggerated.

Probability plots can also be used to check distribution models: (1) a straight line on arithmetic scale suggests a normal distribution, and (2) a straight line on logarithmic scale suggests a lognormal distribution. The practical importance of this depends on whether the predictive methods to be applied are parametric (Fig. 2.3).

There are two common univariate distributions that are discussed in greater detail: the normal or Gaussian and the lognormal distributions. The normal distribution was first introduced by de Moivre in an article in 1733 (reprinted in the second edition of his *The Doctrine of Chances*, 1738) in the context of approximating certain binomial distributions for large *n*. His result was extended by Laplace in his book

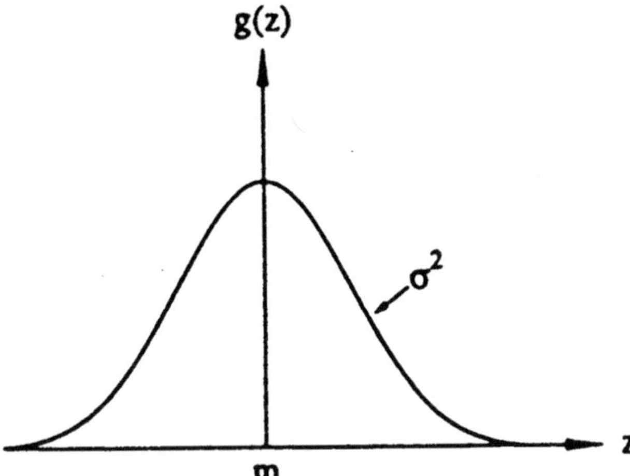

Fig. 2.4 A sketch of a normal or Gaussian distribution

Analytical Theory of Probabilities (1812), and is now called the Theorem of de Moivre-Laplace. Laplace used the normal distribution in the analysis of errors of experiments. The important method of least squares optimization was introduced by Legendre in 1806. Gauss, who claimed to have used the method since 1794, justified it rigorously in 1809 by assuming a normal distribution of the errors.

The Gaussian distribution is fully characterized by its two parameters, the mean and the variance. The standard normal PDF has a mean of zero and a standard deviation of one. The CDF of the Gaussian distribution has no closed form analytical expression, but the standard normal CDF is well tabulated in literature. The Gaussian distribution has a characteristic symmetric bell shaped curve about its mean; thus the mean and median are the same, see Fig. 2.4.

The lognormal distribution is important because of its history in spatial statistics and geostatistics. Many earth science variables are non-negative and positively skewed. The lognormal distribution is a simple distribution that can be used to model non-negative variables with positive skewness. A positive random variable is said to be lognormally distributed if $X = ln(Y)$ is normally distributed (Fig. 2.5). There are many grade distributions that are approximately lognormal. These distributions are also characterized by two parameters, a mean and a variance, although three-parameter lognormal distributions have been used in mining, see for example Sichel (1952). Lognormal distributions can be characterized by either their arithmetic or their logarithmic parameters.

The Central Limit theorem (see for example Lapin 1983) states that *the sum of a great number of independent equally distributed (not necessarily Gaussian) standardized random variables (RV) tends to be normally distributed, i.e. if n RV's Z_i have the same CDF and zero means, the RV tends toward a normal CDF, as n increases towards infinity.* The corollary of this is that the product of a great number of independent, identically distributed RV′s tends to be normally distributed. The theoretical justification of the normal distribution is of little practical importance; however, we commonly observe that the distribution of grades becomes more symmetric and normal-like as the volume of investigation becomes large—the randomness of the grades is averaged and the results tend to a normal distribution.

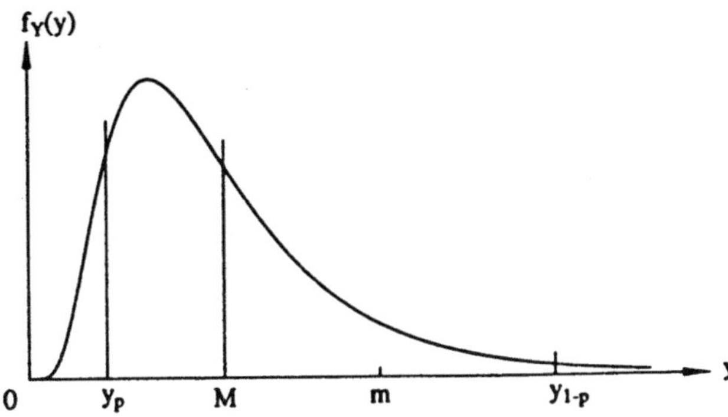

Fig. 2.5 A sketch of a lognormal distribution

2.2.2 Parametric and Non-parametric Distributions

A parametric distribution model has an analytical expression for either the PDF or the CDF, as for the Gaussian density function and the Lognormal distribution. Parametric distributions sometimes relate to an underlying theory, as does the normal distribution to the Central Limit Theorem. There are many parametric distributions that are used in different settings including the lognormal, uniform, triangular, and exponential distributions. Modern geostatistics makes extensive use of the Gaussian distribution because of its mathematical tractability. The lognormal distribution is important as well, but mostly from an historical perspective. In general, however, modern geostatistics is not overly concerned with other parametric distributions because data from any distribution can be transformed to any other distribution including the Gaussian one if needed. Adopting a parametric distribution for the data values may be the only option in presence of very sparse data; a non-parametric distribution is used when there are sufficient data.

There is no general theory for earth science related variables that would predict the parametric form for probability distributions. Nevertheless, certain distribution shapes are commonly observed. There are statistical tests to judge whether a set of data values follow a particular parametric distribution. But these tests are of little value in resource estimation because they require that the data values all be independent one from another, which is not the case in practice.

Parametric distributions have three significant advantages: (1) they are amenable to mathematical calculations, (2) the PDF and CDF are analytically known for all z values, and (3) they are defined with a few parameters. The primary disadvantage of parametric distributions is that, in general, real data do not conveniently fit a parametric model. However, data transformation permits data following any distribution to be transformed to any other distribution, thus capitalizing on most of the benefits of parametric distributions.

Most data distributions are often not well represented by a parametric distribution model. Sometimes distributions are characterized as non-parametric, that is, all of the data are used to define the distribution with experimental proportions; a parametric model for the CDF or PDF is not required. In this case, the CDF probability distribution may be inferred directly from the data, and therefore non-parametric distributions are more flexible. The CDF is inferred directly as the proportion of data less than or equal to the threshold value z. Thus, a proportion is associated to a probability.

A non-parametric cumulative distribution function is a series of step functions. Some form of interpolation may be used to provide a more continuous distribution $F(z)$ that extends to arbitrary minimum $z_{\min}$ and maximum $z_{\max}$ values. Linear interpolation is often used. More complex interpolation models could be considered for highly skewed data distributions with limited data.

2.2.3 Quantiles

Quantiles are specific Z values that have a probabilistic meaning. The p-quantile of the distribution $F(z)$ is the value z_p for which: $F(z_p) = \text{Prob}\{Z \leq z_p\} = p$. The 99 quantiles with probability values from 0.01 to 0.99 in increments of 0.01 are known as percentiles. The nine quantiles at 0.1, 0.2, …, 0.9 are called deciles. The 3 quantiles with probability values of 0.25, 0.5 and 0.75 are known as quartiles. The 0.5 quantile is also known as the median. The cumulative distribution function provides the tool for extracting any quantile of interest. The mathematical inverse of the CDF function is known as the quantile function:

$$z_p = F^{-1}(p) = q(p)$$

The *interquartile Range* (IR or IQR) is the difference between the upper and the lower quartiles: IR = q(0.75) − q(0.25) and is used as a robust measure of the spread of a distribution. The s*kewness* sign is the sign of the difference between the mean and the median (m-M) that indicates positive skewness or negative skewness.

Quantiles are used for comparing distributions in various ways. They can be used to compare the original data distribution to simulated values, compare two types of samples, or compare assay results from two different laboratories. A good way to do this is with a plot of matching quantiles, that is, a quantile-quantile (Q-Q) plot (Fig. 2.6). To generate a Q-Q plot, we must first choose a series of probability values pk, k = 1, 2, …, K; then, we plot $q_1(p_k)$ versus $q_2(p_k)$, k = 1, 2, …, K.

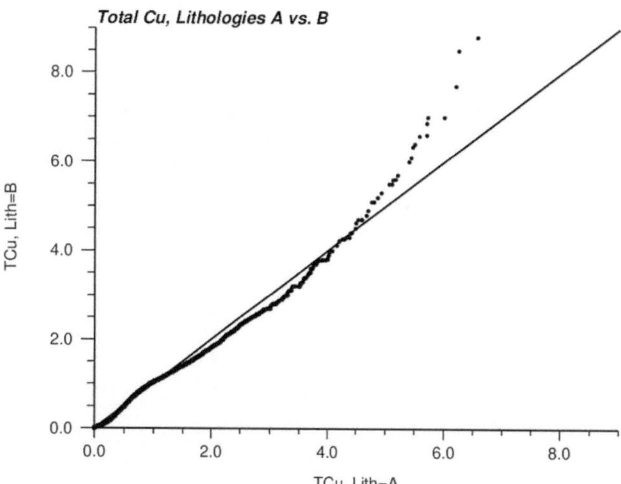

Fig. 2.6 An example of a Q-Q plot. The data is total copper, corresponding to two different lithologies

If all the points fall along the 45° line, the two distribution are exactly the same; if the line is shifted from the 45°, but parallel to it, the two distribution have the same shape but different means; if the slope of the line is not 45°, the two distributions have different variances, but similar shapes; and if there is a nonlinear character to the relationship between the two distributions, they have different histogram shapes and parameters.

The P-P plot considers matching probabilities for a series of fixed Z values. The P-P plot will vary between 0 and 1 (or 0 and 100%), from minimum to maximum values in both distributions. In practice, Q-Q plots are more useful because they plot the values of interest (grades, thicknesses, permeabilities, etc.), and it is therefore easier to conclude how the two distributions compare based on sample values.

2.2.4 Expected Values

The expected value of a random variable is the probability weighted average of that random variable:

$$E\{Z\} = m = \int_{-\infty}^{+\infty} z dF(z) = \int_{-\infty}^{+\infty} z f(z) dz$$

The expected value of a random variable is also known as the mean or the first moment. The expected value can also be considered as a statistical operator. It is a linear operator.

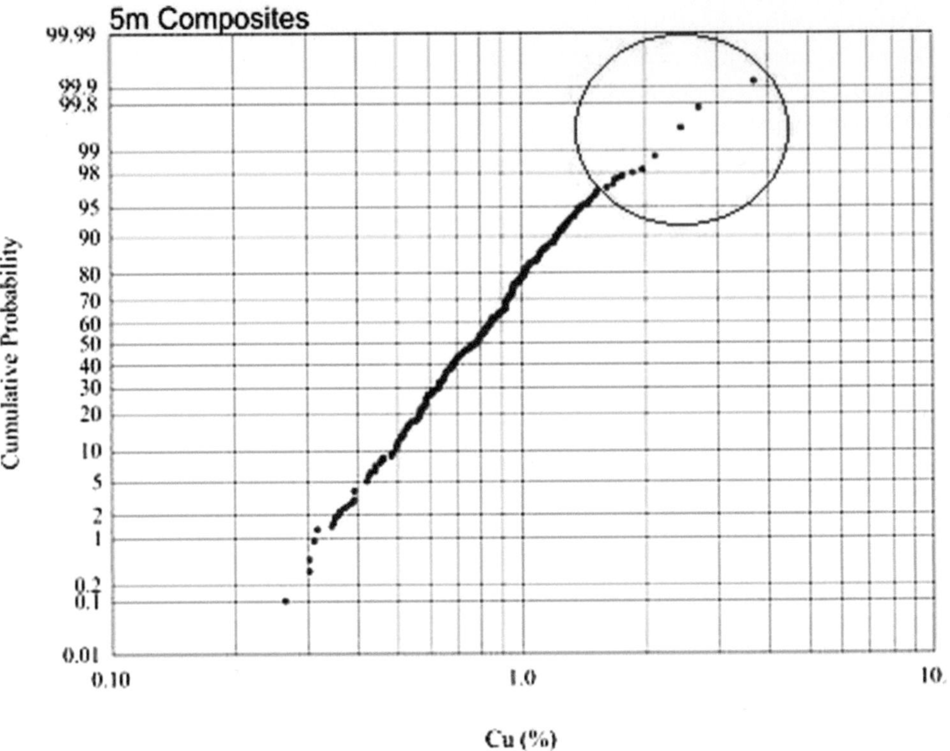

Fig. 2.7 Probability plot with outliers identified

The expected value of the squared difference from the mean is known as the variance (σ^2). It is written:

$$
\begin{aligned}
Var\{Z\} &= E\{[Z - m_z]^2\} = \sigma^2 \\
&= E\{Z^2 - 2Zm_z + m_z^2\} \\
&= E\{Z^2\} - 2m_z E\{Z\} + m_z^2 \\
&= E\{Z^2\} - m^2
\end{aligned}
$$

The square root of the variance is the standard deviation (σ or s). The standard deviation is in the units of the variable. It is common to calculate a dimensionless coefficient of variation (CV), that is, the ratio of the standard deviation divided by the mean.

$$CV = \sigma / m$$

As an approximate guide, a CV less than 0.5 indicates a fairly well behaved set of data. A CV greater than 2.0 or 2.5 indicates a distribution of data with significant variability, such that some predictive models may not be appropriate.

There are additional measures of central tendency aside from the mean. They include the median (50 % of the data smaller and 50 % larger), the mode (the most common observation), and the geometric mean.There are also measures of spread aside from the variance. They include the range (difference between the largest and smallest observation), the interquartile range (described above), and the mean absolute deviation (MAD). These measures are not used extensively.

2.2.5 Extreme Values—Outliers

A small number of very low or very high values may strongly affect summary statistics like the mean or variance of the data, the correlation coefficient, and measures of spatial continuity. If they are proven to be erroneous values, then they should be removed from the data. For extreme values that are valid samples, there are different ways to handle them: (1) classify the extreme values into a separate statistical population for special processing, or (2) use robust statistics, which are less sensitive to extreme values. These options can be used at different times in mineral resource estimation. As a general principle, the data should not be modified unless they are known to be erroneous, although their influence in spatial predictive models may be restricted.

Many geostatistical methods require a transformation of the data that reduces the influence of extreme values. Probability plots can sometimes be used to help identify and correct extreme values, see Fig. 2.7. The values in the upper tail of the distribution could be moved back in line with the trend determined from the other data. An alternative consists of *capping* whereby values higher than a defined outlier threshold are reset to the outlier threshold itself. The high values could be interpreted as a separate population altogether (see for example Parker 1991). There are a number of methods to deal with outliers at the time of variography and resource estimation.

In general, outliers or extreme values are considered on a case-by-case basis with sensitivity studies and considering their impact on local and global resource estimates.

2.2.6 Multiple Variable Distributions

Mineral resource estimation commonly considers multiple variables. The multiple variables could be geometric attributes of the deposit or grades such as thickness, gold, silver, or copper grades. They could be the same grade sampled at different locations. Bivariate and multivariate statistics are used in these cases. There are many references to multivariate statistics, such as Dillon and Goldstein (1984).

The cumulative distribution function and probability density function can be extended to the bivariate case. Let X and Y be two different RVs. The bivariate cdf of X and Y, $F_{XY}(x, y)$ and the pdf of X and Y $f_{XY}(x, y)$ are defined as

$$F_{XY}(x,y) = Prob\{X \leq x, \text{ and } Y \leq y\}$$

and

$$f_{XY}(x,y) = \frac{\partial^2 F_{XY}(x,y)}{\partial x \partial y}$$

We could also define a bivariate histogram, that is, divide the range of the X and Y variables into bins and plot bivariate frequencies. It is more common to simply plot a scatterplot of paired samples on arithmetic or logarithmic scale. Figure 2.8 shows an example from the oil sands in Northern Alberta, Canada, after transformation to a Gaussian variable.

The means and variances of each variable are used as summary statistics. The covariance is used to characterize bivariate distributions:

$$Cov\{X,Y\} = E\{[X - m_X][Y - m_Y]\} = E\{XY\} - m_X m_Y$$
$$= \int_{-\infty}^{+\infty} dx \int_{-\infty}^{+\infty} (x - m_X)(y - m_Y) f_{XY}(x,y) dy$$

The unit of the covariance is the product of the units of the two variables, for example, g/t Au multiplied by thickness in meters. Since these units are hard to understand or interpret, it is common for the covariance to be standardized.

The covariance describes whether the bivariate relationship is dominated by a direct or an inverse relationship, see Fig. 2.9. The product of $[X-m_X][Y-m_Y]$ is positive in quadrants II and IV; it is negative in quadrants I and III. The expected value is the average of the product over all pairs. The example of Fig. 2.9 has a positive covariance, while the example in Fig. 2.8 has a negative covariance because the relationship is dominated by an inverse relationship.

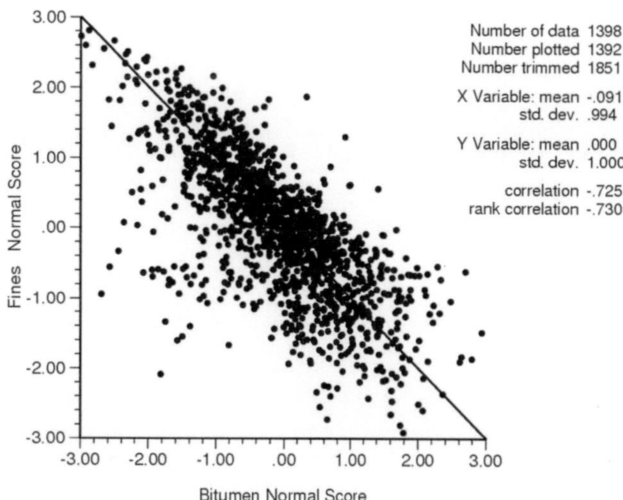

Fig. 2.8 Scatterplot of Bitumen vs. Fines Gaussian variables

The correlation coefficient between random variables X and Y is defined as the covariance between X and Y divided by the standard deviations of the X and Y variables:

$$\rho_{XY} = \frac{Cov\{X,Y\}}{\sigma_X \sigma_Y}$$

The correlation coefficient is a dimensionless measure between -1 (a perfect inverse linear relationship) and $+1$ (a perfect direct linear relationship). Independence between the two variables means that the correlation coefficient is zero, but the reverse is not necessarily true. A covariance or correlation coefficient of zero means there is no dominant direct or inverse relationship, but the variables may be related in a nonlinear manner.

Second order moments like the variance and covariance are significantly affected by outlier data. Some outlier pairs can destroy an otherwise good correlation or enhance an otherwise poor correlation, see Fig. 2.10. The sketch on the left illustrates a case where some outliers would make an otherwise good correlation appear low; the sketch on the right shows a case where a few outliers make an otherwise poor correlation appear high.

The *rank* correlation is more robust with respect to outliers, and is obtained by calculating the correlation coefficient on the rank order of the data. Each data variable is replaced by its rank position in the dataset, and then the correlation coefficient is calculated using the rank positions.

It is common for both correlation coefficients to be shown on experimental cross plots as in Fig. 2.8 where a direct comparison of the two correlation coefficients can be made. Their difference highlights whether there are data features,

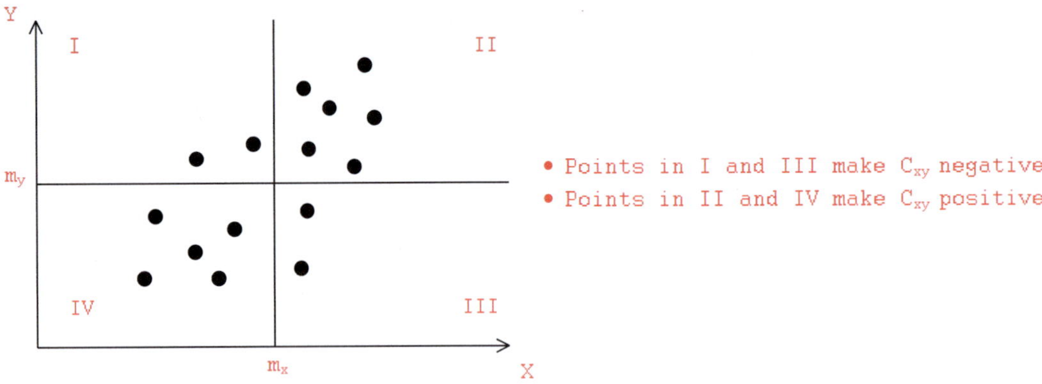

Fig. 2.9 Schematic cross plot with the mean of X drawn as a vertical line, the mean of Y drawn as a horizontal line and the four quadrants numbered

Fig. 2.10 Two schematic scatterplots where outlier data damage a good correlation or enhance a poor correlation

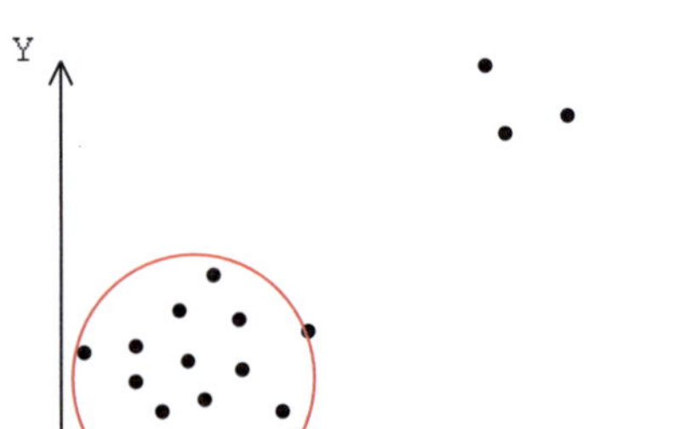

such as outliers, that render the linear correlation measure less useful. Classical least-squares regression requires traditional covariances and not those calculated on a transform of the data. Therefore, rank correlations should only be used for data exploration.

As with the univariate case, scatterplot smoothing is possible and sometimes necessary if the amount of original information is insufficient to characterize the bivariate distribution.

2.3 Spatial Data Analysis

This section describes a series of tools used to better understand spatial distributions. There are several tools that can be used, and are applied in the process called Exploratory Data Analysis, see for example Isaaks and Srivastava (1989).

Posting the data on a variety of cross-sectional or projection views provides clues as to the collection of the data and potential clustering. Posting the values colored differently for values above and below different grade thresholds provides an assessment as to the continuity of high and low grade trends.

Contour maps are used for understanding trends. These can be made by hand or the computer and are used to help in the description of trends. Contouring is typically done on two-dimensional planes defined according to the grid coordinates in plan, cross-sectional, and longitudinal views. It is common to rotate the locations to a local coordinate system prior to any spatial analysis, such that the main coordinate axes are approximately matched with the general orientation of the deposit.

Symbol maps may be more convenient than grade posting maps. A symbol represents some significant aspect of the data, for example drill hole data obtained in different campaigns, by different drilling methods, or at different points in time.

Indicator Maps are a particular form of a symbol map, where a binary variable is used to observe the presence or absence of certain characteristics such as data above and below certain thresholds or presence or absence of specific geologic variables.

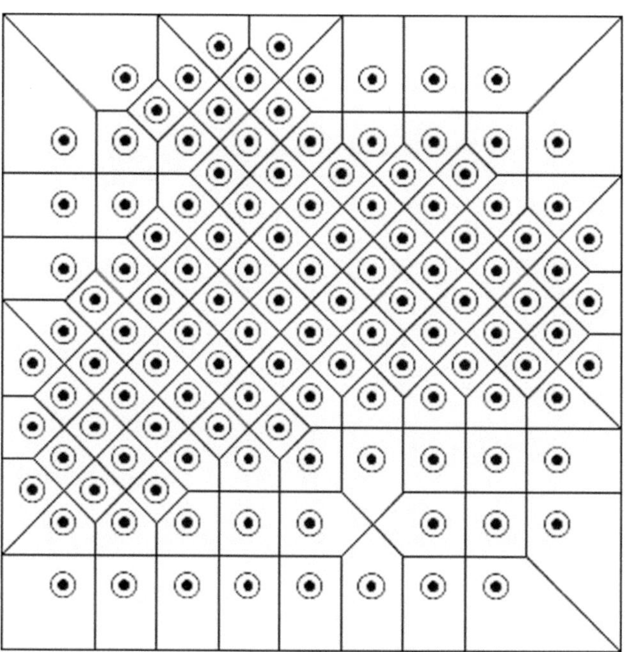

Fig. 2.11 An example of 122 samples with their polygonal areas of influence

2.3.1 Declustering

Data are rarely collected randomly. Drill holes are often drilled in areas of greatest interest, for example high grade areas that will be mined early in the production schedule. This practice of collecting more samples in areas of high grade should not be changed because it leads to the greatest number of data in portions of the study area that are the most important. There is a need, however, to adjust the histograms and summary statistics to be representative of the entire volume of interest.

Declustering techniques assign each datum a weight based on closeness to surrounding data wi, $i=1, ..., n$. These weights are greater than 0 and sum to 1. The experimental distribution and all summary statistics are calculated with the weights instead of a constant $1/n$.

The polygonal declustering method (Fig. 2.11; Isaaks and Srivastava 1989) is perhaps the simplest, and assigns each weight proportional to the area or volume of interest of each sample. Studies have shown that this approach works well when the limits to the area of interest are well defined and the ratio of the largest to smallest weight is less than 10 to 1.

The nearest-neighbor declustering technique is commonly used in resource estimation, and is like the polygonal method. The difference is that it is applied to a regular grid of blocks or grid nodes. The closest datum of the set being declustered is assigned to each block. Because it works on the same blocks that are used to estimate resources, it is more practical in resource estimation.

The technique of cell declustering is another commonly used declustering technique (Journel 1983; Deutsch 1989). Cell declustering works as follows:

1. Divide the volume of interest into a grid of cells $l=1, ..., L$.
2. Count the occupied cells L_o and the number of data in each occupied cell n_{lo}, $lo=1, ..., L_o$.
3. Weight each data according to the number of data falling in the same cell, for example, for datum i falling in cell l, the cell declustering weight is:

$$w_i = \frac{1}{n_l \cdot L_o}$$

The weights are greater than zero and sum to one. Each occupied cell is assigned the same weight. An unoccupied cell simply receives no weight.

Figure 2.12 illustrates the cell declustering procedure. The area of interest is divided into a grid of $L=36$ cells, with $L_o=33$ occupied cells. The number of data in each occupied cell is established by arbitrarily moving data on the grid boundaries to the right and down.

The weights depend on the cell size and the origin of the grid network. It is important to note that the cell size for declustering is *not* the cell size for geologic modeling; it simply defines an intermediate grid that allows assigning a declustering weight.

When the cell size is very small, each datum is in its own cell and receives an equal weight. When the cell size is very large, all data fall into one cell and are equally weighted. Choosing the optimal grid origin, cell shape, and size requires some sensitivity studies. It is common to choose the cell size so that there is approximately one datum per cell in the sparsely sampled areas or, if available, to choose it according to an underlying, quasi-homogeneous sampling grid.

The sensitivity of the results to small changes in the cell size should be checked. If the results change by a large amount, then most likely the declustering weights are changing for one or two anomalously high or low grades.

Since it is generally known whether over-sampling occurs in high- or low-valued areas, the weights can be selected such that they give the minimum or maximum declustered mean of the data. The declustered mean versus a range of cell sizes should be plotted, and the size with the lowest (Fig. 2.13, data clustered in high-valued areas) or highest (data clustered in low-valued areas) chosen. Care should be taken not to over-fit the minimum. The correct cell size should be approximately the spacing of the data in sparsely sampled areas. This qualitative check can be used to ensure that a too-large or too-small cell size is not chosen.

The shape of the cells depends on the geometric configuration of the data, as it is adjusted to conform to the major directions of preferential sampling. For example, if the samples are more closely spaced in the X direction than in the Y direction, the cell size in the X direction should be reduced.

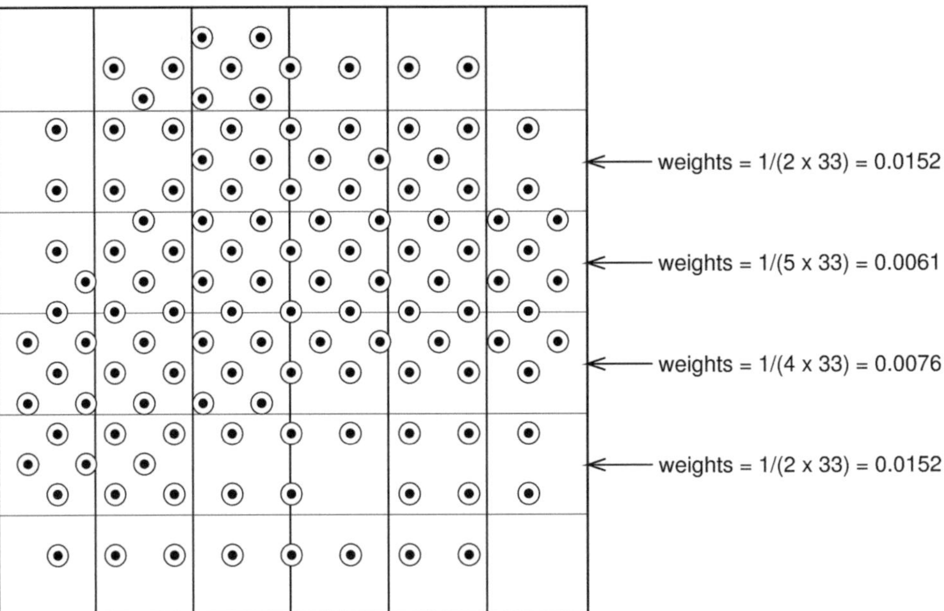

Fig. 2.12 Illustration of the cell declustering method

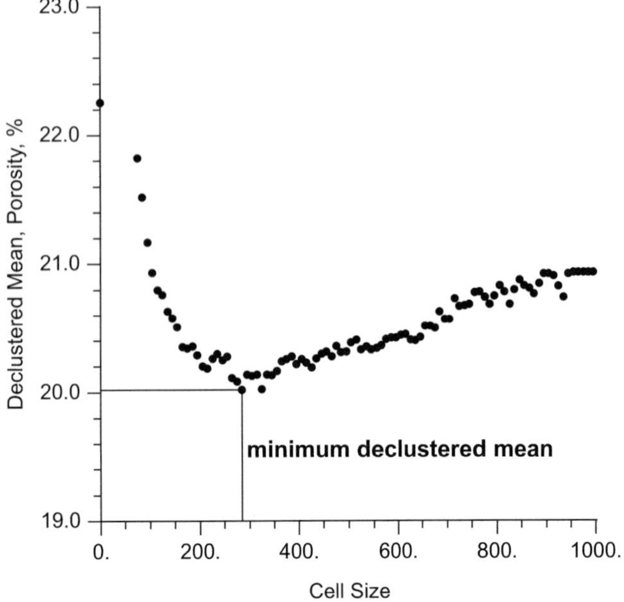

Fig. 2.13 Plot of declustered mean vs. cell size

The origin of the cell declustering grid and the number of cells L must be chosen such that all data are included within the grid network. Fixing the cell size and changing the origin often leads to different declustering weights. To avoid this artifact, a number of different origin locations should be considered for the same cell size. The declustering weights are then averaged for each origin offset.

Declustering assumes that the entire range of the true distribution has been sampled. If this is not the case, then the data is biased and debiasing techniques may be required. These techniques include trend modeling for debiasing and debiasing using qualitative data, subjects that are not covered in this book.

2.3.2 Declustering with Multiple Variables

Declustering weights are determined on the basis of the geometric configuration of the data; therefore, only one set of declustering weights is calculated in presence of multiple variables that have been equally sampled. However, different declustering weights will need to be calculated when there is unequal sampling. For example, there are sometimes different sets of Copper and Molybdenum samples in a Cu-Mo porphyry deposit, which would require two sets of declustering weights.

Declustering weights are primarily used to determine a representative histogram for each variable; however, we also require the correlation between multiple variables. The same set of declustering weights can weight each pair contributing to the correlation coefficient (Deutsch 2002).

2.3.3 Moving Windows and Proportional Effect

Moving windows are used to understand the local spatial behavior of the data, and how it may differ from global statistics. The process is to lay over the volume of interest a grid of cells, which may or may not be partially overlapping, moving them over the entire domain or deposit, and obtaining statistics within them. Overlapping windows are typically used when there are few data within the window to provide reliable statistics (Goovaerts 1997; Isaaks and Srivastava 1989).

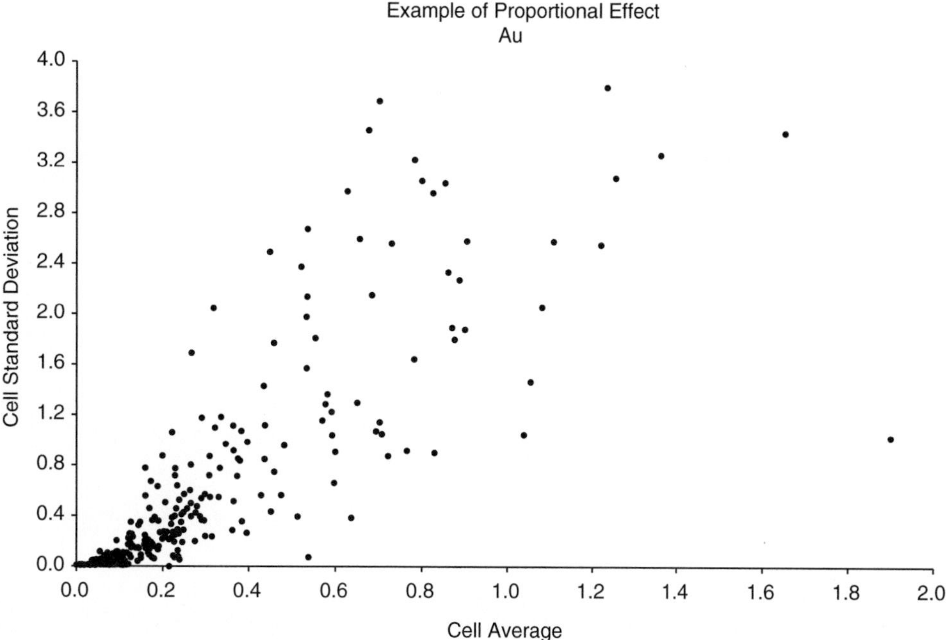

Fig. 2.14 An example of proportional effect from a West-African gold deposit. Cell averages and standard deviations are both in g/t

The most common statistics analyzed are the mean and standard deviations of the data within the windows.

A plot of the mean versus standard deviation calculated from moving windows of data can be used to assess changes in local variability, see Fig. 2.14 for an example. Generally, positively skewed distributions will show that windows with higher local mean usually exhibits higher local standard deviation. This is the proportional effect described by various authors, for example David (1977) and also Journel and Huijbregts (1978). The proportional effect is due to a skewed histogram, but it may also indicate spatial trends or a lack of spatial homogeneity. Proportional effect graphs are sometimes used to help determine homogeneous statistical populations within the deposit (see Chap. 4).

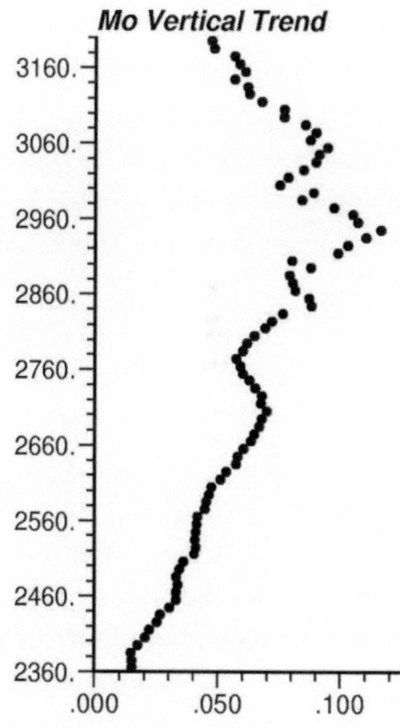

Fig. 2.15 Example of a molybdenum vertical trend

2.3.4 Trend Modeling

Trend modeling is applied when a trend has been detected and is assumed to be well understood. While some geostatistical estimation methods are quite robust with respect to the presence of trends, such as Ordinary Kriging (Chap. 8; Journel and Rossi 1989), there are many others, most notably simulation (Chap. 10) that are quite sensitive to trends.

The trend is modeled as a deterministic component plus a residual component. The deterministic component is removed and then the residual component is modeled either through estimation or simulation techniques. Finally, the deterministic trend is added back. In such a model, the mean of the residual and the correlation between the trend and the residual should be close to 0.

The drill hole data is typically the source for trend detection. In some cases where the geological environment is well understood, trends can be expected and modeled without the drill hole data, but this should only be attempted when there is no other option. Large scale spatial features can be detected during several stages of data analysis and modeling. Sometimes a simple cross-plot of the data against elevation may show a trend, as in the example of Fig. 2.15. In other cases, simple contour maps on cross-sections, longitudinal sections, or plan views are enough to identify and model trends. Moving window averages can also provide an indication of whether or not the local means and variances are stationary. If there are notable changes in the local

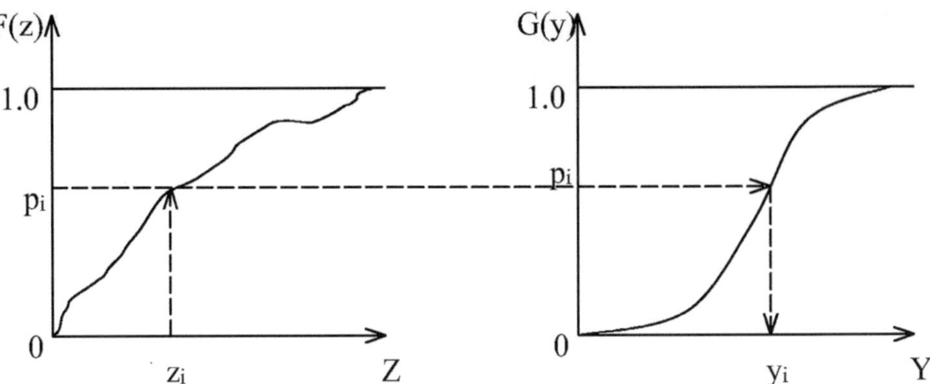

Fig. 2.16 Data transformation using Cumulative Distribution Functions

mean and variance of reasonably large subdivisions within the domain, as in Fig. 2.14, then a spatial trend model may be required.

Although the identification of a trend is subjective, it is generally accepted that the trend is deterministic and should not have short scale variability. It should be identified from features that are significantly larger than the data spacing, i. e., domain-wide. This sometimes can be evident from the experimental variogram that may show a trend in any one or more directions. The experimental variogram continues to increase above the variance of the data as the lag distance increases (Chap. 6; Journel and Huijbregts 1978). This usually indicates that the decision of stationarity should be revisited, and consider whether the domain should be subdivided or a trend considered.

2.4 Gaussian Distribution and Data Transformations

Gaussian distributions are commonly used due to their convenient statistical properties. The Gaussian distribution is derived from the Central Limit Theorem, which is one of the most consequential theorems in statistics.

A univariate Gaussian distribution is fully characterized by its mean (m) and standard deviation (σ). The probability density function is given by:

$$g(z) = \frac{1}{\sigma\sqrt{2\pi}} \exp\left[-\frac{1}{2}\left(\frac{z-m}{\sigma}\right)^2\right]$$

It is common to transform data to a Gaussian distribution. There are many instances where the prediction of uncertainty at un-sampled locations becomes much easier with a Gaussian distribution.

The simplest method to transform any distribution into a Gaussian distribution is a direct quantile-to-quantile transformation, whereby the CDF of each distribution is used to perform the transform. This is known as the Normal Scores (NS) transform, see Fig. 2.16. The NS transform is achieved by quantile transformation:

$$y = G^{-1}(F(z))$$

which is back-transformed by

$$z = F^{-1}(G(y))$$

The expected values should not be back transformed unless the distribution is symmetric.

A variable Z is non-standard Gaussian when the standardized variable Y is standard Gaussian. A non-standard Gaussian value is easily converted to/from a standard Gaussian value.

$$y = \frac{z - m_Z}{\sigma_Z} \qquad z = y \cdot \sigma_Z + m_Z$$

The normal score transform is rank preserving and reversible. The disadvantages of performing such a transform are that the significance of the numbers themselves is less clear, more difficult to interpret, and also that the distribution parameters cannot be back transformed directly due to the nonlinearity of the process.

Spikes of constant values in the original distribution can cause problems. Gaussian values are continuous and ties (equal values) in the original distribution must be resolved prior to transforming the data. There are two different methods commonly used to break the ties or despike. The simpler method is to add a small random component to each tie, which is the most common approach used in popular software packages, such as the GSLIB programs (Deutsch and Journel 1997). A better alternative is to add a random component based on local averages of the data (Verly 1984), which ranks the ties based on the local grades of nearby data. Although more onerous in terms of time and computer effort, it is justified when the proportion of original data with the same values is significant. Typical drill hole data from Au epithermal deposits can show a significant number of values at or below the laboratory's detection limit, sometimes as much as 50 or 60 %, in which case despiking is better accomplished using the local averaging method. Of course, an alternative is to separate the barren or un-mineralized material into its own stationary population.

This is reasonable when the spatial arrangement of the barren material is predictable.

2.5 Data Integration and Inference

The prediction of spatial variables requires consideration of multivariate distributions of values at different locations. Inference requires the combination of sample data to estimate at an unknown location. The calculation of conditional distributions is accomplished by application of Bayes' Law, one of the most important laws in statistical theory.

Bayes' Law provides the probability that a certain event will occur given that (or conditional to) a different event has already occurred. The mathematical expression for Bayes' Law can be written as:

$$P(E_1 \mid E_2) = \frac{P(E_1 \text{ and } E_2)}{P(E_2)}$$

with E_1 and E_2 being the events, and P representing probabilities.

If E_1 and E_2 are independent events, then knowing that E_1 occurred does not give additional information about whether E_2 will occur:

$$P(E_1 \mid E_2) = P(E_1)$$
$$P(E_1 \text{ and } E_2) = P(E_1) \cdot P(E_2)$$

Direct inference of multivariate variables is often difficult, which leads us to use the multivariate Gaussian model, mostly because it is straightforward to extend to higher dimension. The bivariate Gaussian distribution is defined as:

$$(X,Y) \rightarrow N(0,1,\rho_{X,Y}),$$

$$f_{X,Y}(x,y) = \frac{1}{2\pi\sqrt{1-\rho^2}} e^{\left[-\frac{1}{2\left(1-\rho^2\right)}\left(x^2-2\rho xy+y^2\right)\right]}$$

The relationship between the two variables is defined by a single parameter, the correlation coefficient, and in the *XY* cross-plot the probability contours are elliptical. The conditional expectation of *Y* given an event for *X* is a linear function of the conditioning event:

$$E\{Y \mid X = x\} = m_Y + \rho_{X,Y} \frac{\sigma_Y}{\sigma_X}(x - m_x)$$

The conditional expectation follows the equation of a line, $y = mx + b$, where m is the slope (correlation coefficient) and b is the intercept (mean).

The conditional variance is independent of the conditioning event(s). This is an important consideration that will influence some of the geostatistical methods to be described later, and is written as:

$$Var\{Y \mid X = x\} = \sigma_Y^2\left(1 - \rho_{X,Y}^2\right)$$

For a standard bivariate Gaussian distribution (that is, both variables, *X* and *Y* have a mean = 0 and variance = 1.0) the parameters are:

$$E\{Y \mid X = x\} = \rho_{X,Y} \cdot x$$
$$Var\{Y \mid X = x\} = 1 - \rho_{X,Y}^2$$

The extension to multivariate distributions is straightforward, and can be written as:

$$\mathcal{N}(\mathrm{x};\mu,\Sigma) = \frac{1}{(\sqrt{2\pi})^d \left|\Sigma\right|^{1/2}} \cdot \exp\left[-\frac{1}{2}(\mathrm{x}-\mu)^T \Sigma^{-1} (\mathrm{x}-\mu)\right]$$

where d is the dimensionality of x. Note that μ is a (d × 1) vector and Σ is a (d × d) positive definite, symmetric variance-covariance matrix. The expression |Σ| is the determinant of Σ. μ is the mean of the distribution and Σ is the covariance matrix. The *i*-th element of μ expresses the expected value of the *i*-th component in the random vector x; similarly, the (i, j) component of Σ expresses the expected value of $x_i\, x_j$ minus $\mu_i\, \mu_j$. The diagonal elements of Σ are the variances of the corresponding component of x.

The multivariate (N-variate) Gaussian distribution possesses some extraordinary properties (Anderson 1958; Abramovitz and Stegun 1964):

1. All lower order N-k marginal and conditional distributions are Gaussian.
2. All conditional expectations are linear functions of the conditioning data:

$$E\{X_i \mid X_j = x_j, \forall j \neq i\} = \sum_{j \neq i} \lambda_j x_j$$
$$= \varphi(x_j, j \neq i) = [x_i]^*_{SK}$$

3. All conditional variances are homoscedastic (data-values-independent):

$$E\left\{\left[X_i - \varphi(x_j, j \neq i)\right]^2 \mid X_j = x_j, \forall j \neq i\right\}$$
$$= E\left\{\left[X_i - \varphi(x_j, j \neq i)\right]^2\right\}$$

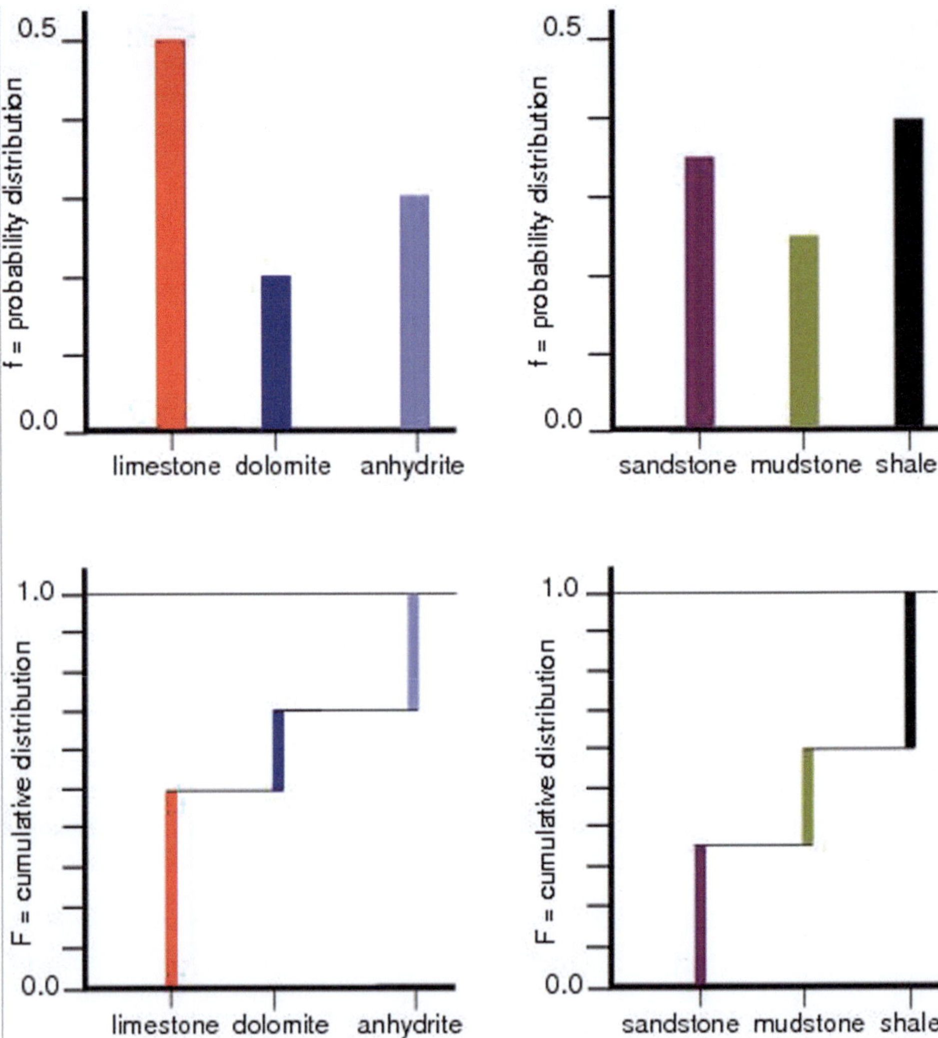

Fig. 2.17 PDFs and CDFs for categorical variables

Conditional expectations are linear functions of the data. All linear combinations of Gaussian variables are also Gaussian, and in particular, averages are Gaussian. Also, conditional variances are data-values-independent, a property called *homoscedasticity*.

In geostatistics, it is common to assume that the normal scores of grade variables are multivariate Gaussian within geologically defined domains. This is done for convenience since the simple (co)kriging method provides exactly the mean and variance of all conditional distributions, as described in Chaps. 8–10.

Performing a univariate normal score transformation guarantees a univariate Gaussian distribution, but there is no guarantee of a multivariate Gaussian distribution. The transformation does not remove nonlinearity or other constraints. The proportional effect and heteroscedasticity is largely removed by the transformation, but then it is reintroduced by the back transformation. Transforming a multivariate distribution is rarely done in mineral resource estimation because of the complexity and requirement for many data.

Categorical Variables The probability distribution of a discrete or categorical variable is defined by the probability or proportion of each category, that is, p_k, $k=1, \ldots, K$, where there are K categories. The probabilities must be non-negative and sum to 1.0. A table of the p_k values completely describes the data distribution. Sometimes, however, it is convenient to consider a histogram and cumulative histogram as shown below (Fig. 2.17):

The cumulative histogram is a series of step functions for an arbitrary ordering of the discrete categories. Such a cumulative histogram is not useful for descriptive purposes but is needed for Monte Carlo simulation and data transformation. In general, but not always, the ordering does not matter. The cases where the ordering affects the results will be discussed later in the book.

Consider K mutually exclusive categories s_k, $k=1, \ldots, K$. This list is also exhaustive; that is, any location $\boldsymbol{u}$ belongs to one and only one of these K categories. Let $i(\mathbf{u}; s_k)$ be the indicator variable corresponding to category s_k, set to 1 if location $\boldsymbol{u}$ in s_k, zero otherwise, that is:

$$i(\mathbf{u}_j;s_k)=\begin{cases}1, & \text{if location } \mathbf{u}_j \text{ in category } s_k\\ 0, & \text{otherwise}\end{cases}$$

Mutual exclusion and exhaustivity entails the following relations:

$$i(\mathbf{u};\ s_k)\cdot i(\mathbf{u};\ s_{k'})=0,\quad \forall k\neq k'$$

$$\sum_{k=1}^{K} i(\mathbf{u};\ s_k)=1$$

The mean indicator for each category s_k, $k=1, \ldots, K$ is the proportion of data in that category:

$$p_k=\overline{i(\mathbf{u};\ s_k)}=\frac{\sum_{j=1}^{N} w_j i(\mathbf{u}_j;\ s_k)}{\sum_{j=1}^{N} w_j}$$

The variance of the indicator for each category s_k, $k=1, \ldots, K$ is a simple function of the mean indicator (Journel 1983; Deutsch 2002):

$$Var\{i(\mathbf{u};s_k)\}=\frac{\sum_{j=1}^{N} w_j[i(\mathbf{u}_j;s_k)-p_k]^2}{\sum_{j=1}^{N} w_j}=p_k(1.0-p_k)$$

The variance would be used to standardize variograms for quicker interpretation and comparison across different categories.

2.6 Exercises

The objective of this exercise is to review some mathematical principles, become familiar with some notation, work with some common probability distribution models and gain some experience with declustering. Some specific (geo)statistical software may be required. The functionality may be available in different public domain or commercial software. Please acquire the required software before beginning the exercise. The data files are available for download from the author's website—a search engine will reveal the location.

2.6.1 Part One: Calculus and Algebra

Question 1: Consider the following function $(aX+bY)(X+Y)$. Calculate the derivative of this function with respect to X and Y.

Question 2: Calculate the integral for the function below:

$$\int_0^5 \frac{1}{2}x^2+x^3-\frac{1}{4}x^5 dx$$

Question 3: Consider the three matrices below:

$$\mathbf{A}=\begin{bmatrix}5 & 2\\ 2 & 3\end{bmatrix}\quad \mathbf{B}=\begin{bmatrix}1\\ 4\end{bmatrix}\quad \mathbf{C}=\begin{bmatrix}2 & 3\end{bmatrix}$$

What is the result of **AB**, $\mathbf{AC}^\mathrm{T}$, and **(AB)C**?

2.6.2 Part Two: Gaussian Distribution

Consider the standard Gaussian or normal distribution that is of extraordinary importance in statistics and geostatistics because it is the limit distribution of the central limit theorem and is mathematically tractable.

Question 1: Verify that the sum of independent random variables tends toward a normal distribution. Consider (1) setting up a grid of 100 rows by 10 columns in Excel with uniform random numbers between 0 and 1, (2) create an 11th column with the sum of the 10 first columns, (3) plot a histogram of the 11th column, and (4) comment.

Question 2: What is the mean and variance of a probability distribution that is uniform between 0 and 1? The central limit theorem tells us that the mean of 10 values added together should be this mean multiplied by 10—check against Question 1 and comment. The central limit theorem would also tell us that the variance is multiplied by 10—check against Question 1 and comment.

Question 3: Create a 12th column in your spreadsheet with the sum (the 11th column) minus the mean divided by the standard deviation, that is, $y12=(y11-m)/\sigma$. Plot a histogram and calculate the statistics of this *standardized deviate*. Comment on the results.

2.6.3 Part Three: Uniform Distribution

Consider the uniform distribution specified below:

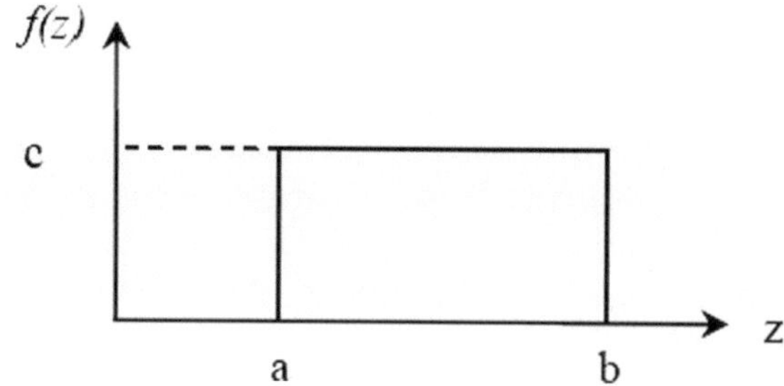

Question 1: Write the definition and equation for the cumulative distribution function (cdf) of

the uniform distribution above. Draw the corresponding cdf to the probability density function (pdf) above.

Question 2: What is the value of *c* that makes *f(z)* a licit probability distribution? Write your answer in terms of *a* and *b*.

Question 3: What is the expected value (or mean) of the variable *Z* in terms of *a*, *b*, and *c*? Solve the integral.

Question 4: What is the variance of the variable *Z* in terms *a*, *b*, and *c*? Solve for the expected value of Z_2 and solve for the variance using $\sigma 2 = E\{Z2\} - [E\{Z\}]^2$.

Question 5: What is the 90% probability interval? Write out the function corresponding to the cdf and solve for the 5th and 95th quantiles.

The objective of this exercise is to become familiar with the different ways to use declustering to infer a representative probability distribution. Declustering software and the specified datasets are required.

2.6.4 Part Four: Small Declustering

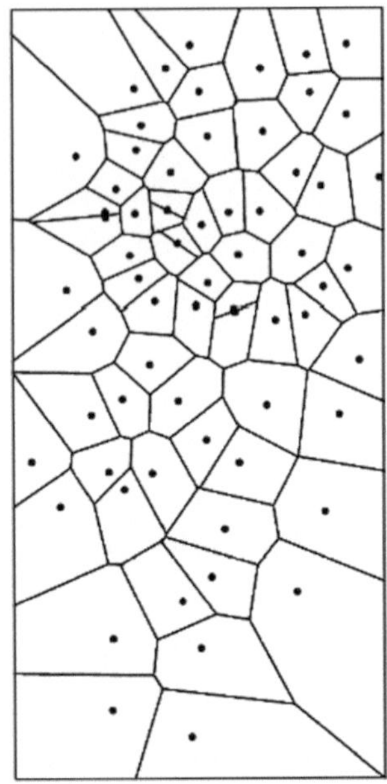

Consider the 2-D data in `red.dat` (see right). The 67 drill hole intersections have a hole ID, location, thickness, four grade values, and a rock type. The area is from 20,100 to 20,400 in the northing direction and –600 to 0 in elevation. The rock type is simply a flag that specifies below or above –300 m. There is a difference below that elevation that warrants our attention.

Question 1: Plot a location map of the thickness and the gold grade. Plot a histogram of all the gold grades without any declustering weight.

Question 2: Setup and run polygonal declustering to get a map that looks like the one to the right. Plot a declustered histogram of the gold grades.

Question 3: Cell declustering is widely used because it is robust in 3-D and is less sensitive to edge effects. Run cell declustering for a range of cell sizes—explain your choice of parameters. Plot the declustered mean versus cell size, choose a cell size, and justify your choice. Compare results to those obtained above.

2.6.5 Part Five: Large Declustering

Consider the 3-D Au/Cu data in `largedata.dat`. This data will be used in some subsequent exercises. We need declustered distributions for the two variables in all rock types.

Question 1: Consider cell declustering on a by-rock type basis and considering all of the data together. Compare the results and comment on the preferred approach. Prepare a reasonable set of plots to support your conclusions including the declustered mean versus cell size plot(s) and tables of declustered mean and standard deviation values.

Question 2: Assemble the reference distributions for subsequent modeling (based on your chosen method).

References

Abramovitz M, Stegun I (1964) Handbook of mathematical functions. Dover Publications, New York, p 1046

Anderson T (1958) An introduction to multivariate statistical analysis. Wiley, New York

Borradaile GJ (2003) Statistics of earth science data. Springer, Heidelberg

David M (1977) Geostatistical ore reserve estimation. Elsevier, Amsterdam

Davis JC (1986) Statistics and data analysis in geology, 2nd edn. Wiley, New York, p 646

de Moivre A (1738) The doctrine of chances: or, a method for calculating the probabilities of events in play, 2nd edn. printed by H. Woodfall, London

Deutsch CV (1989) DECLUS: A FORTRAN 77 program for determining optimum spatial declustering weights. Comput Geosci 15(3):325–332

Deutsch CV (2002) Geostatistical reservoir modeling. Oxford University Press, New York, p 376

Deustch CV, Journel AG (1997) GSLIB: geostatistical software library and user's guide, 2nd edn. Oxford University Press, New York, p 369

Dillon W, Goldstein M (1984) Multivariate analysis: methods and applications. Wiley, New York, p 587

Gauss CF (1809) Theoria Motus Corporum Coelestium in sectionibus conicis solem ambientium. English translation by C. H. Davis, reprinted 1963, Dover, New York

Goovaerts P (1997) Geostatistics for natural resources evaluation. Oxford University Press, New York, p 483

Isaaks EH, Srivastava RM (1989) An introduction to applied geostatistics. Oxford University Press, New York, p 561

Journel AG (1983) Non-parametric estimation of spatial distributions. Math Geol 15(3):445–468

Journel AG, Huijbregts ChJ (1978) Mining geostatistics. Academic Press, New York

Journel AG, Rossi ME (1989) When do we need a trend model? Math Geol 22(8):715–739

Koch G, Link R (1986) Statistical analysis of geological data, 2nd edn. Wiley, New York

Lapin LL (1983) Probability and statistics for modern engineering. PWS Publishers, Boston, p 624

Laplace PS (1812) Théorie analytique des probabilités. Printed in 1814 by Mme. Ve. Courier, Paris

Legendre AM (1806) Nouvelles Methodes pour la Determination des Orbites des Cometes. F. Didot, Paris

Parker HM (1991) Statistical treatment of outlier data in epithermal gold deposit reserve estimation. Math Geol 23:125–199

Ripley BD (1987) Spatial statistics, 2nd edn. Wiley, New York

Rohatgi VK, Ehsanes Saleh AK Md (2000) An introduction to probability and statistics. Wiley, New York

Sichel HS (1952) New methods in the statistical evaluation of mine sampling data. Trans Inst Min Metall Lond 61:261

Verly G (1984) Estimation of spatial point and block distributions: the multigaussian model. PhD Dissertation, Department of Applied Earth Sciences, Stanford University

Geological Controls and Block Modeling

3

Abstract

Mineral deposition is governed by complex processes. The structure of mineral deposits is partly deterministic and partly stochastic. Large scale deterministic geological control must be accounted for explicitly. Block models are commonly used to discretize a deposit because they provide a spatial representation of geologic variables and a useful format to store other important attributes, including the estimated grades.

3.1 Geological and Mineralization Controls

The geology used to support resource estimation is understood from the analysis of the recorded information gathered through detailed exploration work, including drill holes. Surface mapping, underground mapping and sampling, and geochemical and geophysical investigations may also contribute, particularly in the early stages of project development (Peters 1978). This chapter assumes that mapped drill hole information is the basis for geologic modeling, while acknowledging that all geological interpretations are the result of a pool of quantitative and qualitative information.

Figure 3.1 shows an example of a geologic log of the Spence copper Project in Northern Chile. The log sheet shows the from—to interval; mapped lithology (in character and graphical codes); mineralization type; structures; description and percentages of the minerals found; alteration; gangue minerals; and presence and type of veinlets. The specific information collected will vary from one deposit to the next.

The ultimate goal of mineral resource estimation is a numerical model that will accurately predict the tonnages and grades that will be extracted from a mining operation. The geologic variables that controlled the mineral deposition are modeled to help with this. Certain geologic variables are of greater interest to resource estimation, that is, the ones that have a stronger or more direct relationship with the mineralization. One example is the fracturing and permeability of certain rocks in strata-bound or sedimentary-type deposits (for example uranium, gold, or copper in sandstones and/or breccias); the fluids carrying the ore minerals would move preferentially through the fractured permeable rock. Other examples include the geochemical stability of certain minerals and the fracture density in low-grade bulk precious metals deposits. These specific geologic variables should be the focus of geologic investigation and modeling for improved resource estimation.

The information shown in Fig. 3.1 and other site-specific information is the basis for analyzing the relationships between the geologic variables and the grade distribution. Not all mapped geologic information will be a significant mineralization control, and thus it may not aid in estimating grades. The geologic description shown in Fig. 3.1 is too detailed for practical use in defining mineralization controls. An important challenge is to identify the important geologic variables that need to be interpreted, modeled, and carried into the block model. These may include variables needed to build metallurgical and geotechnical models, such as concentrations of certain types of clays, rock hardness indices, metallurgical recoveries, and fracture densities.

The level of geologic detail that can be considered in a block model is limited. It depends on the size of the deposit and the amount of drill hole information available. There is a compromise between the level of detail achieved and the robustness of the statistical analysis within each geologic population defined. A resource model with no geologic support is inadequate because geologic factors highly constrain the distribution of grades. But too much detail is undesirable, since it creates estimation domains with too few data for reliable statistical inference.

Although there are no hard rules that can be used to determine the amount of data required, the general guideline is to

M. E. Rossi, C. V. Deutsch, *Mineral Resource Estimation,* DOI 10.1007/978-1-4020-5717-5_3,

COMPAÑIA MINERA RIOCHILEX S.A.
PROYECTO SPENCE - Logeo Sondajes DDH

SONDAJE	N188
GEOLOGO	F D.bona
Fecha	29/09/01

N°Muestra	Tramo Desde	Tramo Hasta
	0	114
	114	118
	118	126
	126	132
	132	135
	135	138
	138	141
	141	144
	144	147
	147	150
	150	153
	153	156
	156	159
	159	162
	162	165
	165	168
	168	171
	171	174
	174	177
	177	180
	180	183
	183	186
	186	189
	189	192
	192	195

Fig. 3.1 Geologic Log used by BHP Billiton at their Spence deposit, northern Chile. Courtesy of BHP Billiton

obtain a reliable model of spatial continuity for each domain, as well as sufficient drill hole data to obtain a robust grade estimate.

Specifying too many geological factors/domains does not help with estimation. One such case was the Chuquicamata copper mine. During the mid-1990s there were 64 estimation domains defined in the sulphide zone in support of the resource model. Many of these units, although clearly distinguishable and properly characterized from a geologic standpoint, were only weak mineralization controls and in some cases did not have sufficient data support to allow for robust inference. Thus, they were poorly estimated. In the late 1990s, after a review of the production data against the predicted data, the estimation domains used in the resource model were reduced to less than 30. There were other improvements to the overall resource modeling process; however, it was recognized that reducing the number of estimation domains resulted in an improved resource model.

As the project moves from early exploration into resource definition and pre-feasibility, the gathering of geologic information for exploration and resource delineation requires more planning and control (Hartman 1992, p. 30). Some aspects to be considered include:

a. Development of detailed written protocols for data collection and geologic work. While they need to be constantly updated, they are to be used throughout the project. The protocols should describe the procedures used by the field geologist to control drilling; the sampling protocol and equipment to be used at the drill rig; the corresponding mapping and logging protocols; procedures for using and maintaining hand-held logging devices used (if available); quality control and quality assurance (QA/QC) programs for laboratory assay controls; and proper chain of custody for all samples. The consistent use of protocols and procedures will help create more reliable databases.
b. Active supervision and continued training of all personnel involved, ensuring correct and consistent application of procedures and protocols. Protocols are of little value if there is no training and supervision.
c. Correct management of logged and mapped geologic information including handling, storage, electronic input, and interpretation/evaluation. This may include a description of drill hole information according to prescribed conventions; adequate documentation of the work done; adequate storage of the same documentation, and specific descriptions as a documentation trail for future audits.
d. Working sections with (hand) drawn interpretations allow for a dynamic understanding of geologic controls, and better management of future data gathering campaigns.
e. It is critical to properly store and retain half cores, field sample rejects, coarse and pulp rejects from the sample preparation and assaying processes. Sufficient and proper storage of the excess sample material from drilling should be planned for early in the project's life. Storage areas should be covered, clean, and well organized.
f. Information loss because of bad quality control or lack of quality assurance procedures is serious. One often-

Table 3.1 Lithology mapped, logged, and modeled, October 2001 Resource Model (used with permission from BHP Billiton)

LITHOLOGY	Mapped/Logged codes	Modeled (Oct 2001 Model)
Feldspar Porphyry (Escondida Porphyry)	PF	Modeled as such
Rhyolite	PC	Modeled as such
Undifferentiated Porphyry	PU	Modeled either as PF or AN depending on its spatial location
Andesites	AN	Modeled as such
Igneous Breccias	BI	Modeled as Breccias (single unit)
Hydrothermal Breccias	BH	Modeled as Breccias (single unit)
Tectonic Breccias	BT	Modeled as Breccias (single unit)
Gravel	GR	Modeled as such
Late Dacites	DT	Absorbed by the major unit that contains it
Diorites	DR	Absorbed by the major unit that contains it
Tuffs	TB	Absorbed by the major unit that contains it
Pebble Dykes	PD	Absorbed by the major unit that contains it

encountered problem is that information related to a drill hole (geologic mapping; topographic survey of the drill hole collars; down the hole survey of the drill hole inclinations; laboratory assay certificates, etc.) is disorganized and misplaced. It may be located in different file cabinets, different offices, or in different parts of the world. This results in a high probability of losing costly information. The recommended solution is to keep individual binders (one per drill hole) with all its relevant information, and a backup in a different location. Another common oversight is improper backup procedures for computerized information.

g. Determination and modeling of the mineralization controls. Lithology, alteration, mineralization, structural, and other relevant information must be analyzed and interpreted. This data must be maintained even if not all these variables result in interpretable mineralization controls. This process should utilize a combination of field observations, plausible genetic theories, and should make extensive use of statistical tools (see Chaps. 2 and 4) to identify geologic controls. The process is iterative and should be started as soon as there is enough information to statistically describe relationships between grades and geology.
h. Development of a geologic model that adequately captures the mineralization controls for estimation domains and grade estimation. This is in addition to the working geologic model used for exploration.
i. Effective presentation and communication of the model should be considered an essential part of the work itself (Peters 1978). The use of visualization tools such as three-dimensional models, two-dimensional cross sections and plan views are essential. Appropriate scales typically range from 1:200 to 1:1,000. Plots should show color-coded drill hole information (geology and assay), working or final polygons representing the interpreted geologic variables, and topographic and/or bedrock surfaces. All drill holes should be properly identified. A plan view at the top of the plot showing the drill hole trace is also convenient if showing a cross or longitudinal section. Three-dimensional visualization tools should be routinely used both for validation and presentation purposes.

Tables 3.1, 3.2, and 3.3 show an example taken from the Escondida mine in northern Chile operated by BHP Billiton for lithology, alteration and mineralization type variables. The tables show the variables mapped, logged, and then modeled as of 2001.

There can be several reasons why a given unit is not modeled. For example, Dacites, Diorites, Pebble Dykes, and Tuffs are absorbed into the major unit that surrounds them at the time of modeling, because their spatial extent is not significant compared to the scale of mining (Table 3.1). Undifferentiated Porphyry is either a transition between an Andesitic-type rock and the feldspar porphyry, or it is logged as such because the sample is too altered or broken up to be properly recognized. In either case, Undifferentiated Porphyry is usually located along the Andesite—Escondida Porphyry contact; therefore, it is assimilated to one or the other according to which one is closest.

The mineralization types (Table 3.2) are considered the most important mineralization controls. More of the mapped and logged units are actually modeled. Also, different mineralization types are routed to different processing plants. Oxide mineralization is recovered using an acid leaching process with a solvent extraction and electro-winning (SX-EW) recovery plant, while sulphide mineralization (high enrichment, low enrichment, and primary mineralization) is processed in a floatation plant. The units with cuprite are modeled together because they are quite small.

More grouping is done for alteration zones because they are more difficult to accurately map. There are a number of transitional units with mixtures of different alteration events, which complicate their mapping and modeling. Therefore, the tendency of geologists is to model only the major units, see Table 3.3.

This example shows that some variables are mapped and logged, but not necessarily modeled. This example is spe-

Table 3.2 Mineralization mapped, logged, and modeled, October 2001 Resource Model (used with permission from BHP Billiton)

Mineralization types	Mapped/Logged abrev. codes	Modeled (Oct 2001 Model)
Leach	LX	Modeled as such
Green Copper Oxides	OX	Modeled as such
Cuprite	CP	Modeled as Cuprite
Cuprita + Copper Oxides	CPOX	Modeled as Cuprite
Cuprite + Mixed	CPMX	Modeled as Cuprite
Cuprite + Chalcocite + Pyrite	CPCCPY	Modeled as Cuprite
Partial Leach	PL	Modeled as such
Mixed Copper Oxides + Sulfides	MX	Modeled as such
Chacolcite + Pyrite	HE1	Modeled as High Enrichment
Chacolcite + Covelite + Pyrite	HE2	Modeled as High Enrichment
Covelite + Pyrite	HE3	Modeled as High Enrichment
Chalcocite + Chalcopyrite + Pyrite	LE1	Modeled as Low Enrichment
Chacolcite + Covelite + Chalcopyrite + Pyrite	LE2	Modeled as Low Enrichment
Covelite + Chalcopyrite + Pyrite	LE3	Modeled as Low Enrichment
Pyrite	PR1	Modeled as Primary Mineralization
Chalcopyrite + Pyrite	PR2	Modeled as Primary Mineralization
Bornite + Chalcopyrite + Pyrite	PR3	Modeled as such

Table 3.3 Alteration mapped, logged, and modeled, October 2001 Resource Model (used with permission from BHP Billiton)

Alteration zones	Mapped/Logged abrev. codes	Modeled (Oct 2001 Model)
Unaltered	F	Not modeled
Propilitic	P	Not explicitly modeled
Chlorite-Sericite-Clay	SCC	Modeled as such
Quartz-Sericite	S	Modeled as QSC
Potassic	K	Modeled as K-B
Biotitic	B	Modeled as K-B
Advanced Argillic	AA	Modeled as QSC
Clays	AS	Modeled as QSC
Silicified	Q	Modeled as QSC
Potassic-Sericite Transition in Porphyry	QSC	Modeled as QSC
Silicified Sericite-Chlorite-Clay in Andesites	SSCC	Modeled as SCC
Silicified Quartz-Sericite-Clay in Porphyry	SQSC	Modeled as QSC

cific for a porphyry-copper type orebody, but the process of mapping, logging, and modeling geologic variables is general and applies to other types of mineral deposits.

Figures 3.2 and 3.3 show a plan view and a cross section of the resulting interpretation of lithology. Cross sections used to model lithology in this deposit are located 50 m apart, while benches are 15 m high. Only the abundant high-volume units can be represented. Both figures use the same color codes, with Fig. 3.2 showing the correspondence between the unit names and the colors.

3.2 Geologic Interpretation and Modeling

A traditional approach to create a geologic model is to interpret the geologic variables on cross sections and plan maps, then, extend the interpretations to three-dimensional volumes. This is sometimes called deterministic geologic modeling, since it does not carry a measure of uncertainty. The interpreted models are assumed to be exact and accurate.

Geologic interpretation and modeling uses the data and general geologic knowledge gained from other studies in context of the type of deposit. This outside information may include geological knowledge, a plausible theory about the genesis of the deposit, and past experience with similar deposits. Deterministic interpretations are preferred because they are unique and easy to manage, although sometimes difficult and time-consuming to build.

Some basic guidelines for creating good sectional or plan views are worth noting. First, features of interest must be properly drawn and clearly labelled. These features include coordinate axes and a reference datum. The map should also include a title block which includes the title of the drawing, who it was drawn by, and the date it was drawn. Any third party should be able to easily figure out what they are looking at and from what view angle.

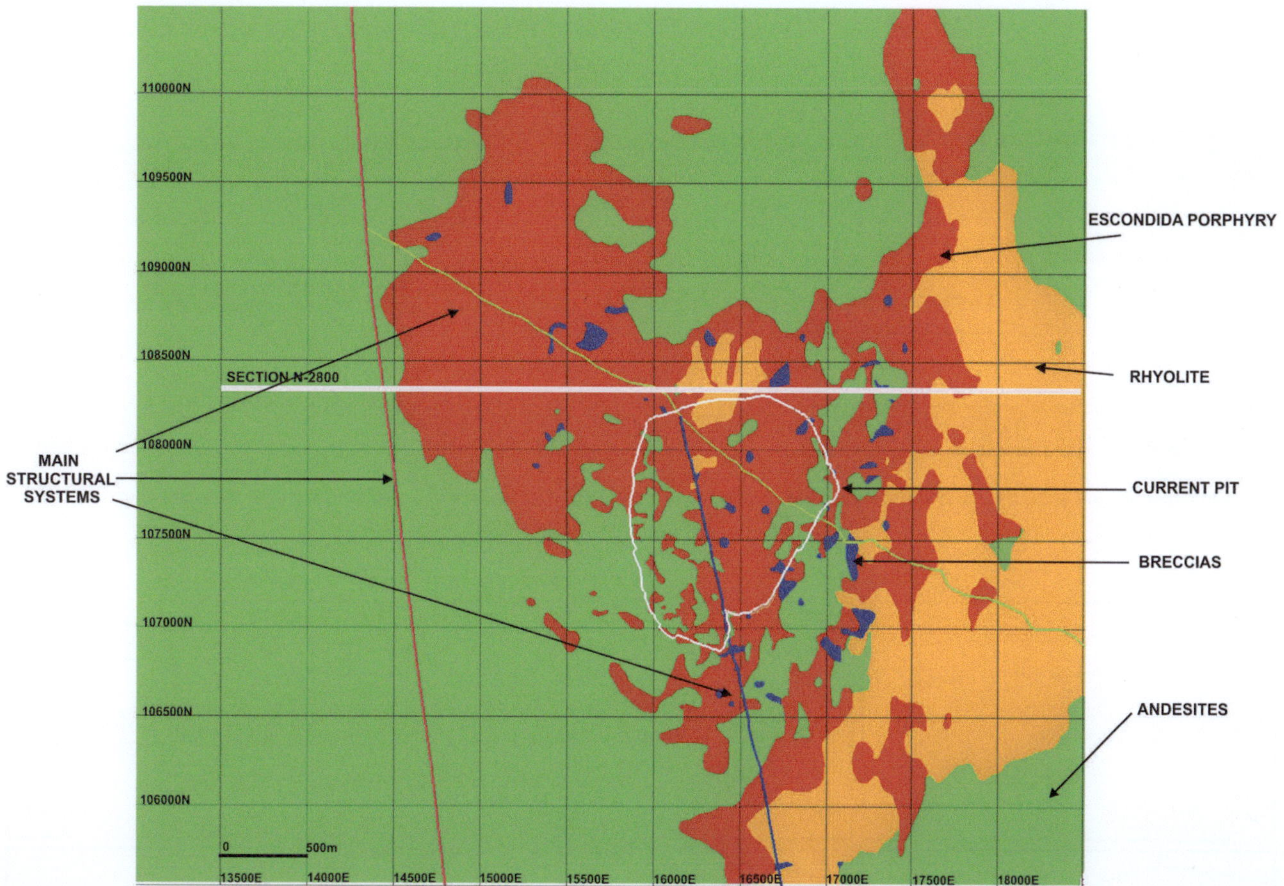

Fig. 3.2 Plan view of the Escondida Lithology model, October 2001, Bench 2800. Note the position of Section N104800, shown in Fig. 3.3

The drawn geological shapes should be based on a sufficient amount of drill hole information and other geologic knowledge that could include a model for ore deposition, surface mapping, and structural and radiometric information. Surfaces are sometimes referred to as Digital Terrain Models (DTM), while wireframes are used to define three-dimensional volumes corresponding to a geologic variable. An alternative to wireframes is to simply extrude or extend the two-dimensional polygonal shapes a fixed distance to either side of the plan of interpretation. Although simpler and quicker, this option often results in an overly simplistic and sometimes inconsistent model.

The shapes should be drawn with a degree of confidence related to the data density. If continuity cannot be established between sections, then the confidence in the interpretation will be poor. The required drill hole spacing for adequate interpretation is related to the mineralization type. A small vein type deposit may have drill hole information spaced 20 m or less, while a massive bulk tonnage deposit may have drill holes spaced 50, 70, or more meters apart. In either case, the geological variable being modeled should be continuous for two or three sections of drill hole spacing units, implying a reasonable degree of confidence in continuity. However, although it should never become part of the published, official resources, it is sometimes necessary to allow for extrapolation of geologic features to aid in future exploration.

Rules about extrapolating at depth or laterally past the last points of information should be clearly stated. A safe option is to avoid excessive extrapolation by creating an outer envelope around the drill hole information to constrain interpretation and modeling.

Drill hole data and a plan view of drill hole locations provide a starting point for geological interpretation. Sections are chosen based on the drill hole distribution. Multiple sections are combined to form a 3-D model that is consistent with information from all sections.

The simplest methods used to obtain models of geologic attributes are based on two-dimensional interpretation, generally done on cross sections. Then, the resulting polygons representing the interpreted shapes are refined on a second set of longitudinal sections. Finally, the model can be refined on plan views. This was done for the lithology model shown in Figs. 3.2 and 3.3.

The order of two-dimensional interpretation depends on the geometry of the deposit. For open-pit, disseminated-type deposits, the final stage should be plan views, because the

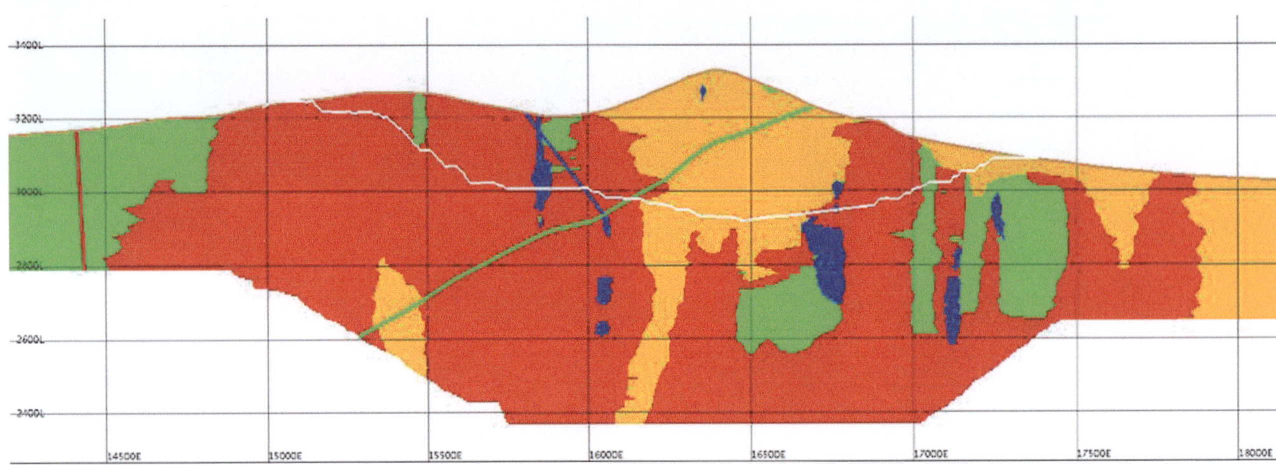

Fig. 3.3 Cross Section of the Escondida Lithology model, October 2001, N108400, looking north. Corresponds to the section line shown in Fig. 3.2

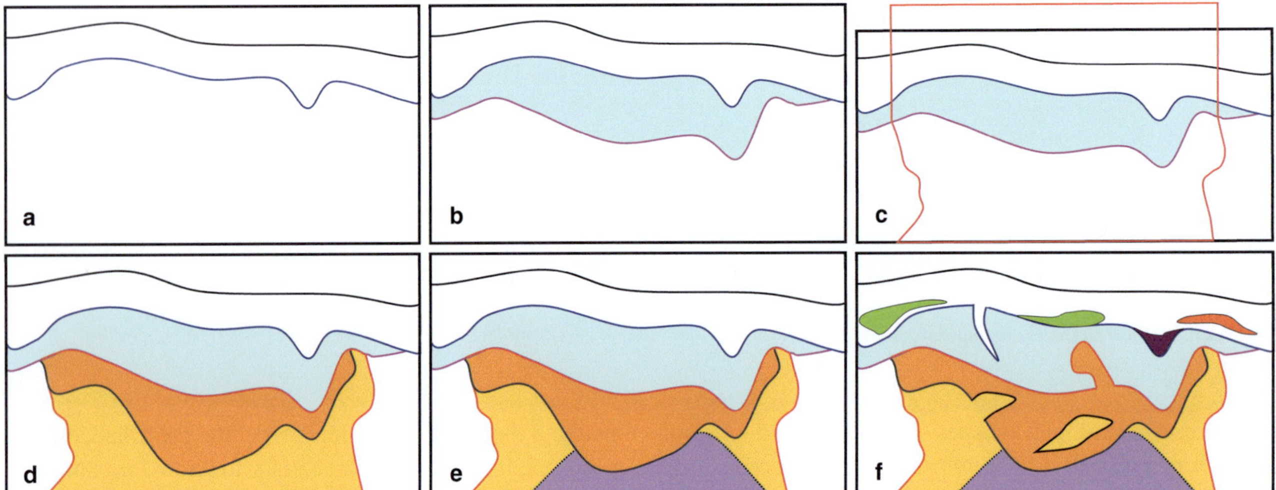

Fig. 3.4 Steps used in developing the Escondida Norte geologic model. Courtesy of Minera Escondida Ltda. **a**) Topography and the interpretation of the Top of Dominant Sulfides (TDS) using prior mineralization models, an alteration model, and structural interpretations. **b**) The Top of Dominant Chalcopyrite (TDCpy) is interpreted next. **c**) The Lateral Limits of the TDCpy establish the inner limit of the outer Pyrite shell. **d**) The Top of Primary Mineralization (TPr) is interpreted next. Three main sulphide volumes are thus generated **e**) The Top of Bornite (TBn) is interpreted next. **f**) Additional "patches" are added on, including above-TDS mineralization, Leached and Oxide remnants within the enrichment blanket, and other isolated volumes

mining engineers will usually plan the operation based on benches. For vein-type deposits, it is likely that the most important view will be cross sectional, although sometimes along longitudinal views. A simplified interpretation would skip one of the two sectional views, for example, the longitudinal section. This may be acceptable for deposits in an early stage of exploration.

An example of surface and volume modeling as done for the Escondida Norte deposit is reproduced here with permission from Minera Escondida Ltda and BHP Billiton. Figure 3.4a–f shows the sequence of steps involved in producing the mineralization model. A combination of surfaces and three-dimensional shapes are used to define the different volumes representing mineralization zones, including patches and remnants of one type of mineralization within other. Although only one set of sections is shown, this process is repeated for orthogonal sections.

The minerals shown in these figures are OxCu (Copper Oxides), Cup (Cuprite), OxFe (Iron Oxides), Chalcosite (Cc), Covelite (Cv), Chalcopyrite (Cpy), Pyrite (Py), and Bornite (Bn). The surfaces shown are Topography; TDS (Top of Dominant Sulfides); TDCpy (Top of Dominant Chalcopyrite), TPr (Top of Primary Mineralization); and TBn (Top of Bornite).

The surfaces were interpreted individually from the top down. In the case of the units above TDS, they were interpolated probabilistically using indicator kriging (see details in Sect. 14.3). Care must be taken to ensure that the surfaces do not cross each other as the interpretation progresses from one section to the next. The proto-ore limit TDCpy shown above the topographic surface (Fig. 3.4c) is only an interpretational technique to produce a fully enclosed volume.

There are different manual and semi-automatic techniques that can be used to produce an interpretation such as this. The specific procedure depends in part on the software available to do the work. Regardless of the details, thorough checking and validation is necessary to ensure that the modeling process occurred as intended.

3.2.1 Distance Functions and Tonnage Uncertainty

Geologic modeling with extensive interpretation and digitization is recommended; trained professionals can understand a great deal about the geometry of the deposit. Stochastic geostatistical techniques such as indicator simulation, truncated (pluri)Gaussian or other techniques often create models that are very random. A relatively recent approach to geologic modeling is to use a signed distance function (DF) that maps the location of boundaries and at the same time allows for an assessment of the uncertainty. This uncertainty is represented spatially by a zone (or bandwidth) that is quantifiable and needs to be calibrated. The DF is calculated directly from individual drillhole samples coded with a distance, rather than a wireframe model. This approach is currently applicable to binary geologic systems, with only two geologic domains, as in for example vein-type deposits, although further development into multivariate systems is ongoing.

Changing the DF impacts the size and shape of the zone of uncertainty. Two parameters, the distance function uncertainty component, *C*, and the distance function fairness component beta, β, are used to modify the DF. The *C* parameter controls the bandwidth and therefore the uncertainty. The β parameter controls the position of bandwidth. With proper calibration, values of *C* and β can result in models that are both accurate and precise.

The DF is the Euclidean distance between different types of samples. The distance is the shortest distance to a sample with a different rock type (vein or non-vein). The distance is given a positive sign in one rock type and a negative sign in another. The contact between samples has a distance function of zero. An isoline connecting successive 'zero' points defines the iso-zero surface or shell.

The vein geometry (and tonnage) uncertainty cannot be calculated directly using an Euclidean distance because it produces a single boundary. This is much like doing a traditional interpretation and wireframing. However, a modified distance function, DF_{mod} can consider the uncertainty using parameters *C* and β, creating a range of probable boundaries. The vein geometry and corresponding tonnage uncertainty can be calculated using these different vein boundaries.

To calculate the distance function, assume that a first sample is non-vein and has an indicator of 0, VI=0. The distance function is the distance to the nearest sample with indicator of 1, VI=1. This sample could exist next to the original sample if located at the contact between vein and non-vein or in a nearby drillhole if located at some distance from the vein, see Fig. 3.5. The actual distance is then modified depending on the value of the indicator VI. Consider the DF:

$$DF = \begin{bmatrix} \left(\sqrt{dx^2 + dy^2 + dz^2} + C\right) & \forall \text{ VI} = 0 \\ \left(\sqrt{dx^2 + dy^2 + dz^2} + C\right) \cdot -1 & \forall \text{ VI} = 1 \end{bmatrix}$$

where, $\sqrt{dx^2 + dy^2 + dz^2}$ is the Euclidean distance between the current point and the closest point with a different VI, *C* is the uncertainty parameter. When the indicator VI is 0, or non-vein, the DF returns a positive value equal to the distance plus the uncertainty parameter *C*. If the indicator I_V is 1 signalling the presence of vein, the DF returns a negative value equal to the distance plus the uncertainty parameter C. The distance from $-C$ to $+C$ is defined as the uncertainty bandwidth.

3.2.1.1 Uncertainty Parameter C

The parameter *C* must be calibrated so that the width of uncertainty is neither too large nor too small. Consider two drill holes (Fig. 3.6), one with a vein intercept, the other without, separated by a distance ***ds*** that represents the typical drill hole spacing. The true vein boundary, or iso-zero boundary of the vein must exist at some location between the two drillholes. The drill hole distance, ***ds***, is the maximum geologically reasonable distance that can be assigned to *C* and is equal to the drill hole spacing. For example, in Fig. 3.6, the mid-point between the holes could be a likely position of the contact, the iso-zero boundaries. However, the vein could extend to almost any point in between, with higher or lower probability depending on the local geology.

The uncertainty parameter *C* is not designed to define the location of the iso-zero boundaries but rather to define a reasonable bandwidth of uncertainty associated with it. The uncertainty bandwidth cannot be greater than the drill hole spacing.

Widely spaced drill holes would suggest a large bandwidth, which would produce large vein boundary and ton-

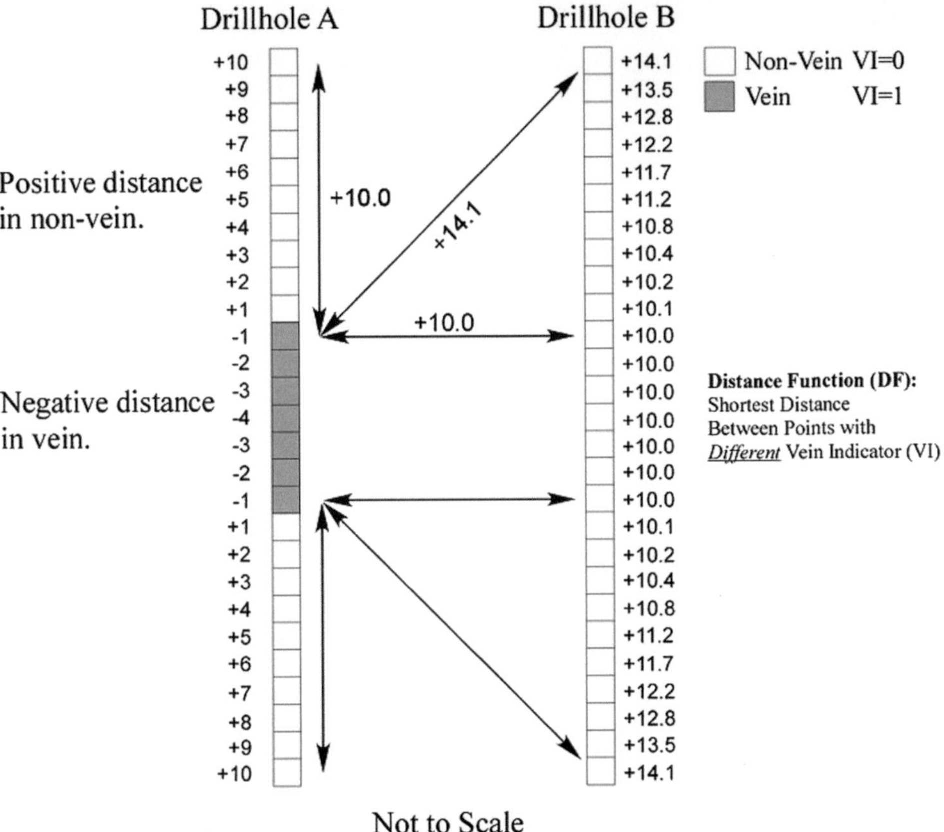

Fig. 3.5 Schematic of distance function. Numbers indicate the distance assigned by the DF. (Taken from Munroe 2012)

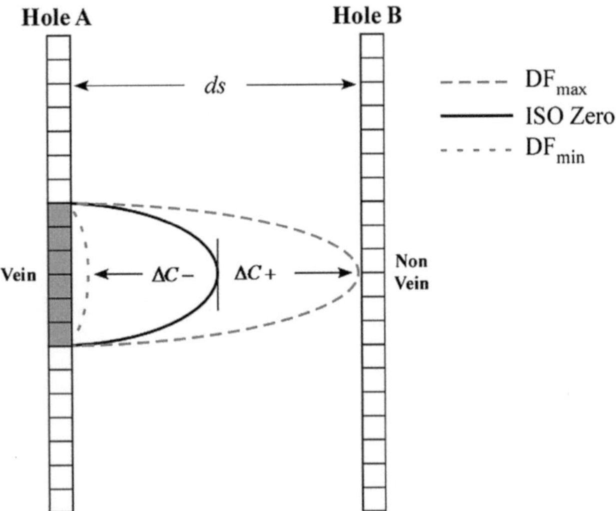

Fig. 3.6 Schematic of the uncertainty bandwidth defined by C, from Munroe (2012)

nage uncertainty. The contrary would be true for closely spaced drill holes. A symmetric variation on a constant C (a traditional "plus or minus") could possibly lead to biases, since it does not incorporate geologic knowledge. A second parameter is defined to center the uncertainty width, beta (β), providing more flexibility in the modeling of the vein geometry.

3.2.1.2 Modified Distance Function (DF_{mod})

The DF is modified in a second step by applying a bias parameter, β, used to center the distribution of estimates. The bias parameter is applied to the original DF:

$$DF_{mod} = \begin{cases} (dist + C)/\beta & \forall \text{ VI} = 0 \\ (dist + C)\cdot\beta & \forall \text{ VI} = 1 \end{cases}$$

The fairness parameter β divides or multiples the original DF depending whether VI indicates non-vein or vein, respectively. A value of $\beta = 1$ returns the original DF value.

The iso-zero surface is the contact between vein and non-vein. This zero point is known and honoured by the data. But at locations away from sampled locations, however, there will be uncertainty as to where the actual position of the contact surface is located. The shape and size of the iso-zero surface is controlled by β; it has the effect of dilating it for the larger values of β (outer dashed ellipse in Fig. 3.7), or eroding it for decreasing values of β (corresponding inner dotted line).

The β parameter is a number typically between 0.1 and 2 and is dependent on drill hole spacing. For example, if the drill hole spacing tends to inflate the vein geometry and overestimate tonnage, then β values greater than 1 are used to decrease it. The parameter β imposes a control on the final surface and makes it possible to adjust the iso-zero surface so that fair and unbiased estimates can be obtained. The calibration of β needs careful consideration.

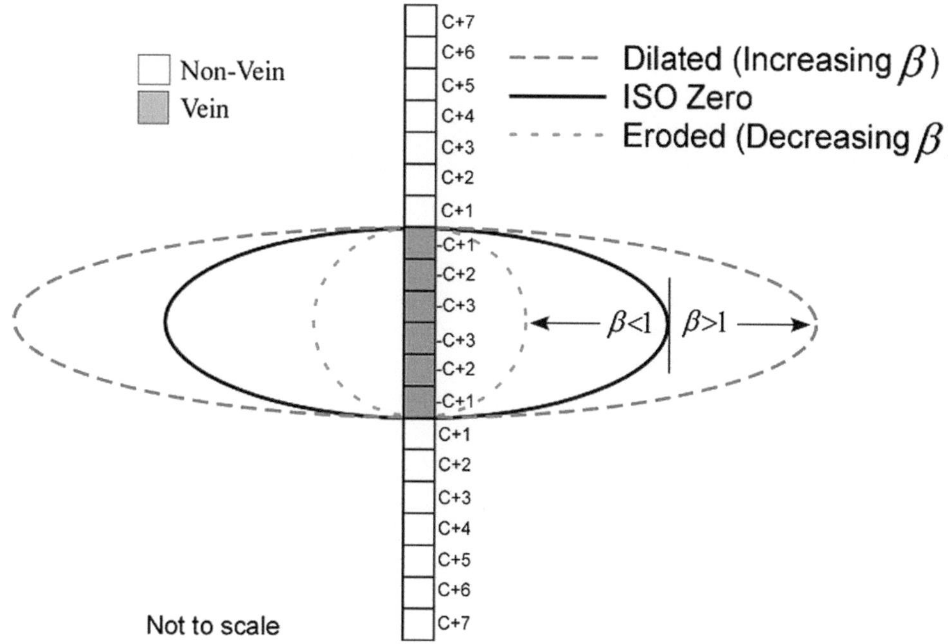

Fig. 3.7 Effect of β on the iso-zero surface. With increasing β, the surface expands; with decreasing β the surface contracts. (Munroe 2012)

3.2.1.3 Distance Function Thresholds

Tonnages can be calculated from the uncertainty bandwidth, the size of which is determined by the uncertainty constant C and the minimum and maximum limits of the bandwidth determined from both C and β. The inner limit of the uncertainty band, DF_{min} is calculated as;

$$DF_{min} = -\frac{1}{2} C \cdot DS \cdot \beta$$

where DS is the drill spacing and is the lower limit defined as one half the distance function of the portion inside the vein structure.

The outer limit of the uncertainty band, DF_{max} is calculated as:

$$DF_{max} = \frac{1}{2} \frac{C \cdot DS}{\beta}$$

and is the maximum limit defined as one half the distance function of the portion outside the vein structure (Fig. 3.8).

The probability thresholds within the bandwidth are defined as a p—probability value. The bandwidth interval is rescaled to [0,1] so that $DF_{min}=0$ and $DF_{max}=1$. The dashed line is the p_{50} and has a p value of 0.5.

The p values are used to define the possible vein geometries and extract tonnages for defined probability intervals by converting individual model cell values into p values. The p value is calculated as:

$$p = \frac{z - DF_{min}}{DF_{max} - DF_{min}}$$

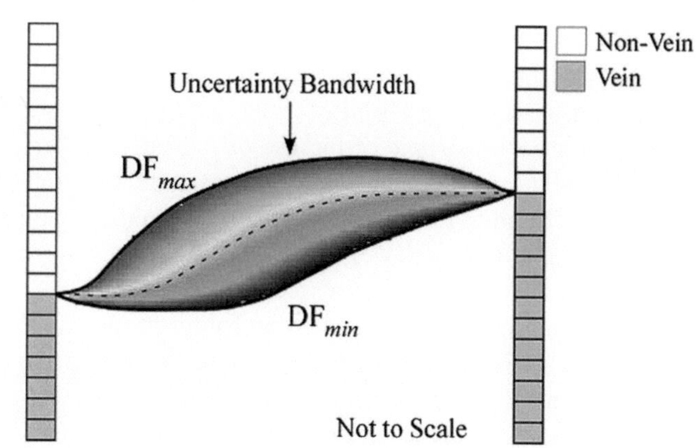

Fig. 3.8 Schematic of uncertainty bandwidth between drillholes. (Munroe 2012)

where z is the estimated DF value. The total tonnage for a particular probability interval p_i, is the total number of cells where $p \leq p_i$. Recall the zone of uncertainty is located between DF_{min} and DF_{max}. If $z < DF_{min}$ then z is certainly located within the vein structure. If $z > DF_{max}$ then z is most certainly located outside the vein structure. By dividing the space between DF_{min} and DF_{max} into a [0,1] interval, we can readily extract tonnages from a mapped distance function for any probability interval.

In order to define those probabilities and vein geometries, the DF is mapped using Simple Kriging on a regular 3D grid. Although other options are available, simple kriging is the preffered option because it allows more control over areas with less or uneven drilling.

3.2.2 Geostatistical Geologic Modeling

A geologic model can also be obtained by using a polygonal (or nearest-neighbor) method, or a geostatistical algorithm. These techniques can sometimes be used if the data is too sparse to confidently draw the interpreted shapes and extend them between sections. One of two basic approaches could be used: (a) a deterministic model that estimates the geological category at each location, or (b) a model that provides a probability of each category at each location.

The polygonal (or nearest-neighbor) technique assigns the geologic attribute to points or blocks in the three-dimensional space according to a fixed rule; for example, each block is assigned the geologic attribute from the drill hole data that is closest to its centroid (Stone and Dunn 1996). It is similar to a geologist drawing the interpreted shapes, except that the computer will not use additional geologic knowledge or judgement to guide the assignment of geology in the three-dimensional space. Often, nearest-neighbor models are used to check the global volumes of interpreted models.

Indicator-based techniques describe the discrete distribution by assigning an indicator to each geologic attribute. The indicators can be kriged simultaneously (Multiple Indicator Kriging) or sequentially one at a time. In either case indicator kriging provides a probability of the geologic variable being present. An example of the use of indicator kriging to sequentially estimate mineralogical zones is presented in Sect. 14.3. The preferred MIK technique has been used with success at multiple sites as well.

More advanced geostatistical techniques include conditional simulation which provide a probabilistic model of the geologic variables. There are three common techniques: (1) sequential indicator simulation, (2) truncated Gaussian simulation, and (3) object-based modeling. These are discussed in Chap. 10.

The basic idea of indicator-based techniques is to calculate the probability of each geologic code at an un-sampled location using all nearby data (Journel 1983). A specific realization is drawn by Monte Carlo simulation and added to the set of conditioning data. All un-sampled locations are visited sequentially. Multiple realizations of the geologic codes are calculated by repeating the procedure with a different random number seed.

Truncated Gaussian simulation requires the geological codes to be assigned to a range of the Gaussian distribution (Delfiner and Chilès 1999). The Gaussian variable is simulated with a geostatistical procedure such as sequential Gaussian simulation (SGS, see Sect. 10.3) and then truncated to get back to geological codes. This approach enforces a particular ordering in the geological codes.

Object-based modeling stochastically positions objects of arbitrary shape and size within a matrix geological code (Deutsch 2002). This approach is applicable when the geological codes are organized in physical shapes that can be parameterized or digitized. Object-based models have been used extensively for siliciclastic sedimentary facies models, for example.

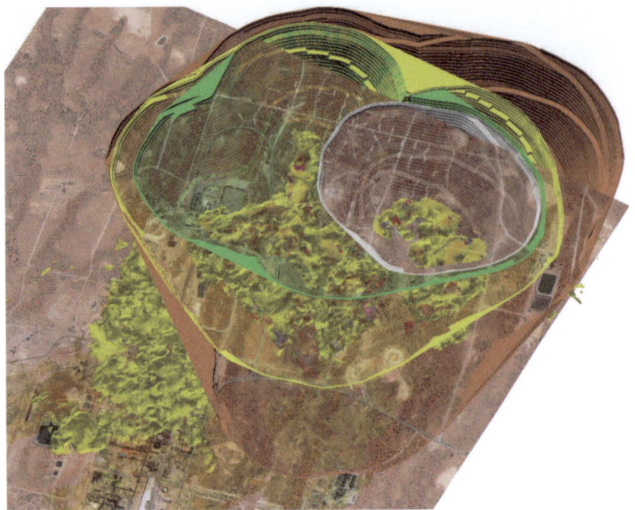

Fig. 3.9 View of Olympic Dam's Au resources with respect to surface infrastructure and showing four different Mining Phases (pits). Courtesy of BHP Billiton

3.3 Visualization

The visualization of geologic and block models is an important part of resource modeling. Visualization of three-dimensional bodies and their relative positions and interactions in space can be used to better understand the proposed geologic and resource model, oftentimes in relation to a planned open pit or underground operation.

Visualization provides qualitative evidence of anisotropic behavior of the attributes as well as an easier grasp of the scales of variability. Geologic variables and grades distributions can be visualized in conjunction with some of the tools described in Sect. 3.3.2. This allows a better understanding of the continuity of the mineralization.

Figure 3.9 shows the Olympic Dam Au resource model model in relation to surface infrastructure and Mining Phases (pits), looking to the North-Northwest. The blocks have been made into a solids model using a grade indicator. Note that some of the resources are left out of all the pits shown.

Another example is shown in Fig. 3.10 where a set of tourmaline veins are shown with topography, the location of the cross sections used to interpret the geometry of the veins, and several existing underground workings and accesses. The view is from underneath. If the visualization is developed within an interactive three-dimensional environment, the relation of the underground workings and the interpreted veins can be used to design future stopes and accesses.

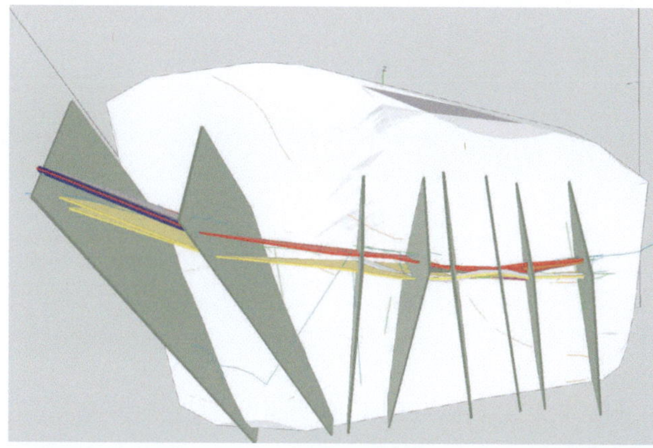

Fig. 3.10 Tourmaline Veins in *Red* and *Yellow*, with interpretative cross sections in *green*, and underground workings and accesses. View is from underneath, courtesy of HRK International

Visualization of geologic models in relation to mine phases or stopes, and in relation to production can also be used as validation. Figure 3.11 shows a detailed view of an area in Mine Area A of the Olympic Dam Underground mine as of March 2006.

Another example is shown in Fig. 3.12, where a conceptual exploration program is described. The geology corresponds to the Lince-Estefanía mine (owned by Minera Michilla S.A., a subsidiary of Antofagasta Minerals, S.A., Chile) and the exploration target is the Susana area where manto-type Oxide Cu mineralization is present. Figure 3.12 shows the Lince pit (as of early 2002) and some important underground workings corresponding to the Estefanía mine.

Another purpose for visualization is to transfer an understanding of geologic knowledge to individuals less familiar with the deposit as well as highlighting mining and engineering details that are relevant to the overall work. Figure 3.13 shows the ultimate pit of Escondida Norte according to its final Feasibility study, looking NW, and showing main haul ore and waste roads.

It should be noted that for all the examples shown, the images are a static version of a three-dimensional rendering of block models, geology, topography, or infrastructure. In practice, geologists and engineers have the ability to move, rotate, and incorporate different information into their visualization exercise, according to the objectives.

There is often a distinction between visualization tools for presentations and visualization tools of actual models. In all cases, there must be some kind of model built to produce the spatial information for the display. Mathematical tools are commonly used to smooth surfaces and volumes to produce more pleasant views; however, these views do not necessarily represent the most accurate model, just as contour lines are generally overly smoothed and do not provide the most accurate estimate of the variable of interest.

The main difference between geologic and engineering plots and model presentation tools is the level of technical work that is based on them. Visualization tools often allow

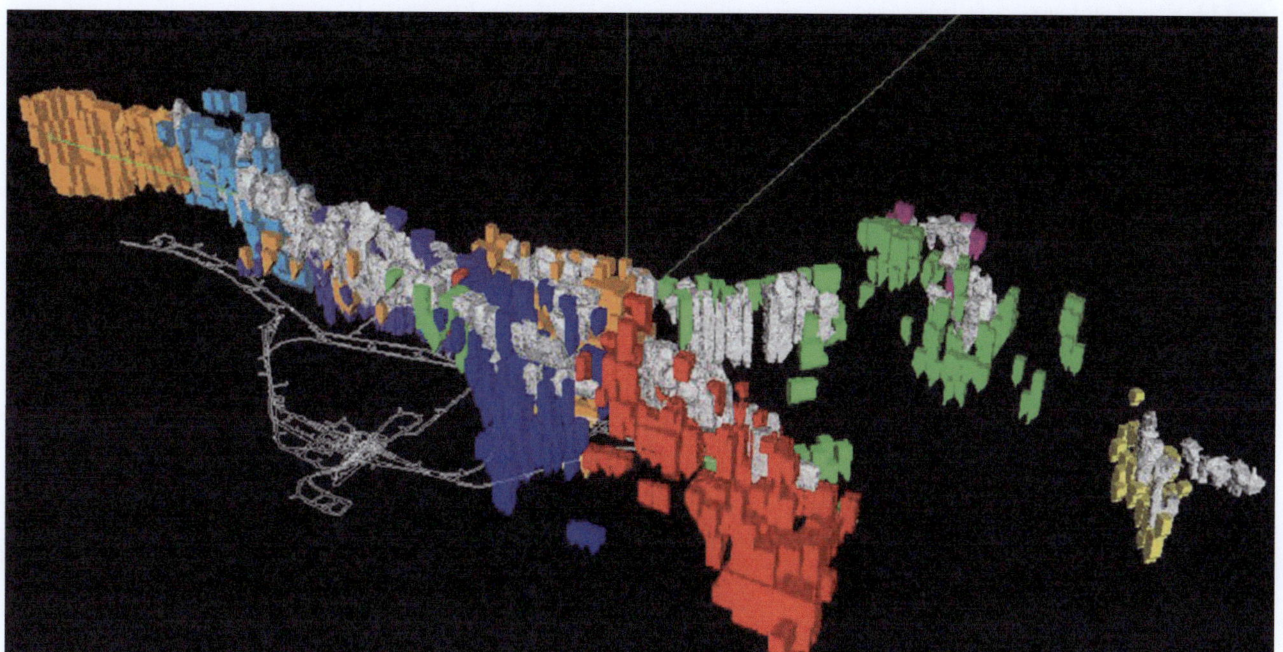

Fig. 3.11 Mined out stopes (in *white*) and planned stopes, Olympic Dam Underground Mine, as of mid-2007. Planned stopes are color-coded by Mine Area, and generally are 30 × 30 × 100 m, and up to 40 × 40 × 200 m, and can be used for scale. Mine workings are also shown. Courtesy of BHP Billiton/Olympic Dam

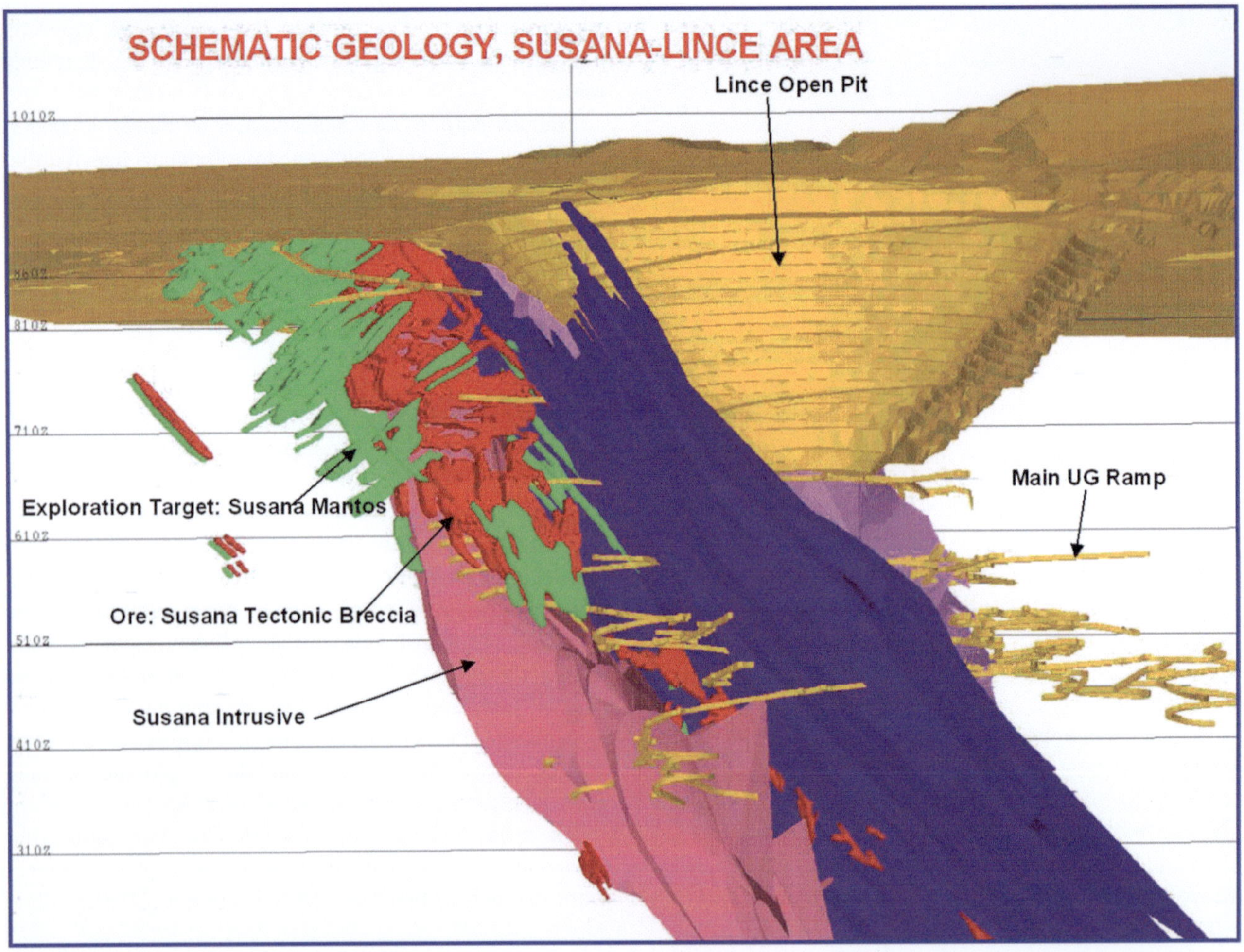

Fig. 3.12 View of the Lince-Estefanía Mine and Mineralization, with Susana Mantos as an Exploration Target area. Topography, the Lince Open Pit and some of the Estefanía Underground workings are shown in light brown. Courtesy of Minera Michilla S.A.

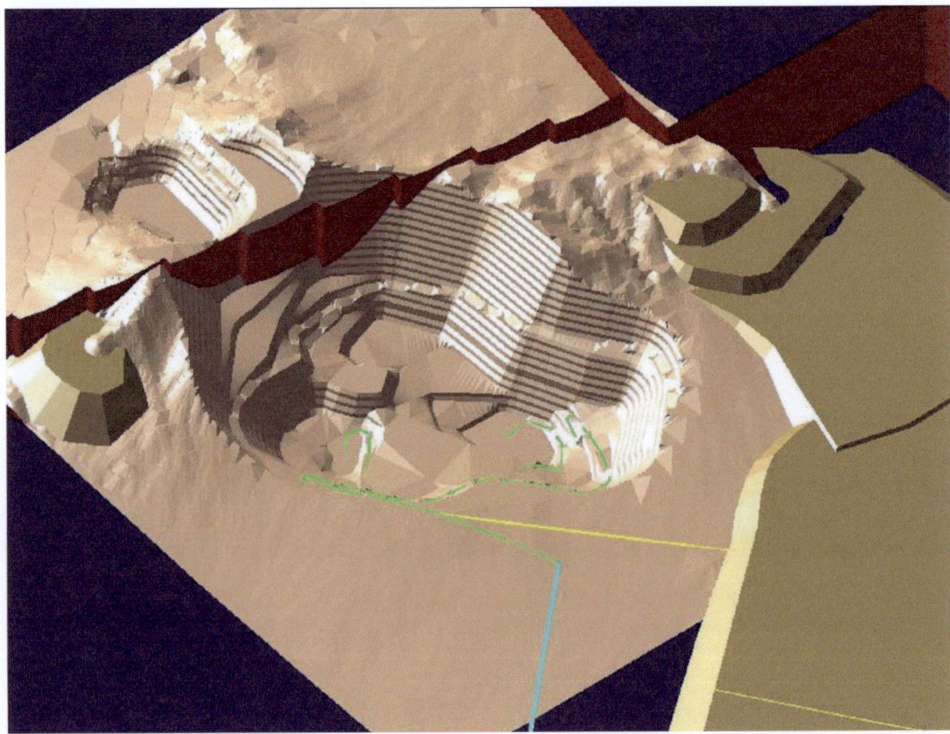

Fig. 3.13 Ultimate Pit for Escondida Norte (Final Feasibility Study), looking NW. Waste Haul Roads are shown in *yellow*, while Ore Haul roads are shown in *green*. Courtesy of Minera Escondida Ltda

little interaction. Most major mining software packages allow for interactive geological and engineering work in three dimensions, and also visualization and presentation of the results. The required basic elements of geologic and engineering plots vary with the objective of the work, but some elements are essential and common to most of them.

3.3.1 Scale

Showing data at appropriate scales is a fundamental attribute of any plot. Accurate information on locations, distances, and volumes should reflect the level of detail and accuracy desired. Scales of 1:2.5 million may be appropriate for metallogenic maps of entire continents, but will not show individual deposits. Regional geologic maps at 1:500,000 up to 1:100,000 are typically used in mineral reconnaissance work. Large, open pit mines may use scales of 1:1,000 to 1:500 for the geologic cross sections and plan views used for interpretation. Smaller underground mines will generally require larger scales, and may work on 1:200 or 1:100 scales for better definition of smaller orebodies and mine stopes. There are no fixed rules about the appropriate scale to use in each circumstance, but the scale should allow for the best accuracy that is practical.

3.3.2 Data

Virtually all work involved in developing geologic models and mineral resource estimations require geologic and assay data from drill hole data, surface trenching or underground stope sampling. It is necessary to plot all relevant information for analysis, interpretation, and checking. Drill hole names and traces (in three dimensions or represented in a two-dimensional section), down-the-hole geologic codes and assays, topography, underground workings, and other relevant surfaces and volumes should be included in plots used for model development and engineering. These may be cross sections, longitudinal sections, or plan maps.

In the case of model-checking, the data presented needs to be relevant, legible, and accurate, and, in the case of block models, for example, should include estimated grades, assigned rock types, assigned resource classification codes, and metallurgical properties. Additional specifics related to model-checking are described in Chap. 11.

Color-coding is used as an aid to visualization, interpretation and engineering work. Suggestions for color-coding include: (a) standardization to common practice, whereby warm colors (yellow, orange, red, magenta) indicate high values, and colder colors (grey, blue, green) indicate lower values; (b) consistency in the definition of the set of color codes (legends) to be used for the different attributes, always applying the same codes to the same attributes and data types; and (c) considerations about legibility; for example, it is very difficult to read numbers plotted using yellow; shades of brown can be used for different surfaces, but they should plot in a distinct way; the point size of the numbers and letters chosen should be such that they do not overprint or abut against each other.

Although it may seem that some of these details are obvious and common-sense, a well-thought out plotting scheme can save significant time and effort, lead to better modeling and validating practices, and should be considered an important part of the work plan.

3.4 Block Model Setup and Geometry

3.4.1 Coordinate Systems

Spatial data, block models and other mineral resource estimation data require the use of coordinate systems to locate relevant attributes. Appropriate coordinates have to be defined based on the available sources of information. The largest scales of work for mapping and map projections are dealt within the field of cartography, and are outside the scope of this book. Geographic coordinates are measured in terms of *longitude and latitude*. Longitudes are referenced to meridians defined by a line joining the North and South Pole and latitude are parallels that result from planes that intercept the idealized spherical earth. Latitude measures the angle between any point in the earth's surface and the Equator. Longitude measures the corresponding angle between the meridian that passes through that same point and the central meridian, arbitrarily defined as that which passes through Greenwich, England. For further discussion on the subject, see Maling (1992); Snyder (1987); and Bonham-Carter (1994).

Geographic coordinates (latitudes and longitudes) are converted for use at smaller scales into planar coordinates, based on projection transformations from a quasi-spherical globe (the earth) onto a plane. These projections introduce some geometric distortion. The type of distortion is used to classify the projections into equal angle, equal area, or equal distance, depending upon whether the projection preserves angular, area, or distance relationships between features of interest.

The most commonly used system is the Universal Transverse Mercator (UTM) system that is an equal angle system established in 1936 by the International Union of Geodesy and Geophysics (Snyder 1987). The UTM system is widely used by national geographic surveys and offices of countries around the world as well as mining companies. The popularity of the UTM system is mostly because regional-scale maps available from geological surveys and governmental offices around the world are UTM-based, generally at a 1:250,000 scale or larger. The world is divided into 60 UTM zones, numbered from west to east and starting with 1° at 180° west.

The UTM projection is such that at scales smaller than 1:250,000 significant distorsions can occur. Also, if the zone of interest straddles two UTM zones, then the conversion from one zone to the next is not straightforward. This is significant because, occasionally, a local triangulation (survey) point may be assigned a UTM coordinate that has not been corrected for geometric distortion. Mining companies will generally define local grids based on truncated UTM coordinates, which, although sometimes suffering from geometric distorsion, have little impact on relative positions at the mining project scale.

Another common practice in geology and mining is to use local, arbitrary "mine" grids to more easily characterize the geometry of the ore deposit. Many ore deposits have principal features aligned in directions other than the Easting and Northing of the UTM coordinates. Common examples of this include vein-type deposits, and deposits where the mineralization is structurally or shear-hosted, or otherwise controlled by elongated features. One specific case study can be found in Barnes (1982).

The drill hole information and the corresponding geological cross sections used to model the deposit should be perpendicular to the main directions of mineralization (strike and dip). The more anisotropic the geometry of the deposit, the more important this aspect becomes. In these situations it is common to rotate the coordinates to align the mine grid north with the main strike or dip direction. Additionally, the conversion often includes a translation to an arbitrary origin, selected such that (a) the Easting, Northing, and Elevation coordinates have different number of digits to avoid confusion and (b) there are no negative coordinates in the new, rotated grid.

The rotations can be done using any convention desired. For example, and using the GSLIB convention (Deutsch and Journel 1997), the first initial clockwise rotation of the original $(x_0; y_0; z_0)$ coordinates by an azimuth angle α can be written as, see Fig. 3.14:

$$\begin{bmatrix} x_1 \\ y_1 \\ z_1 \end{bmatrix} = \begin{bmatrix} \cos\alpha & -\sin\alpha & 0 \\ \sin\alpha & \cos\alpha & 0 \\ 0 & 0 & 1 \end{bmatrix} \cdot \begin{bmatrix} x_0 \\ y_0 \\ z_0 \end{bmatrix}$$

Following the GSLIB convention, the new $(x_1; y_1; z_1)$ can then be rotated a dip angle β counter-clockwise on the Y'-Z plane around the X' axis:

$$\begin{bmatrix} x_2 \\ y_2 \\ z_2 \end{bmatrix} = \begin{bmatrix} 1 & 0 & 0 \\ 0 & \cos\beta & \sin\beta \\ 0 & -\sin\beta & \cos\beta \end{bmatrix} \cdot \begin{bmatrix} x_1 \\ y_1 \\ z_1 \end{bmatrix}$$

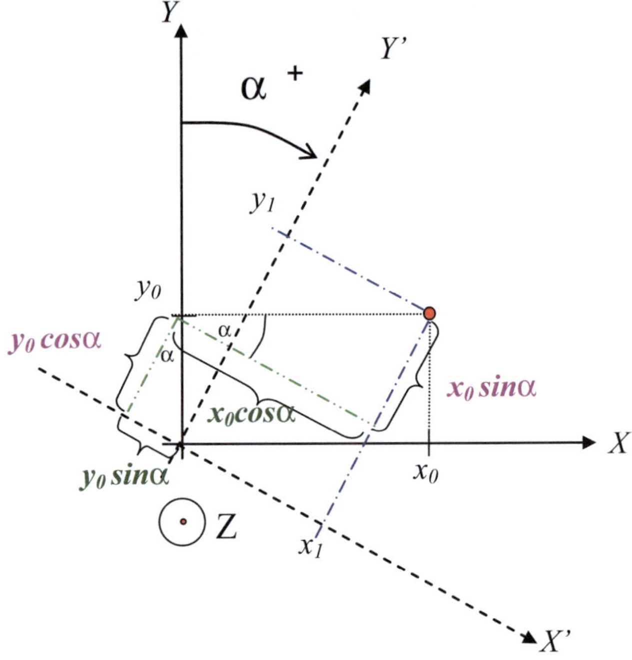

Fig. 3.14 Azimuth rotation on the X-Y plane around the Z axis, GSLIB convention

The third rotation of the $(x_2; y_2; z_2)$ coordinates of a plunge angle φ is counter-clockwise on the X'-Z' plane around the Y' axis:

$$\begin{bmatrix} x_3 \\ y_3 \\ z_3 \end{bmatrix} = \begin{bmatrix} \cos\varphi & 0 & \sin\varphi \\ 0 & 1 & 0 \\ -\sin\varphi & 0 & \cos\varphi \end{bmatrix} \cdot \begin{bmatrix} x_2 \\ y_2 \\ z_2 \end{bmatrix}$$

The three-step rotation described can be summarized in a single step:

$$\begin{bmatrix} x_3 \\ y_3 \\ z_3 \end{bmatrix} = \begin{bmatrix} \cos\alpha\cos\varphi - \sin\alpha\sin\beta\sin\varphi & -\sin\alpha\cos\varphi - \cos\alpha\sin\beta\sin\varphi & \cos\beta\sin\varphi \\ \sin\alpha\cos\beta & \cos\alpha\cos\beta & \sin\beta \\ -\cos\alpha\sin\varphi - \sin\alpha\sin\beta\cos\varphi & \sin\alpha\sin\varphi - \cos\alpha\sin\beta\cos\varphi & \cos\beta\cos\varphi \end{bmatrix} \cdot \begin{bmatrix} x_0 \\ y_0 \\ z_0 \end{bmatrix}$$

At any stage of the rotation, a translation may be applied, as needed. For example, translating the (X, Y) coordinates of the origin of the model can be represented according to the equations shown below and Fig. 3.15:

$$x' = x + x^o$$
$$y' = y + y^o$$
$$z' = z$$

3.4.2 Stratigraphic Coordinates

Specific non-linear transformations are sometimes used to better model certain types of deposits. The use of stratigraphic coordinates is convenient in the cases of sedi-

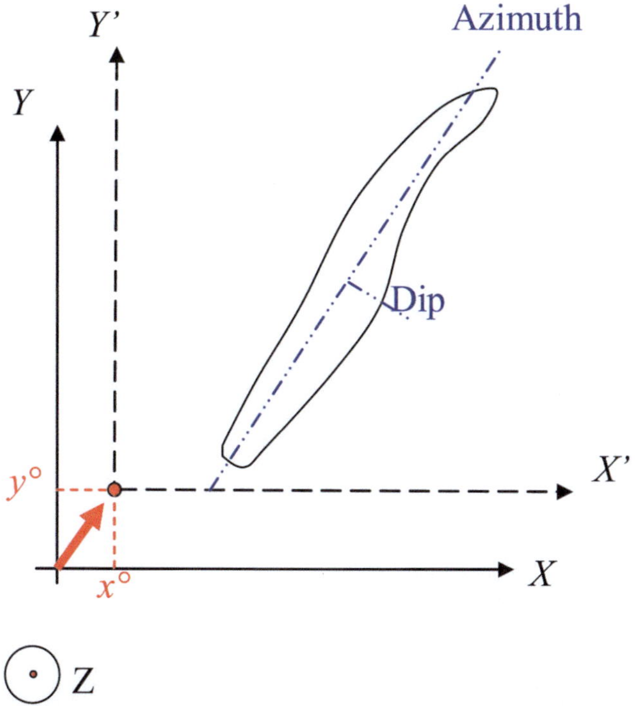

Fig. 3.15 Translation of a (X^0;Y^0) origin to (X';Y')

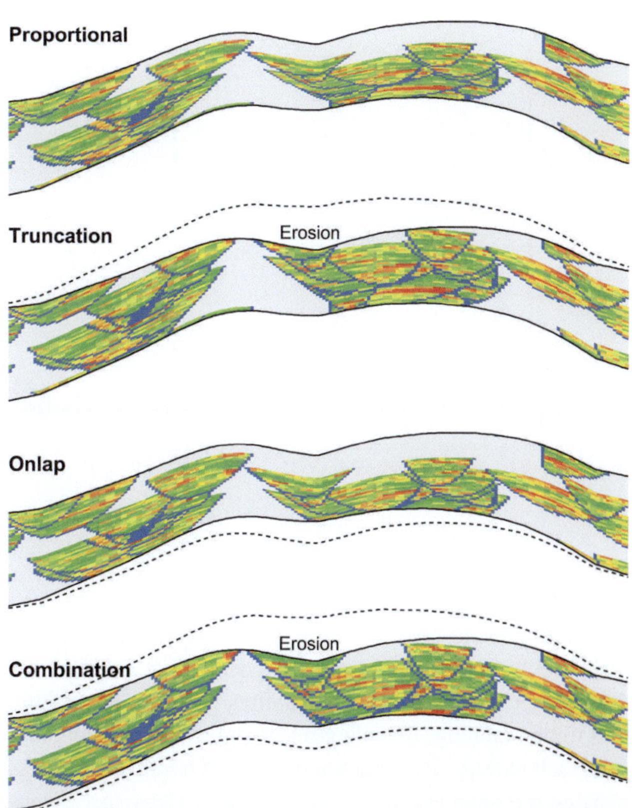

Fig. 3.16 Examples of different correlation styles. (Deutsch 2002)

mentary strata-bound deposits where the mineralization has continuity along the stratigraphic unit of interest. The strata may be folded (plastic) or fractured with little lateral displacements of the units, such that it appears to be continuous. Then, the original Cartesian coordinates X, Y, Z are transformed into a Stratigraphic or Unfolded Coordinate Systems (SCS or UCS). All calculations being carried out in the SCS and then back-transformed to real-world Cartesian coordinates. Examples and details of this type of transformation are shown in David (1988); Dagbert et al. (1984); Sides (1987), or Deutsch (2005).

Modeling 2-D stratigraphic or multiple layered deposits often requires the vertical coordinate to be transformed to a stratigraphic coordinate. Figure 3.16 shows schematic illustrations of different basic correlation types for stratigraphic layers. Each layer is modeled independently with a relative stratigraphic coordinate z_{rel} derived from four surface grids:

$$z_{rel} = \frac{z - z_{cb}}{z_{ct} - z_{cb}} \cdot T$$

where the cb refers to the correlation base and ct refers to the correlation top. The T parameter is the average thickness of the layer under consideration. The relative stratigraphic coordinate is 0° at the base and T at the top.

This transform may be reversed by

$$z = z_{cb} + \frac{z_{rel}}{T} \cdot (Z_{ct} - Z_{cb})$$

Converting all depth measurements to z_{rel} permits modeling of each stratigraphic layer in regular Cartesian X, Y, Z_{rel} coordinates. The locations of all data (geologic variables and grades) are converted back to real Z coordinates before visualization, volumetric calculations, or planning calculations. There will be no back-transformed z values outside the existing interval $z \notin (z_{et}, z_{eb})$ because these locations are known ahead of time and excluded from modeling.

Another transformation could be considered in cases where there is large scale undulation or curvature. There are many variants used of this type of transformation commonly called unfolding. One procedure is to transform the X coordinate to be the distance from an arbitrary center line Y coordinate, which is along the primary direction of continuity and is left unchanged, see Fig. 3.17:

$$x = x - f^{c}(y)$$

where $f^{c}(y)$ is the deviation of the undulating centerline from a straight constant X reference line.

Straightforward normal faults can be handled by correlation grids and stratigraphic coordinate transformation. Reverse faults cause problems because there are multiple surface values at same x, y locations, see Figs. 3.18a and b. The grid can be expanded to avoid overlap. Specialized software is used for more complex gridding schemes.

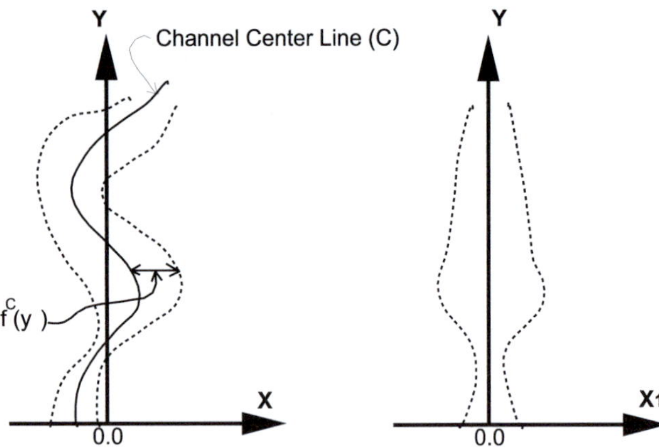

Fig. 3.17 Straightening function

3.4.3 Block Models

Block models describe the three-dimensional volumes with (relatively) small-sized parallelepipeds. Block models are convenient tools for mine evaluation, resource estimation and mine planning, including pit or stope optimization, and mine scheduling. The vast majority of mineral resource estimates are obtained using block models. There are exceptions, particularly in early exploration stages and where no computerized modeling techniques are used. Preliminary estimates can be obtained by manual calculations, typically on cross sections, where areas of influence are drawn on paper, and then projected in between sections. For each of these areas, an estimated grade is obtained by a weighted arithmetic average which, along with the dimensions of the area and an assumed specific gravity, can be used to estimate the tonnage and grade of each section (Stone and Dunn 1995; Sinclair and Blackwell 2002).

The geometry of the block model depends on the characteristics of the deposit, the geological features being modeled, and mine planning requirements, such as equipment size and type to be used by the operation. Block size and geometry is an important decision in resource modeling.

Two-dimensional models may be appropriate for stratigraphically controlled sedimentary or evaporitic deposits including coal, heavy sands deposits, oil sands, some uranium deposits, and some industrial minerals, such as nitrates, iodine, and boron mineral deposits. Typical applications of two-dimensional models relate to modeling surfaces such as topography and surfaces that define different geologic characteristics. Sometimes block models are defined as a stacked set of two-dimensional models in the presence of sequences of mineralized strata.

Three-dimensional grids are used to model massive deposits such as porphyry copper deposits, massive sulphide deposits, skarn, vein-types, and other types of tabular and

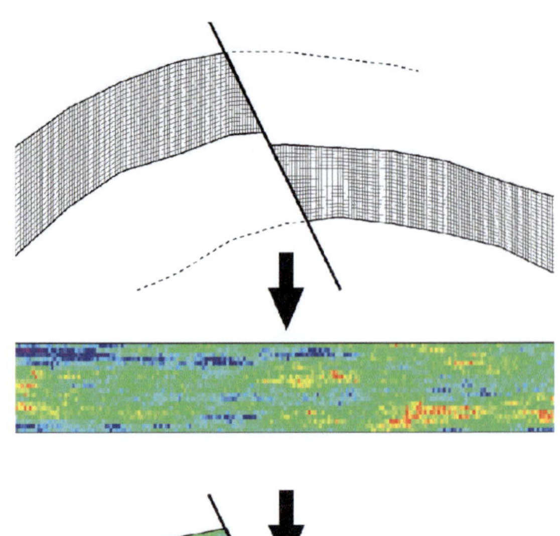

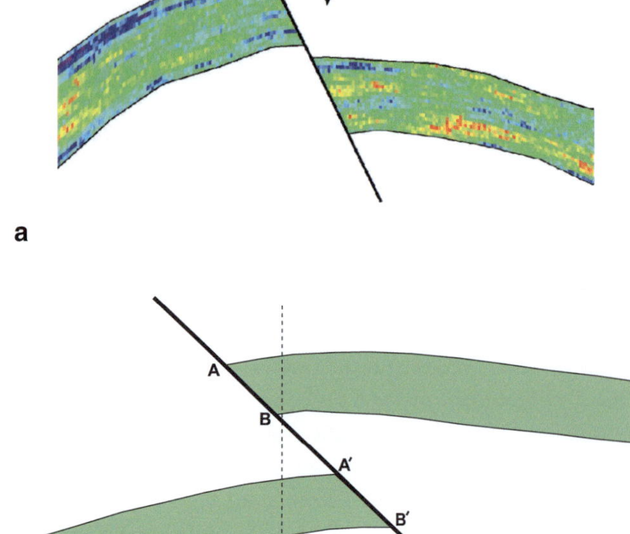

Fig. 3.18 **a** Normal fault transform **b** Reverse fault transform leads to multiple Z values at some locations

sedimentary or pseudo-sedimentary deposits with significant development in the third dimension.

3.4.4 Block Size

The block size should be decided based on the drill hole data spacing and other engineering considerations. Larger blocks are easier to estimate than smaller blocks in the sense that the predicted grades are more likely to be close to the actual grade of the block. On the other hand, too large a block size is not useful for pit optimization and mine planning. Typical mine planning packages work on smaller blocks that discretize to the time interval on which the mine plan is based. For example, for long-term planning, incremental monthly planning units are frequently used. The block size should represent a suitable incremental tonnage. Mining engineers will sometimes sub-divide large panels into smaller units if necessary; the easiest choice is to assign the same panel grade to all smaller units. This was, for example, the scheduling practice at AngloGold's Cerro Vanguardia gold mine in the Argentinean Patagonia in 2002. The practice resulted in erroneous predictions of monthly tonnages and grades to the mill, because the reserve model forecasted smooth variations in grade.

The block size should be less than the data spacing. Journel and Huijbregts (1978, Sect. 5) propose a block size from 1/3 to 1/2 of the drill hole data spacing as an approximate guideline. The logic behind this recommendation is that smaller block sizes will produce artificial smoothing of the model. Adjacent small blocks will receive about the same grade if the same drill hole data are used to estimate them. Too large a block size with respect to the drill hole data grid will not fully utilize the resolution available from the drill hole data. In the context of geostatistical simulation, the grid node spacing and final block size do not depend on data spacing.

Another important aspect that could impact the decision on block size is related to recoverable resource estimation. This involves the concepts of mine selectivity and the Selective Mining Unit (SMU). This is a particularly useful concept in open pit mines, although also applicable to underground stopes as well.

An SMU is defined as the smallest volume of material that can be selectively extracted as ore or waste. In practice, mining selectivity along boundaries is better (smaller) than the chosen SMU size. Moreover, it is rare that an isolated pod of marginal ore or mineralized waste the size of an SMU would be selectively mined. The SMU size is partially subjective, and is based on mining experience, calibration to production, and other characteristics of the operation itself, such as equipment size, anticipated grade control practices and the data available for the final decision. For example, the height of the block is typically related to the mining method: it coincides with the bench height in open pit mines, or with mining lifts in the case of the more common underground methods (cut-and-fill, open stope, and sub-level stoping).

3.4.5 Block Model Geometry

The block geometry should be adapted to the geometry of the deposit, which in turn dictates the geometry of the mining operation. Tabular or vein-type deposits will generally be estimated as two-dimensional deposits, considering that one of its dimensions may be an order-of-magnitude smaller than the other two. There is little opportunity to select material types from within the vein. Massive deposits will commonly be represented by a three-dimensional block network, which can be single- or variable-sized blocks.

For deposits with simple geometry, and in cases where there is limited amount of geologic and drill hole information (early exploration or pre-feasibility level), regular single-size blocks are generally used. The attributes of interest and estimated grades are generally assigned on a whole-block basis.

The blocks are locally subdivided when the deposit has a more complex geometry, or with multiple estimation domains that have large and multiple contact areas between them. The idea is to gain resolution in representing the interpreted geologic contacts into the block model. This is reasonable only when the data density warrants the refinement. Geologic contact modeling is important for incorporating contact dilution in the block model (see Chap. 7).

Care must be taken in defining the sub-blocks (or sub-cells) along contacts, since some commercial software allow for unrealistically small sub-block sizes to be used. This introduces increased computational cost and creates a false sense of resolution. Figure 3.19 shows an example of a block model section with sub-blocks and full size blocks. The full block size is $25 \times 25 \times 15$ m, with sub-blocks defined at a minimum of $5 \times 5 \times 5$ m. Note that in this case the software will automatically define full-size blocks if there are no contacts within the block, and will re-combine the sub-blocks into larger sub-blocks whenever possible to optimize the block model. The grades should be estimated into the main blocks and assigned to the sub-blocks; i.e., grade estimation should be based on the parent block.

An alternative to using sub-blocks in the model is to define a single-size block model, but adding the capability of calculating percentages of geologic and other attributes in each block. Most commercial mining software has the capability of calculating the percentages of material in each block from the interpreted solids or wireframes of geologic attributes.

The grades of the attributes of interest are then estimated for each geologic unit defined in the block, and the final (diluted) block grade is obtained as a weighted averaged of the grade of all the units present in the block. In this case, there are less blocks defined, but with more variables defined within each block.

3.4.6 Block Model Volume and Variables

Block models should cover the entire volume of interest. The units used to define the grid must be consistent with the

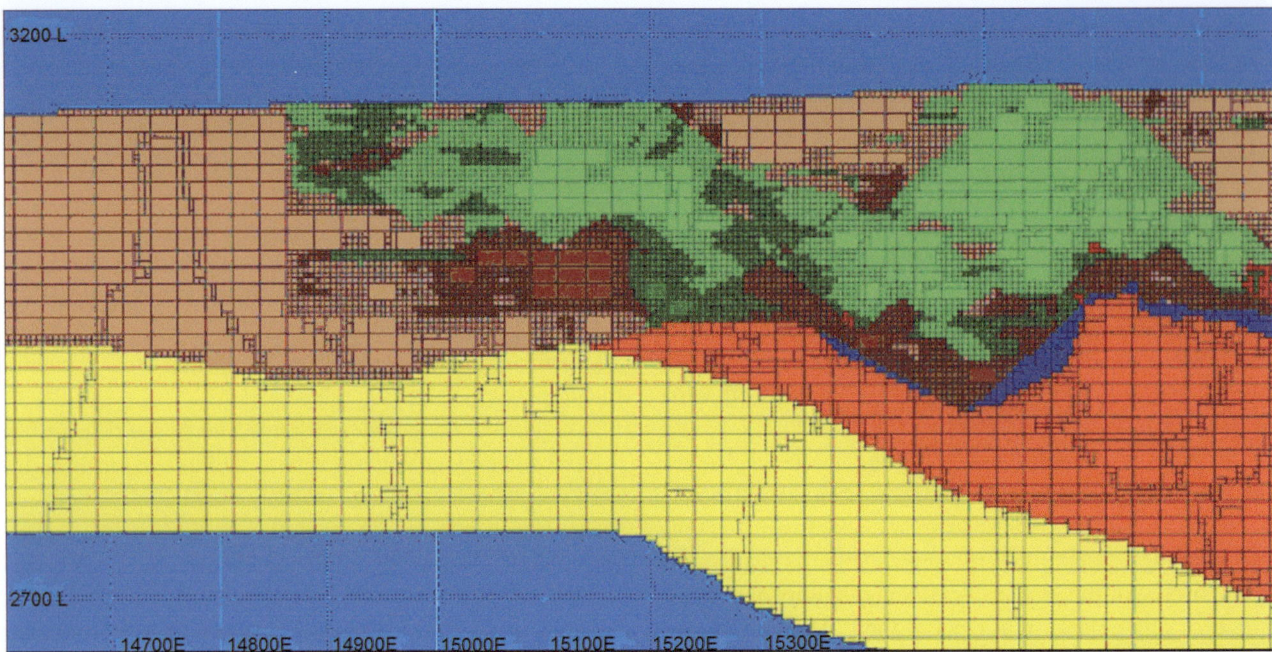

Fig. 3.19 A section of a Block Model showing full size blocks and sub-blocks. Colors represent different mineralization types and full block size is 25 × 25 × 15 m for scale

drill hole database. Open pit mine planning requires that the block model include the ultimate pit that is often larger than the mineralized zones.

The variables defined within a block model should be all of those necessary for mine planning including grades and in situ bulk density of the different geologic units. It may commonly involve percentages, indicators, or other auxiliary variables, such as percentage of air in the block (due to the block being near surface or underground workings) and contacts with certain geologic units that could be either waste or contaminants.

Variables that may be included in the block model, in addition to grid indexes and block coordinates and block sizes (if variable) are geologic attributes such as codes for lithology, mineralization type, degree of oxidation, alteration, structural information, and estimation domains. The estimated grades for all constituents of interest (ore and contaminants) will be in the block model. Other variables include the presence of clays and other consequential units, rock hardness, bond mill indexes, crushing plant throughput prediction, and metallurgical recoveries.

All these variables require storage. The storage requirements can be large for many variables in large block models. Given current computer hardware capabilities the block model size generally is kept to a few million blocks. Clearly this number will continue increasing.

Appropriate procedures should be in place to ensure the quality of all the information developed and contained in the block model, since it forms the basis of the subsequent mine planning and economic decisions. This subject is developed in more detail in Chap. 11.

3.5 Summary of Minimum, Good and Best Practices

Minimum practice related to geologic mapping, logging, and interpretation include the following aspects:

a. A series of written protocols and procedures should be developed, clearly specifying the mappable geologic attributes and the quality control procedures. The protocols should specify the units that need to be mapped and the general description for each unit. A physical example (a rock chip) should be retained as a reference specimen for the particular characteristics. In some instances, color photographs are appropriate. The purpose is to limit the inconsistencies that occur when different geologists map the same rock. The protocols and procedures should be updated on a yearly basis, based on the experience gained in prior campaigns.
b. There should be clearly specified procedures for logging the mapped information and entering this information into the computerized database. At a minimum, the specifics of the procedure, supervisors involved, and timing of the process should be included. Quality control procedures such as double-entry (if manually entering the

data) or external reviews of the information should be specified.

c. All original drill hole information should be properly stored for possible future use: laboratory checking, re-mapping, or for other types of reviews. The information should be electronically stored in no less than two different locations, one of them away from the project or operation site. In addition, all information relating to each drill hole, including topography, drill hole deviation measurements, mapped geologic attributes, and a copy of the assays returned should be available in hardcopy with a single folder for each drill hole.
d. Explicit definition and modeling of at least the most important mineralization controls is required. In the case where no strong geologic controls are evident, the geometry of the deposit and the data extents should be used to limit the resource model. If the deposit has no natural limits, then some form of grade outline may be required to constrain grade interpolation.
e. The geologic interpretation should be completed on two orthogonal planes (cross sections and plans, for example), or, if modeling a vein or tabular deposit, a single cross sectional set of views, orthogonal to the mineralization strike, should be used. Plan or longitudinal views should be used for proper control of the main features.
f. The geologic interpretation should be used to code the block model. The interpreted geology can be converted into solids using simple techniques such as extrusion adjacent sections to or with areas of influence. At the earliest stages of exploration, a computerized geologic modeling (using a nearest-neighbor technique, for example) could be used to assign geologic codes to blocks, instead of explicitly interpreting geology.
g. Simple visual checks should be made to ensure that the blocks have been correctly flagged with the geologic attributes. These checks may include plotting sections and plans of drill hole information against assigned codes for all geologic variables modeled in different geologic units.
h. The block model should consider the deposit geometry and the characteristics of the (potential) mining operation. The geometric considerations should include volume, shape of the blocks, the block sizes adequate for the drill hole density, the use or not of sub-blocks for better geologic contact definition, and the mining selectivity envisioned.

In addition to the above, good practice should include the following:

a. Periodic internal checks should be made on all aspects of the database, including the database geology against the original log sheets. This review should include at least 20 % of the incremental drill holes, and should be done after each major drilling campaign. Most mining exploration projects and operations work with a yearly budget that defines the periodicity of the drilling campaigns.
b. Detailed definition of demonstrable geologic controls should be available, with properly documented geologic and statistical supporting evidence. This should include reports or memoranda describing the different controls, and which geologic variables should be explicitly modeled. Scatterplots and Q-Q plots (Chap. 2) should be used to describe the relationships between grades and each geologic variable.
c. Whenever possible, two-dimensional surfaces should be used to interpret planar geologic features, and three-dimensional solids should be created to represent volumes, including vein and tabular deposits of significant width.
d. A clear audit trail and documentation should be left to facilitate third party reviews of the work. This includes working and final sections and plans in paper (typically at a 1:500 scale), as well as procedures, reports and memoranda documenting the internal checks, and the step-by-step process followed.
e. The geologic model should be checked for volumetric biases. The declustered drill hole database represents the proportions or relative volumes among the different modeled geologic attributes. These relationships should be maintained in the model. Statistics and proportions of each geologic variable (code) in the database and in the resulting block model should be compared. If the observed differences are larger than 10 %, then a clear justification should be made for accepting such a difference.
f. The block model should characterize the contact between the different geologic attributes to the level of detail warranted by the data. This can be done through the use of partial blocks or sub-blocks. The re-blocked model should preserve the information obtained at the better resolution.

Additionally, best practice includes the following:

a. Multiple procedures and checks should be in place to ensure constant quality of the geologic database. Automatic mapping, logging and input into the computer database in the field is recommended through the use of computerized logging. The database where the information is stored should be relational and a sufficient number of automatic basic checks should be in place to facilitate database maintenance.
b. Exhaustive and detailed work on the definition of geologic controls, including a ranking of importance, if appropriate, should be done and available as part of the audit trail. The geologic descriptions and the statistical methods used should be detailed, including analysis such as Classification and Regression Trees (CART), possibly varying from one sector of the deposit to another.
c. The geology should be interpreted on all three orthogonal planes (cross sections, longitudinal section, and plan views). The geological controls and limits should be mod-

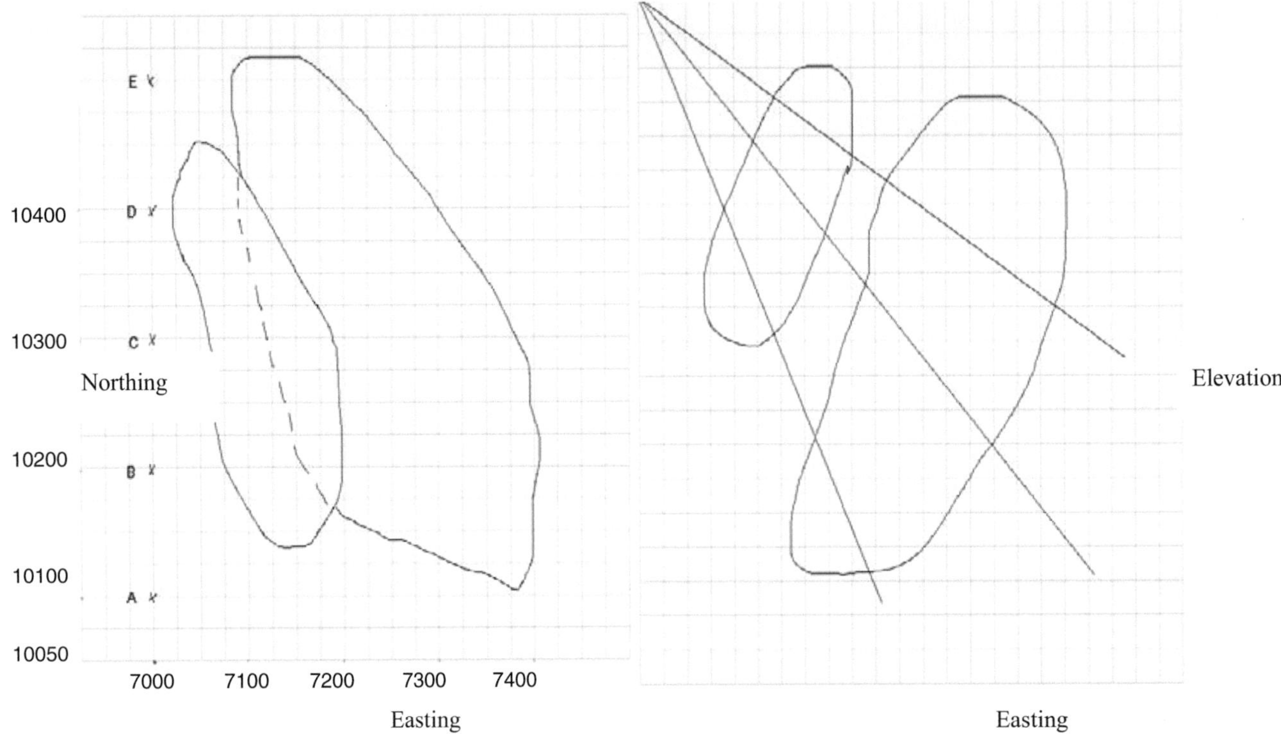

Plan view of the deposit

Sectional view (looking north)

eled in true 3-D avoiding projecting to the plane of interpretation. Thorough manual and visual checks should be performed on each set including plotting at an appropriate scale the working and final sections and plans.

d. The full set of working and final sections should be available for auditing. These should be stored as historical documentation with the full drilling program report. The modeling procedures and criteria, reports of internal checks and external audits, and production reconciliation information (if available) should be archived in the same place.
e. Wireframes should be used to build solids (as opposed to volumes obtained from sectional extrusions), and a full set of visualization tools should be used in the presentation of the modeled geology.
f. The interpreted geology should be dynamically updated as new drill hole information becomes available. This is particularly important for operating mines where infill drilling is done on a regular basis and production data also becomes available on a regular basis.
g. The geologic model should be checked against geologic flagging obtained through nearest-neighbor estimation. Volumetric differences and biases should be explained or resolved, as appropriate.

3.6 Exercises

The objective of this exercise is to perform some basic geometric modeling and work with conventional Cartesian coordinate systems. Some specific (geo)statistical software may be required. The functionality may be available in different public domain or commercial software. Please acquire the required software before beginning the exercise. The data files are available for download from the author's website—a search engine will reveal the location.

3.6.1 Part One: Vein Type Modeling

Consider 15 drill holes—3 drill holes from each of 5 drill stations (labeled as A, B, C, D, and E) to delineate a deposit with two mineralized zones. Drill stations are aligned N-S and spaced 100 m, see Figure below.

The deposit is in the shape of two steeply dipping en echelon ore lenses where the strike direction of the lenses is approximately northwest. The lower limb is not always mineralized. An estimate of the total volume of mineralized resource is required. The exercise is to develop sectional drawings from drillhole data, derive level maps from the sectional drawings, and generate volumetric estimates.

Question 1: Using the drill hole information given below, generate cross sectional views (easting-elevation) for 10100N, 10200N, 10300NE, 10400N and 10500N looking north. Label each drawing appropriately including the location of the drill station, the trace of the drill holes, an interpreted outline of the ore body, and other labels and notes. Your interpretation of the ore body limits should consider the drillhole data, the conceptual model briefly described above, and realistic limits for extrapolation.

Question 2: To check the consistency of the five cross sections, generate two N-S sections at 7200E and 7400E, and two level maps at elevations 700 and 500 m. On each drawing, include the location of the drill stations, the trace of the drill holes, your interpreted outline of the ore body, and appropriate labels and notes. Revise your interpretation if there are abrupt discontinuities in the sectional or level drawings.

Question 3: Using the five cross sections from Question 1, estimate the volume of ore contained within the deposit. Critically evaluate your results. In particular, consider the potential for error, how to minimize such error, and how you would assess uncertainty.

Drill Data:

Drill Station Data

Station A	**Collar:** 10100N 7000E		**Elevation:** 1000m		
Hole #	From	To	Dip	Azimuth	Rock Type
DH1	0	595	-33	90	waste
DH2	0	725	-51	90	waste
DH3	0	800	-66	90	waste

Station B	**Collar:** 10200N 7000E		**Elevation:** 1000m		
Hole #	From	To	Dip	Azimuth	Rock Type
DH4	0	331	-29	90	waste
	331	390	-29	90	ore
	390	474	-29	90	waste
	474	609	-29	90	ore
	609	701	-29	90	waste
DH5	0	302	-44	90	waste
	302	407	-44	90	ore
	407	462	-44	90	waste
	462	634	-44	90	ore
	634	725	-44	90	waste
DH6	0	320	-60	90	waste
	320	408	-60	90	ore
	408	512	-60	90	waste
	512	629	-60	90	ore
	629	712	-60	90	waste

Station C	**Collar:** 10300 N 7000E		**Elevation:** 1000m		
Hole #	From	To	Dip	Azimuth	Rock Type
DH7	0	323	-30	90	waste
	323	407	-30	90	ore
	407	491	-30	90	waste
	491	617	-30	90	ore
	617	659	-30	90	waste
DH8	0	310	-47	90	waste
	310	420	-47	90	ore
	420	487	-47	90	waste
	487	646	-47	90	ore
	646	682	-47	90	waste
DH9	0	344	-61	90	waste
	344	445	-61	90	ore
	445	512	-61	90	waste
	512	604	-61	90	ore
	604	701	-61	90	waste

Station D	**Collar:** 10400N 7000E		**Elevation:** 1000m		
Hole #	From	To	Dip	Azimuth	Rock Type
DH10	0	487	-28	90	waste
	487	601	-28	90	ore
	601	623	-28	90	waste
DH11	0	302	-46	90	waste
	302	390	-46	90	ore
	390	466	-46	90	waste
	466	604	-46	90	ore
	604	688	-46	90	waste
DH12	0	327	-59	90	waste
	327	373	-59	90	ore
	373	470	-59	90	waste
	470	621	-59	90	ore
	621	645	-59	90	waste

Station E	**Collar:** 10500N 7000E		**Elevation:** 1000m		
Hole #	From	To	Dip	Azimuth	Rock Type
DH13	0	445	-33	90	waste
	445	562	-33	90	ore
	562	593	-33	90	waste
DH14	0	470	-47	90	waste
	470	588	-47	90	ore
	588	632	-47	90	waste
DH15	0	478	-58	90	waste
	478	600	-58	90	ore
	600	658	-58	90	waste

3.6.2 Part Two: Coordinate Systems

Consider a tabular deposit with a strike direction at an azimuth of 70° (East of North). The dip is at −35° (downward) to the Northwest. The original X, Y, and Z coordinate system is aligned toward the East, North and Elevation, respectively.

Question 1: Write the matrix equation to rotate the $\boldsymbol{x}$, $\boldsymbol{y}$ system to the $\boldsymbol{x}'$, $\boldsymbol{y}'$ system (so that $\boldsymbol{x}'$ is in the dip direction and $\boldsymbol{y}'$ in the strike direction). Write the matrix equation to rotate the $\boldsymbol{x}'$, z system to the $\boldsymbol{x}''$, z' system (so that $\boldsymbol{x}'$ is down the dip direction). Show the matrix equations in terms of an azimuth angle (α) and dip angle (β) and in numeric form. Define all your terms and comment on your results.

Question 2: Write the single matrix equation to perform the complete data transformation from x, y, z to x'', y', z'. Show the matrix equation in terms of an azimuth angle (α) and dip angle (β) and in numeric form. Define your terms and comment.

Question 3: A vertical drillhole intersects an apparent thickness of 11.5 m. What is the true thickness measured perpendicular to the deposit? Describe your methodology and comment.

References

Barnes TE (1982) The transformation of coordinate systems to model continuity at Mt. Emmons. In: Proceedings 17th APCOM, AIME, New York, pp 765–770

Bateman AM (1950) Economic mineral deposits. Wiley, New York, pp 316–325

Bonham-Carter GF (1994) Geographic information systems for geoscientists: modeling with GIS. Elsevier Science Ltd., New York, p 398

Dagbert M, David M, Corcel D, Desbarats A (1984) Computing variograms in folded strata controlled deposits. In: Verly et al. (ed) Geostatistics for natural resource characterization 1, Reidel, Dordrecht, Holland, pp 70–90

David M (1988) Handbook of applied advanced geostatistical ore reserve estimation. Elsevier, Amsterdam

Delfiner P, Chilès (2012) Modeling spatial uncertainty, 2nd edn. Wiley, Hoboken, p 699

Deutsch CV (2002) Geostatistical reservoir modeling. Oxford University Press, New York, p 376

Deustch CV (2005) Practical unfolding for geostatistical modeling of vein type and complex tabular mineral deposits. In: Dessureault S, Ganguli R, Kekojevic, Dwyer (eds) Proceedings of the 32nd international APCOM symposium, published by the Taylor and Francis Group, London, pp 197–202

Deustch CV, Journel AG (1997) GSLIB: geostatistical software library and user's guide, 2nd edn. Oxford University Press, New York, p 369

Guilbert JM, Park CF Jr (1985) The geology of ore deposits. W.H. freeman and Co, New York, p 985

Hartman H (ed) (1992) SME mining engineering handbook, 2nd ed, vol 1. Society for Mining, Metallurgy, and Exploration, Littleton, CO

Journel AG (1983) Non-parametric estimation of spatial distributions. Math Geol 15(3):445–468

Journel AG, Huijbregts ChJ (1978) Mining geostatistics. Academic, New York

Maling DH (1992) Coordinate systems and map projections, 2nd edn. Pergamon Press, Oxford, p 476

Munroe MJ (2012) A methodology for calculating tonnage uncertainty in vein type deposits. MSc Thesis, University of Alberta, p 142

Peters WC (1978) Exploration and mining geology, 2nd ed. Wiley, New York

Sides EJ (1987) An alternative approach to the modeling of deformed stratiform and stratabound deposits. In: Proceedings of the 20th APCOM symposium, vol 3, SAIMM, Johannesburg, pp 187–198

Sinclair AJ, Blackwell GH (2002) Applied mineral inventory estimation. Cambridge University Press, New York, p 381

Snyder JP (1987) Map projections – a working manual. United States Geological Survey Professional Paper 1395, U.S. Government Printing Office, p 386

Stone JG, Dunn PG (1996) Ore reserve estimates in the real world. Soc Econ Geol Special Publication (3), p 150

4 Definition of Estimation Domains

Abstract

Estimation of grades proceeds within domains defined on the basis of geological and statistical considerations. The definition and modeling of these domains is an important step in mineral resource estimation. This Chapter presents practical aspects of the development of estimation domains, the limitations faced when defining these domains, details of the modeling of estimation domains, and the most commonly used methods to assign estimation domains to a resource block model.

4.1 Estimation Domains

Estimation domains are the geological equivalent to geostatistical stationary zones and are defined as a volume of rock with mineralization controls that result in approximately homogeneous distributions of mineralization. The spatial distributions of grade exhibit consistent statistical properties. This does not mean that the grades are constant within the domains; however, the geological and statistical properties of the grades facilitate its prediction.

The concept of statistically homogeneous populations is termed stationarity. Stationarity is a two-fold decision. First, there is a choice of the data to pool together for common analysis. Second, there is a choice of how statistics such as the mean vary by location within the domain. Stationarity is a property of the random function model (Isaaks and Srivastava 1989) and is not an intrinsic characteristic of the variable. It is a decision made by the resource estimator and is necessary to make inferences. Stationarity was formally defined by Matheron (1962–1963) in the context of geostatistics and is also discussed in Chap. 6.

Exploratory data analysis (EDA) may indicate the existence of several populations with significantly different summary statistics. The understanding of the statistical characteristics of the data, coupled with geologic knowledge, leads to subdividing the deposit into domains for estimation. This is considered more reasonable than taking the entire deposit at one time. Domain definition depends on the availability of enough data to reliably infer statistical parameters within each domain. Moreover, the domains must have some spatial predictability and not be overly mixed with other domains.

A good definition of estimation domains is very important. The consequences of defining inadequate estimation domains are rarely evaluated. It is common to confuse the concepts of geologic and estimation domains. Geologic domains are commonly described by a single geologic variable. Estimation domains are defined by a set of mineralization controls and may contain more than one geological domain.

In multi-element deposits it is common to assume that the estimation domains defined for the main element/mineral of interest applies to all secondary elements that may be present. In practice, different grades are controlled by different geologic variables, and thus they may be predicted using different estimation domains.

For example, porphyry deposits with copper and gold mineralization may exhibit an inverse spatial relationship, that is, gold may not leach through weathering as copper does. Gold may form a cap on the upper part of the deposit. In such cases, copper and gold should be modeled using different estimation domains. In epithermal deposits, gold and silver mineralization may exhibit little correlation since they are deposited differently. Estimating gold and silver using the same estimation domains would lead to suboptimal results.

Estimation domains must make spatial and geologic sense (Coombes 2008). The combination of geologic variables used to define the domains must have spatial and geologic characteristics that are recognizable in drilling and/or production data. The estimation domain must be sufficiently

M. E. Rossi, C. V. Deutsch, *Mineral Resource Estimation,* DOI 10.1007/978-1-4020-5717-5_4,

represented in the database and in the deposit. These conditions provide constraints on what can realistically be modeled in practice.

4.2 Defining the Estimation Domains

A thorough stepwise approach is suggested here. It is based on a combination of geological and statistical analyses. This approach is more detailed and time-consuming, but it provides better support for estimation. The concept is based on decomposing the problem by describing and modeling the relationships between each geologic variable. The combination of variables results in a matrix that ranks the most critical grade controls as identified by the data. These should be explained in terms of plausible natural processes, to ensure that the controls derived from the data are consistent with known geology.

Development of the grade domains begins and ends with geologic knowledge. The first step is to define the geologic variables that are used as the building blocks for the estimation domains definition. Typical variables mapped from drill hole data include lithology, alteration, mineralogy, weathering (oxide/sulfide, for example), and structures or structural domains. Not all these variables are always mapped; some may not be relevant for a particular deposit type.

The second step is to decide the specific geologic variables that are the most important. This is based on geologic considerations, overall abundance within the deposit, and drill hole information.

Third, estimation domains based on all reasonable combinations of the geologic attributes are defined. Consider, for example, 3 geologic attributes each with 4 variables, and thus a total of 64 theoretically possible estimation domains. For example, porphyry, andesites, breccias, and dacites could be the 4 variables of lithology in a porphyry copper-type deposit. Data abundance will filter out a number of these. Consideration of practical aspects will further reduce the number of theoretical domains, such as existing or planned mineral processing facilities. In copper, gold, and many other precious and base metal deposits, for example, it is not advisable to mix oxide and sulfide mineralization, since they are frequently treated at separate processing plants, or, if one of the two metallurgical types is small in volume or low grade, it may be simply stockpiled. Another criterion often used is proximity: certain units may be at the periphery of the deposit, and therefore should not be mixed with units at the central portion of the deposit.

The fourth step involves a statistical description of the initial domains. The main purpose is to remove or group domains according to geologic considerations. Variables that have little representation in the database should be removed, regardless of whether they represent a strong mineralization control or not. A rule of thumb threshold is 1 % of the total number of intervals in the database, although this is dependent on the total size of database.

Next, statistical comparisons between the initial domains accepted will often lead to grouping. Statistical tools such as histograms, probability plots, box plots, scatterplots, quantile-quantile (Q-Q) plots, proportional effect plots, and variograms are used. They allow comparisons of grade distributions within each of the domains proposed. Analysis of the statistics requires a degree of subjectivity, since an acceptable degree of similarity needs to be defined. Once two variables are shown to provide a similar degree of mineralization control, and assuming it makes geologic sense, they are grouped, and the statistical analysis repeated.

This iterative process can be labor-intensive, and is usually repeated until a group of geologic variables and elements have been defined that clearly separates different types of mineralization. Some of the variables will be grouped even though there are clear differences in the spatial characteristics of the mineralization. This is often done because of practical limitations, including data quantity, metallurgical considerations, and other economic and technical factors.

Alternative Statistical Techniques Other multivariate statistical techniques could be used to describe the relationships between geology and grade. For example, some practitioners have proposed the use of Classification and Regression Tree analysis (CART, Breiman et al. 1984) to determine and categorize relationships between geology and grade distributions. Techniques such as Principal Component Analysis and Cluster Analysis have also been proposed. A common problem, however, is that these techniques are often used to classify the relationships based on statistical parameters without geological consideration.

The proportional effect may also be used to define domains. The proportional effect appears in the presence of positively skewed distributions. It indicates that, as the average of the variable increases, so does its variability. These plots, when comparing means and standard deviations of groups of data defined according to geologic variables, may show clusters of data. The assumption is that data within each cluster belong to a quasi-stationary population, thus defining estimation domains. These data clusters should be correlated to specific geologic controls.

The iterative process using simple statistics described is recommended. An important by-product is that the more labor-intensive process leads to a more thorough understanding of the geology. It ensures that the estimation domains are a group of quasi-stationary domains that make spatial and geologic sense, as opposed to only statistical groupings.

4.3 Case Study: Estimation Domains Definition for the Escondida Mine

The process of defining estimation domains is best illustrated with an example. The following has been taken from a definition of Total Copper (TCu) estimation domains at BHP Billiton's Escondida copper deposit. It is reproduced here courtesy of BHP Billiton, Base Metals Division.

Not all the aspects of a given geologic variable are valid or useful at the time of defining estimation domains. The first step in the process is to define those aspects that will be considered. This initial selection of important geologic attributes should be decided by the geologists who know the deposit well. An understanding of how geologic variables may impact resource estimation is also required.

The definition of estimation domains at Escondida was greatly assisted by the operating mine. The open pit afforded the opportunity for confirmation by direct observation of the assumed relationships described by the drill hole data. At Escondida, the geologic variables considered were mineralization type, alteration, lithology, and structural domains.

In the case of mineralization types, all high enrichment mineralization (HE1, HE2, and HE3) would be modeled as below Top of Sulfides (TDS) and above Top of Chalcopyrite (TDCpy). It was shown through statistical and additional chemical analyses that the Covelite-Pyrite (Cv+Py) unit has the statistical and spatial characteristics of low enrichment mineralization. Also the Chalcocite-Chalcopyrite-Pyrite (Cc+Cpy+Py) unit has characteristics of high enrichment mineralization, particularly for higher benches where the proportion of Cc in this unit is more significant. This is to be expected, since these mineral assemblages are transitions from higher to lower enrichment mineralization.

Following similar reasoning for all alteration, lithology and structure categories, the original codes in the database were translated into a simplified version, and are shown in Table 4.1. The most important characteristics of the resulting mineralization codes are the following:

- All mineralization with some cuprite described was grouped into a single code (Cuprite, Cuprite+Ox, Cuprite+Mx, and Cuprite+Cc+Py into Cuprite). This is because Cu cannot be recovered from cuprite using the existing processing facilities, and is detrimental to the overall Cu recovery in a flotation plant.
- High Enrichment is defined as of Cc+Py and Cc+Cv+Py.
- Low Enrichment groups the units Cc+Cpy+Py, Cc+Cv+Cpy+Py, Cv+Cpy+Py, and Cv+Py.
- All primary mineralization is lumped into one category (Py, Cpy+Py, and Bn+Cpy+Py), because, at the time of the study, the bulk of the processed ore will come from enriched mineralization units.
- All other elements used are the original codes: Leach (code 0), green Oxides (code 1), Partial Leach (code 4), and Mixed (code 5).

A similar process of developing new variables for lithology and alteration was completed. The grouping of initial mapped elements resulted in three alteration codes, QSA, SCC, and K-B (white, green and potassic-biotite alteration, respectively), and three lithologies: porphyry, andesite, and rhyolite.

With respect to lithology, the following characteristics are noted:

- Tuffs were grouped with Rhyolite, (PC).
- The following codes were ignored due to lack of spatial representation: Dacites, Gravels, Tectonic Breccias, Undifferentiated Porphyry (9), Diorites, and Pebble Dykes.
- Hydrothermal Breccias and Igneous Breccias were grouped with the main Escondida Porphyry unit.

With respect to alteration, the following groupings were made:

- A new code QSA was formed grouping all Quartz, Sericite, and Clays (Sericite, Clays, Silicified, and Advanced Argillic). This also known as white alteration.
- Similarly, a new SCC code was formed by grouping Propilitic, Sericite-Chlorite-Clays, K-S Transition in Porphyry, Silicified in Andesites, and Silicified in Porphyry. This is sometimes referred to as "green alteration" because of the presence of chlorite. Propilitic alteration is very different from the other components of this SCC grouping described. However, it is deemed pertinent here because there are very few intervals coded as propilitic alteration. Normally, in other porphyry deposits, propilitic alteration is observed as a halo on the outskirts of the deposit, and would be advisable to model it separately.
- A third alteration K-B was formed by grouping Potassic (K) and Biotite alterations (B).
- The fresh, unaltered rock is volumetrically unimportant and was ignored.

With the simplification of the original codes, completed by Escondida geologists, the basic elements of the geological model have been defined, and combinations of these elements define the initial set of estimation domains.

Five structural domains were identified based on observations in the pit and drill hole data. At Escondida, like most mineral deposits, structures control the spatial distribution of TCu grades in different areas of the deposit. Figure 4.1 shows the domains and the current pit projection as modeled by structural geologists. Domain 5 (to the West of the deposit-bounding Ferrocarril fault, in brown) was not considered, since there is no evidence of mineralization. The basic building blocks for defining the estimation domains were defined with the remaining four structural domains.

4.3.1 Exploratory Data Analysis of the Initial Database

The database consisted of 2,140 drill holes with 215,681 assays, lithology, mineralization, and alteration records. Histo-

Table 4.1 Original and simplified geologic codes, Escondida database

LITHOLOGY	GROUPING	Alpha CODES	Numeric Codes
K-Porphyry	OK	PF	1
Quartz-Porphyry	Grouped with Tuffs	PC	2
Undifferentiated Porphyry	Ignore	PU	-99
Andesite	OK	AN	3
Igneous Breccias	Grouped with PF	BI	4 (1)
Hydrothermal Breccias	Grouped with other PF	BH	7 (1)
Tectonic Breccias	Ignore	BT	-99
Gravel and Pebble Dykes	Ignore	GR/PD	-99
Late Dacite	Ignore	DT	-99
Diorite	Ignore	DR	-99
Tuff	Grouped with PC	TB	2
MINERALIZATION TYPES	**GROUPING**	**Alpha Codes**	**Numeric Codes**
Leach	OK	LX	0
Green Cu Oxides	OK	OX	1
Cuprite	Grouped with other Cuprite	CP	2
Cuprite + Ox Cu	Grouped with other Cuprite	CPOX	2
Cuprite + Mixto	Grouped with other Cuprite	CPMX	2
Cuprite + Cc + Py	Grouped with other Cuprite	CPCCPY	2
Partial Leach	OK	PL	4
Mixed Oxide and Sulfides	OK	MX	5
Chalcocite/Pyrite	Grouped with HE2	HE1	6
Chalcocite/Covelite/ Pyrite	Grouped with HE1	HE2	6
Covelite/Pyrite	Grouped with LE	HE3	7
Chalcocite/Chalcopyrite/ Pyrite	Grouped with LE	LE1	7
Chalcocite/Covelite/ Chalcopyrite/Pyrite	Grouped with LE	LE2	7
Covelite/Chalcopyrite/ Pyrite	Grouped with LE	LE3	7
Pyrite	Grouped with other Primary	PR1	8
Chalcopyrite/Pyrite	Grouped with other Primary	PR2	8
Bornite/Chalcopyrite/ Pyrite	Grouped with other Primary	PR3	8
ALTERATION	**GROUPING**	**Alpha Codes**	**Numeric Codes**
Fresh rock	Ignore	F	-99
Propilitic	Grouped with SCC	P	2
Clorite-Sericite-Clay	OK	SCC	2
Quartz-Sericite	OK	S	1
Potassic	OK	K	3
Biotitic	Grouped with K	B	3
Advanced Argilic	Grouped with S	AA	1
Clay	Grouped with S	AS	1
Silicified	Grouped with S	Q	1
K-S Transition in Porphyry	Grouped with SCC	QSC	2
SCC Silicified in Andesite	Grouped with SCC	SCC-An	2
QSC Silicified in Porphyry	Grouped with SCC	SCC-Pf	2

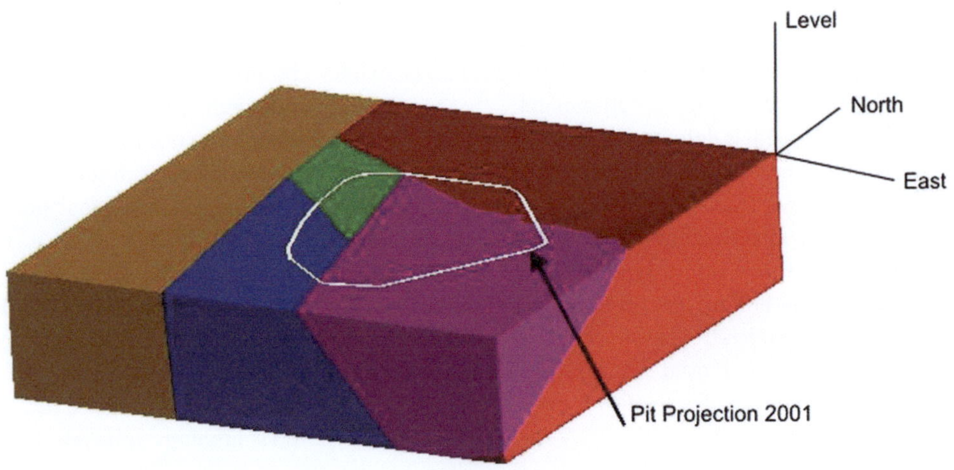

Fig. 4.1 Diagram showing the four structural domains. Domain 5 (in brown, west of the Ferrocarril fault) is non-mineralized, and outside the area of interest. For scale, the projection of the 2001 pit to the surface has an approximate dimension of 3 × 3 km, and no vertical exaggeration

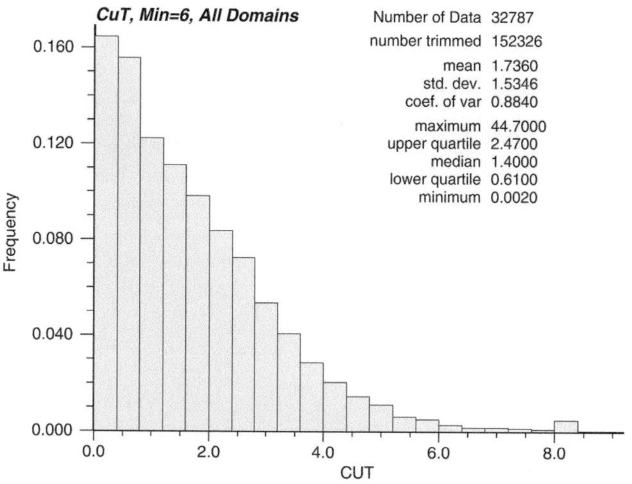

Fig. 4.2 Histogram and basic statistics of TCu (%), Cc+Py unit

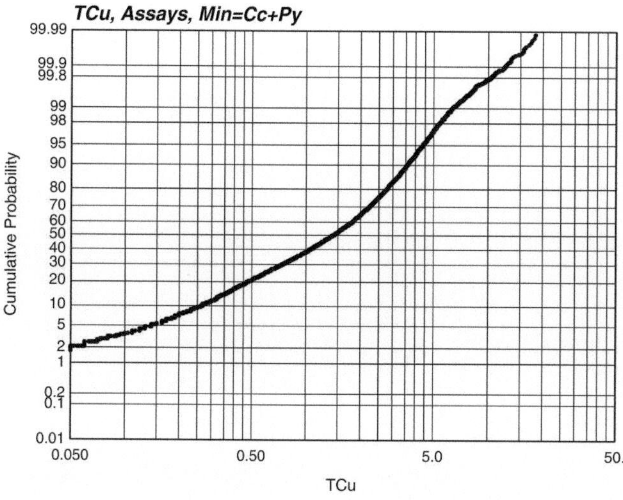

Fig. 4.3 Probability plot of TCu (%), Cc+Py unit

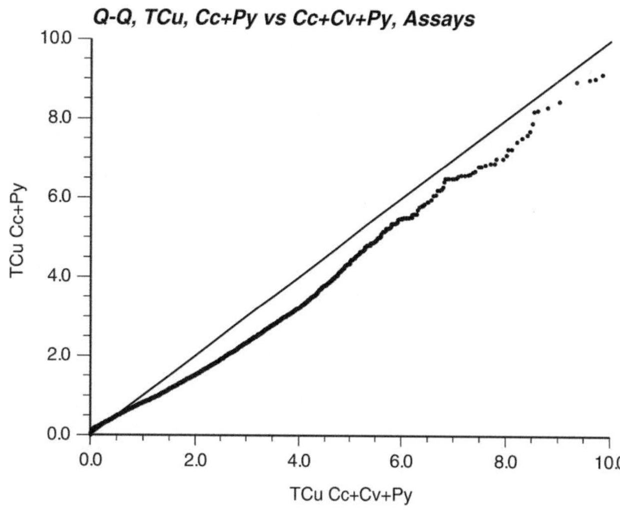

Fig. 4.4 Quantile-Quantile plot of TCu (%), Cc+Py vs. Cc+Cv+Py mineralization

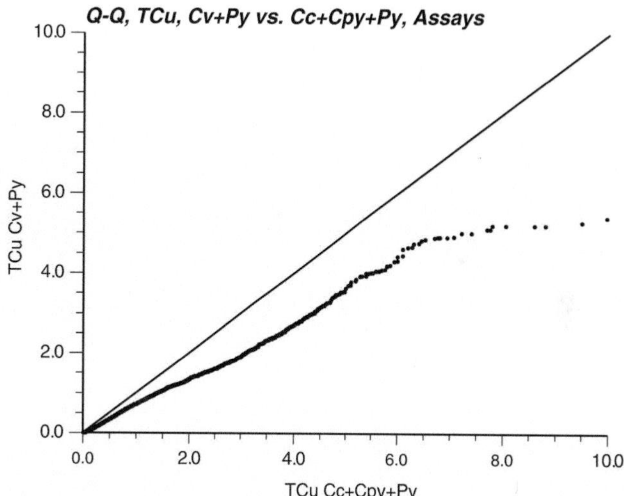

Fig. 4.5 Quantile-Quantile plot of TCu (%), Cc+Cpy+Py and Cv+Py mineralization

grams were used to provide a global description of the variable, along with summary statistics. Figure 4.2 shows the histogram and summary statistics for TCu, all assays logged as chalcocite plus pyrite (Cc+Py, HE1 in Table 4.1). The histogram shows a positively-skewed distribution with an average grade of 1.74 % TCu and a coefficient of variation of 0.88, which is considered low for assay data.

The cumulative frequency plot is often used to describe important characteristics of the distribution, such as looking for breaks along an expected continuous line. Figure 4.3 shows the probability plot corresponding to the data in Fig. 4.2 (TCu, Cc+Py). Note how the curve has inflection points, one at approximately 2 % TCu, and the other at about 6 % TCu, suggesting a mixture of populations in the domain.

Two distributions can be compared using quantile-quantile (Q-Q) plots. Figure 4.4 shows a Q-Q plot comparing Cc+Py and Cc+Cv+Py mineralization, while Fig. 4.5 shows the comparison for Cc+Cpy+Py and Cv+Py. These and other similar figures illustrate the similarity of the grade distributions based on mineralization types alone.

4.3.2 Initial Definition of Estimation Domains

The definition of preliminary estimation domains was done by analyzing all geologically feasible combinations of the four variables: mineralization, lithology, alteration, and structural domains.

Table 4.2 Initial estimation domains

Estimatoin Domain	*Mineralization*	*Lithology*	*Alteration*	*Structural Domain*	*Comments*
0	Leach	ALL	ALL	ALL	Mostly barren.
1	Oxides	ALL	ALL	ALL	Defined by Interpreted Oxide Envelope
2	Cuprite	ALL	ALL	ALL	Cannot be processed, mined as waste regardless of grade
3	Partial Leach	ALL	ALL	ALL	Small bodies, difficult to model
4	Mixed	ALL	ALL	ALL	Small bodies, difficult to model
5	ALL	Rhyolites	ALL	ALL	Eastern edge of the deposit, low grade, little development in the near future.
6	Cc+Py; Cc+Cv+Py	ALL	QSA	1+4	High Enrichment
7	Cc+Py; Cc+Cv+Py	ALL	SCC	1+4	High Enrichment
8	Cc+Py; Cc+Cv+Py	ALL	QSA	3	High Enrichment
9	Cc+Py; Cc+Cv+Py	ALL	SCC	3	High Enrichment
10	Cc+Cpy+Py; Cv+Py Cc+Cv+Cpy+Py; Cv+Cpy+Py	ALL	QSA	1+4	Low Enrichment
11	Cc+Cpy+Py; Cv+Py Cc+Cv+Cpy+Py; Cv+Cpy+Py	ALL	SCC	1+4	Low Enrichment
12	Cc+Cpy+Py; Cv+Py Cc+Cv+Cpy+Py; Cv+Cpy+Py)	ALL	QSA	3	Low Enrichment
13	Cc+Cpy+Py; Cv+Py Cc+Cv+Cpy+Py; Cv+Cpy+Py	ALL	SCC	3	Low Enrichment
14	Cpy+Py; Py; Bn+Cpy+Py	Porphyries + Breccias	K+B	1+4+2	Primary
15	Cpy+Py; Py; Bn+Cpy+Py	Andesites	K+B	1+4+2	Primary
16	Cpy+Py; Py; Bn+Cpy+Py	Porphyries + Breccias	K+B	3	Primary
17	Cpy+Py; Py; Bn+Cpy+Py	Andesites	K+B	3	Primary
18	Cc+Py; Cc+Cv+Py	ALL	ALL	2	High Enrichment
19	Cc+Cpy+Py; Cv+Py Cc+Cv+Cpy+Py; Cv+Cpy+Py	ALL	ALL	2	Low Enrichment

Table 4.2 shows the 20 initial estimation domains defined. The initial six estimation domains are defined based on mineralization alone, and is due to two factors: as these are all non-sulfide units (with the exception of Partial Leach), their economic importance is minor if compared to the supergene sulfide mineralization. Also, the spatial distribution of these mineralization units, with the exception of the leached cap, is complex and difficult to model. Typical sizes of oxide and mixed bodies are at best approximately equal to the better drilling spacing available (50–70 m of lateral extension). Subdividing these small domains even further is likely to result in poor grade estimates.

It was found that within the supergene enrichment zone the lithological control is redundant with alteration. Lithology is an important control for mineralization types, but in the supergene areas alteration overprints and obliterates the Lithologic control.

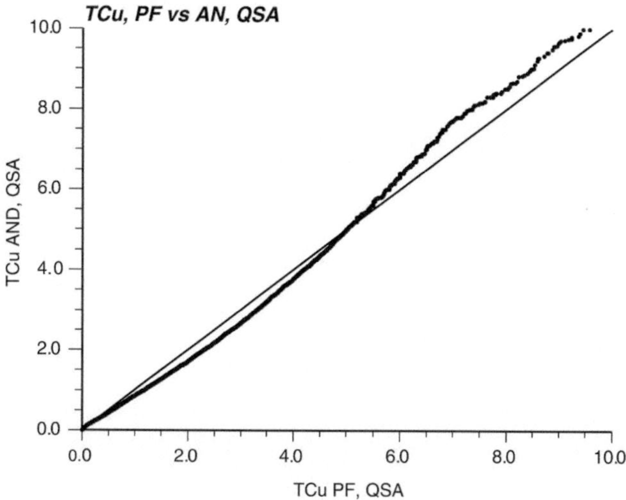

Fig. 4.6 Q-Q plot of TCu (%), Porphyry vs. Andesite, QSA alteration

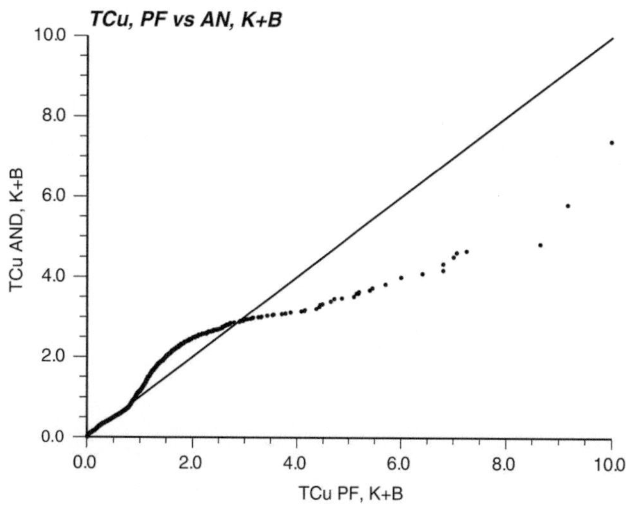

Fig. 4.8 Q-Q plot of TCu (%), Porphyry vs. Andesite, Potassic+Biotite alteration

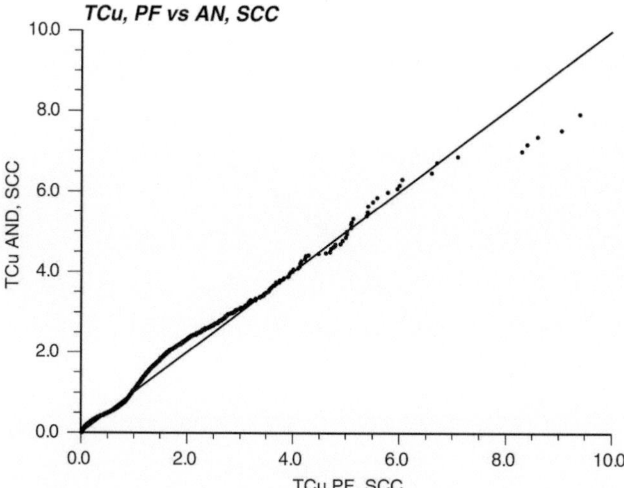

Fig. 4.7 Q-Q plot of TCu (%), Porphyry vs. Andesite, SCC alteration

Figures 4.6 and 4.7 show the Q-Q plots of all Escondida Porphyry vs. andesite lithologies, conditioned to the two main alterations, QSA and SCC, respectively. Note how the plots are close to the 45° line, which implies similar statistical distributions. Therefore, TCu grades do not change much in andesite or Escondida porphyries, as long as the alteration remains the same. Lesser grades can be expected if the alteration is SCC, regardless of whether lithology is andesite or Porphyry. Approximately 18 % of the total assay intervals are andesite with QSA alteration, while there are approximately 4 % of Escondida porphyry assays with SCC alteration.

This is not the case for primary mineralization where there are significant differences in the statistical characteristics of TCu grades when comparing andesites with the Escondida porphyry. Figure 4.8 shows the Q-Q plot of both lithologies for alteration K+B; note how the distributions are quite different. The number of assays available in primary mineralization with potassic and biotite alterations is relatively small since drilling targets the supergene enriched mineralization. This is why they were grouped. Primary mineralization is not as important economically as the upper part of the deposit, so it appears reasonable, mostly for pragmatic reasons, to group the primary mineralization units.

Structural domains 1 and 4 present a clear difference in terms of TCu grades, compared to structural domains 2 and 3. Domain 3, in particular, is the most different. This is evident both from descriptive statistics and TCu correlogram models for the different domains.

Figures 4.9, 4.10, 4.11, and 4.12 show the Q-Q plots of HE versus LE mineralization (Cc+Py vs. Cc+Cpy+Py) for Domains 1 through 4, respectively.

Figure 4.9 (structural domain 1) shows that the global Cc+Py distribution has significantly more grade for the 1–4 % TCu range. The quantile values for higher grades tend to be similar, which implies that both distributions have a significant high grade tail.

Figure 4.10 (structural domain 2) shows that the low enrichment material (Cc+Cpy+Py) has a higher-grade distribution. This is an indication that there is less chalcopyrite in structural Domain 2, probably due to a deepening of the enrichment process in a down-thrown structural block. Therefore, it would be reasonable to combine HE and LE into a single group. Structural domain 2 is the smallest in volume of the four domains considered.

The grade distributions in structural Domain 3 (Fig. 4.11) behave as expected, with the HE distribution consistently showing higher grade, while the grade distributions for structural domain 4 (Fig. 4.12) are very similar, again prob-

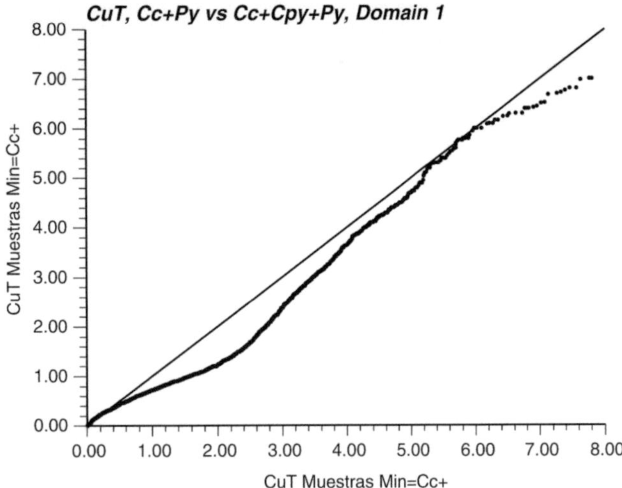

Fig. 4.9 Q-Q plot of TCu (%), Cc+Py vs. Cc+Cpy+Py, Structural Domain 1

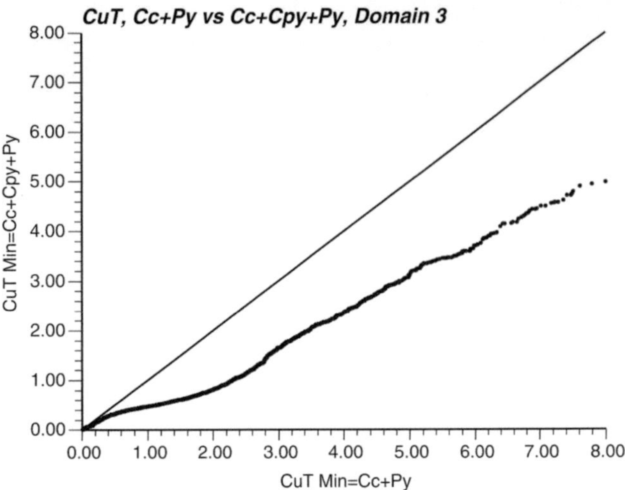

Fig. 4.11 Q-Q plot of TCu (%), Cc+Py vs. Cc+Cpy+Py, Structural Domain 3

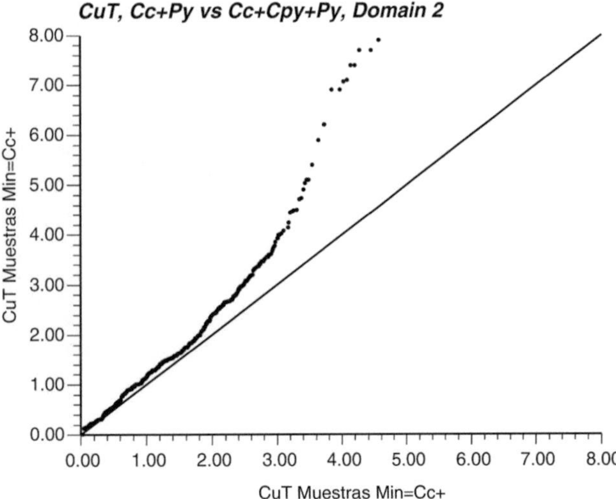

Fig. 4.10 Q-Q plot of TCu (%), Cc+Py vs. Cc+Cpy+Py, Structural Domain 2

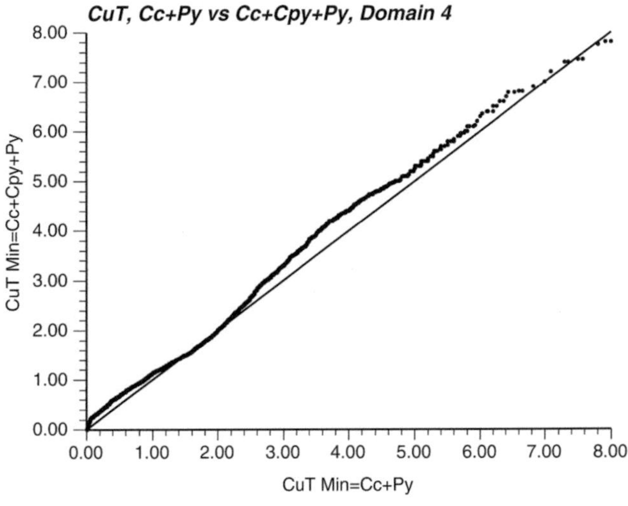

Fig. 4.12 Q-Q plot of TCu (%), Cc+Py vs. Cc+Cpy+Py, Structural Domain 4

ably due to the relative abundance of chalcocite vs. chalcopyrite in the LE unit. The analysis of the relative movements of each structural block explains this observation, since the enrichment process also reached deeper levels for structural Domain 4.

In conclusion, the TCu grade distribution shows different statistical characteristics in each structural domain. The structural control on mineralization explains the relationship between high enrichment and low enrichment mineralization for different parts of the deposit.

In developing Table 4.2 it was assumed that supergene enrichment mineralization (HE and LE) do not show potassic or biotite alteration. This is based on a geologic assumption. Assay intervals logged as HE or LE with K-B alteration were dismissed as incorrectly logged intervals.

4.3.3 Tcu Grade Correlogram Models by Structural Domains

Another perspective of the differences between domains can be gained by analyzing the spatial continuity of the TCu grade distribution, considering again HE mineralization (Cc+Py) as an example. Correlograms (Chap. 6) were run and modeled for all main geologic variables and for each structural domain.

There are practical aspects that need to be considered when analyzing correlogram models within the scope of estimation domain definition. Correlograms and other spatial continuity models are affected by the amount of data available. At Escondida, this implies that the models for structural domain 2, primary mineralization, and some of the low enrichment mineralization units are less reliable compared to the more populated units.

The correlogram models developed showed the following:

- The prevalent anisotropy directions are NE and NW as expected, but not in the horizontal plane. The main axes of continuity are dipping 20–50° towards the center of the deposit, depending on the mineralization unit and domain. This is not a simple, layered deposit that it is sometimes envisioned when dealing with porphyry type deposits.
- Structural Domain 3 consistently presents a much higher nugget effect than the other domains. The grade distribution is more erratic and discontinuous. More dilution can be expected at the time of mining, relative to other domains, which indeed has been the operation's experience.
- Correlograms from structural Domains 2 and 4 show evidence of a deeper enrichment process, consistent with field observations. A NW trending zone of deeper enrichment results in better mineralization as observed in the pit. Correlograms from structural Domain 1 tend to plunge towards the W-SW, while correlograms from Domains 3 and 4 tend to plunge towards the S-SE.
- Structural Domains 1 and 4 show a stronger NE anisotropy, with less emphasis on the NW or SW dipping structures. Structural Domain 2 shows also significant (long-range) NE anisotropy overprinting the expected NW short range anisotropy. The longer-range N-NE anisotropies observed correspond to the general orientation of the two main intrusive bodies that are thought to be the mineralization source.

4.3.4 Final Estimation Domains

Several simplifications were made to the original proposed estimation domains since additional constraints need to be considered to obtain the final estimation domains. First, both enrichment mineralization units in structural Domain 2 (18 and 19) were joined into a single estimation domain, partly because of the similarity of the grade distribution, and partly because of lack of data. Estimation Domains 7 and 11 were merged into a single domain (HE and LE, with SCC alteration, for Domains 1+4), again because of statistical similarity and lack of data. All primary mineralization was combined into a single domain because of lack of data; low TCu grades, and also because production of Cu from primary mineralization will not happen until much later in the mine life.

The final estimation domains are shown in Table 4.3. Descriptive statistics, clustering analysis, contact analysis, and variography are used to confirm the statistical characteristics of TCu within each domain. The results of the domain definition study can be summarized as follows:

1. Fourteen estimation domains (GUs) were defined for TCu. These include the GUs defined for the upper portion of the deposit.
2. Two unexpected features at the time were the use of structural domains and the lesser role that lithology plays as mineralization control in the supergene enrichment zone.
3. The correlogram models obtained for the different datasets and conditioned to different geologic attributes and the GUs show a pattern of anisotropies consistent with geologic knowledge and observations in the pit.
4. There are important details in terms of correlogram models that result from the addition of the structural domains. The most important one is that in Domain 3 the relative nugget effect is significantly higher than for the other domains. This is a result of a local mixture of phyllic (QSA) and SCC alterations, with a corresponding increase in grade variability.
5. The anisotropies detected confirm that the shorter-range, higher-grade mineralization trends mostly NW, but with significant N-NE long-range anisotropies. Also, for units to the south and west of the deposit, the dips and plunges of the ellipsoids of continuity generally will dip to the SW and plunge towards the NE; for units to the north and North East of the deposit, the dip may still be SW, but the plunge is more commonly to the SE.

4.4 Boundaries and Trends

The geological interpretations and modeling of estimation domains produce boundaries that often carry significant uncertainty. The treatment and definition of boundaries have implications on resource estimation such as dilution, lost ore or a mixture of geological populations. The treatment of boundaries at the time of grade estimation is of practical importance. The terms *hard* and *soft* boundaries are used to describe whether the change in grade distribution across the contact is abrupt or not, respectively. Conventional grade estimation usually treats the boundaries between geological units as hard boundaries, whereby no mixing occurs across the boundary. Soft boundaries allow grades from neighboring domains to be used. Sometimes, soft and hard boundaries can be predicted or expected from geological knowledge, but should always be confirmed with statistical contact analysis (Ortiz and Emery 2006; Larrondo and Deutsch 2005).

Contact analysis helps determine whether the grade estimation for any given unit should incorporate characteristics of a neighboring unit. It is a practical tool to describe grade trends and behavior near contacts and define the data to be used in the estimation of each unit.

The behavior of grades across contacts can be analyzed by finding pairs of data in the two estimation domains of interest at pre-defined distances. There are different methods

Table 4.3 Estimation domains for total copper, Escondida 2001 resource model

Estimation Domain	Mineralization	Lithology	Alteration	Structural Domain
0	LIX (0)	ALL	ALL	ALL
1	OXIDE (1)	ALL	ALL	ALL
2	CUPRITE	ALL	ALL	ALL
3	PARTIAL LEACH	ALL	ALL	ALL
4	MIX	ALL	ALL	ALL
5	ALL	PC+TB (Rhyolite+Tuffs)	ALL	ALL
6	6+9 (Cc+Py; Cc+Cv+Py)	ALL	QSA (1)	1+4
7	6+9+7+10+13+14 (Cc+Py; Cc+Cv+Py; Cc+Cpy+Py; Cv+Py; Cc+Cv+Cpy+Py; Cv+Cpy+Py)	ALL	SCC (2)	1+4
8	6+9 (Cc+Py; Cc+Cv+Py)	ALL	QSA (1)	3
9	6+9 (Cc+Py; Cc+Cv+Py)	ALL	SCC (2)	3
10	7+10+13+14 (Cc+Cpy+Py; Cv+Py Cc+Cv+Cpy+Py; Cv+Cpy+Py)	ALL	QSA (1)	1+4
11	7+10+13+14 (Cc+Cpy+Py; Cv+Py Cc+Cv+Cpy+Py; Cv+Cpy+Py)	ALL	QSA (1)	3
12	7+10+13+14 (Cc+Cpy+Py; Cv+Py Cc+Cv+Cpy+Py; Cv+Cpy+Py)	ALL	SCC (2)	3
13	8+10+12 (Cpy+Py; Py; Bn+Cpy+Py)	ALL	K+B (3)	ALL
14	6+9+7+10+13+14 (Cc+Py; Cc+Cv+Py; Cc+Cpy+Py; Cv+Py;Cc+Cv+Cpy+Py; Cv+Cpy+Py)	ALL	ALL	2

to define the pairs, but a true three-dimensional method is preferred to avoid directional biases. In this method, pairs within pre-specified distances are found through a three-dimensional search of nearby assay intervals belonging to a different unit.

Figure 4.13 shows the grade averages at either side of the contact between the Cc+Py and the Cc+Cpy+Py units from the Escondida case study. Each point in the figure corresponds to the TCu average grouped at 2 m distance classes from the contact. Despite the high variability in the averages, the grade transition is smooth, from higher grades in the Cc+Py unit to lower grades in the Cc+Cpy+Py unit, and as would be expected from units that are defined as transitional mineralogical assemblages. A trend could be modeled as a function of distance from the contact.

Another example (Fig. 4.14) shows that the profile of average TCu grades at the contact between the final estimation Domains 6 and 7 at Escondida (see Table 4.3) is hard. In this case, the TCu grades change significantly crossing from one unit to the other in a very short distance. Therefore, it is not advisable to use composites from estimation Domain 7 to estimate TCu grade in estimation Domain 6.

Considering stationary domains in the presence of soft boundaries is often inappropriate. In general, soft boundaries as the one shown in Fig. 4.13 are characterized by a non-stationary behavior near the contact. The mean, variance or covariance are not constant within a zone of influence of one rock type into the other and their values depend on the location relative to the boundary, as illustrated by Fig. 4.15.

The correct reproduction of soft boundaries in resource models improves dilution and mineral resource estimates. The areas close to contacts are usually areas of higher uncertainty, as shown by the abundance of red colors in Fig. 4.16.

In the presence of complex contacts and multiple boundaries, it may be appropriate to model the non-stationary features present in the local neighborhood. The non-stationary features of the mean, variance, and covariance can be parametrized into a local model of coregionalization (Larrondo and Deutsch 2005). Estimation of the grades can be performed using a form of non-stationary cokriging (Chap. 8).

Trends within estimation domains are also common. In certain circumstances, trends need to be explicitly modeled or taken into account, particularly when simulating grade distributions (Chap. 10). In other instances, such as grade es-

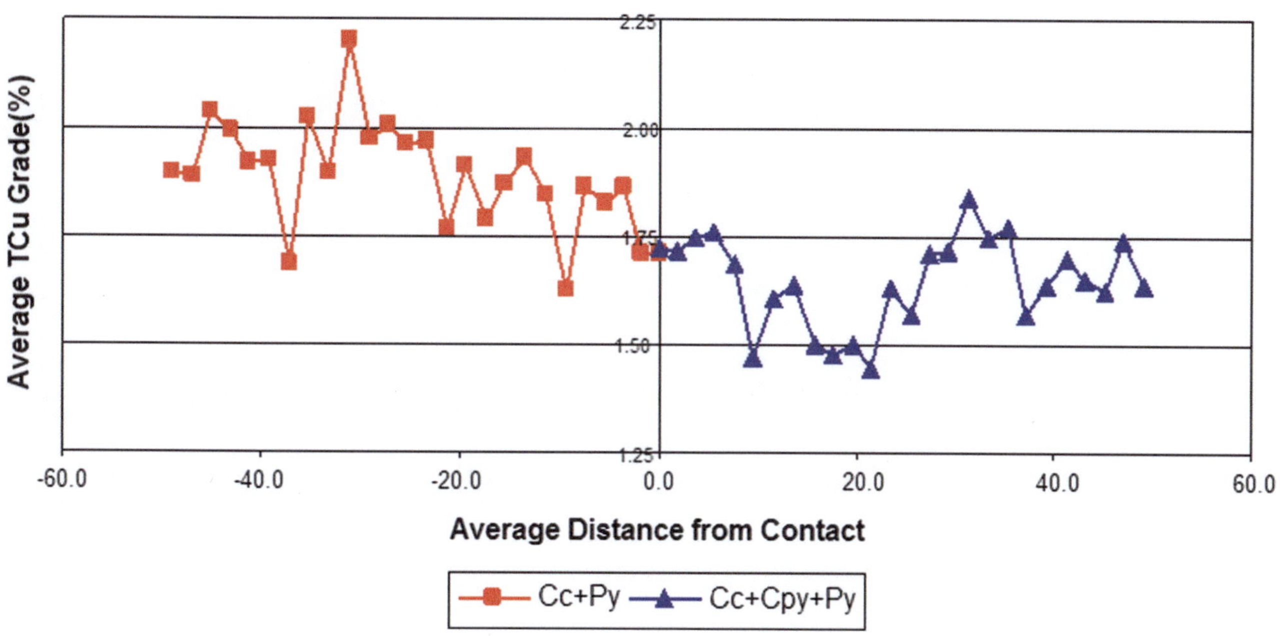

Fig. 4.13 TCu grade transition at the contact between mineralization units Cc+Py and Cc+Cpy+Py, 2 m assays

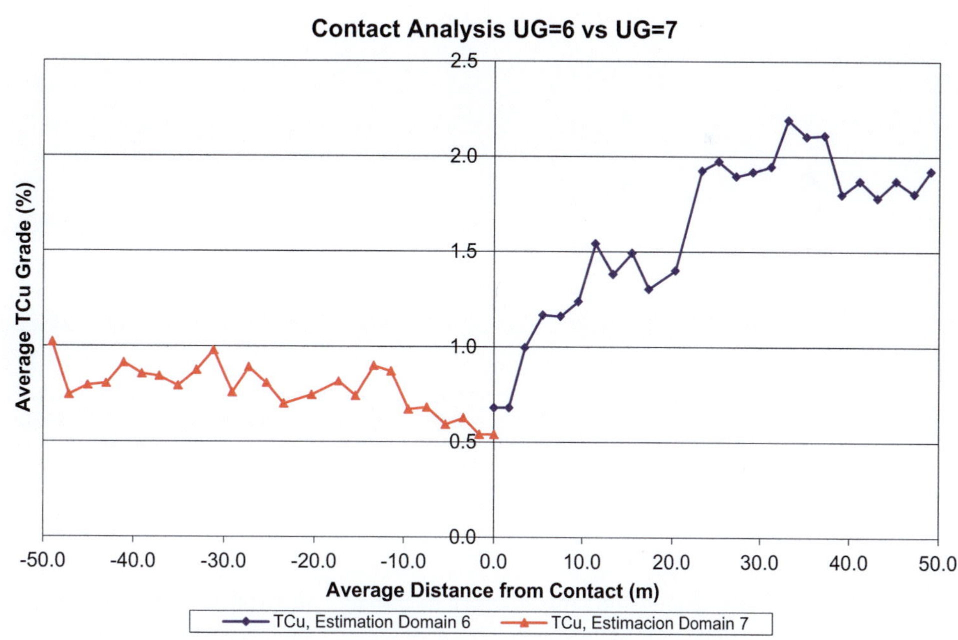

Fig. 4.14 Contact Analysis, Estimation Domains 6 vs. 7 (see Table 4.3)

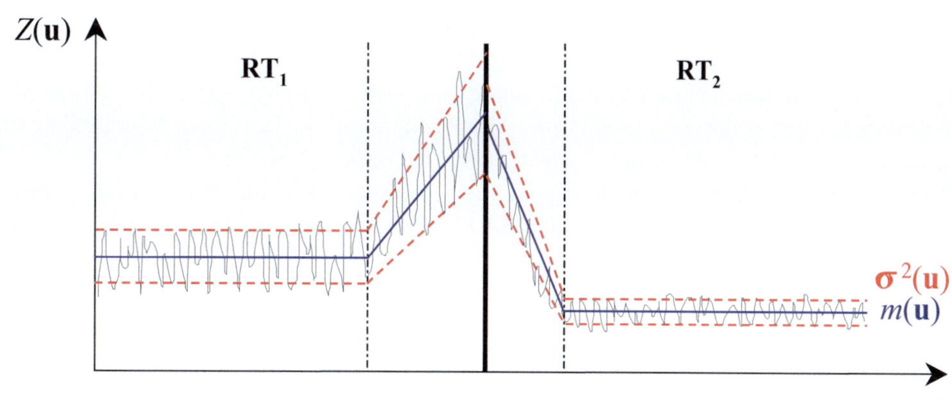

Fig. 4.15 Schematic illustration of trends near boundaries

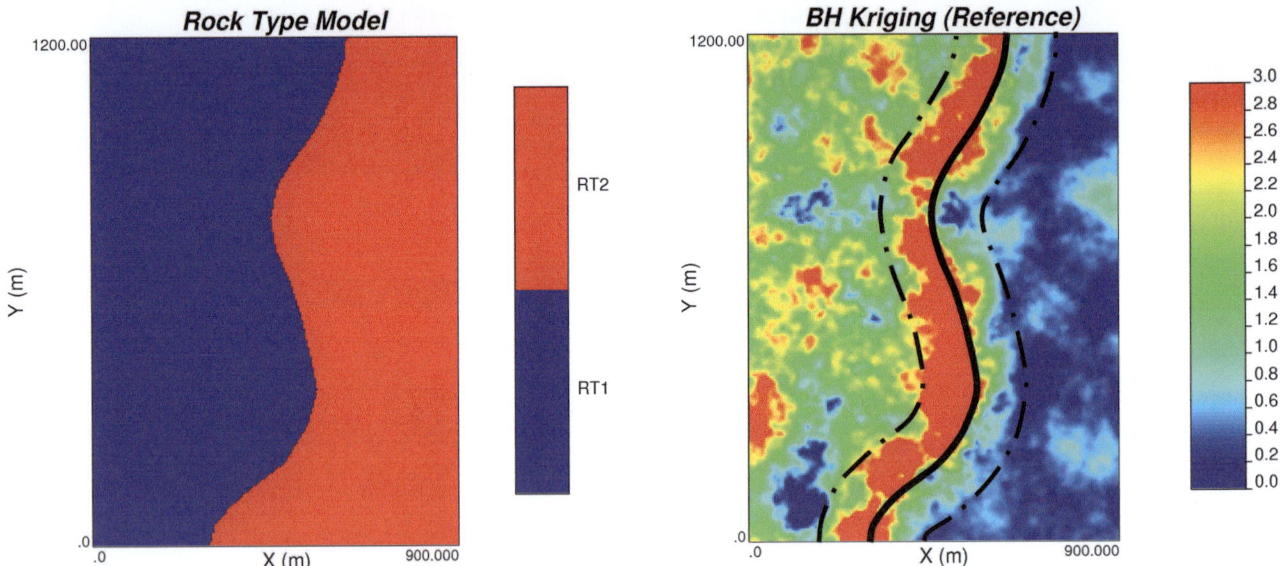

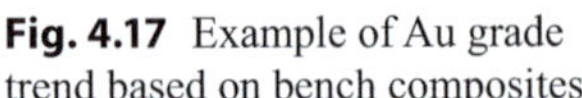

Fig. 4.16 An example of higher uncertainty (and higher grade) near contacts

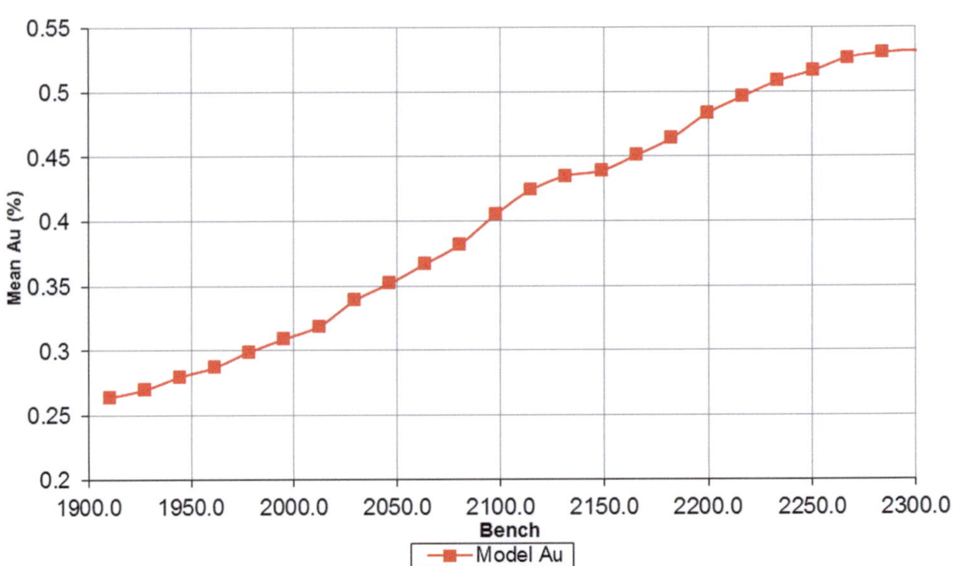

Fig. 4.17 Example of Au grade trend based on bench composites

timation using ordinary kriging and limited search neighborhoods, trends are accounted for by the implicit re-estimation of the mean within the search neighborhood (see Chap. 8 and Journel and Rossi 1989).

Some trends can be inferred from geological knowledge. For example, the distribution of nitrate, borate, and iodine in evaporitic-type deposits is predictable. More commonly, trends are detected and modeled directly from the data. Trends can be described using plots of grade versus distance along a relevant coordinate direction. Figure 4.17 shows the gold grade trend in the vertical direction in a low-grade porphyry Au deposit. The data show that the Au grade declines for lower elevations at an approximate rate of about 0.1 g/t per 100 m. This trend may persist even after defining the final estimation domains. If not taken into account, the trend may result in overestimation of the Au resource for the lower benches.

If trends must be accounted for explicitly, then the following approach is commonly applied in presence of a trend:

- Develop a deterministic trend model and remove it from the data;
- Model the residual component; and
- Add the deterministic trend to obtain the final model.

There are some common deterministic methods for building a trend model. They include hand or computerized contouring, and fitting simple polynomial models. In practice, we might consider 1-D vertical trends and 2-D areal trends that are then merged into a 3-D trend model. There is no unique way to merge 1-D and 2-D trends into a 3-D trend model, but a simple approach is to merge these trends by assuming conditional independence of vertical and areal trends:

$$m(x,y,z) = \frac{m_z(z) \cdot m_{x,y}(x,y)}{m_{global}}$$

Where $m_z(z)$ = mean from vertical trend, $m_{x,y}(x,y)$ = mean from areal trend, m_{global} = global mean from histogram, and $m(x,y,z)$ = mean at location (x,y,z). This equation effectively rescales the vertical trend curve by the areal trend. Other probability combination schemes such as permanence of ratios could be used in situations where assuming conditional independence leads to extreme mean values too close to zero or too high.

4.5 Uncertainties Related to Estimation Domain Definition

The definition of estimation domains is an important prerequisite in the application of most geostatistical tools used in resource modeling. The domains determine the mineralized volume available, and thus is a major factor in the estimated tonnage above economic cutoffs.

The definition of estimation domains is subjective and limited by data and practical considerations. There are many sources of uncertainty contributing to the uncertainty in the definitions of contacts and volumes.

Some of the more typical sources of uncertainty include geologic data: errors, omissions, or imprecise mapping and logging are common. For example, in highly altered rock, the precise description of lithology types can be difficult, more so if diamond drilling is not used. Porphyries of different kinds are difficult to differentiate and different lithologies may not be easy to distinguish. Human perceptions and errors are important since many geologic attributes are subject to visual estimations and interpretations in the field. For example, the alteration intensity or the percentage of sulfides may have to be estimated by the geologist.

Limited data also may be a significant source of uncertainty. It is common that two domains with clearly different mineralization controls have to be combined into one domain because one of them does not have enough drill hole information. This results in a mixture of populations that cannot be resolved until more data are collected. The domain with more data will influence the statistics, the variogram models, and the kriging plans applied to estimate the grades of the combined units.

There is also the uncertainty carried over from the geologic interpretation and modeling which is more significant in sparsely drilled areas. The geologic model can be another important source of uncertainty that, when combined into estimation domains, can result in serious flaws in the resource model.

All these sources of uncertainty combine with the fact that mineralization will be naturally varying from one location to another. This natural variability within the estimation domains exists at different scales and should be considered at the time of estimation.

4.6 Summary of Minimum, Good and Best Practices

At a minimum, the methodology used to define estimation domains should consider the most evident mineralization controls, and include the basic tools needed to demonstrate the relationships between geologic attributes and grade. The main mineralization controls can often be described through mapped geology and a working hypothesis of the genesis of the deposit. Basic exploratory data analyses characterize mineralization controls.

Good practice considers all available geologic information and the relationship between grades and each geologic variable. This process involves a first phase, in which the individual mapped geology, such as mineralization, lithology, alteration, or others, is grouped in part by applying geologic knowledge and common sense, in part applying numeric and statistical constraints.

A new set of descriptive statistics is then developed in a second phase of the study, from which an initial set of estimation domains may be proposed. An iterative process that includes further statistical analysis supported by geologic knowledge results in the final definition of the estimation domains.

The definition of estimation domains is an imperfect process, characterized by compromises between the estimation domains that should be defined (according to geology and statistical analysis) and the amount of data available to define them. Sometimes, limitations in the coding of the original database may also affect the definition of the estimation domains.

Best practice is to define the estimation domains and accompany it by an assessment of its uncertainty and the limitations and assumptions used to define it. The definition should include limitations related to data quality and quantity, geologic information used, and the type of statistical analysis used to assess whether the domains contacts are hard or soft. The better tool to assess geologic uncertainty is simulation.

4.7 Exercises

The objective of this exercise is to construct trend models for a 2-D example and a larger 3-D example. Some specific (geo)statistical software may be required. The functionality may be available in different public domain or commercial software. Please acquire the required software before beginning the exercise. The data files are available for download from the author's website—a search engine will reveal the location.

4.7.1 Part One: Basic Statistics

Consider the 2-D data in `red.dat`. A small exploratory data analysis is required for the five different variables in this dataset: thickness, gold grade, silver grade, copper grade and zinc grade.

Question 1: Tabulate the key statistics for each variable: number of data, minimum, maximum, mean and variance. Plot histograms of the different variables and comment on the results.

Question 2: Plot probability plots of the variables on arithmetic or logarithmic scaling as appropriate. Comment on outliers, inflection points or any other interesting features.

Question 3: Plot scatterplots between all pairs of variables and create a matrix of correlation coefficients to summarize how the variables relate to one another.

Question 4: Repeat the previous question with normal scores of all the variables.

4.7.2 Part Two: 2-D Trend Modeling

Consider the 2-D data in `red.dat`. There is a significant trend with lower thickness at depth (below about −250 m) to the North and South.

Question 1: Create a contour map that represents the trend. Take care that the contours do not too closely match short scale variations. The general rule is to match large scale variations at a scale of greater than 2–3 times the drillhole spacing.
Kriging or inverse distance (or some other gridding algorithm) can be used as well; however, hand contouring is robust and gives an improved understanding of the data. Post the thickness data with the thicknesses posted on the map. Hand contour the map. Choose your own contour intervals; however, you could take 0.5, 1.0, 2.0, 5.0, or 10.0 if you are unsure.

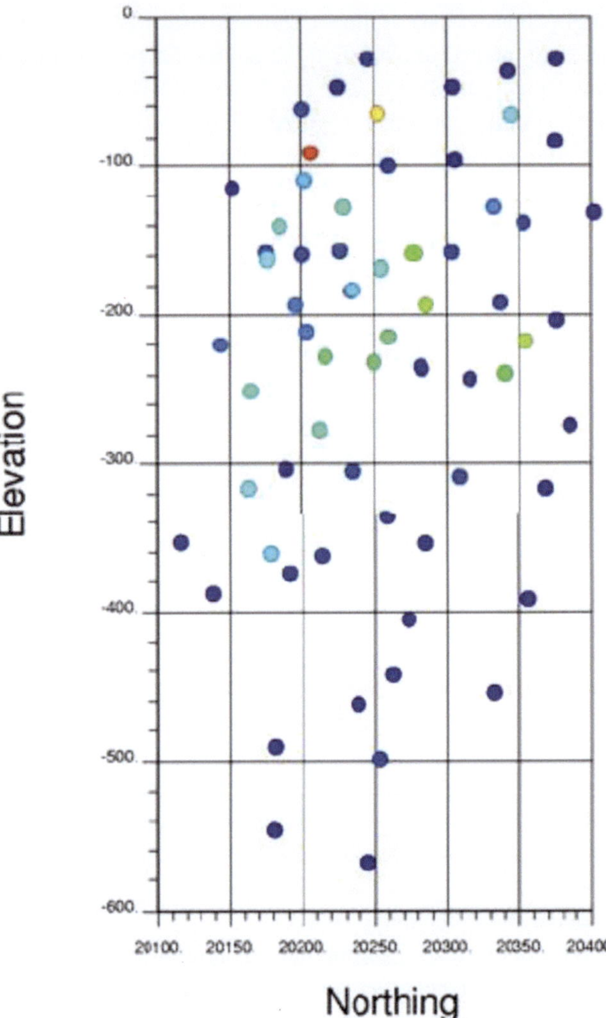

Question 2: There are a number of programs to get the contour lines in a "point-data" format for gridding algorithms. Create a gridded model of your contour map. Ensure that the map is smooth with no artifacts from your chosen gridding algorithm.

Question 3: Calculate residuals as *res* = *thickness-thicknesstrend*. Plot a histogram of the residuals. Plot a cross plot of the residuals versus the *thicknesstrend* values. Comment on any features that would make it awkward to simulate the thickness residuals independently of the thickness trend.

4.7.3 Part Three: 3-D Trend Modeling

Consider the 3-D data in `largedata.dat` for 3-D trend modeling. Build a trend model for the copper grade.

Question 1: Build a smooth vertical average of the grades by averaging the grades in vertical slices. The 1-D averaging program can be used for this purpose. Consider a number of sensitivity runs with different slice thicknesses and other parameters. Plot the results. Comment on the presence of a vertical trend and the importance of considering it in the simulation model. Present the final result that you choose.

Question 2: Calculate the vertical average of the drillhole data, make a map of the vertical averages, and comment on the need for modeling the areal trend.

Generate a smooth areal trend using kriging, inverse distance, or the contouring approach used in Part One.

Question 3: Construct a 3-D trend model by combining the 1-D vertical trend and the 2-D areal trend. Comment on the practical implications of the conditional independence assumption implicit to the combination approach commonly used. Also comment on the alternatives to construct a 3-D trend.

Question 4: Calculate residuals as *res = grade − grade trend.* Plot a histogram of the residuals. Plot a cross plot of the residuals versus the *grade trend* values. Comment on any features that would make it awkward to simulate the grade residuals independently of the grade trend.

References

Breiman L, Friedman JH, Olsen RA, Stone CJ (1984) Classification and regression trees. Wadsworth and Brooks/Cole Advanced Books and Software, Monterey, p 368

Coombes J (2008) The art and science of resource estimation. Coombes Capability, Subiaco

Isaaks EH, Srivastava RM (1989) An introduction to applied geostatistics. Oxford University Press, New York, p 561

Journel AG, Rossi ME (1989) When do we need a trend model? Math Geol 22(8):715–739

Larrondo P, Deutsch CV (2005) Accounting for geological boundaries in geostatistical modeling of multiple rock types. In: Leuangthong O, Deutsch CV (eds) Geostatistics Banff 2004, vol 1, Springer, November, pp 3–12

Matheron G (1962–1963) Tome I: Traité de Géostatistique Appliquée, Tome II: Le Kriegeage I: Mémoires du Bureau de Recherches Géologiques et Minières, No 14 (1962), Editions Technip, Paris; II: Mémoires du Bureau de Recherches Géologiques et Minières, No 24 (1963), Editions B.R.G.M, Paris

Ortiz JM, Emery X (2006) Geostatistical estimation of mineral resources with soft geological boundaries: a comparative study J South Afr Inst Min Metall 106(8):577–584

5 Data Collection and Handling

Abstract

Estimates of mineral resources are dependent on the available data. This Chapter reviews the main challenges related to data collection and handling. Special attention is given to data representativity, extreme high grades, and compositing the data to practical and consistent lengths. A summary of sampling theory is also presented.

5.1 Data

The mining industry collects more data than other natural-resource industries. This provides an opportunity to better understand local variations and obtain robust local estimates. The abundance of data play a major role in defining the modeling techniques used and their implementation, and has historically influenced the development of geostatistical techniques. This is in contrast with, for example, some petroleum and environmental modeling applications, where the amount of data collected is limited, and the final results are more model-dependent.

The quality of the mineral resource estimate is dependent on the quality of the data collection and handling procedures used (Erickson and Padgett 2011; Magri 1987). A number of technical issues affect the overall quality of the data, but only the most important ones are discussed here. The concept of data quality is used in a pragmatic way, that is, with a view to how the data affect the tonnage and grade estimates in the resource model.

Sample data will be used to predict tonnages and grades. Statistical analyses will be used with geological and other technical information to make inference decisions. The sample database has to provide for sound and robust decision-making. Although there may be many data, only a small portion of the deposit is actually sampled; often less than one billionth of the mass of a deposit is drilled.

The samples should be representative of the material intended for sampling which means that the sample obtained should result in a value that is similar to any other sample that could have been obtained for the same volume or material.

The samples should also be representative in a spatial sense, which means that the spatial coverage of the deposit is adequate. For example, the samples may have been taken in an approximately regular or quasi-regular sampling grid, such that each sample represents a similar volume or area within the orebody of interest. In practice, this is rarely the case and some declustering may be requried.

To ensure sample representativity, strict quality assurance and quality control programs should be put in place. If the samples are not representative, then there may be sample bias that will directly affect the final resource estimate. A number of issues need to be considered in relation to sample collection, handling, preparation, and analysis.

5.1.1 Location of Drill Holes, Trenches, and Pits

The geostatistical tools used to predict the tonnages and grade of ore material are based on knowledge of the location of the samples. An exception is Random Kriging that is designed for those cases where only imprecise sample locations within a defined domain are available, see Journel and Huijbregts (1978, pp. 352–355) and Rossi and Posa (1990) for a case study. The location of each sample is expressed as a two or three dimensional coordinates (X, Y, and Z) and is obtained by surveying its position in space. There are several surveying methods that can be used. The location of the drill hole collar as well as the deviations down the hole are surveyed. The location information can be handled using different coordinate systems, see Chap. 3, but one system should be used for the project to avoid errors.

M. E. Rossi, C. V. Deutsch, *Mineral Resource Estimation,* DOI 10.1007/978-1-4020-5717-5_5,

The location of the drill hole collars is typically surveyed with total stations tied to a local triangulation point. High-precision GPS systems are increasingly common. It is also common to develop a local topographic map from a topographic satellite or fly-over (aerial) image.

All surveys should be checked against other information such as the general topography map of the area. The elevation of the drill holes should coincide with the available topographic surface within an acceptable tolerance. A discrepancy of more than half a bench or stope height is considered a problem. Two meters maximum error in elevation is generally acceptable.

Down-the-hole surveys measure drill hole deviations after the drill hole is completed. Commonly used measuring devices are based on photographs of a bubble ring and related to an original orientation, such as single or multi-shot photos, a magnetic compass, or small gyroscopes, from which azimuth and dip measurements are taken. For additonal details of measuring devices see for example Skinner and Callas (1981) and Peters (1978).

The device is lowered into the hole, taking azimuth and dip measurements at pre-specified intervals, typically every 20 to 50 m down the hole. The measurements are later used to determine the X, Y, Z location of each sample. The measured azimuths and dips are particularly important for long, inclined holes. The deviation of a drill hole is a function of the rock it traverses, the drilling technique used, and the depth and initial inclination of the hole. If the hole is drilled close to the schistosity or the natural fabric of the rock, it will tend to follow the weaker planes in the rock. If the drill hole is drilled at a higher angle, it will tend to deviate normal to planes of weakness.

If the hole is expected to deviate significantly, then more frequent measurements should be taken. The composition of the rock being drilled through is another consideration, since some of the instruments used are affected by natural magnetism, such as the reflex system and single-shot devices. The presence of magnetic iron ore minerals, such as magnetite, phyrrotite and quartz-magnetite alterations may affect the readings. Other factors that increase the likelihood of down-the-hole deviations are sudden and periodic changes in rock hardness. Finally, the measured azimuths should be corrected for magnetic declination, particularly in high latitudes.

5.1.2 Sampling Methods and Drilling Equipment Used

In addition to drilling, samples can be obtained directly on the surface or from underground workings through trenching, channel samples, or chip samples. Samples chipped from a rock exposure are generally not used in resource estimation. Although a properly done channel sample provides good information, in practice, it is very difficult to obtain consistently representative samples.

Fig. 5.1 Boart Longyear's LF-140–2 Core Drill Rig (diamond drilling)

Representative channel samples will correspond to limited spatial coverage along exploration adits or underground workings. In underground mines, where channel samples are routinely gathered for grade control, the spatial coverage is more significant, but the sample quality tends to be poorer, because of the shorter times allowed for the sampling and assaying cycle. Most channel samples in this case become chip samples, where only a small amount of rock is taken from the face, with a high probability of being biased.

Drilling is the most common and important method for obtaining representative samples. Drilling allows sampling an unexposed orebody. The most common types of drill holes include conventional rotary (percussion), reverse circulation, and diamond drill holes (see Fig. 5.1). Each drilling method has its own characteristics and variants that affect the quality of the samples collected. Although other methods exist, they are either for special applications, or have been replaced because they are slower and more expensive. W.C. Peters provides a clear discussion on different drilling and sampling methods (Exploration and Mining Geology 1978, pp. 435–443).

5.1.3 Relative Quality of Each Drill Hole or Sample Type

It may not be appropriate to use samples from a percussion rig and from a diamond drill hole simultaneously to obtain a resource estimate. One set of samples may be biased with respect to the other. When more than one sample type is available, it is necessary to make comparisons of each set of samples and their statistical properties. Ideally, it is better to compare sets of twins or duplicate samples, but they are not always available. It is common that channel or chip samples will also show significant differences with nearby drill hole data. Data from biased drill hole or channel samples should be discarded or used cautiously solely for the construction of a geological model. In some cases secondary poorer quality data could be used in some form of cokriging (Journel and Huijbregts 1978).

Samples from percussion drilling commonly suffer from significant loss of material and little control during the drilling operation; high or low grades may be preferentially lost. Also, significant mixing of material occurs as the samples come up the hole; thus the exact location of the sample is uncertain. In most cases, data from percussion drilling is not acceptable for resource estimation.

Reverse circulation drilling is cheaper than diamond drilling, and so, for a given budget, may provide more information. If done carefully and under good sampling conditions, it can provide good samples. Often, reverse circulation drill holes are of larger diameter than common diamond drill holes. It may be difficult to obtain good geological descriptions, since the material is recovered as broken rock chips.

Diamond drilling is more expensive, although if core recovery is good, it has the advantage of bringing intact rock to the surface. This allows for better geologic mapping, and, after splitting the core in half, provides a representative sample for preparation and assaying. The down-the-hole location is better known than other types of drilling. Diamond drilling is generally considered to provide the best sample quality. Figure 5.2 shows a partial view of the very large core farm for BHP-Billiton's Olympic Dam multimetal deposit in South Australia.

Fig. 5.2 Partial view of the core farm for Olympic Dam's mine and expansion project, courtesy of BHP Billiton, Roxby Downs, South Australia

5.1.4 Sampling Conditions

The quality of samples also depends on the material being sampled and the conditions under which the samples are taken. For example, the presence of underground water or very fractured rock requires careful and sometimes much slower and expensive sampling methods to minimize possible biases.

Reverse circulation drilling and sampling could be particularly difficult below the water table or in the presence of significant amounts of water. Down-the-hole contamination, washing, and cave-ins are concerns, as well as loss of mineralization in the slimes produced. In these situations, and to avoid losing fines, the output water from the hole can be redirected to a large decantation barrel before final discharge. The fine material that decants in the barrel can be collected and analyzed, providing an indication of whether the loss of fines results in a mineral grade bias. In practice, the amount of material that can be decanted is limited and it is difficult to assign an exact down-the-hole location to the analyzed fines.

Diamond drill holes can also have problems with excess water during drilling. For example when the mineral is lodged in veinlets that can be washed away before the core is recovered. A multiple-tubing system is sometimes used to achieve better core recoveries in weak, fractured rock. Core cutting is sometimes a source of concern, particularly when a diamond-blade saw cutter is used in the presence of schistose or friable material; a hydraulic press may be preferable in these cases.

5.1.5 Core and Weight Sample Recoveries

For drill types where the samples are taken from drill cuttings, it is useful to record the total weight of the material extracted from the hole for the given sampling interval. This total weight should be compared to the theoretical weight of the sample:

$$Sample\,Weight = \pi * d^2 * length * \delta$$

where d is the hole radius, *length* is the sample interval length, and δ is the density of the material. This slows down the drilling process and may be difficult or cumbersome; however, it is a recommended quality control step.

The actual weight of the sample should be reasonably close (within 10–15 %) of its theoretical weight. There may have been sample loss or additions (from cave-ins or down the hole contamination) that may affect the final sample weights. In the case of diamond drill core, measuring the length of the piece(s) recovered allows a *% recovery* value to be assigned to each sample. Systematically low recoveries are generally a cause for concern. It is important to analyze the relationship between core recovery and grades since better grades may occur in intervals with less recovery, such as fractured rock.

5.1.6 Sample Collection and Preparation Procedures

Sample collection methods depend on the drill rig used. Generally, automatic samplers such as the one shown in Fig. 5.3 are preferred for cuttings since they allow for a more systematic splitting of the material recovered from the drill hole. Particular attention should be paid to the potential loss of fines since they often are high grade material.

A portion of the sampled material should be kept for historic records and future re-checking. Diamond drill cores are usually split in half with one half being used for preparation and assaying and the other half stored as a historic record of the drill hole. The preparation of the sample prior to assaying is a series of size reduction and splitting steps until a fine powder is obtained for assaying.

5.1.7 Geologic Mapping and Logging Procedures

It is important to describe the geologic characteristics of the rocks at different scales. This is done by geologists in the field or in the core shack and requires knowledge of the rock formations, mineralization and alteration types of the deposit. These observations are made visually with the help of a hand-lens. There is always a degree of subjectivity in the geologic descriptions depending on the experience of the geologists and knowledge of the local geology. A mature deposit with a large number of drill holes will probably have well-established procedures for identifying geologic attributes.

Figure 5.4 shows an example of a logging sheet taken from one of the early exploration holes of the Escondida Cu porphyry project, owned at the time by Minera Utah de Chile, Inc. For each mapped interval, information such as

Fig. 5.3 AusDrill Blast Hole Sampling System. Courtesy of AusDrill Ltd

lithology, alteration type and intensity, presence of fractures, identified minerals, and other relevant comments are registered, and eventually entered into a computer database.

The geologic information is used to better understand and predict the mineralization of interest. There are several references that describe procedures to map geologic information, such as the already mentioned Peters (1978). The geologic information gathered from drill hole is as important as the sample grades obtained.

5.1.8 Sample Preparation and Assaying Procedures

Sampling heterogeneous materials will incur some error, regardless of the care and technique used to obtain the sample. It is important to understand the possible impact of these sampling errors on a resource estimate. Sampling errors are also related to the distribution of the variable being sampled, which in some cases is particularly difficult, see Rombouts

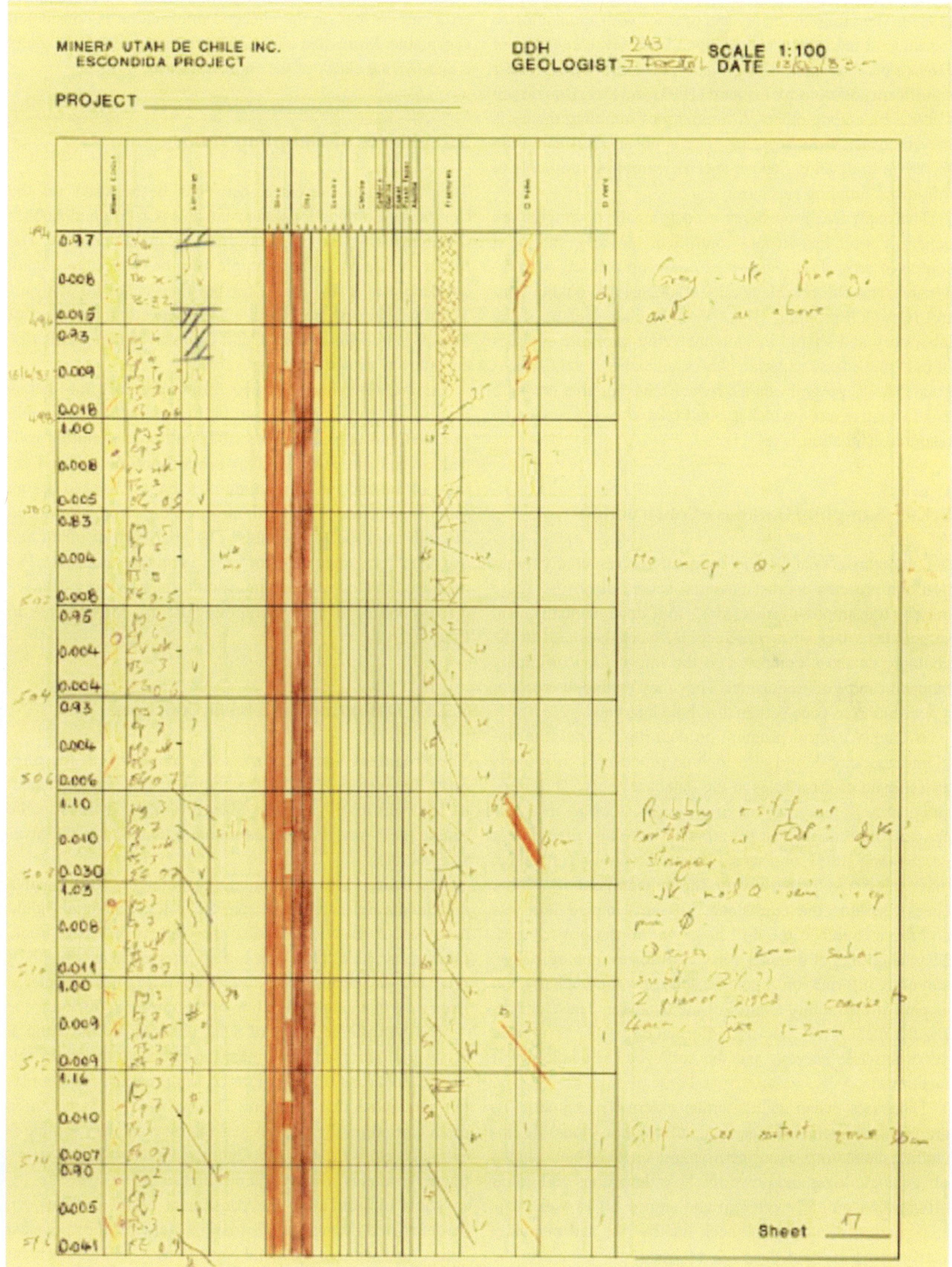

MINERA UTAH DE CHILE INC.
ESCONDIDA PROJECT

DDH 243
GEOLOGIST
SCALE 1:100
DATE

PROJECT

Sheet 17

Fig. 5.4 Historic Geologic Logging Sheet, courtesy of BHP Billiton

(1995). The sampling theory and practice with application to the mineral industry was developed by P. Gy, (*Sampling of Particulate Materials, Theory and Practice* 1982), with later significant additions by F. Pitard (1993) and D. F. Bongarçon (1998a, b), among others. A summary of sampling theory is presented below in Sect. 5.2. The application of sampling theory is to develop deposit-specific sampling protocols to minimize sampling variances.

The analytical procedures used to analyze samples are generally well known and controlled, but may still be a source of error. It is necessary to develop a strict, comprehensive, and enforceable quality assurance and quality control (QA/QC) program. This should be independent of the laboratory and should include analyzing duplicate samples of pulp and coarse material, blanks, and known standards. A good QA/QC program should reduce the laboratory errors to 2–5 % relative error, which is small relative to other errors in resource estimation.

5.1.9 Sampling Database Construction

A computerized database is required for resource estimation and presents another potential source of errors. There may be transcription errors (more so if done manually) and sometimes a lack of record-keeping. Inconsistencies in the geologic database compared to the information originally mapped can be consequential. They may be due to errors or a decision to re-code certain drill hole intervals.

A quality control program on data input should be implemented and should also include procedures to provide an estimate of error rates in the database. There should be safeguards in place against gross errors, such as in a percentage of grade in rock not allowed to be less than 0 % or greater than 100 %, or greater than the maximum percentage that can exist in the rock according to its mineralogy. Other checks include the consistency of the sampled intervals and the location of the drill holes within the project area. Manual checks of original assay certificates, geologic logs, and other information should be done on a routine basis and as part of the quality control of the database. These checks should include all relevant information, such as grades, down-the-hole surveys and the surveyed drill hole collar locations.

Databases, when audited either externally or internally, are checked against the original information available, including laboratory assay certificates; checked and signed off geologic logs; and down the hole deviations and collar information properly checked and signed off as well. It is customary for auditors to check line by line and manually verify about 10 % of the total information available in the database; although actual practice varies, generally a 1 % error rate or less on the checked information is considered acceptable. More than a 2 or 3 % error rate generally triggers a line-by-line check of the entire database.

5.2 Basics of Sampling Theory

The discussion presented here has been based on the Centre for Compuational Geostatistics (CCG) Guidebook 2 (Neufeld 2005). Perfect measurements are not possible. The relatively large mass of a sample must be reduced to a small subsample of a few grams for the final chemical analysis. There will always be a discrepancy between the content of the lot, the original sample, and the assay sample. This discrepancy is termed the sample error.

In sampling there are two forms of error: one that is present due to the intrinsic properties of the material being sampled and one that arises from improper sampling procedures and preparation. This section presents a brief review of the concepts and guidelines that are used to design sampling protocols that will minimize the errors introduced through improper procedures. The goal is to estimate and use the "fundamental error" that is always present. The reader interested in this topic should refer to more detailed discussions of P. Gy's Sampling Theory, for example in Pitard (1993).

5.2.1 Definitions and Basic Concepts

The **Fragment Size, d_α(cm)**, is the actual size of the fragment, or average size of the fragments, in the increment α. The **Nominal Fragment Size, d (cm)**, is defined as the square mesh size that retains no more than 5 % of the oversize material.

The **Lot, L**, is the amount of material from which increments and samples are selected. A lot of material should have well-defined boundaries: the content of a bag, truck, railroad car, ship, etc. A lot is also referred to as a batch of material. An **Increment, I**, is a group of fragments extracted from a lot in a single operation of the sampling device.

The **Sample** is a part of a lot obtained by the reunion of several increments and meant to represent the lot in further calculations or operations. A sample must respect certain guidelines that Sampling Theory lays out. Sampling is often carried out by progressive stages: a primary sample is extracted from the lot, and then a secondary sample is extracted from the primary sample, and so on.

The **Component** is the constituent of the lot that can be quantified by analysis. It may be a chemical or physical component such as: a mineral content, water content, percent fines, sulphur content, hardness, etc.

The **Critical Content**, **a**, is the proportion of a critical component that is to be estimated. The critical component of a lot L is denoted a_L, the critical content of a sample S is denoted a_S, etc.

$$\text{Critical content } a_L = \frac{\text{Weight of a critical component in the lot L}}{\text{Weight all components in the lot L}}$$

Heterogeneity is the condition of a lot where not all elements are identical. There are two types of heterogeneity that we are concerned with: a) the constitution heterogeneity and b) the distribution heterogeneity.

The **Constitution Heterogeneity**, **CH**, represents the differences between the composition of each fragment within the lot. Contributing factors are the fragment shape, size, density, chemical composition, mineralogical composition, etc. Constitution heterogeneity generates the fundamental sampling error.

The **Distribution Heterogeneity**, **DH**, represents the differences from one group to another within the lot. Contributing factors are the constitution heterogeneity, spatial distribution, shape of the lot due to gravity, etc.

The **Sampling Protocol** is a set of steps for sample taking and preparation meant to minimize errors and to provide a representative sample that is within acceptable standards.

5.2.2 Error Basics and Their Effects on Sample Results

Errors may be introduced during the stages required for sampling and sample preparation. They can be random with a mean of zero, random with a non-zero mean, or accidental (ocassional or non-systematic).

The "fundamental error (FE)" is the only error that cannot be eliminated using proper sampling procedures. It will be present even if the sampling operation is perfect. Fundamental error is a function of the constitution heterogeneity of the material being sampled and it can be quantified before sampling. The errors it generates are random with a mean of zero.

The "increment delimitation error (DE)" and "increment extraction error (EE)" are random errors but their mean is typically non-zero. Unlike the fundamental error delimitation and extraction error can be eliminated through proper sampling procedures.

Delimitation error occurs when the shape of the volume for the increment extracted is not correct; for example not taking the entire cross section of a conveyor belt. Extraction error occurs when all of the fragments that belong in the correctly delimited volume for the increment do not end up in that volume. The mean of these errors is typically non-zero, so bias can be introduced to the sampling procedure.

Accidental errors that occur during sampling or preparation cannot be analysed using statistical methods as they are non-random events. Prevention of accidental errors is crucial for reliable sampling, and have little to do with Sampling Theory, more to do with good sampling practices.

Since the errors are random variables and are independent the following relationships are true:

- Total sampling error:

$$TE = FE + DE + EE + \ldots$$

- Average error:

$$E\{TE\} = E\{FE\} + E\{DE\} + E\{EE\} + \ldots$$

- Total error variance:

$$\sigma^2\{TE\} = \sigma^2\{FE\} + \sigma^2\{DE\} + \sigma^2\{EE\} + \ldots$$

Thus, individual errors do not cancel out, but are compounded. This compounding effect emphasises the care and attention that sampling requires.

When the mean of the sampling error, $E\{SE\}$, approaches zero the sample is accurate, or non-biased. A sampling selection is said to be precise when the variance of the sampling error, $\sigma^2(SE)$, is less than the standard required for a given purpose. It is not related to the sample average or accuracy of the sample. Accuracy and precision, the two measures of sample quality, can be combined, leading to the notion of representativeness:

$$r^2\{SE\} = m^2\{SE\} + \sigma^2\{SE\} \le r_0^2$$

When the mean square of the sampling error, $r^2(SE)$, is less than a standard threshold, r_o^2, the sample is considered representative.

5.2.3 Heterogeneity and the Fundamental Error

Sampling theory differentiates between two types of heterogeneity: (1) the distribution heterogeneity and (2) the constitution heterogeneity.

The constitution heterogeneity could be considered in two different ways: the heterogeneity between the fragments making up a sample, or the heterogeneity within the fragments of the sample. For our purposes, and sampling in general, the heterogeneity between the fragments is more consequential.

The constitution heterogeneity is defined based on the number of fragments within the lot and still requiring the characteristics of the material within the lot. But in practice this is difficult, and thus the constitution heterogeneity is multiplied by the average fragment mass. Doing this simplification defines the constant factor of constitution heterogeneity, also called the intrinsic heterogeneity, IH_L, which can be expressed and estimated using simple factors. The required factors to define the intrinsic heterogeneity are:

- **d:** the nominal fragment size; d has units in centimetres.
- **f:** the shape factor accounts for the shape of the fragments and is a measure of the fragment shape deviation from a cubic shape. It is a dimensionless number, used to estimate the volume of the fragment. Since the fragment volume is equal to the product of the shape factor and the cubed fragment size, $f_\alpha d_\alpha^3$, the shape factor is a correction factor to determine its volume.
  If the fragments are perfect cubes, $f_\alpha = 1$. If the fragments are perfect unit spheres with $r = 1$, the volume of the sphere would be $4/3\ \pi r^3 = 0.523$, therefore with shape factor $f = 0.523$.
  It is dimensionless and is experimentally determined with most minerals having a shape factor approximately equal to 0.5: coal=0.446; iron ore=0.495 to 0.514; pure pyrite=0.470; quartz=0.474; etc. Flaky materials, such as mica, have a shape factor around 0.1; soft solids submitted to mechanical stresses, such as gold nuggets, have a shape factor around 0.2; and acicular minerals, such as asbestos, have a shape factor greater than 1 and may be as large as 10.
- **g:** the granulometric factor accounts for the size differences between the fragments, also a dimensionless number. Using the granulometric factor, g, and the nominal fragment size, d, the fragment size distribution can be accounted for. The granulometric factor is a measure of the range in fragment sizes in the sample:
  - Noncalibrated material, crusher product, it is around 0.25.
  - Calibrated material, between two screens, it is around 0.55.
  - Naturally calibrated material, cereals or beans, it is around 0.75.
- **c:** the mineralogical factor accounts for the maximum heterogeneity condition that can be present in the sample; units is in grams per centimetre cubed or specific gravity.
- **ℓ:** the liberation factor accounts for the degree of liberation in the sample; it is a dimensionless number. When the material is perfectly homogenous $\ell = 0$ and when the mineral is completely liberated $\ell = 1$. Most materials can be classified according to their degree of heterogeneity. The liberation factor varies considerably and it is difficult to assign an accurate average to it. The calculation of the liberation factor has evolved over time, and has changed since the second edition of *Pierre Gy's Sampling Theory and Sampling Practice* by F. Pitard (1993) was published. François-Bongarçon and Gy wrote a paper (2001) presenting an improved method for estimating the liberation factor. This method corrected some of the problems associated with the previous calculation and use of the liberation factor.

The constant factor of constitution or intrinsic heterogeneity has units of mass (grams), and is used to relate the fundamental error to the mass of the sample:

$$IH_L = c\ell fgd^3$$

5.3 Liberation Size Method

5.3.1 Fundamental Sample Error, FE

The fundamental sampling error, FE, is defined as the error that occurs when the selection of the increments composing the sample is correct. This error is generated entirely by the constitution heterogeneity. Gy has demonstrated that the mean, m(FE), of the fundamental error is negligible and that the variance, σ_{FE}^2, can be expressed as:

$$\sigma_{FE}^2 = \frac{1-P}{PM_L} IH_L$$

where P is the probability of selection for any one fragment within the lot and:

$$M_S = PM_L$$

Substituting this into the variance equation gives us:

$$\sigma_{FE}^2 = \left(\frac{1}{M_S} - \frac{1}{M_L}\right) IH_L$$

and when $M_L >> M_S$:

$$\sigma_{FE}^2 = \left(\frac{1}{M_S}\right) IH_L$$

These are very practical and useful formulae for designing and optimizing sampling protocols.

5.3.2 The Nomograph

To make use of the constitution heterogeneity and quantify/show its effects on the sampling process we have to relate the state that the sample is in, fragment size and mass, to the state that we want the sample to be in, that is, with a smaller mass and a smaller fragment size. The error produced during

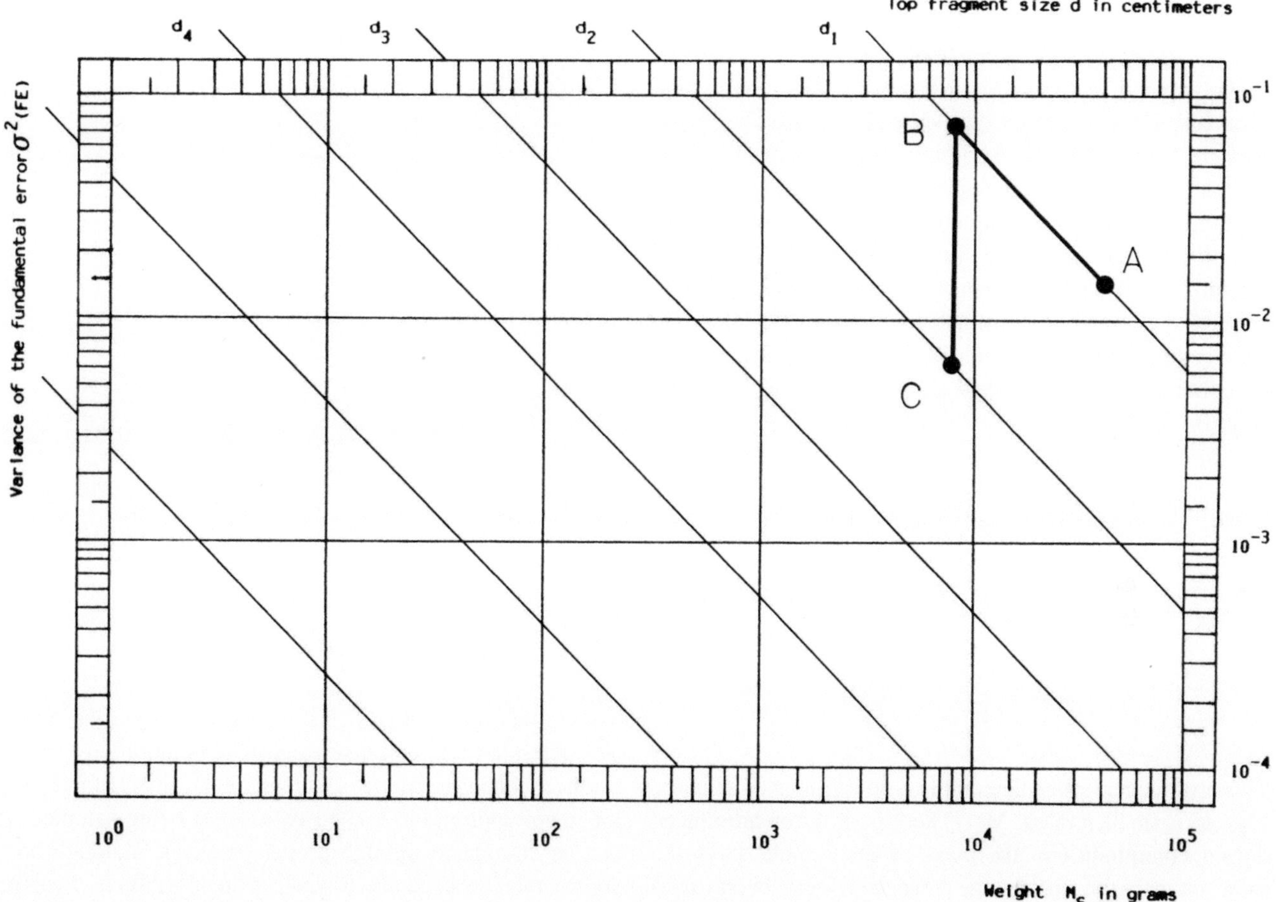

Fig. 5.5 Example nomograph showing the nominal size lines, a sample splitting cycle, and a comminution step. (Hartmann 1992)

the stages to get a smaller sample are measured and minimized using the constitution heterogeneity of the material being sampled and the nomograph.

5.3.3 Nomograph Construction

A nomograph is a base 10 log-log plot with the sample variance on the ordinate axis and the sample mass on the abscissa axis. To plot the variances versus sample size the formula must be converted to logarithmic space:

$$\begin{aligned}\sigma_{FE}^2 &= \left(\frac{1}{M_S}\right) IH_L \\ &= \frac{1}{M_S} c\ell fgd^3\end{aligned}$$

$$\begin{aligned}\log(\sigma_{FE}^2) &= \log\left(\frac{1}{M_S} c\ell fgd^3\right) \\ &= \log(c\ell fg) + 3\log(d) - \log(M_S) \\ &= C + 3\log(d) - \log(M_S)\end{aligned}$$

Plotting changes to the sample using this method allows easy visualization of changes made to the sample. These changes are actual steps in the sample preparation and can be either a reduction in fragment size through comminution, crushing or grinding, or a reduction in the sample mass through splitting.

When the sample is split there is no change in the nominal fragment size so all of the terms in the sample variance equation are constant and the variance becomes directly proportional to $-\log(M_S)$. Changes to the mass of the sample, and the variance of the sample, by splitting will follow a line on the nomograph with a negative one slope. This allows lines representing the different nominal fragment sizes during the sample preparation to be plotted on the nomograph. During comminution the mass of the sample stays constant and the other terms of the equation will change. Comminution results in a reduction of the sample variance due to the reduced fragment size, and on the nomograph this will be a vertical line from the larger nominal fragment size line straight down vertically to the lower nominal fragment size line.

Figure 5.5 shows the size lines for six different nominal fragment sizes, a sample mass reduction step, and a comminution step. The size lines, for the nominal fragment sizes d_1 decreasing to d_6, are shown as the thin lines that extend beyond the upper and left edges of the nomograph boundary. When the sample was split the nominal size stayed the same but the mass decreased resulting in an increase in the sample variance. This is shown as the thick line from point A to point B. During the comminution phase the sample mass stayed the same and the nominal fragment size decreased resulting in the sample variance dropping from its position on the larger size line down to a point on a lower size line corresponding to the nominal fragment size produced from the comminution cycle. This is shown as the thick line from point B to point C.

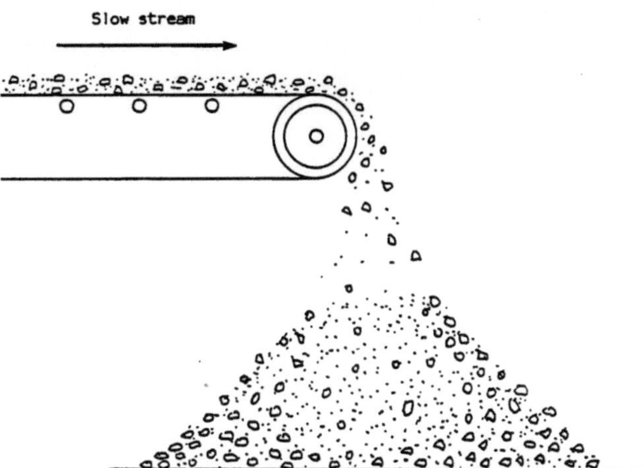

Fig. 5.6 Segregation due to differences in the fragment size (coarse fragments roll down the pile concentrating themselves at the edge of the pile while the fine fragments remain near the centre) (Hartmann 1992)

5.3.4 Sampling Fundamental Error

The fundamental error that occurs during the sampling protocol is:

$$\sigma^2_{FE} = \left(\frac{1}{M_{S2}} - \frac{1}{M_{S1}}\right) IH_L$$

This error occurs when the sample is split from a large mass, M_{S1}, to a smaller mass, $M_{S2.}$ There is no error introduced during comminuition—the mass of the sample stays the same and only the particle size is reduced. Over several sample preparation stages the fundamental error is the sum of the error variances during the individual stages:

$$\begin{aligned}\sigma^2_{FE} &= \sum_i \sigma^2_{(FE)i} \\ &= \sigma^2_{(FE)1} + \sigma^2_{(FE)2} + \cdots + \sigma^2_{(FE)n}\end{aligned}$$

This is a simple method for calculating the fundamental error introduced during sample preparation. The nomograph can thus be used for sampling protocol optimization.

5.3.5 Segregation or Distribution Heterogeneity

Segregation is the heterogeneity between groups of fragments within the sample lot. Segregation can occur during handling of the material once it has been mined or sampled. This segregation of material in a lot is due to differences in the fragment size, shape, density, mass, angle of repose, etc. Figure 5.6 shows an example of segregation and how the material properties contribute to the segregation. The sampling protocol must be designed to remove any influence that the segregation may have.

5.3.6 Delimitation and Extraction Errors

These two errors are related in that if one is present, probably the other one will also occur. Delimitation error is introduced when the shape of the increment selected from the lot is not appropriate for the type of lot being sampled. It can be eliminated through proper practices, although how controlable it is depends on whether the lot is three-, two-, or one-dimensional. The ideal samples being a uniform sphere, a cylinder, or two parallel planes, respectively.

Extraction error is introduced when the method used to obtain the sample is incorrect. Along with the Delimitation Error, it can be very detrimental to correct sampling.

The errors produced generally have non-zero means and introduce a bias to the results. After the correct shape of the increment has been determined its extraction must follow the centre of gravity rule. This rule states that all fragments with their centre of gravity inside the increment belong in that increment and all of the fragments with their centre of gravity outside the increment do not belong.

5.3.7 Preparation Error

As soon as the insitu material is disturbed to the moment that it is assayed it is possible to introduce a wide range of possible preparation errors. Due to the non-random nature of these errors and the long time frame over which they can occur, it is important to have a good sampling protocol that outlines handling and preparation of the sample material so that the samples are representative. The errors can come from contamination, losses, changes in the chemical composition, unintentional mistakes, and fraud or sabotage.

Contamination of the sample can be in the form of dust falling on the sample, material left over in the sampling circuit, abrasion on the sampling equipment, or corrosion. These errors can either increase or decrease the critical content for the component of interest and once the sample has been contaminated there is nothing that can be done remove the contamination.

Losses may occur as fines in dust, material left in the sampling circuit, or smearing of the critical component. Usually the component being sampled is a minor portion of the overall material and it is significantly different in composition from the gangue material. This means that losing a specific fraction of the sample, the fine or coarse fraction, will result in the sample no longer being representative.

Changes in the chemical composition of the sample must be avoided. Some potential changes are oxidation, or fixation of water or carbon dioxide. Alterations in the chemical nature of the material will impact the assay results and the expected recovery of the material as it is processed.

One time human errors have the potential to be significant and it is very hard to determine the source of the error. Dropping the samples, mixing of samples, improper labelling, poor maintenance of the equipment, contamination, etc, are mistakes that may introduce error. Care and attention to detail as well as following the sampling protocol will ensure that errors of this type are minimal.

Fraud and sabotage are the intentional alteration of the samples used to increase, or decrease, the value of a sample for personal or corporate gain. Inflation of the mineral content in the deposit to increase share value has occurred in the past and as a result the requirements for sampling quality and documentation of the sampling undertaken have become more stringent.

5.4 Sampling Quality Assurance and Quality Control

The process of mineral resource estimation requires a strict program of quality assurance and quality control (QA/QC) to provide confidence about the precision and accuracy of the drill hole data used for estimation. The QA/QC program implemented may have somewhat different characteristics if implemented for mine operations (production sampling), but it will have the same general objective.

A QA/QC program of an appropriate standard is required by most international resource reporting standards. Published resource estimates should be accompanied by a description and statement of the data quality. This is also a basic item in any third-party review of resource models, and may have a significant impact on the overall perception of the model quality.

There are no universally accepted procedures for QA/QC, although certain basic steps are always recommended. An outline of a recommended set of procedures, derived mostly for gold sampling, will be presented. Good general references are Long (1999) and Roden and Smith (2001).

5.4.1 General Principles

The main objective of the QA/QC program is to minimize errors introduced due to sampling, sample preparation, and sample assaying procedures. The QA/QC program is a continuous process providing information necessary to correct defects in the shortest amount of time possible.

Accuracy and precision are two terms used to evaluate the quality of the information provided by analytical laboratories. Accuracy is a measure of the degree of agreement of the assayed sample value to the true unknown value of that sample. An indication of accuracy can only be obtained through re-assaying samples of known values such as standards or reference materials.

Precision is a measure of the reproducibility of the sample value, which can be estimated by re-assaying the same sample a number of times. Precision and accuracy are different concepts. A laboratory could have any combination of good or bad precision and accuracy.

Figure 5.7 illustrates the concepts of accuracy and precision using the common analogy of the shooter's bullseye. The left image shows a precisely inaccurate set of three shots; the center figure shows an accurate but imprecise series, while the right image shows the case where the shooter has been both accurate and precise.

All QA/QC check samples sent for analysis to the laboratories should be blind, meaning that the laboratory should not be able to differentiate a check sample from a regular submission. The internal checks that analytical laboratories often implement are performed with the technicians being aware of the fact that they are assaying duplicate samples. These internal checks, often reported by the laboratories as measures of their sampling precision and accuracy, should never be considered as part of a formal QA/QC program. This applies both to company-owned and external laboratories.

The minimum control unit should be the batch of samples sent originally to the laboratory. The batch concept derives from the fact that gold fire assays are done by oven batches. Typically a set of 40 samples is introduced into the oven. It is a useful concept that has been extended to other types of assaying in the context of QA/QC.

Any check sample that fails implies that the complete batch to which the control sample was incorporated should be re-assayed. This applies to drill hole samples, but not necessarily for production samples, since there is no time to

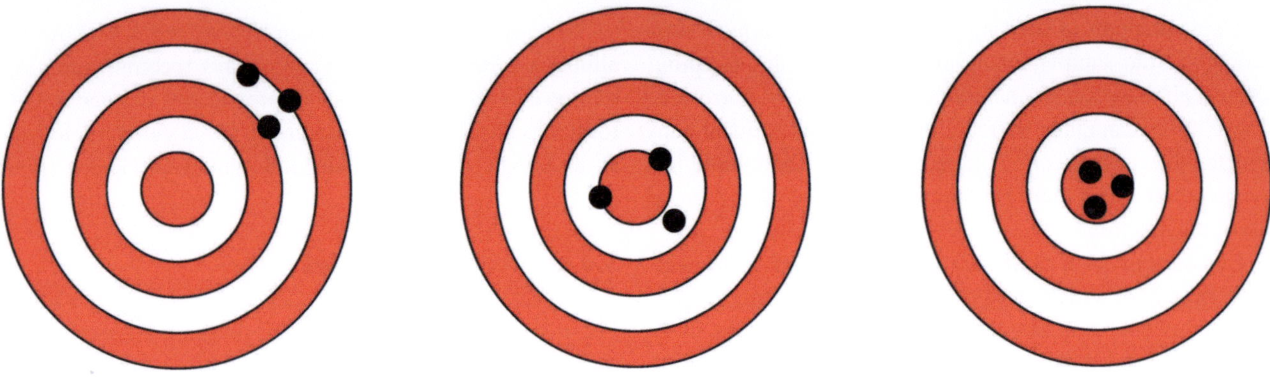

Fig. 5.7 Accuracy and precision. *Left* is precise but inaccurate; *center* is accurate but imprecise; and *right* is both precise and accurate

re-assay them. In this case, failed check samples trigger corrective measures for future assaying.

The sampling QA/QC program should cover (a) sampling conditions in the field; (b) sample preparation; (c) analytical accuracy and precision; and (d) correctness of the laboratory reports and transfer of the information to the database(s).

The materials to be used in the QA/QC program include (a) standards, or reference material; (b) blanks, which are samples with no grade; (c) field duplicates, taken at the drill hole site or core box; (d) coarse duplicates, taken as the first reject at the sample preparation stage, typically −10 mesh in size; and (e) pulp duplicates, which are taken from the last size-reduction and splitting at the end of the sample preparation process.

There are generally two or more laboratories involved that would include a primary or principal laboratory for routine work, and a secondary or check laboratory. Occasionally, a referee laboratory is needed when discrepancies between the primary and secondary laboratories cannot to be resolved.

Sampling and assaying protocols are established prior to processing the samples from the field. These protocols should cover all aspects of sample processing and handling including chain of custody. The sampling theory originally developed by P. Gy (1982) can be used to determine optimum sample preparation protocol, such that errors introduced in the preparation and assaying procedures are minimized (see above).

The sample preparation and assaying protocols should be identical for the primary and secondary laboratory. The mining company should have a staff person in charge of the overall QA/QC program whose duties include ensuring that the protocols used at the different labs are consistent. That person should inspect the facilities on a regular basis.

5.4.2 Elements of a QA/QC Program

5.4.2.1 Blanks

Blanks are samples with no grade of interest whose purpose is to check laboratory contamination and to verify correct handling of the samples. There should be both pulp and coarse blanks prepared and inserted into the sample preparation stream. In the case of core samples, the coarse blank is introduced after the first crushing stage, while the pulp blank should be inserted as a separate envelope in the sample batch.

It is advisable that the blank have the same matrix (mineralogy) and result in a pulp with the same characteristics as the main samples so it is not obvious to the laboratory that the sample is different in any way. This is sometimes difficult to achieve, although at least the main characteristics such as color should be as similar as possible. Very low grade samples should not be used as proxies for blanks.

5.4.2.2 Standards

Standards are samples for which its grade is known within a certain precision. They are used to check the accuracy of the analytical laboratories, by comparison of the re-assays to the reference value.

There are commercial standards that can be purchased. They provide samples with known grades for some types of ore deposits. This material can be purchased from laboratories and institutions around the world. The standards are delivered with certificates stating the accepted value and its precision, in addition to a full description of the procedures used to analyze them.

Alternatively, the mining company has the option of developing its own standards. The material used to obtain the standard is typically from the same deposit as the main sample stream, which ensures that differences in the sample matrices will be minimal. The certification of standards requires major analytical work, which can be done through a round robin analysis using no less than 6 laboratories, and more commonly 8. This work can be managed by the mining company or outsourced to an external analytical laboratory, which would also be part of the round robin laboratories, and would provide the final certification for the samples and their corresponding tolerance limits.

If the standards are done using a commercial laboratory, it should be involved in the assaying of the project samples. The most probable value for the standard should be reported with the $\pm 2\sigma$ (2 standard deviations of the distribution of all assayed values). These or alternative upper and lower limits should be used as acceptance criteria for the re-assayed standard.

5.4.2.3 Coarse and Field Duplicates

The purpose of the coarse duplicates is to quantify the variances introduced into the assayed grade by errors at different sample preparation stages. They provide a measure of the sample precision. There will commonly be more than one size reduction and splitting steps in the preparation stage. These coarse duplicates should be inserted into the primary laboratory stream, providing an estimate of the sum of the assay variance plus the sample preparation variance, up to the first crushing stage.

An alternative is to obtain a field duplicate. In the case of diamond drilling, a duplicate from the core box (i.e., a quarter core or the other half core) is sent to the laboratory, most commonly with the intention of replacing the coarse duplicate. The advantage is that the variance observed in field duplicates includes the actual sampling and the first size reduction step. The price of leaving the interval without core may be too high. Also, a quarter core may be too small a volume for the duplicate to be representative. In the case of reverse circulation drilling, it is more likely that field duplicates do not exhaust the sample as generally there are abundant chips available.

In the case of blast hole sampling, it is also possible to take a duplicate sample in the field from the cuttings pile or the reject from the hydrocyclone if an automatic sampler is used. These field duplicates can be used to check the first stage crushing and sampling process.

5.4.2.4 Pulp Duplicates

Pulp duplicates provide a measure of precision of the analytical procedures used. They are taken at the final stage of sample preparation, and generally are a second envelope with the 100 or 200 g final sample sent for assaying, inserted blindly into the sample batch. Pulp duplicates sent to the same primary laboratory provide an estimate of the analytical variance of that laboratory. When sent to the second, check laboratory, the pulp duplicates quantify the precision (analytical variance) between the two laboratories.

5.4.3 Insertion Procedures and Handling of Check Material

The basic unit is a batch. This can be defined for drill hole samples, blast hole samples, or any other type of production samples. A batch should contain sufficient samples to allow the insertion of control samples. At the same time it cannot be too large to become too difficult to manage, evaluate, or re-assay. For drill hole samples, it is generally recommended that a batch be no less than 20 samples, and preferably 40 samples. For blast holes and crusher samples, the batch is generally larger and composed of samples from a fixed timeframe (work shift or day).

For each 40-sample batch, and assuming that the mining company has full control of the sample preparation stages, the following are the suggested control samples to be sent to the primary laboratory:

1. For drill hole samples:
   a. Two blanks (5 % of total number of samples). Of these, insert one coarse blank for every 4th blank inserted (25 % of the total number of blanks inserted).
   b. Two pulp duplicates (5 % of total number of samples).
   c. Two coarse duplicates (5 % of total number of samples).
   d. Two standards appropriate to the expected grade of the batch samples (5 % of total).
2. For blast hole and crusher samples:
   a. One blank (2.5 % of total number of samples).
   b. One pulp duplicate (2.5 % of total number of samples).
   c. One coarse duplicate (2.5 % of total number of samples).
   d. One standard appropriate to the expected grade of the batch samples (2.5 % of total).

This implies that there will be 20 % check samples for exploration data, and 10 % additional control samples for production data. A second check laboratory should be used for the drill hole samples, but it is not necessary for production samples. Since production samples are normally processed in-house, an additional 2.5 % of pulp control samples should be sent out for re-assaying at a different laboratory as routine check.

For drill hole samples, the control samples sent to the second (check) laboratory should be from pulp duplicates in all cases and should include one blank, two sample pulps, and one standard for every 40-sample batch. This implies an additional 10 % sent to a second laboratory.

There are cases where the mining company or project development team does not have a sample preparation facility, and is not able to control the sample preparation process. For these cases when the sample preparation is done by the laboratory itself, coarse duplicates should be sent for preparation and assaying by the second laboratory.

All control samples should have a pre-defined logical sequence of numbers, such that the flow of samples is easy to control, and the control samples are inserted into the stream in a disguised fashion.

5.4.4 Evaluation Procedures and Acceptance Criteria

The unit of acceptance/rejection should be a complete batch. Individual samples should not be sent out for re-assaying. The numbers of samples in a batch is variable, but 40 is commonly used. Laboratories tend to process samples by batches so whatever problem caused the check sample to fail is likely to have affected the remaining samples in the batch.

All acceptance/rejection criteria should be enforceable as part of the agreements with the laboratories, including re-assaying if the batch is not compliant. The exploration or mining company should perform, as part of its QA/QC program, the acceptance/rejection tests consistently using the same procedures. Transparency and good relations with the laboratory are always necessary to ensure a successful QA/QC program.

The exploration or mining company should not shortcut the QA/QC program, and should allow sufficient time, budget, and contractual arrangements for a significant number of re-assays. The program should be implemented on an ongoing basis and not at the end of drilling campaigns or predetermined time periods.

The expected accuracy and precision depend on the type of mineralization being sampled. Sampling gold is particularly difficult especially if there are coarse particles. Certain base metal deposits may be easier to sample. The acceptance criteria will change according to the type of mineralization being sampled. In the case of gold, commonly accepted criteria include:

1. ***Coarse blanks*:** 80% or more of these samples should return with a value less than or equal to *three times* the detection limit. Thus, at least four control samples are required to make a decision, which implies 8 batches (one coarse blank in every other batch). Another way to express it, is to say that 1 in every 5th blank may fail the criterion.
2. ***Pulp blanks*:** 90% or more of these samples should return with a value less than or equal to *three times* the detection limit. Therefore, 1 in every 10 can be above the accepted limit.
3. ***Standards*:** In all cases, standards have to fall within the accepted tolerace limits of the certified reference value. This can be 2 or 3 times the standard deviation or some intermediate value, depending on the round robin results.

In the case of duplicates, the suggested criteria for pairs of samples (original-duplicate) with averages equal or to greater than 5 times the laboratory's detection limit (DL), the following formulas are suggested:

4. ***Pulp duplicates*:** 90% of the pairs' absolute relative differences equal to or smaller than 10%. The absolute relative difference for each pair is defined as:

$$Rel.Error = \left| \frac{Original - Dup}{\left(\frac{Original + Dup}{2} \right)} \right| < 0.10 \qquad (5.1)$$

1. ***Coarse duplicates*:** Using the same Eq. 5.1, 90% of the pairs' absolute relative differences equal to or smaller than 20%.
2. ***Field duplicates*:** Using the same Eq. 5.1, 90% of the pairs' absolute relative differences equal to or smaller than 25%.
3. Additionally, for pulp duplicates, if the absolute value of the difference [Original-Duplicate] (numerator in Eq. 5.1) is equal to or less than two times the detection limit (DL), the pair is accepted.
4. For coarse duplicates, if the absolute value of the difference [Original-Duplicate] (numerator in Eq. 5.1) is equal to or less than three times the detection limit (DL), the pair is accepted.

The standards should be plotted to verify the laboratory's performance over time. The graph typically shows the expected value, the upper and lower acceptance limits, and the assay results for each control sample inserted. This allows for trends to be detected. For example, if the control samples are consistently above the expected values (but still within the acceptance limit), there may be a small persistent bias that the laboratory should be notified of and correct.

In performing these tests, the practical detection limit (DL) should be considered. The practical DL may be different (and generally higher) than the nominal or theoretical DL as stated by the laboratory, because it takes into account the lower precision of analytical methods when working at or near their nominal DL. A higher DL should not cause a problem as long as it is much lower than the mineralized or economic cutoffs.

The sample preparation and the analytical laboratories should be supervised constantly. Ideally, the responsible person should make surprise visits to each laboratory on a regular basis. These informal inspections should result in a brief report describing the laboratories operating conditions, cleanliness, orderliness, sample handling procedures, and, most importantly, the extent to which they are correctly implementing the prescribed sample preparation and analytical protocols. Photographs should be used to record and document each visit.

5.4.5 Statistical and Graphical Control Tools

There are several tools that can be used in analyzing and describing QA/QC information. Among the alternatives, the suggested basic tools are:

- Error histograms and basic statistics: these should include the relative errors defined in the equations above for

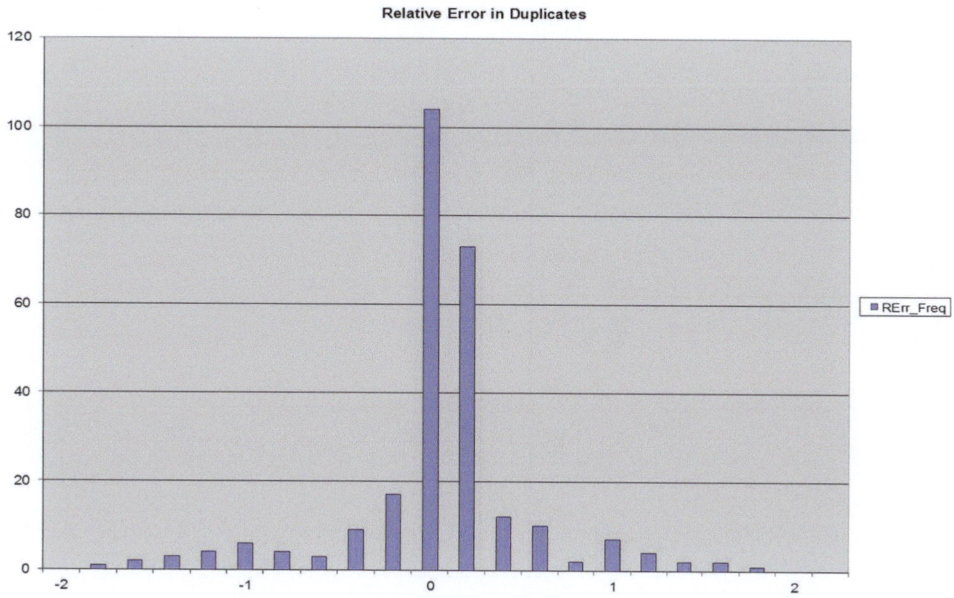

Fig. 5.8 Histogram of the absolute differences of duplicate sample pairs

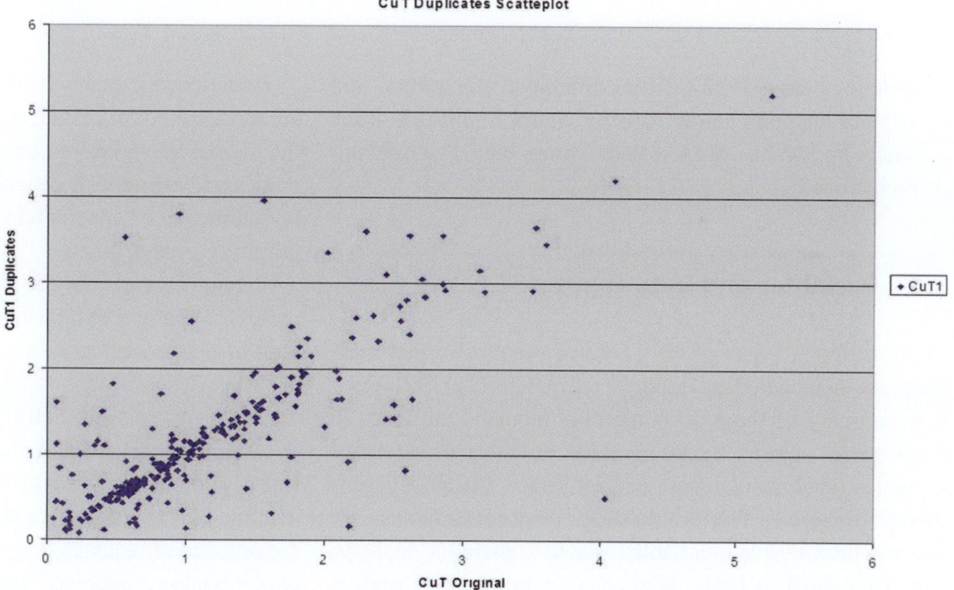

Fig. 5.9 Scatter plot of the original vs. the duplicate samples

pulp and coarse duplicates; the differences between the expected values of the standard and the control samples; histograms and basic statistics of the percentages of deviation of pulp and coarse duplicates; etc., see, for example, Fig. 5.8.

- A scatterplot of the original versus duplicate samples (pulp and coarse) along with correlation statistics. This plot provides a visual and numerical analysis of the correlation of the two variables, see Fig. 5.9.
- Control graphs for standards versus time including lines representing the expected value of the standard and the upper and lower limits. These graphs are also useful for duplicate samples because it may help identify periods when the quality of the laboratory work is poor. It is often seen that the laboratory work performed immediately after holidays or long weekends is of lower quality. The quality may also be poor when the laboratory is working at capacity or overloaded.
- Another time-dependent control chart, for both standards and duplicates, is to plot a moving average of results that would include 20 to 40 control samples at a time, i.e., several batches. This control chart would be useful to detect longer-term trends, which are sometimes disguised or difficult to detect when looking at individual batches.
- Another useful graph is the cumulative frequency of the absolute value of Eq. 5.1, see Fig. 5.10. It shows if the control samples comply with the prescribed 90 % of the

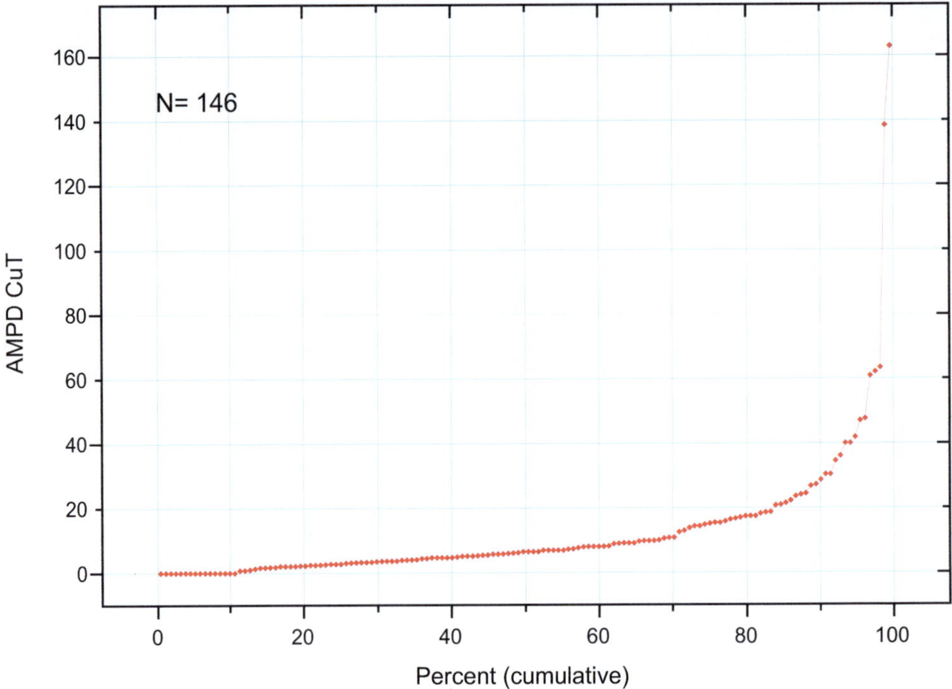

Fig. 5.10 Cumulative frequency plot of the absolute value of pair differences

samples below 10% relative difference (for pulps) and 90% of the pairs below 20% for coarse duplicates. Pairs where both values are less than 5 times the DL should not be included.

5.5 Variables and Data Types

Many different types of raw, derived or transformed variables are considered in resource estimation. The following is a summary of the most commonly modeled variables in the minerals industry including grade variables, transformed variables (such as Gaussian and indicators variables); geological properties; fracture densities; texture variables; dry and wet densities; and variables related to metallurgical performance, such as bond index; SemiAutogenous Grinding (SAG) power index; hardness; plant throughput; mill recoveries; and mineralogical species.

5.5.1 Raw and Transformed Variables

The most common variables are grade variables (mass fractions) since they directly measure the resource being estimated. Different types of grade variables will lead to different modeling techniques. Typical examples include precious metals, such as gold, silver, platinum, and palladium.; base metals, such as Cu, Pb, Zn, Fe, and Ni; iron ores such as magnetite and hematite; coal, uranium, potash, and nitrates. Most metal variables will be positively skewed, although there are noteable exceptions such as Fe.

Experience has shown that certain modeling techniques are more suited for certain types of deposits. This is based on the degree of variability commonly associated with certain grades and the deposit geometry. There are tabular vein-type or sedimentary deposits and more massive disseminated or porphyry-type deposits.

Precious metals tend to have grade distributions that are more skewed and exhibit higher spatial variability. The impact of outlier values is more significant and thus it is appropriate to consider modeling techniques better suited to control erratic distributions.

An important aspect of grade variables is that they upscale linearly, that is, the grades of larger volumes is the arithmetic average of the constituent smaller volumes. This simplifies the practice of block kriging (Chap. 8) and the application of conditional simulations (Chap. 10). Change of support methods (Chap. 7) are based on the linear averaging of variables.

In-situ bulk density (dry and wet) is a critical component in resource modeling. The resource estimates should normally be reported using dry density values, since all grade assays are normally done on a dry basis; however, estimates of moisture content (or wet densities) is necessary for mine planning, since it provides a more realistic estimate of true tonnage that will be mined, and therefore is an important component in the scheduling of trucks and other equipment. This is particularly important in tropical or wet environments. For example, the moisture content in Nickel laterite deposits can be greater than 15% of total weight.

Bulk density and moisture values can be modeled through arithmetic averages by rock types or other geologic variable types or estimated through kriging or inverse distance methods. Bulk density is sometimes forgotten and few good samples are available. Parrish (1993) has a good discussion on bulk densities and their importance to resource estimation.

Another group of raw variables that are commonly modeled are auxiliary variables. Some examples include thicknesses of formations, typical of sedimentary deposits; elevations of the top or bottom of particular surfaces of interest, such as bedrock contact; the geometry of diamondiferous pipes; top of sulfides enrichment; footwall and/or hanging wall positions in tabular deposits. Sometimes, variables such as grade multiplied by thickness (proportional to metal content) is used in tabular deposits. This grade times thickness variable transforms a 3-D modeling problem into a 2-D exercise, because the third spatial dimension, usually much smaller than the other two, is incorporated into the variable being estimated.

Another important group of variables are metallurgical performance variables. Resource and reserve models should include predictions of rock characteristics, crushing/grinding throughputs, final product recoveries, and other variables as a more realistic basis for cash flow predictions. There is a trend in the industry to model geometallurgical variables and include them in resource models. They can be ore and gangue mineralogical variables, useful for better predicting plant performance, concentrate grade, and heap or vat leach performances. Some of these variables do not average linearly, and thus require special consideration.

Resource models must consider a breadth of issues that were simplified in the past including all types of dilution (Chap. 7) and geologic variables that affect mine and plant performance including geo-mechanical and geo-metallurgical variables. A resource model is much more than a geological in-situ model.

5.5.2 Soft Data

Soft data is a term used in reference to information that provides imprecise or indirect measurements of the variables of interest. Some specific examples include geophysical readings such as magnetic anomalies associated with an iron (magnetite) deposit. Another example is the use of radiometric readings (a gamma probe) to obtain the ratio of parent-to-daughter products of Uranium$_{238}$ (U_{238}) decay chain, from which, and after proper calibration, the U_{238} grade may be estimated.

Soft data does not have the same quality of information as compared to hard data, which is typically the assayed mineral grade. Depending on the specifics of the indirect measurements, there are two general characteristics that determine the methodology used to analyze and apply the soft data.

First, the quantity of soft data may be significantly larger than the hard data, but of poorer quality. The data may be close-spaced while the hard data will tend to be located at a much larger spacing. This is characteristic of geophysical data, where a dense grid of data is available.

Second, the soft variable may be a simple condition, such as "within this rock type, the Au grade will be no larger the 1.0 g/t". The soft information is qualitative in nature and somehow must be expressed in a numeric format before it can be used explicitly in the modeling process to any advantage.

A common procedure is to turn the soft information into an indicator or a series of indicators that allows merging it with the hard data. The details of the modeling techniques are discussed in Chap. 9.

5.5.3 Compositional Data

The following discussion is a summary of CCG (Centre for Computational Geostatistics) Guidebook 7 (Manchuk 2008). Compositional data are multivariate data where the variables or components represent some part of a whole (Pawlowsky, 1989; Pawlowsky et.al., 1995). All variables from a composition are measured on the same scale and unit system and are constrained by a constant sum property. The sum depends on the measurement scale. Some common ones are 1 for fractional data, 100 for percentages, and 10^6 for parts per million or ppm. A set of variables summing to a constant is also referred to as a closed array (Chayes 1962). The concept of a compositional data set **X** with D components and N observations can be written as:

$$\mathbf{X}^D \equiv \{\mathbf{x}_1,\ldots,\mathbf{x}_D : \mathbf{x}_1 \geq \mathbf{0},\ldots,\mathbf{x}_D \geq \mathbf{0}; \mathbf{X}\mathbf{1}^D = \mathbf{1}^N\}$$

Note that $\mathbf{1}^D$ is a column vector of ones of size D and $\mathbf{1}^N$ of size N.

Two issues for statistical analysis are raised by this equation: (1) variables are not free to range in $(-\infty,+\infty)$, thus the relationships are not free to vary independently, and (2) the constant sum constraint must force at least one covariance or correlation to be negative; when one component gets large, the others must necessarily decrease. Correlations are not free to range in $[-1,1]$ causing spurious correlations (Aitchison 1986).

5.5.3.1 Compositions in Natural Resources

In a natural resources context, compositions are geochemical, geophysical, or lithological. Whole rock geochemistry is an example, and may come in various forms depending on the depositional environment and target resource. Consider mining to extract metal products, such as copper suplhides. Several mineralogical species may be responsible for all the Cu in a sample, such as chalcopyrite, chacolsite, covelite,

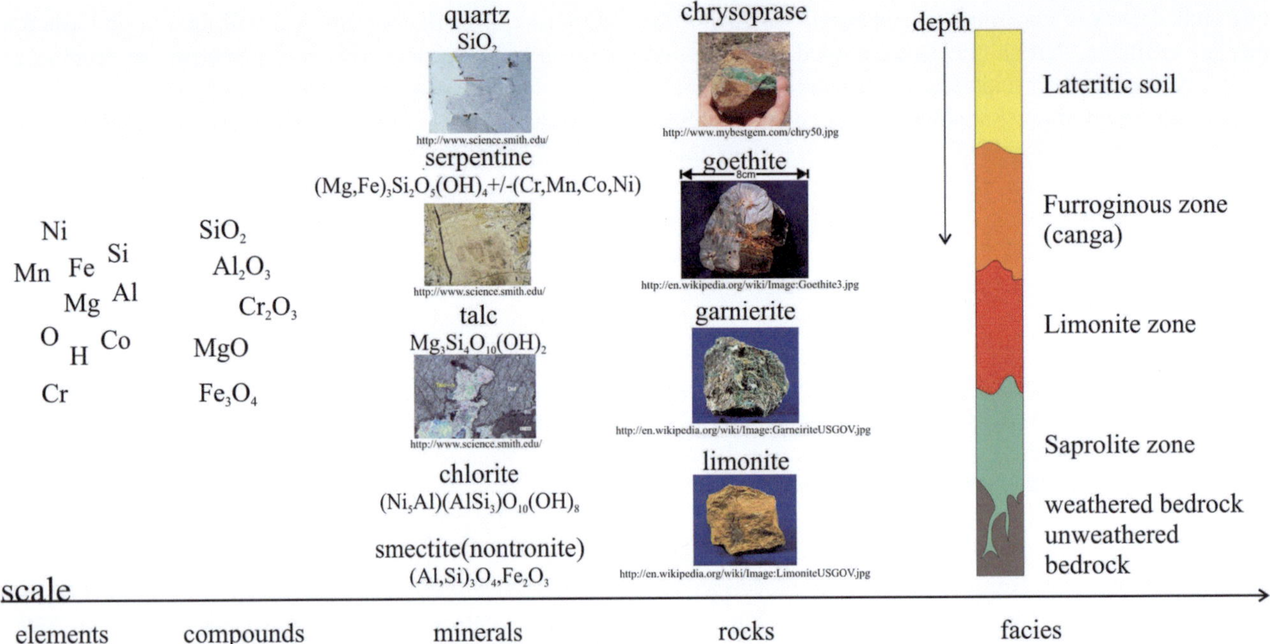

Fig. 5.11 Various scales of components of a nickel laterite deposit

and bornite. In addition, several other minerals may have smaller amounts of Cu.

Lithological data comes in two forms: continuous like geochemical data or categorical. The continuous component, which can be considered on a microscopic scale, is concerned with proportions of elements that define a particular type of lithology, for example the percentages of sand, silt and clay that are indicative of a sediment specimen. The macroscopic or categorical component deals with actual lithology or facies types. Natural resources are characterized by a set of facies types, which can be considered as compositional data on a larger scale.

5.5.3.2 Compositional Scale

Compositional data can be defined at various scales. Geochemical data for example exist at the atomic scale for specific elements, the microscopic scale for compounds of those elements, the mesoscopic scale where different combinations of compounds and elements define different minerals, and the macroscopic scale defining facies types and lithology zones.

An example showing the various scales is a nickel laterite deposit. This type of deposit is typically characterized by a very high grade of limonite clay containing 1–2 % nickel, high iron concentrations, and trace cobalt. Other constituents are present including silica, magnesium and aluminum. Starting with element contents, Fig. 5.11 shows the hierarchy of clustering into compositions with increasing scale.

The nickel laterite data set consists of just over 30,000 samples with 9 variables: 2 indicator type variables and 7 continuous. All continuous variables are measured in percent and are: nickel (Ni), iron (Fe), silica (SiO_2), magnesium oxide (MgO), cobalt (Co), aluminum oxide (Al_2O_3) and chromium oxide (Cr_2O_3). Indicator variables are ore-type and rock-type. Ore has been classified into seven different types and rock into two different types.

A unit mass of material from this deposit contains some proportion of the compounds and elements mentioned above as well as other media, which will be denoted as Z. This forms the full composition of the deposit with the other media being a filler variable to achieve 100 % of a unit mass:

$$\mathbf{X} = \{Ni, Fe, SiO_2, MgO, Co, Al_2O_3, Cr_2O_3, Z\}$$
$$100\,\% = Ni + Fe + SiO_2 + MgO + Co + Al_2O_3 + Cr_2O_3 + Z$$

5.5.3.3 Ternary Diagrams

Due to the high dimensionality of the nickel laterite data set, only subcompositions can be visualized. Subcompositions of size 3 plotted in a ternary diagram must undergo a transformation to add up to 100 %. Figure 5.12 shows the ternary diagrams (called also simplexes in the literature) for Ni-Fe-SiO_2 and Ni-MgO-Co colored by ore-type. Notice that the data are poorly distributed in the simplex. This is a common issue when one element of the composition is typically high valued or low valued with little variability. A method called data centering is used to redistribute data more appropriately.

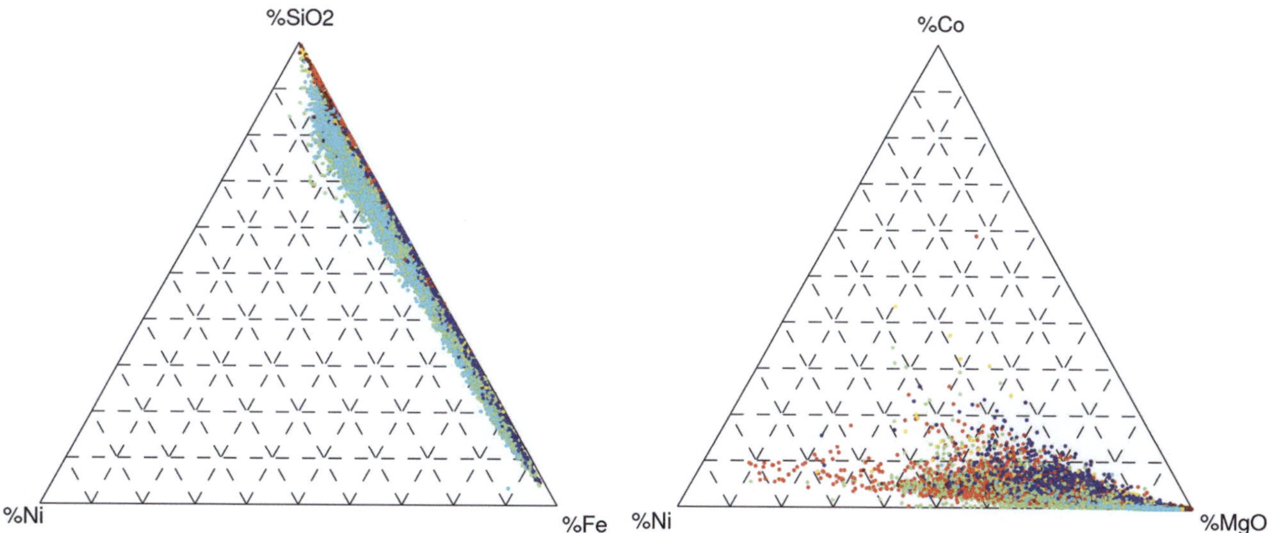

Fig. 5.12 Ni-Fe-SiO_2 and Ni-MgO-Co ternary diagrams

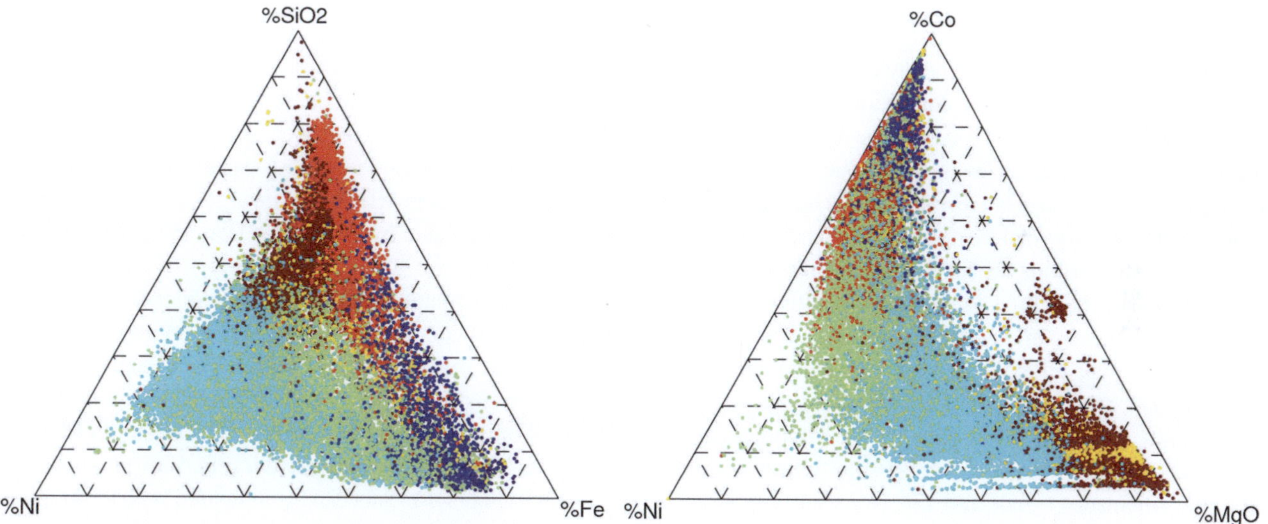

Fig. 5.13 Centered Ni-Fe-SiO_2 and Ni-MgO-Co simplexes

5.5.3.4 Operations

Two common operations for manipulating data and for displaying various geometric images within the ternary diagram are powering and perturbation (Aitchison 1986; Pawlowsky-Glahn and Olea 2004; Pawlowsky-Glahn and Egozcue 2006). Pertinent geometric images may be zones discriminating compositional observations into classes and ellipses denoting confidence intervals of distributions.

A particular form of perturbation called data centering is used when data are compressed to a small region of the ternary diagram (simplex) offering poor visual analysis of the data. The perturbing vector is set as the inverse geometric mean of the compositional data and such that it obeys the unit sum property.

$$\mathbf{u} = \zeta\left\{\left(\prod \mathbf{x}_1\right)^{-\frac{1}{N}}, \ldots, \left(\prod \mathbf{x}_D\right)^{-\frac{1}{N}}\right\}$$

Centered data for subcompositions Ni-Fe-SiO_2 and Ni-MgO-Co (Fig. 5.13) give a much better picture of the data distribution. The segregation of ore type classes (color coded points) is more visible in both plots. An important problem to note here is the occurrence of zeros and missing values in the sample data. Because the geometric mean is used, these values must be handled with care. For the purposes of centering in the nickel laterite example, missing values were ignored and zeros were set to 1 such that the products did not resolve to zero.

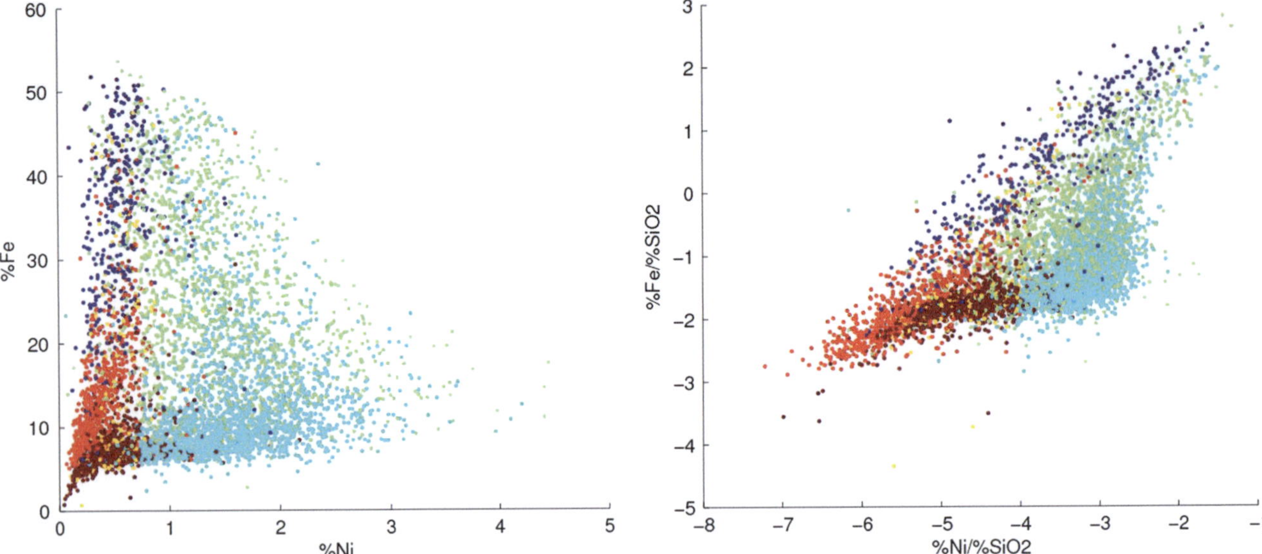

Fig. 5.14 Crossplots of Ni-Fe and Ni/SiO_2-Fe/SiO_2

5.5.3.5 Transformations

The majority of analyses concerned with compositions apply logratio transformations to the data prior to any exploratory analysis. Aitchison et al. (2002) argue that compositions provide data on a relative level, rather than on an absolute level and logarithms of relative values expressed as ratios are easier to statistically analyze than ratios themselves. Also, logratio transformations do not affect the information content of the data. The reason one must work with ratios rather than raw variables as in the ternry diagrams shown above is those variables are not scale invariant: the subcompositions shown are not coherent with the full composition.

The more common transforms are the additive logratio (alr), the centered logratio (clr), the multiplicative logratio (mlr, Aitchison 1986), and the isometric logratio (ilr, Egozcue et al. 2003). Choice of transformation depends on the problem being considered and the targeted properties of results. The outcome of these transformations is a set of vectors that exist in real space not constrained to the simplex. Each component of the vectors refers to a coordinate. These transforms are also one-to-one in that they map distinct values from one sample space to distinct values in the transformed sample space.

5.5.3.6 Additive Logratio Transform

Forward and inverse alr transformations are expressed respectively by the following equations:

$$y_i = \log\left(\frac{x_i}{x_D}\right), \qquad i = 1, \ldots, d$$

$$x_i = \frac{\exp(y_i)}{\sum_{i=1}^{d} \exp(y_i) + 1}, \qquad i = 1, \ldots, d$$

The denominator x_D can be any one of the components of $\mathbf{x}$, but it must remain consistent when applying this transform to a complete set of compositions, and $\mathbf{x}$ must be greater than 0. The advantage is a new space free of constraints where classical multivariate analysis methods can be applied; however, the space is not isometric. Coordinate axes are not orthogonal, but are separated by 60 degrees (Pawlowsky-Glahn and Egozcue 2006).

This transformation is applied to the nickel laterite data by dividing all components by the filler component Z of the compositions, resulting in 7 variables. For visualization purposes, the transformation was applied to the Ni-Fe-SiO_2 subcomposition with silica as the divisor. A crossplot of the resulting variables, $\log(Ni/SiO_2)$ and $\log(Fe/SiO_2)$ is shown on the right in Fig. 5.14. This can be compared with scatterplots of the original variables Ni and Fe on the left. Prior to transformation the data were constrained to positive real space since they are expressed as percentages. Post-transformation shows the data are unconstrained and that division by SiO_2 imposes a relationship.

5.5.3.7 Centered Logratio Transform

Unlike the alr transformation, clr results in orthogonal axes which simplifies further multivariate computations. The nature of this transformation results in vectors with a zero sum meaning the subspace is actually a plane. This zero sum property results in singular covariance matrices (Pawlowsky-Glahn and Egozcue 2006), but there are methods to overcome this limitation (Quintana and West 1988). The forward clr transform with $g(\mathbf{x})$ the geometric mean of $\mathbf{x}$ is:

$$y_i = \log\left(\frac{x_i}{g(\mathbf{x})}\right), i = 1, \ldots, D$$

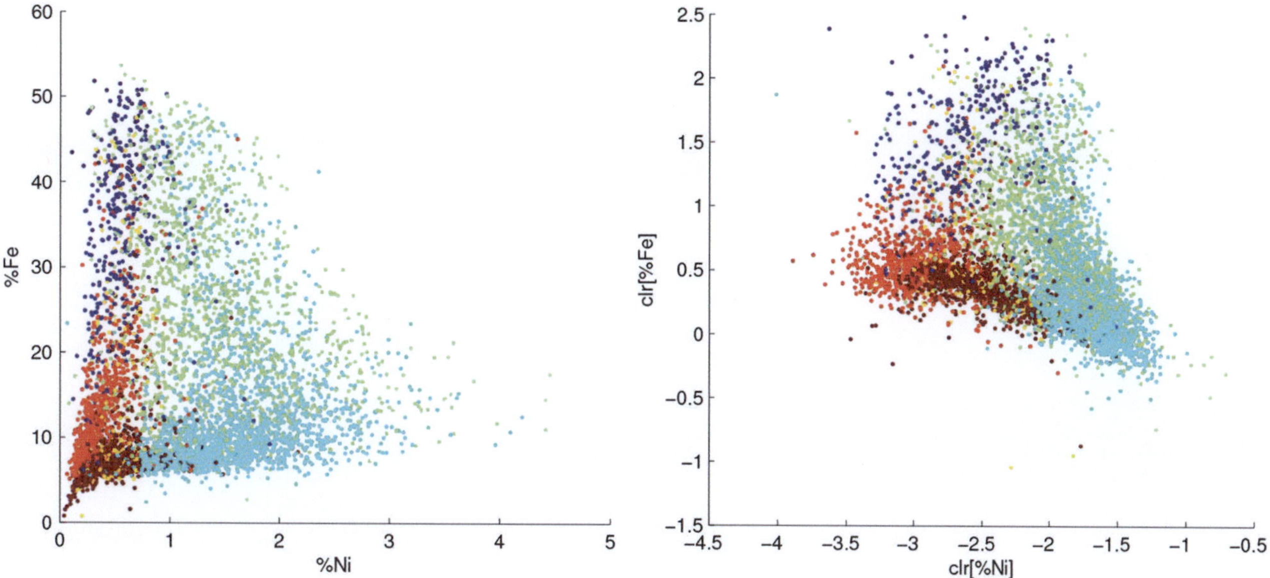

Fig. 5.15 Crossplots of Ni-Fe and clr(Ni)-clr(Fe)

Figure 5.15 shows the original and transformed scatterplots. Compared to the alr transform, the data is now centered by the geometric mean. This is better shown with a simplification of the transformation on the Ni-Fe-SiO_2 subcomposition.

$$y_i = \log\left(\frac{x_i}{g(\mathbf{x})}\right) = \log(x_i) - \log(g(\mathbf{x}))$$

5.5.3.8 Multiplicative Logratio Transform

This transformation is similar to the alr method except that the divisor cannot be any one of the elements in the composition. Rather it is the filler required for a composition to meet the unit sum constraint similarly to how Z was used in the nickel laterite data. This transformation is appropriate for exploring relationships between single divisions of a composition (Aitchison 1986). For example, analyzing how the elements of Ni-Fe-SiO_2 relate to MgO-Co-Al_2O_3-Cr_2O_3. The forward and inverse transformations are shown below, with Fig. 5.16 showing graphically this transformation to the Ni-Fe-SiO_2 subcomposition.

$$y_i = \log\left(\frac{x_i}{1 - x_1 - \cdots - x_i}\right), i = 1,...d$$

$$x_i = \frac{\exp(y_i)}{(1+\exp(y_1))\cdot\cdots\cdot(1+\exp(y_i))}, i = 1,...,d$$

$$x_D = \frac{1}{(1+\exp(y_1))\cdot\cdots\cdot(1+\exp(y_d))}$$

5.5.3.9 Isometric Logratio Transform

The isometric transform relies on orthonormal bases to transform from the simplex to real coordinates and has the property of conserving the metric properties in both spaces. Geometric notions such as angles and distances in the simplex are associated with angles and distances in the real space of transformed data. Applying this transformation requires that an orthonormal basis be defined in the Aitchison metric (Egozcue et al. 2003; Tolosana-Delgado et al. 2005). The Aitchison metric is the simplex sample space and geometry. The orthonormal basis is defined by a set of vectors $\mathbf{e}_1,\ldots,\mathbf{e}_d$ and the interested reader is directed to Egozcue et al. (2003) for a complete derivation. The ilr transform is defined given these vectors, where $\langle \cdot,\cdot \rangle_a$ defines the Aitchison inner product.

$$y_i = \langle \mathbf{x}, \mathbf{e}_i \rangle_a, i = 1,...,d$$

$$\langle \mathbf{x}_1, \mathbf{x}_2 \rangle_a = \langle \text{clr}(\mathbf{x}_1), \text{clr}(\mathbf{x}_2) \rangle = \sum_{i=1}^{D} \text{clr}(\mathbf{x}_1)_i \cdot \text{clr}(\mathbf{x}_2)_i$$

The Aitchison inner product is the Euclidean inner product (dot product) applied to alr or clr transformed data. Transformation using ilr calculates inner products of clr transformed data.

Despite the increased complexity of this transformation, it has the advantage of generating vectors in unconstrained orthogonal space (Fig. 5.17). Any multivariate analysis technique can be applied to the result.

5.5.4 Service Variables

The use of accumulation (or service) variables in geostatistics is fairly common in industry when developing block

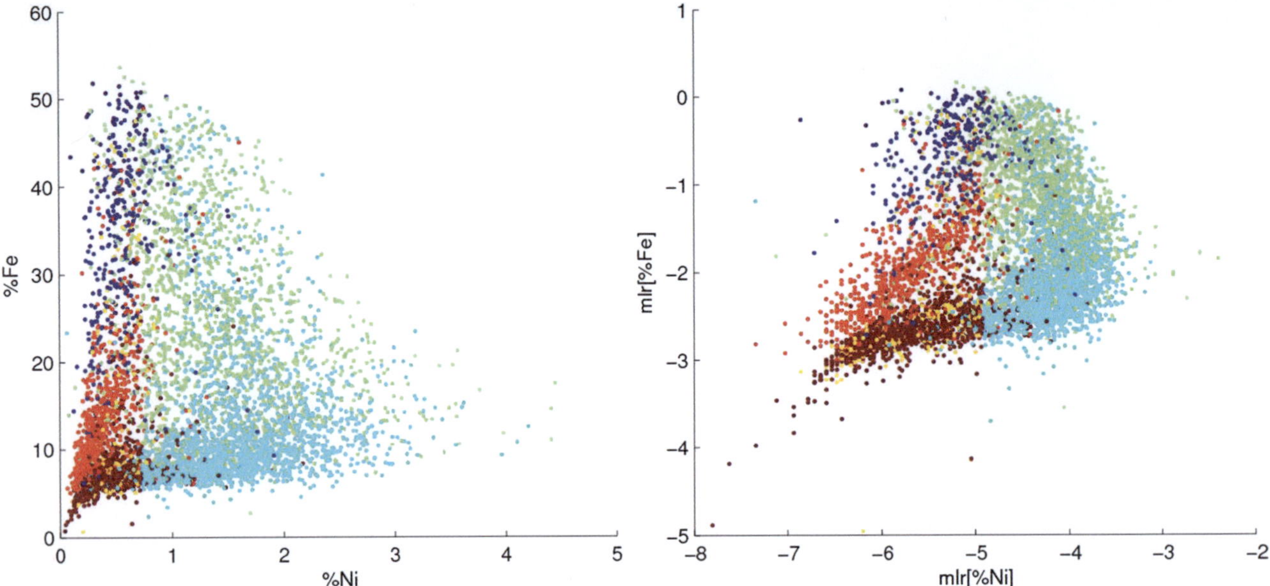

Fig. 5.16 Crossplots of Ni-Fe and mlr(Ni)-mlr(Fe)

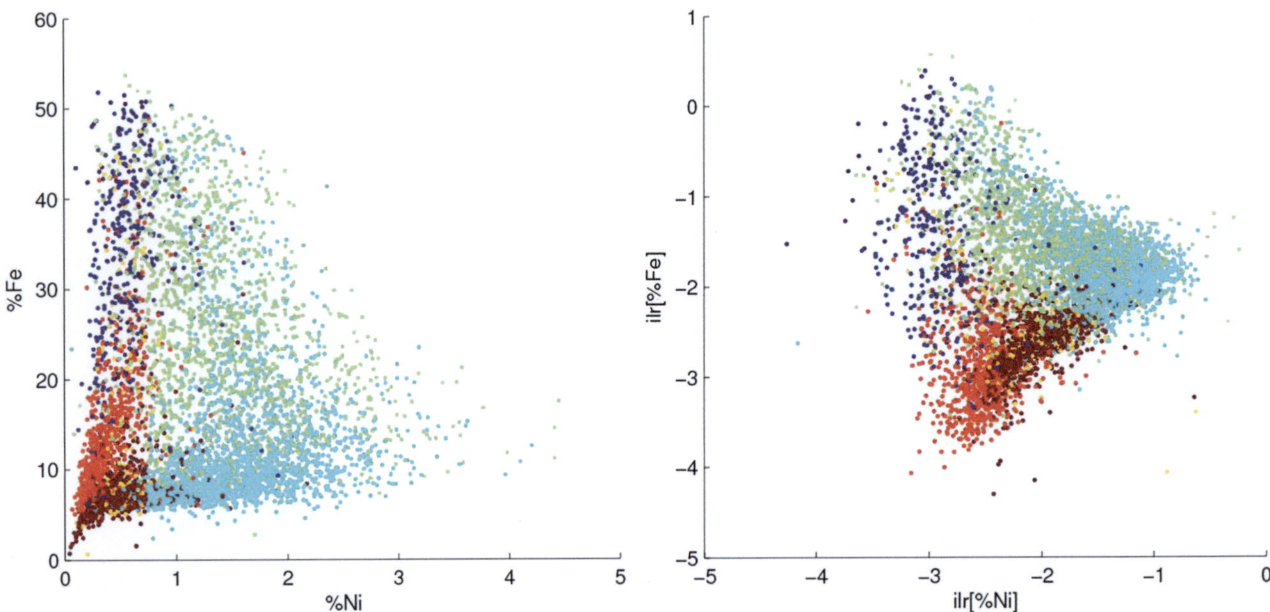

Fig. 5.17 Crossplots of Ni-Fe and ilr(Ni)-ilr(Fe)

grade models and geologic resource estimates (David 1977; Glacken and Snowden 2001). This is particularly true in the case of sedimentary and stratabound-type deposits.

Estimating accumulation (grade-thickness (GT) and thickness (T) variables) came about early in the history of resource estimation. It has been applied mostly to narrow orebodies, and initially as an adaptation of the polygonal approach. Grade multiplied by thickness is proportional to metal content, which is estimated independently from thickness. The final grade is obtained at each point or block by dividing the accumulation by the thickness.

The accumulation approach works well when there is little or no correlation between grade and thickness. Also, thickness should be true thickness, that is, the samples need to be considered as true (normal to dip) mineralized thicknesses. It is worth noting that the accumulation and thickness variables can be estimated using any technique, including geostatistical methods.

The motivation for using accumulations is generally that, for thin tabular or vein-type deposits, the mineralized intercepts are of irregular length; there are very few intercepts (or just one) per drill hole; and that grade variability is high, with somtimes the thinner intercept having the highest grades. In these situations, multiplying the grade with the thickness will result in a variable with lower coefficient of variation (variability), and thus easier to estimate without having to risk over-spreading grade. Thickness, while sometimes variable, will generally be a smoothly varying attribute in space, and thus even easier to estimate.

Deposits in which grade-thickness variables have been used include tabular and vein-type gold, platinum reefs, ultramafic deposits (chromite, for example), some nickel laterites, and uranium roll-front type deposits.

5.6 Compositing and Outliers

5.6.1 Drill Hole Composites

The original grade values (assays) in the database are usually averaged to pre-specified lengths: a procedure referred to as compositing. This is not strictly a requirement for resource estimation; however, the homogenization of the data scale or support and correction for incompletely sampled intervals motivates compositing in almost all cases. Most resource estimation software assumes the data are at a constant support.

Compositing also incorporates a certain amount of dilution into the raw data prior to estimation or simulation. The mining operation is expected to work at a certain level of selectivity that is larger than the scale of the raw assays. In the case of open pits, selectivity in the vertical dimension is generally fixed by the bench height. In the case of underground mines, selectivity is a function of the mining method. The height of the lift or slice in a cut-and-fill or similar method determines the selectivity in the vertical direction. The composite length may be made equal to the bench or lift height to put data at the same vertical support as the mine selectivity.

The composite is typically calculated by a length-weighted average and may also be weighted by specific gravity and core recovery. Compositing can be done to obtain a representative value for ore body intersections, lithological or metallurgical composites, regular length down-the-hole composites, bench composites or section composites, high grade composites, or minimum length and grade composites.

Each of these types of composites are produced for different purposes and in different situations. Regular length or bench composites are most common in resource estimation. There are geostatistcal models that can provide for mixtures of support sizes in the original data, but estimation software almost always assumes the data are of constant support.

Drill holes of different diameters are commonly used in the same deposit. There will also be partial composites at the end of drill holes or at hard geologic contacts. In practice, slight differences in support size will have little effect on the final resource estimate.

Consider an operation that has both reverse circulation (RC) and diamond drill holes (DDH) in its resource database. For a 1 m interval, a typical RC drill hole sample will represent close to 50 kg, considering that a 5.25 inch diameter hole is drilled. The corresponding HQ-size DDH, for the same 1 m length, will represent close to 28 kg. Although the two sample weights are quite different, the difference and its impact is negligible. This is because the sample is (a) usually composited, and (b) multiple composites are used to estimate blocks. The blocks may be as small as $5 \times 5 \times 5$ m (for smaller, selective deposits), equivalent perhaps to 350 metric tons, and up to $25 \times 25 \times 15$ m or larger blocks for massive deposits, equivalent to almost 25,000 tons considering an in-situ bulk density of 2.65 t/m^3. This will depend on the specific resource modeling case.

Further justification for compositing is that small scale assays may be highly variable, which can be mitigated by compositing. Compositing to an appropriate length will show less variability making the corresponding geostatistical analysis, including variography, more robust.

In particular, compositing has a dramatic affect on the nugget effect, that is, the completely random portion of the variability. The decrease in the nugget variance will be inversely proportional to the degree of compositing.

The composited dataset is important to the overall quality of the resource model. Several decisions must be made in practice, including whether or not to use composites; the most appropriate length; the compositing method used; whether to truncate the composites at geologic boundaries; how to handle missing intervals within the composite (with no assay information); and the minimum acceptable composite length.

5.6.2 Composite Lengths and Methods

The composite length chosen is commonly a function of the anticipated mine selectivity. Shorter composite lengths can be used to increase the statistical population available for variography and estimation; however, the final variograms should be based on the composites that are going into estimation. Shorter composites permit more accurate representation of geological contacts.

Compositing can be done using two basic methods: down-the-hole, or by bench. Bench composites are common in open pit mines with nearly vertical drilling. The method involves defining the top and bottom elevations of each bench and then compositing all sample intervals that fall within those elevations. The mid-bench elevation is assigned as the centroid of the composite which is assumed to be vertical.

Although convenient, the bench compositing method commonly used in open pits has some shortcomings. If the drill hole is inclined, the actual composited length will be different than the bench height. For example, for a drill hole with a 45° inclination, a 10 m bench composite will incorporate over 14 m of sample. Bench compositing should be restricted to cases where all drill holes have no more than a 70° inclination, although this choice is subjective. This method should not be used when there is a significant mixture of vertical, subvertical, and inclined drill holes.

Down-the-hole compositing usually begins at the top of the hole, although most mining software packages will allow for other options such as truncating at important geological contacts. The length of material composited is always the same, with the exceptions noted below. The inclination of the drill hole is no longer a factor. The centroid of the composite corresponds to the exact location in space of the combined samples.

An important decision is whether to truncate the composite at geologic boundaries or not. The decision amounts to incorporate some contact dilution (mixtures of grades from either side of the contact) or to avoid such mixing. If the composites are truncated at the boundaries, the estimation domains will be more sharply defined and contact dilution better controlled. The cost is a larger number of composites of a shorter length. The most appropriate decision will depend on the characteristics of the deposit and the amount of information available.

Missing sample intervals within a composite may be problematic if the voids are significant because there will be a significant difference between the nominal length of the composite versus the actual length of sample intervals composited. Simple statistics can be length-weighted, using the actual length of material composited, which is often recorded by most mining software packages. A maximum length of voids within any composite is tolerated before discarding the composite altogether. The weighting of mass fractions should be by mass, that is, specific gravity should be considered as well as length.

The decision related to the acceptable minimum length of a composite depends on the representativity and support of the actual composite. This issue is more relevant if composites are truncated at geologic boundaries, because there can be a large proportion of the total number of composites that are shorter than the nominal length. If no boundary truncation is applied, then the only composites shorter than the nominal length will be at the end of the drill holes.

A common choice in industry is to use 50 % of the nominal length as the cutoff for acceptable composite lengths. Discarding all composites less than 50 % of the nominal length is arbitrary. Since the main concern is representativity, a correlation study between composite length and grades can be used to support the choice of minimum acceptable composite length. A much shorter minimum composite length may be acceptable where there is no detectable correlation between composite length and grade. The relationship between grade and composite length will likely be small in base metal and massive deposits, with good core recovery. The opposite is generally true for massive sulfide deposits, some precious metals deposits, and in cases where core recoveries are poor.

5.6.3 Outliers

The term outliers is used to describe extreme high values since many grade distributions are positively skewed. Some distributions have low grade outliers, but this situation is less common. These grades deviate from the general tendency of most other grades and can be spatially and statistically isolated. In the discussion that follows, outliers are valid assayed samples, not a consequence of spurious or erroneous data collection. It is assumed that all potential database or analytical errors have been checked for and all possible errors discarded or corrected in the database. Outliers are defined in terms of geological and statistical populations.

Extreme grades are consequential in precious metal deposits, but not as much in base metal deposits. A statistical analysis is always warranted to quantify how much impact the outliers have on the final resource estimate.

The importance of outliers is often described in terms of their contribution to the overall metal content of the deposit. This is because not handling them appropriately can lead to overestimation of the recoverable resource. Two aspects must be resolved: (a) what assays should be considered outlier values, and (b) how to deal with them at the time of estimating the resources. In all cases, the analysis should be done on the original, assayed samples. If performed on the composited data, the outlier values may already be smoothed depending on the type and length of the composites.

The presence of extreme grades is particularly problematic if they have little spatial connectivity, that is, they are located within a small spatially restricted volume. The more skewed the grade distribution, the larger the potential impact of outliers on the resource estimation process.

The determination of what values are considered outliers is subjective. Outlier values are commonly examined on a log-normal cumulative frequency plot. Breaks at the high end of the distribution may represent outlier populations. For example, Fig. 5.18 shows the log-normal probability plot of Au grade in a copper-gold porphyry deposit. Note how, for grades higher than 5.0 g/t, the distribution appears to break up and exhibits a sudden slope change, represented by less than 0.1 % of the total samples. Outliers are also sometimes defined as those values that are outside a specific interval, such as plus or minus 2 or 3 standard deviations ($\pm 2\sigma$ or $\pm 3\sigma$) with respect to the mean or median of the distribution. There will always be outliers according to this definition; professional judgement is required.

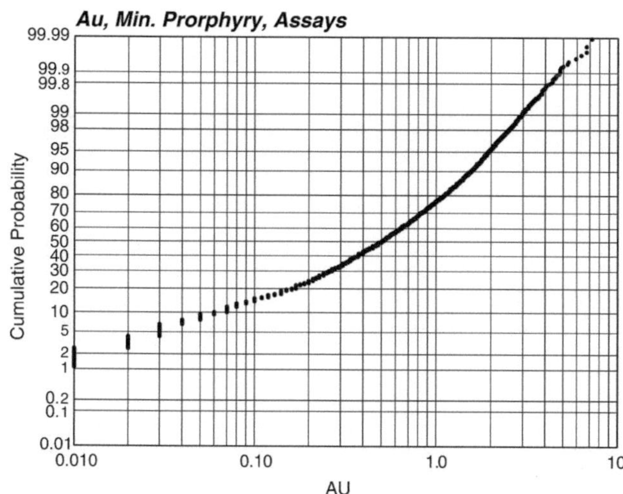

Fig. 5.18 Probability plot, Au grade in a porphyry copper deposit

It is preferable to define a range of possible cutoff values for studying outliers. An example, taken from the Chang Shan Hao Au deposit in Inner Mongolia, China, is shown in Table 5.1. The impact in terms of quantity of metal (QM) of samples above a series of grade cutoffs is presented. Note that, for example at the 4.0 g/t Au cutoff, just over 40 m of samples, representing about 0.49 % of the total meterage in the database, is responsible for over 5.7 % of the total quantity of metal. Although this is not a very extreme case if compared to other Au deposits, it indicates that outliers must be considered to avoid overestimation when estimating the resources for this deposit.

Statistical methods can also be used to determine the impact and modeling of outlier grades. One such method, proposed originally by Parker (1991), is based on assuming specific distributions for the upper tail of the grade distribution including a log-normal distribution.

Another method, also proposed originally by Parker (personal communication), is to assume that, above a certain grade threshold, the grade values are uncorrelated and independent of each other. In this case, a Monte Carlo method is used, whereby the high grade distribution is simulated. The amount of metal that must be removed from the database is estimated based on the simulated high grade distribution and for a specified condition. The condition is generally that the predicted metal production on a yearly basis (for example) can be assured with a given confidence level. This concept of analyzing the problem from the perspective of mining risk is appealing, but it has the caveat of the data independence related to the Monte Carlo simulation, which is not always applicable. Also, the distribution of outlier grades must be fairly homogeneous for the metal to be accurately predicted by mining period.

To limit the influence of the outlier data, the most common procedure is to define a cutting or capping grade whereby all samples above the specified grade cutoff are either ignored (not used) or reset to the top value defined, respectively. Ignoring the outlier values altogether is not recommended, since it tends to be overly conservative. More commonly, practitioners will reset all assays above the specified cutoff to that value.

For some estimation and simulation methods, the treatment of outlier values is accomplished within the method itself. For example, if using multiple indicator kriging, the impact of high values may be dealt through the definition of a more conservative value for the upper class mean, see Chap. 9.

For most cases, it is preferable to restrict the spatial influence of the outlier values at the time of estimation or simulation. This is implemented in some software packages. The assumption is that extreme values are valid and should be used to estimate resources, but their spatial influence should be limited. The high grade may be constrained to small-size veinlets, or represent a nugget with little or no spatial extension.

Capping the grades removes metal from the distribution and limits the influence of the outliers. There may still be a region of high estimates around the outliers, yet there may be isolated high grade. The local estimates are checked on a case-by-case basis.

5.7 Density Determinations

In-situ density must be modeled at the time of resource estimation. The predicted tonnage of the ore deposit is directly dependent on the tonnage factor or density applied to the modeled volumes.

A geologic model is used to predict the mineralized volume, and this volume in turn is multiplied by its in-situ density to obtain an estimated tonnage for the deposit. Any error made in density determination and estimation is directly incorporated into tonnage estimates. A good discussion on this issue can be found in Parrish (1993).

Several factors affect bulk density determinations such as heterogeneity of the materials being sampled, the method of determination, the practice of determining dry or wet densities, rocks with voids in them ("vuggy" breccias, for example), the material consolidation, relationships between ore grade and densities, such as in massive sulfide deposits, and so on.

Immersion methods are commonly used to determine the density of rock samples. The sample is weighed in air and then in water. Density is then determined as:

$$Density = \left(\frac{W_{air}}{W_{air} - W_{water}} \right) \quad (5.2)$$

where W_{air} is the dry weight of the sample, and W_{water} is the weight of the submerged sample. In practice, since it is

Table 5.1 Example of the impact extreme assay values in terms of quantity of metal (QM) for a gold deposit

Au Cutoff (g/t)	Total DDH Meters	Grade above cutoff (g/t)	Metal above cutoff (m*g/t)	% QM above cutoff	% Total Meters
4.00	40.37	6.515	263	5.74%	0.49%
4.20	40.37	6.515	263	5.74%	0.49%
4.40	36.37	6.757	246	5.37%	0.44%
4.60	32.37	7.042	228	4.98%	0.39%
4.80	26.00	7.624	198	4.33%	0.32%
5.00	20.00	8.436	169	3.68%	0.24%
5.20	20.00	8.436	169	3.68%	0.24%
5.40	18.00	8.789	158	3.45%	0.22%
5.60	15.99	9.192	147	3.21%	0.19%
5.80	12.00	10.347	124	2.71%	0.15%

advisable to obtain dry density measurements, a commonly used procedure involves:

1. Carefully weigh the sample as it was received from the field.
2. Completely dry the sample on a conventional oven at (typically) 105 °C, and weigh again. The difference between the two weights provides an estimate of moisture content in the rock.
3. Submerge the sample in water (density 1.0 g/cm^3) and record its weight.
4. Apply Eq. 5.2 above to determine the rock density.

The impact of water absorption by the sample is a function of how porous the rock is, but it will always increase the density value, since water is replacing air within the specimen. To avoid measurement errors derived from the water absorbed by the rock during immersion, a more robust alternative for density measurements is to coat the sample with a wax of known density to seal the internal voids in the rock. Figure 5.19 shows a photograph of the waxing step, bulk density determination using the wax-coating method.

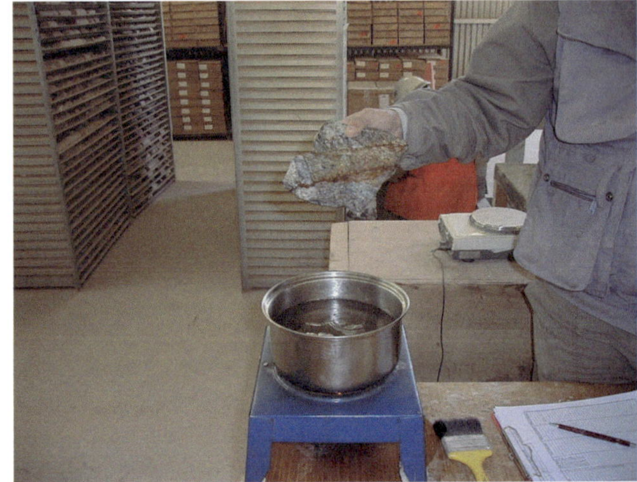

Fig. 5.19 Wax-coating density determination method, waxing step. Courtesy of BHP Billiton, Chile

The density determination now is a function of the dry and wet coated weights, dry uncoated weight, and the known wax density:

$$Density = \left(\frac{W_{air}}{(W_{Cair} - W_{Cwater}) - \left[\frac{W_{Cair} - W_{air}}{\delta_{wax}} \right]} \right)$$

where W_{air} is the dry weight in air, prior to wax coating; W_{Cair} is the dry weight of the coated sample; W_{Cwater} is the weight of the coated sample in water, and δ_{wax} is the known density of the wax.

The spatial distribution of density measurements is also an important consideration. The density samples obtained must be representative in a spatial and geologic sense.

Sample density may be correlated to sample grade. This is because most metals have a higher specific gravity than the host rock. If there is a sufficiently high concentration of metals (grade), then the density of the rock will be influenced by its metal content. Typical examples of deposits where there is usually a clear correlation between grade and density are uranium deposits, massive sulfides, iron deposits, and some high grade gold deposits.

Certain geologic factors will also have an impact on the expected density. For example, in an open pit mine, barren material that must be removed to access the deposit may be lighter than consolidated rock (bedrock), since it may consist of gravels or loose highly weathered material. Certain types of host rocks are porous such as breccias with significant voids, which despite being favorable hosts for mineralization and having generally good grade, will tend to be lighter. The type and intensity of alteration will also affect the in situ density values. Highly altered rock is friable, powdery in extreme cases, and generally lighter than unaltered rock.

It is usually convenient to define density domains based on a combination of lithology, alteration, mineralogy, and average grades of the rock. There should be an adequate amount of density samples available to model density, which may be from 100 to 1000 per domain, depending on the size and type of the deposit.

5.8 Geometallurgical Data

Geometallurgy is a relatively new field in Mining that tries to understand and model variables that are related to metallurgical performance. These may include variables that directly or indirectly measures throughput (hardness, grindability), recovery (liberation, mineral shape/texture, etc) and concentrate quality. The importance of geometallurgy has been understood for quite some time in some deposits, such as limestone, iron, lateritic nickel, bauxite, manganese and coal. In recent years it is increasingly becoming a key component in resource models for all base and precious metal deposits.

Geometallurgical variables can pose challenges. There are at least two aspects that need to be considered: (a) the measured variables are usually indirect measures, or proxies, of the metallurgical performance of interest; in some cases they do not average linearly; and (b) many aspects of metallurgical performance are non-linearly dependent on the measured variables. The issue of non-linearity is relevant because the predictions are required at a very different volume (support) than the original measurement is taken, as will be discussed in Chap. 7. When a variable is non-linear, simply taking the average or the tonnage-weighted average of the variable is not correct. Another aspect that may be significant is that in this type of predictions that extreme values are more relevant than large-volume averages.

The more typical grade estimates can play a part in the material characterization that is sought when predicting metallurgical performance. For specific performance measurements, additional variables may be considered; for example, for hardness and grindability, Drop Weight Index (DWi); Bond Work Index (BWi); in-situ density; and P80, which characterizes the size of the throughput material, may be combined in a linear or non-linear equation to predict tonnage per hours processed; alternatively, SPI (SAG Power Index) may be relevant, depending on the modeling approach taken. Sometimes geotechnical variables, such as RQD (Rock Quality Designation), UCS (uniaxial compressive test), PLT (point load test), etc may be used as proxies to estimate throughput.

Other variables may include grades for payable elements, as well as deleterious elements that may result in penalties for concentrates sold, as well as those impacting on metallurgical recovery. May also include mineral species present, mineral liberation, texture, grain sizes and size distribution curves (p80), etc. Commonly, direct tests of metallurgical recovery are available, which in turn may lead to modeling relationships between grades, mineral species, and metallurgical recoveries.

Challenges that may be posed by these variables at the time of estimation will be discussed in Chaps. 8–10.

5.9 Summary of Minimum, Good and Best Practices

At a minimum, the data must have a demonstrable level of quality such that it adequately supports the resource modeling objectives. Quality requirements will thus increase from low to high as the level of detail of the resource model increases, from initial deposit modeling, pre-feasibility, and feasibility studies, mine planning and mine operations support. Database quality, measured in terms of error rate, should be better than 5 % for geologic codes and assay values. Specific issues to consider include:

a. Written procedures for data collection and handling should be available. They should include procedures and protocols for field work, geologic mapping and logging, quality assurance and quality control, database construction, sample chain of custody, and documentation trail. The procedures should include a QA/QC program for the analytical work, including acceptance/rejection criteria for batches of samples.
b. A detailed review of field practices and sample collection procedures should be performed on a regular basis, to ensure that the correct procedures and protocols are being followed.
c. Similarly, review of laboratory work should be an on-going process, including occasional visits to the laboratories involved.
d. A QA/QC program should be implemented, and should include at least pulp duplicates, standards, and blanks, as discussed above. Samples should be controlled on a batch-by-batch basis, and rejection criteria should be enforced.
e. Information about core recovery and sample weights for RC drilling should be compiled and analyzed as drilling progresses.
f. Sufficient density information should be available to characterize the main geologic units. No less than 30 samples per unit is a suggested minimum.
g. The compositing method should be adapted to the characteristics of the (future) operation, and described and justified in detail. Similarly, a description and analysis of outliers is necessary.

In addition to the above, good practice requires that:

a. Every drill hole campaign should have a quantified degree of uncertainty. This includes discussions and reports of the quality and potential problems of sample collection procedures, data handling, and the overall quality of the computerized database.

b. The QA/QC program should be extended to include not only sample preparation and assaying, but also of the sample collection process, drill hole collar and sample locations, including drill hole deviations, drill hole spatial coverage, geologic mapping and logging, and a clear definition of the variables to be used and their purpose.
c. Issues such as the combined use of soft and hard data should be properly dealt with and justified.
d. Density determinations should be sufficient to adequately provide spatial coverage, and also should be subject to quality control procedures.
e. Detailed, written QA/QC reports should exist for every step of the sampling and data handling process. These procedures should include corrective measures as required.
f. Overall database errors rates, for all elements stored, should be below 2 %.

Best practice additionally includes:

a. The use of all possible (and relevant) geologic, grade, and other data to obtain a resource model. The process of defining the variables and their characteristics should be well documented. If metallurgical information is available, a complete description of the relationships between geology, grade, and metallurgical performance should be available, such as grade-recovery curves, hardness-plant throughput curves, etc.
b. There should be detailed reports on the QA/QC procedures implemented available, including its results and relevant discussions, both for current and historic campaigns. Also, records of internal checks and audits performed for each step of the process should be compiled and archived.
c. A summary of qualitative or quantitative data uncertainty should be available for each data component. These can be derived from the database error rates (for qualitative or categorical variables), from sampling variance studies for sample preparation and analysis protocols, or from other statistical analysis.
d. Database error rates should be below 1 % for all variables being used.

5.10 Exercises

The objective of this exercise is to review some sampling theory and gain some experience with sampling nomographs. Some specific (geo)statistical software may be required. The functionality may be available in different public domain or commercial software. Please acquire the required software before beginning the exercise. The data files are available for download from the author's website—a search engine will reveal the location.

The primary metal in the hypothetical deposit of this exercise is copper. The copper mineral is chalcocite (Cu_2S). Chalcocite has a density, λ_m, of 5.5. The average copper grade is 2.0 % (note that this is not the average chalcocite grade). The host rock is a granite with a density, λ_g, of 2.3. The Liberation size, *d*, of the chalcocite is 50 μm. The shape factor, *f*, for chalcocite is 0.47. The granulometric factor, *g*, is 0.25.

Samples are taken using diamond core drilling. The core has a diameter of 52.0 mm. Half of the core will be sent for assay and the other half will be retained. The nominal sample length is 2.5 m.

The fundamental sample error is defined as:

$$\sigma_{FE}^2 = \left(\frac{1}{M_{S2}} - \frac{1}{M_{S2}} \right) IH_L$$

$$IH_L = clfgd^3$$

where σ_{FE}^2 is the sampling error introduced when splitting the sample from M_{SI} to M_{S2}, *d* is the particle size when the sample is being split, *c* is the mineralogical factor, and *l* is the liberation factor. The mineralogical and liberation factors are calculated in part one of this exercise.

5.10.1 Part One: Prerequisites for the Sampling Nomograph

Question 1: Calculate the amount of material from 1/2 of the drill core for a 2.5 m sample length. This is the starting mass for the sampling protocol.

Question 2: Calculate the chalcocite content, a_L, in the sample. The calculation needs to be done using fractions, not a percentage or ppm. Use the average copper grade for this calculation. The molecular weight for chalcocite is 159.17, for copper it is 63.55, and for sulfur it is 32.70.

Question 3: Calculate the mineralogical factor, *c*, for chalcocite. The result from question 2 is required for this step.

$$c = \lambda_m \frac{(1-a_L)^2}{a_L} + \lambda_g (1-a_L)$$

Question 4: The next step is to calculate the location of the size lines for the nomograph plot. This requires some iteration. The first step is to choose some nominal particle sizes. The second step is to assume a sample mass for each particle size. This mass is not related to the actual sample. The mass is used for plotting lines on the nomograph. Calculate the error for the size and mass pairs using the formulae below. If any of the points do not lie within the nomograph window, modify the size of the sample so that the point is within the window. Use a value of 1 for b in the calculation of l. This is a conservative value for b. Note that the liberation factor cannot go above 1.0.

$$\sigma_{FE}^2 = \frac{clfgd^3}{M_S} \qquad l = \left(\frac{d_l}{d}\right)^b$$

Alternatively, you may use the following table if needed.

d (cm)	l	IH_L	size	σ^2(FE)
2.5000				
0.9500				
0.4750				
0.2360				
0.1700				
0.1000				
0.0710				
0.0425				
0.0250				
0.0150				
0.0106				

5.10.2 Part Two: Nomograph Construction and Fundamental Error

Question 1: Plot the size lines calculated above onto a blank nomograph. The size lines are at a 45° angle that increases to the left.

Question 2: Recall the starting sample mass from part 1. Propose a sampling protocol, including crushing and splitting, such that the total sampling error, σ_{FE}, is less than 7.5 %. A useful constraint is to not introducing more than 5 % error in a single step. An example is:

Question 3: Generate a sampling nomograph and check your results.

References

Aitchison J (1986) The statistical analysis of compositional data: Monographs on statistics and applied probability. Chapman and Hall, London, p 416

Aitchison J, Barcelo-Vidal C, Pawlowsky-Glahn V (2002) Some comments on compositional data analysis in archaeometry, in particular the fallacies in Tangri and Wright's dismissal of logratio analysis. Archaeometry 44(2):295–304

Chayes F (1962) Numerical correlation and petrographic variation. J Geol 70(4):440–452

David M (1977) Geostatistical ore reserve estimation. Elsevier, Amsterdam

Egozcue J, Pawlowsky-Glahn V, Mateu-Figueras G, Barcelo-Vidal C (2003) Isometric logratio transformations for compositional data analysis. Math Geol 35(3):279–300

Erickson AJ, Padgett JT (2011) Chapter 4.1: geological data collection. In: Darling P (ed) SME mining engineering handbook. Society for Mining Metallurgy & Exploration, pp 145–159

François-Bongarçon DM (1998a) Extensions to the demonstrations of Gy's formula. In: Valleé M, Sinclair AJ (eds) Quality assurance, continuous quality improvement and standards in mineral resource estimation. Exp Min Geol 7(1–2):149–154

François-Bongarçon DM (1998b) Error variance information from paired data: applications to sampling theory. In: Valleé M, Sinclair AJ (eds) Quality assurance, continuous quality improvement and standards in mineral resource estimation. Exp Min Geol 7(1–2):161–168

François-Bongarçon D, Gy P (2001) The most common error in applying 'Gy's Formula' in the theory of mineral sampling, and the history of the liberation factor. In: Edwards AC (ed) Mineral resource and ore reserve estimation—the AusIMM guide to good practice. The Australasian Institute of Mining and Metallurgy, Melbourne, pp 67–72

Glacken IM, Snowden DV (2001) Mineral resource estimation. In: Edwards AC (ed) Mineral resource and ore reserve estimation—the AusIMM guide to good practice. The Australasian Institute of Mining and Metallurgy, Melbourne, pp 189–198

Gy P (1982) Sampling of particulate materials, theory and practice, 2nd ed. Elsevier, Amsterdam

Hartmann H (ed) (1992) SME mining engineering handbook, vol 2, 2nd edn. Society for Mining Metallurgy, and Exploration, Littleton, CO

Journel AG, Huijbregts CJ (1978) Mining geostatistics. Academic Press, New York

Position in Protocol	Point on Nomograph	Sample Mass M_S (g)	Fragment Size d (cm)	Fundamental Error σ^2	Fundamental Sample Error σ^2	Sample Error % σ
Primary crushing of core	A	20,000	0.475			
Split for first subsample	A-B	2,000	0.475			
Secondary crushing	B-C	2,000	0.071			
Second split	C-D	50	0.071			
Grinding of sample	D-E	50	0.015			
Split for lab assay	E-F	1	0.015			
Sample FE						
Total sample error (2 times the sample error)						

Long S (1999) Practical quality control procedures in mineral inventory estimation. Explor Min Geol J 7(2) Canadian Institute of Mining and Metallurgy (CIMM)

Magri EJ (1987) Economic optimization of the number of boreholes and deflections in deep gold exploration. J South Afr Inst Min Metal 87(10):307–321

Manchuk JG (2008) Guide to geostatistics with compositional data, Guidebook Series No. 7, Centre for Computational Geostatistics, University of Alberta, p 45

Neufeld CT (2005) Guide to sampling, Guidebokk Series No. 2, Center for Computational Geostatistics. University of Alberta, p 35

Parker HM (1991) Statistical treatment of outlier data in epithermal gold deposit reserve estimation. Math Geol 23:125–199

Parrish IS (1993) Tonnage factor- a matter of some gravity. Min Eng 45(10):1268–1271

Pawlowsky V (1989) Cokriging of regionalized compositions. Math Geol 21(5):513–521

Pawlowsky V, Olea RA, Davis JC (1995) Estimation of regionalized compositions: a comparison of three methods. Math Geol 27(1):105–127

Pawlowsky-Glahn V, Olea RA (2004) Geostatistical analysis of compositional data. Oxford University Press, New York, p 304

Pawlowsky-Glahn V, Egozcue JJ (2006) Compositional data and their analysis: an introduction. In: Buccianti A, Mateu-Figueras G, Pawlowsky-Glahn V (eds) Compositional data analysis in the geosciences: From theory to practice. Geological Society, London, Special Publications, 264, pp 1–10

Peters WC (1978) Exploration and mining geology, 2nd ed. Wiley, New York

Pitard F (1993) Pierre Gy's sampling theory and sampling practice, 2nd ed. CRC Press, Boca Raton

Quintana JM, West M (1988) The time series analysis of compositional data. In: Bayesian statistics 3. pp 747–756

Roden S, Smith T (2001) Sampling and analysis protocols and their role in mineral exploration and new resource development. In: Edwards AC (ed) Mineral resource and ore reserve estimation—the AusIMM guide to good practice. The Australasian Institute of Mining and Metallurgy, Melbourne, pp 67–72

Rombouts L (1995) Sampling and statistical evaluation of diamond deposits. J Geochem Explor 53:351–367

Rossi ME, Posa D (1990) 3-D mapping of dissolved oxygen in Mar Piccolo: a case study. Environ Geol Water Sci 16(3):209–219

Skinner EH, Callas NP (1981) Borehole deviation control in pre-mining investigations for hardrock mines. U.S. Bureau of Mines Information Circular, 8891, pp 79–95

Tolosana-Delgado R, Otero N, Pawlowsky-Glahn V (2005) Some basic concepts of compositional geometry. Math Geol 37(7):673–680

Spatial Variability

6

Abstract

An essential aspect of geostatistical modeling is to establish quantitative measures of spatial variability or continuity to be used for subsequent estimation and simulation. The modeling of the spatial variability has become a standard tool of mineral resource analysts. In the last 20 years or so, the traditional experimental variogram has given way to more robust measures of variability. Details of how to calculate, interpret and model variograms or their more robust alternatives are contained in this chapter.

6.1 Concepts

Mineral grades are generated through a succession of geological processes not always completely known or understood. Necessary conditions for mineral deposition include mineralization sources, pathways, and favorable geological conditions for deposition. The right physical and chemical processes can lead to significant mineral concentrations. The characteristics of mineral deposition invariably impart patterns of spatial correlation that are important for resource evaluation and mine planning.

The description and modeling of these correlation patterns allows better understanding of the depositional processes and improves on the prediction of mineralization and mineral grades at unsampled locations. Statistical tools can be used to describe those correlations within an appropriate theoretical framework.

The material in this section summarizes other geostatistical texts such as Geostatistical Ore Reserve Estimation (David 1977), Mining Geostatistics (Journel and Huijbregts 1978), An Introduction to Applied Geostatistics (Isaaks and Srivastava 1989), or Geostatistics for Natural Resources Evaluation (Goovaerts 1997).

Random Function Concept The uncertainty about an unsampled value z is modeled through the probability distribution of a random variable (RV) Z. The probability distribution of Z after data conditioning is usually location dependent; hence, the notation Z(**u**), with **u** being the coordinate location vector.

A random function (RF) is a set of RVs defined over some field of interest, for example, {Z(**u**), **u** is an element of study area A} also denoted simply as Z(**u**). Usually the RF definition is restricted to RVs related to the same attribute, hence, another RF would be defined to model the spatial variability of a second attribute, say {Y(**u**), **u** is an element of the study area A}.

The use of Random Functions implies that the variables are within a subset of the deposit or an area that is considered stationary. The ability to apply the RF concept is based on the belief that the locations **u** in A and variable Z belong to the same statistical population. The purpose for conceptualizing a RF as {Z(**u**), **u** is an element of study area A} is never to study the case where the variable Z is completely known. If all the z(**u**)'s were known for all **u** in the study area A, there would be neither any problem left nor any need for the concept of a random function. The ultimate goal of a RF model is to make some predictive statement about locations **u** where the true outcome z(**u**) is unknown.

Just as a RV Z(**u**) is characterized by its cumulative distribution function (cdf), a RF Z(**u**) is characterized by the set of all its N-variate cdfs for any number N and any choice of the N locations $\mathbf{u}_i$, i = 1,…, N within the study area A:

$$F(\mathbf{u}_1,\ldots,\mathbf{u}_N;z_1,\ldots,z_N) = Prob\{Z(\mathbf{u}_1) \le z_1,\ldots,Z(\mathbf{u}_N) \le z_N\}$$

M. E. Rossi, C. V. Deutsch, *Mineral Resource Estimation,* DOI 10.1007/978-1-4020-5717-5_6,

The univariate cdf of the RV Z(**u**) is used to characterize uncertainty about the value z(**u**), and the multivariate cdf is used to characterize joint uncertainty about the N values z($\mathbf{u}_1$),…, z($\mathbf{u}_N$).

The bivariate (N=2) cdf of any two RVs Z($\mathbf{u}_1$), Z($\mathbf{u}_2$), or more generally Z($\mathbf{u}_1$), Y($\mathbf{u}_2$), is particularly important since conventional geostatistical procedures are restricted to univariate (F(**u**;z)) and bivariate distributions:

$$F(\mathbf{u}_1,\mathbf{u}_2;z_1,z_2) = Prob\{Z(\mathbf{u}_1)\le z_1, Z(\mathbf{u}_2)\le z_2\}$$

One important statistic of the bivariate cdf F($\mathbf{u}_1$, $\mathbf{u}_2$; z_1, z_2) is the covariance function defined as:

$$C(\mathbf{u}_1,\mathbf{u}_2) = E\{Z(\mathbf{u}_1)Z(\mathbf{u}_2)\} - E\{Z(\mathbf{u}_1)\}E\{Z(\mathbf{u}_2)\}$$

The covariance is a summary statistic that is positive when Z($\mathbf{u}_1$) and Z($\mathbf{u}_2$) are directly related and negative when they are inversely related. The magnitude of the covariance summarizes the strength of the relationship. It is a single number that summarizes the bivariate distribution. When a more complete summary is needed, the bivariate cdf F($\mathbf{u}_1$, $\mathbf{u}_2$; z_1, z_2) can be described by considering binary indicator transforms for thresholds of the Z variable. Then, the previous bivariate cdf at various thresholds z_1 and z_2 appears as the non-centered covariance of the indicator variables:

$$F(\mathbf{u}_1,\mathbf{u}_2;z_1,z_2) = E\{I(\mathbf{u}_1,z_1)I(\mathbf{u}_2,z_2)\}$$

This relation is important for the interpretation of the indicator geostatistics formalism; it shows that the inference of bivariate cdfs can be done through sample indicator covariances.

The probability density (or mass) function (pdf) representation is more relevant for categorical variables. Recall that a categorical variable Z(**u**) may take one of K outcome values $k=1,\ldots,K$, arising from a naturally occurring categorical variable or from a continuous variable discretized into K classes.

Inference of any statistic requires some repetitive sampling. For example, repetitive sampling of the variable z(**u**) is needed to evaluate the cdf through experimental proportions:

$$F(\mathbf{u};z) = Prob\{Z(\mathbf{u})\le z\} = Proportion\{z(\mathbf{u})\le z\}$$

However, in almost all applications at most one sample is available at any single location **u** in which case z(**u**) is known (ignoring sampling errors), and the need to consider the RV model Z(**u**) vanishes. The need remains to infer the statistical parameters at unsampled locations. The paradigm underlying geostatistical inference is to trade the unavailable replication at location **u** for another replication available somewhere else in space and/or time. For example, the cdf F(**u**;z) may be inferred from the sampling distribution of z-samples collected at other locations within the same field.

This trade of replication corresponds to the decision of stationarity. Stationarity is a property of the RF model, not of the underlying physical spatial distribution. Thus, it cannot be checked from data. The decision to pool data into statistics across geologic units is not refutable a priori from data; however, it can be shown inappropriate a posteriori if differentiation of a domain significantly improves the inferences and estimations obtained.

The RF {Z(**u**), **u** in A} is said to be stationary within the field A if its multivariate cdf is invariant under any translation of the N coordinate vectors $\mathbf{u}_k$, that is:

$$F(\mathbf{u}_1,\ldots,\mathbf{u}_N;\, z_1,\ldots,\, z_N) \\ = F(\mathbf{u}_1+\mathbf{l},\ldots,\mathbf{u}_n+\mathbf{l};z_1,\ldots,\, z_n) \text{ for any vector } \mathbf{l}$$

Invariance of the multivariate cdf entails invariance of any lower order cdf, including the univariate and bivariate cdfs, and invariance of all their moments. The decision of stationarity allows inference. For example, the unique stationary cdf

$$F(z) = F(\mathbf{u};z), \text{ for all } \mathbf{u} \text{ in A}$$

can be inferred from the cumulative sample histogram of the *z*-data values available at various locations within A. The stationary mean and variance can then be calculated from that stationary cdf F(*z*), and also the stationary covariance can be inferred.

Stationarity is critical for the appropriateness and reliability of geostatistical methods. Pooling data across geological boundaries may mask important grade differences; on the other hand, splitting the data into too many small stationary subsets may lead to unreliable statistics based on too few data per subset. The rule in statistical inference is to pool the largest amount of *relevant* information to formulate predictive statements (Chap. 4).

Since stationarity is a property of the RF model, the decision of stationarity may change if the scale of the study changes or if more data becomes available. If the goal of the study is global, then local details may be less important; conversely, the more data available approaching final decisions such as grade control or final mine design, the more statistically significant differentiation becomes possible.

Consider a stationary random function Z with known mean m and variance σ^2. The mean and variance are independent of location, that is, $m(\mathbf{u})=m$ and $\sigma^2(\mathbf{u})=\sigma^2$ for all locations $\mathbf{u}$ in the study area. The variogram is defined as:

$$2\gamma(\mathbf{h}) = Var[Z(\mathbf{u}) - Z(\mathbf{u}+\mathbf{h})] = E\{[Z(\mathbf{u}) - Z(\mathbf{u}+\mathbf{h})]^2\} \tag{6.1}$$

In words, the variogram is the expected squared difference between two data values separated by a distance vector **h**. The *semi*variogram $\gamma(\mathbf{h})$ is one half of the variogram $2\gamma(\mathbf{h})$. To avoid excessive jargon we simply refer to the variogram, except where mathematical rigor requires a precise definition. As with the mean and the variance, the variogram does not depend on location; it applies for a separation vector that is translated or scanned over all locations in the chosen area of interest. The variogram is a measure of variability; it increases as samples become more dissimilar. The covariance is a statistical measure that is used to measure correlation or similarity:

$$C(\mathbf{h}) = E\{[Z(\mathbf{u}) \cdot Z(\mathbf{u}+\mathbf{h})]\} - m^2 \tag{6.2}$$

The covariance $C(\mathbf{h})$ is 0.0 when the values **h**-apart are not linearly correlated. At $\mathbf{h}=0$ the stationary covariance C(0) equals the stationary variance σ^2, that is,

$$\begin{aligned} C(0) &= E\{Z(\mathbf{u}+0)\, Z(\mathbf{u})\} - [E\{Z(\mathbf{u})\}]^2 \\ &= E\{Z(\mathbf{u})^2\} - [E\{Z(\mathbf{u})\}]^2 \\ &= Var\{Z(\mathbf{u})\} = \sigma^2 \end{aligned}$$

In certain situations the standardized covariance, the correlation coefficient, is preferred:

$$\rho(\mathbf{h}) = C(\mathbf{h})/C(0)$$

By further expanding Eq. 6.1, the following relation between the semi-variogram and covariance is established for a stationary RF:

$$\gamma(\mathbf{h}) = C(0) - C(\mathbf{h}) \quad \text{or} \quad C(\mathbf{h}) = C(0) - \gamma(\mathbf{h}) \tag{6.3}$$

This relation depends on the model decision that the mean and variance are constant and independent of location. These relations are important in variogram interpretation and in providing covariances to kriging equations.

The principal features of the variogram are the sill, range, and nugget effect. Figure 6.1 shows a variogram with these three important parameters:

1. The "sill" of the variogram is the equal weighted variance of the data going into variogram calculation, which is the variogram value that corresponds to zero linear correlation. The variogram may flatten off at an apparent sill below or above the sill variance.

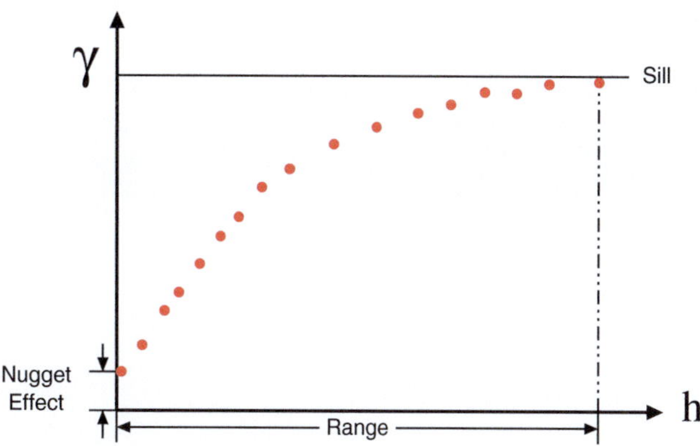

Fig. 6.1 Features of a variogram

2. The "range" is the distance at which this zero correlation is reached. If the variogram reaches the sill multiple times, it is common to consider the range as the first occurrence.
3. The "nugget effect" is the variogram value at a distance just larger than the sample size, which characterizes the very short scale variability. It is common also to use the term short-scale variability when referring to variogram distances less than the smallest spacing between sample points.

The correlation between $Z(\mathbf{u})$ and $Z(\mathbf{u}+\mathbf{h})$ is positive when the variogram value is less than the sill, and the correlation is negative when the variogram exceeds the sill. Figure 6.2 shows three **h**-scatterplots corresponding to three lags on a typical semivariogram. Geostatistical modeling generally uses the variogram instead of the covariance for mainly historical reasons, that is, it was considered more robust to cases where the mean changes locally. In practice, second order stationarity is almost always assumed and this advantage is not practically important.

6.2 Experimental Variograms and Exploratory Analysis

A single variogram value $\gamma(\mathbf{h})$ for a particular distance and direction vector **h** is straightforward to interpret and understand. Practical difficulties arise from the fact that we must calculate reliable values and simultaneously consider many lag vectors **h**. The variogram is a measure of geologic variability versus distance. This variability can be different in different directions; for example, in sedimentary formations there is typically much greater spatial correlation in the horizontal plane. The pattern of spatial continuity is anisotropic when the variable is more continuous in one direction than another.

Some important steps have to be completed before calculating experimental variograms: (1) the data must be visualized and understood from a geological perspective, (2) an appropriate coordinate system must be established, and (3)

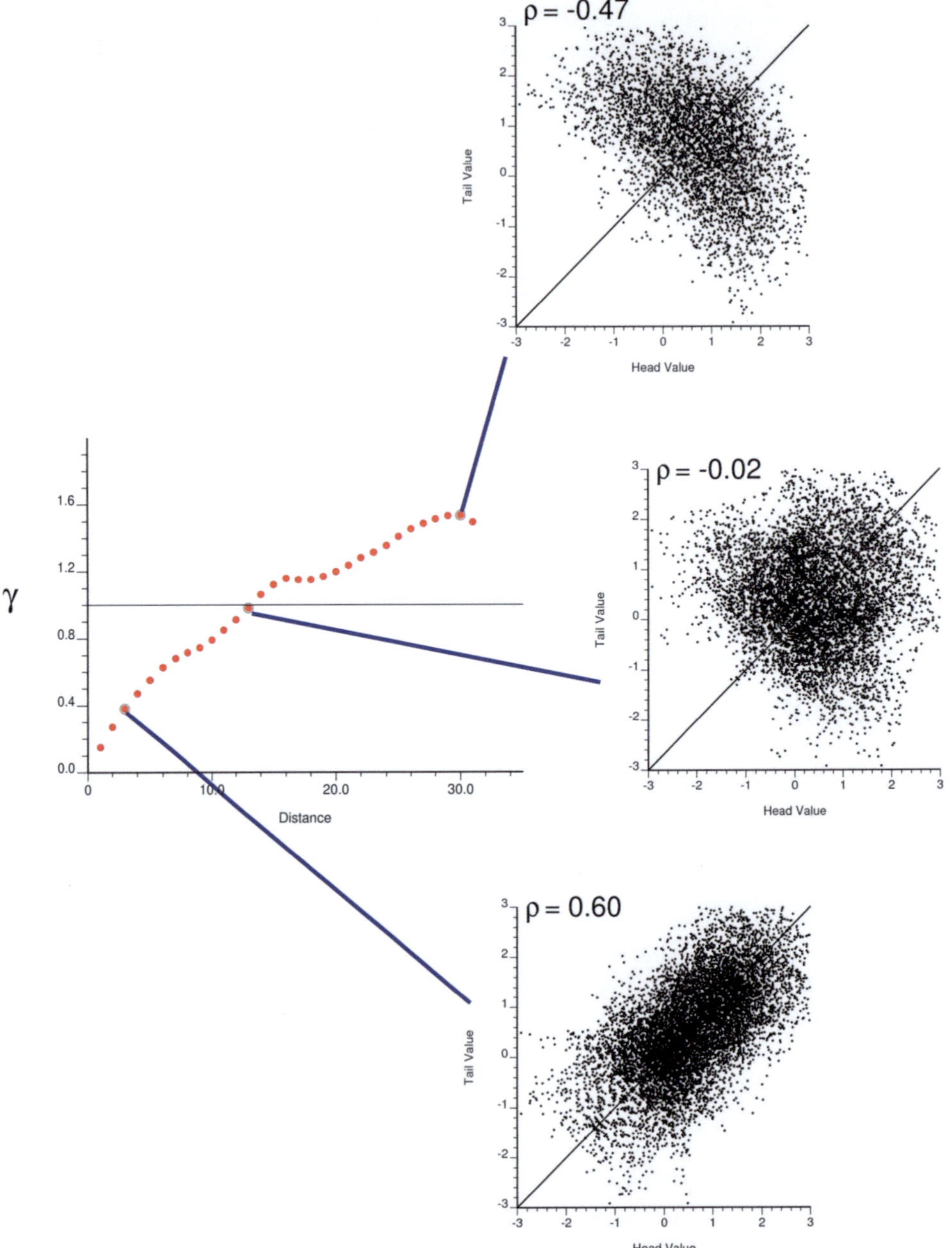

Fig. 6.2 Variogram values from a normal score transformed variable. The data points going into the calculation of three lags are shown as **h**-scatterplots

outliers and data transformations should be considered. In general, the variogram is computed in the coordinate system and in the units that will be used for modeling, that is, composite grades are considered and normal scores are considered if Gaussian uncertainty or simulation techniques are being considered.

The first prerequisite is to understand the data. Although experimental variograms may help understand the geology and spatial variability of a particular variable, misunderstanding and errors could be propagated by calculating variograms without a reasonable understanding of the data, trends, and data configuration. The histogram and univariate statistics of the data should be investigated. Odd data and extremely high and low data values should be questioned. The data should be visualized in many different ways. The complexity of the geologic continuity should be understood relative to the data spacing. The choice of estimation domains and variables should seem reasonable.

A second prerequisite is to perform variogram calculations in an appropriate coordinate system. Standard X/Y/Z coordinates based on elevation and a UTM or a local mine system may be reasonable for a disseminated deposit. Tabular or stratigraphic type orebodies may require the calculation and usage of stratigraphic coordinates (see Sect. 3.4). Weathered deposits may require the use of depth below topographic surface as the vertical Z coordinate.

The variogram is a two-point statistic; curvilinear continuity cannot be represented by a two-point statistic. The study area may need to be subdivided if the direction of continuity changes systematically over the study area; but there will be a trade-off between preserving local precision and maintaining sufficient data for reliable calculations. Some newer tools for locally varying anisotropy are becoming available, but are not commonly used.

Variogram calculation is preceded by selection of the Z variable to use in variogram calculation. The variable should not be transformed for obtaining experimental variograms for conventional kriging applications. The use of Gaussian techniques requires a prior normal score transform of the data and the variogram of those transformed data. Indicator techniques require an indicator coding of the data prior to variogram calculation.

Another aspect of choosing the correct variable is outlier detection and removal. Extreme high and low data values can have a large influence on the variogram value since each pair is squared in the calculation. While erroneous data should be removed, legitimate high values may also mask the spatial structure of the majority of the data. The increased variability of high values combined with preferential sampling in high valued areas can lead to experimental variograms that are noisy and difficult to interpret. Logarithmic or normal score transformation mitigates the effect of outliers, but an appropriate back transform is being considered in later calculations.

The correct variable also depends on how trends are going to be handled in subsequent model building. Sometimes, clear areal or vertical trends are removed prior to geostatistical modeling and then added to geostatistical models of the residual (original value minus trend). If this two-step modeling procedure is being considered, then the variogram of the residual data is required. There is a risk, however, of introducing artifact structures in the definition of the trend and residual data.

The variogram is calculated for distance/direction lags where there are a sufficient number of paired data. The variogram is the average of squared differences from data pairs:

$$2\gamma(\mathbf{h}) \approx \frac{1}{N(\mathbf{h})}\sum_{i=1}^{N(\mathbf{h})}[z(\mathbf{u}_i)-z(\mathbf{u}_i+\mathbf{h})]^2 \tag{6.4}$$

It is rare to find data pairs exactly the same distance apart. This requires that the $N(\mathbf{h})$ pairs be assembled using reasonable distance and direction tolerances (Fig. 6.3). Variogram calculation programs scan over all pairs and assemble the ones that fall into approximate distance/direction **h** lags after applying the specified tolerances. These tolerances define sectors within which separation vectors are defined.

A simple 2-D example of this is shown in Fig. 6.4. The experimental variogram is the average squared difference between the 16 pairs. Notice how some data are not used, some are used once, some twice, and one data is used three times. This depends on the data configuration and the tolerance parameters.

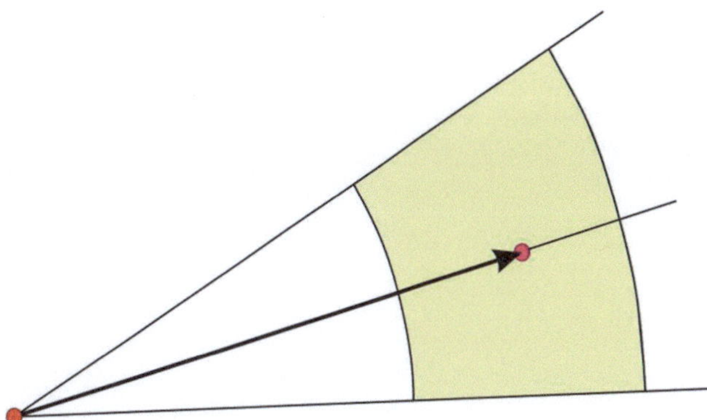

Fig. 6.3 Example tolerance for variogram calculation. The *bold arrow* between the two colored dots represents the vector of interest. Any vector from the right hand side dot to any location in the shaded area will be accepted

Establishing the variogram tolerance parameters is trial and error. If tolerances are too small, then the variogram will be noisy. If tolerances are too large, then the spatial continuity will be averaged out and imprecise. In general, lag and angle tolerances should be set as small as possible to ensure good definition of directional continuity while still obtaining a stable variogram.

The maximum number of lags should be so that the maximum lag distance is less than one half of the domain size. The variogram is only valid for a distance of one half of the field size since for larger distances the variogram begins to leave data out of the calculations. The lag tolerance is usually one half the lag separation distance. In cases of erratic variograms or few data, the lag tolerance can be greater than one half the lag separation to add additional pairs to the variogram calculation and to smooth between lags.

The choice of directions for variogram calculation depends on the anticipated anisotropy of the geological variable, the number of samples available and the software used. The geologic characteristics of the deposit can be understood by looking at sections and plan views to define potential directions of anisotropy. The orientation of the drill holes should also be considered. If there are enough samples, multiple directions are reviewed before choosing a set of three perpendicular directions. These three directions then become the three main axes of the ellipsoid that represents the anisotropy.

An omnidirectional variogram considers all possible directions simultaneously by opening the angle tolerance to 90°; they often yield the best-behaved variograms, but all directional characteristics are lost. A down-the-hole variogram will provide a good estimate of the nugget effect, as well as the short scale continuity, since it is calculated from adjacent data along the drill holes paths, without considering their orientation.

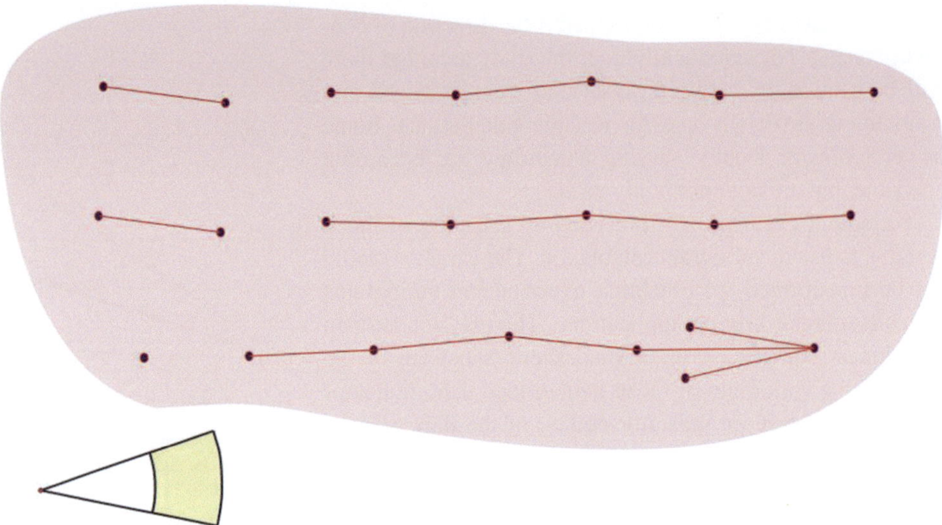

Fig. 6.4 Twenty-three data and 16 pairs that meet the tolerance sketched in the *lower left* of the figure. There are some gaps where the data are too close

In some cases there may be enough data to calculate the variogram values for many different distances and directions and plot a two- or three-dimensional map of variogram values. This is true for both gridded data and dense irregular data. These maps are useful for detecting directions of anisotropy, and to avoid imposing pre-conceived ideas on the variogram.

6.2.1 Other Continuity Estimators

The experimental variogram is often calculated as in Eq. 6.4; this is the traditional tool used to assess variability. However, the presence of a small number of extreme data values can cause the variogram to become very noisy. Normal scores, indicator data, or log-normal transformations often make the variogram more robust, but the nature of the variability is changed. Several other continuity estimators have been proposed to make the variogram function more robust (Journel 1988; Isaaks and Srivastava 1988).

The general relative and the pairwise relative variograms were popularized by M. David (1977). The general relative variogram standardizes the traditional variogram using the squared average of the data points for each lag:

$$\gamma_{GR}(\mathbf{h}) = \gamma(\mathbf{h}) / m(\mathbf{h})^2$$

with the average of data for each lag being

$$m(\mathbf{h}) = \frac{m_{\mathbf{h}} + m_{-\mathbf{h}}}{2}$$

The pairwise relative variogram often produces a more clear spatial continuity function than the general relative variogram. The difference is that the adjustment is done separately for each pair of values, using the average of the two values:

$$\gamma_{PR}(\mathbf{h}) = \frac{1}{2N(\mathbf{h})} \sum_{N(\mathbf{h})} \frac{[z(\mathbf{u}) - z(\mathbf{u}+\mathbf{h})]^2}{\left[\dfrac{z(\mathbf{u}) + z(\mathbf{u}+\mathbf{h})}{2}\right]^2}$$

The variable used should be strictly positive because of the denominators in both relative variogram's definition. Relative variograms were originally proposed to remove the *proportional effect* commonly found in positively-skewed mineral grade distributions. The standard deviation of the samples within sub-zones of our "stationary" domain is likely proportional to their mean grade. Experience has shown that the relative variograms, and in particular the pairwise relative variogram, are significantly more structured and easier to model than the traditional variogram. This makes it a suitable experimental variogram estimator in presence of sparse, clustered and erratic data. A concern is that the sill is not clearly defined—it depends on the shape of the histogram and the coefficient of variation. One useful implementation approach is to transform the data to a lognormal distribution with a mean of 1 and a variance of 1, then the sill of the pairwise relative variogram is 0.44 (Babakhani and Deutsch 2012).

The covariance defined Eq. 6.2 assumes that the domain is stationary, and thus the mean of the data is the same at both ends of the separation vector. A more general spatial covariance definition is the non-ergodic covariance (Srivastava 1987) that does not assume that the averages of the tail and head of the separation vector are the same:

$$C(\mathbf{h}) = \frac{1}{N(\mathbf{h})} \sum_{N(\mathbf{h})} [z(\mathbf{u}) \cdot z(\mathbf{u}+\mathbf{h})] - m_{-\mathbf{h}} \cdot m_{\mathbf{h}}$$

The corresponding non-ergodic correlogram (Srivastava and Parker 1988) is commonly used as a robust alternative to the traditional variogram. The sample correlogram is calculated with:

$$\rho(\mathbf{h}) = C(\mathbf{h})/(\sigma_{-\mathbf{h}} \cdot \sigma_{\mathbf{h}})$$

where C(**h**) is defined above and:

$$m_{-\mathbf{h}} = \frac{1}{N(\mathbf{h})}\sum_{N(\mathbf{h})} z(\mathbf{u}), \quad m_{\mathbf{h}} = \frac{1}{N(\mathbf{h})}\sum_{N(\mathbf{h})} z(\mathbf{u}+\mathbf{h})$$

$$\sigma_{-\mathbf{h}} = \sqrt{\frac{1}{N(\mathbf{h})}\sum_{N(\mathbf{h})} [z(\mathbf{u}) - m_{-\mathbf{h}}]^2},$$

$$\text{and } \sigma_{\mathbf{h}} = \sqrt{\frac{1}{N(\mathbf{h})}\sum_{N(\mathbf{h})} [z(\mathbf{u}+\mathbf{h}) - m_{\mathbf{h}}]^2}$$

This measure is robust because of the use of lag-specific mean and variance values. In practice, it has become the most popular option when dealing with untransformed variables.

6.2.2 Inference and Interpretation of Variograms

The single biggest problem in variogram interpretation is a lack of data to calculate a reliable variogram. Too few data for reliable variogram interpretation does not mean there is no variogram; it is just hidden or masked by data paucity. Geological analogues or expert judgment may be required.

In general, inference is affected by data density; the use of different data types (drill holes, blast holes, trench samples, etc.); influence of outliers; and trends. Also, a high relative variability of the samples as measured, for example, by the coefficient of variation (CV), is an indication that robust measures of continuity are necessary.

Inference is an iterative process and exploratory in nature. Inference generally starts with the initial sample collection, which results in early geologic interpretation and estimation domain definition.

Spatial clustering is common in regions of high grades that will likely be mined sooner. Geologists would naturally seek to confirm and carefully delineate such areas. These clustered data, however, can cause the variogram at short lags to be too high or too low, which could lead to a misinterpretation of the variogram structure. The most common misinterpretations would be a too-high a nugget effect or short-scale structure. There can also appear to be a short-scale cyclic characteristic. The clustered data could provide an improved representation of the higher grades (and higher variance if a proportional effect exists) sub-zone within the stationary domain. Different patterns of anisotropy may emerge from short-scale data, compared to the larger, domain-wide scale.

Decisions as to whether clustering is truly imparting an artifact to the variogram model should be made considering geologic information. A few possible solutions may be to (1) remove the clustered data, leaving an underlying grid; (2) discredit unusually high variogram points for the short distances; or (3) using more robust measures such as the sample correlogram or pairwise relative variogram rather than the traditional variogram.

The interpretation of the variogram consists of explaining the variability over different distance scales as a function of known geological and mineralogical factors. Experimental variograms should always be reconciled with known geology. Potential artifacts introduced by data configurations and sampling practices should be discussed with geologists familiar with the deposit. The degree of continuity observed, the anisotropies observed, and the relative variances of each structure should be discussed. The variogram function should be representative of the expected geologic variability.

Discrepancies that may arise between perceived geologic knowledge and interpretations and inferences from the experimental variograms should be resolved before proceeding with the resource evaluation process.

There are four common cases in variogram interpretation: trends, cyclicity, zonal (or areal) anisotropy, and geometric anisotropy (Fig. 6.5). Trends are common in mineral deposits, and often question the definition of stationarity chosen. For example, Cu grades tend to decrease towards the periphery of a typical Porphyry Cu deposit, while the grades in some base metal mines are a function of the porosity of the host rock; if it is sedimentary or pseudo-sedimentary, a clear trend across the direction of deposition will be found. These trends cause the variogram to increase above its sill variance σ^2, or stationary variance, and show a negative correlation at large distances.

Cyclicity can be a result of the mineral depositional phenomena occurring repeatedly over geologic time and leading to repetitive en echelon variations in the resulting mineral grades. Cyclical behaviors are more common, but not exclusive to, sedimentary or strata-bound deposits. This behavior is observed in the variogram as cycles of positive and negative correlation at the length scale of the geologic cycles. These cyclic variations often dampen out over large distances as the size or length scale of the geologic cycles is not perfectly regular. These cycles are usually modeled with sinusoidal functions, often referred to as a *hole effect*. In early mining geostatistics, cyclicity was observed in "down-hole" variograms; hence the name *hole effect*.

Anisotropic variograms are extremely common in mining and other geostatistical applications. Occasionally the pattern of continuity will be similar in all directions, and thus the variogram is said to be isotropic. But by far the most common occurrence is for variograms to be anisotropic.

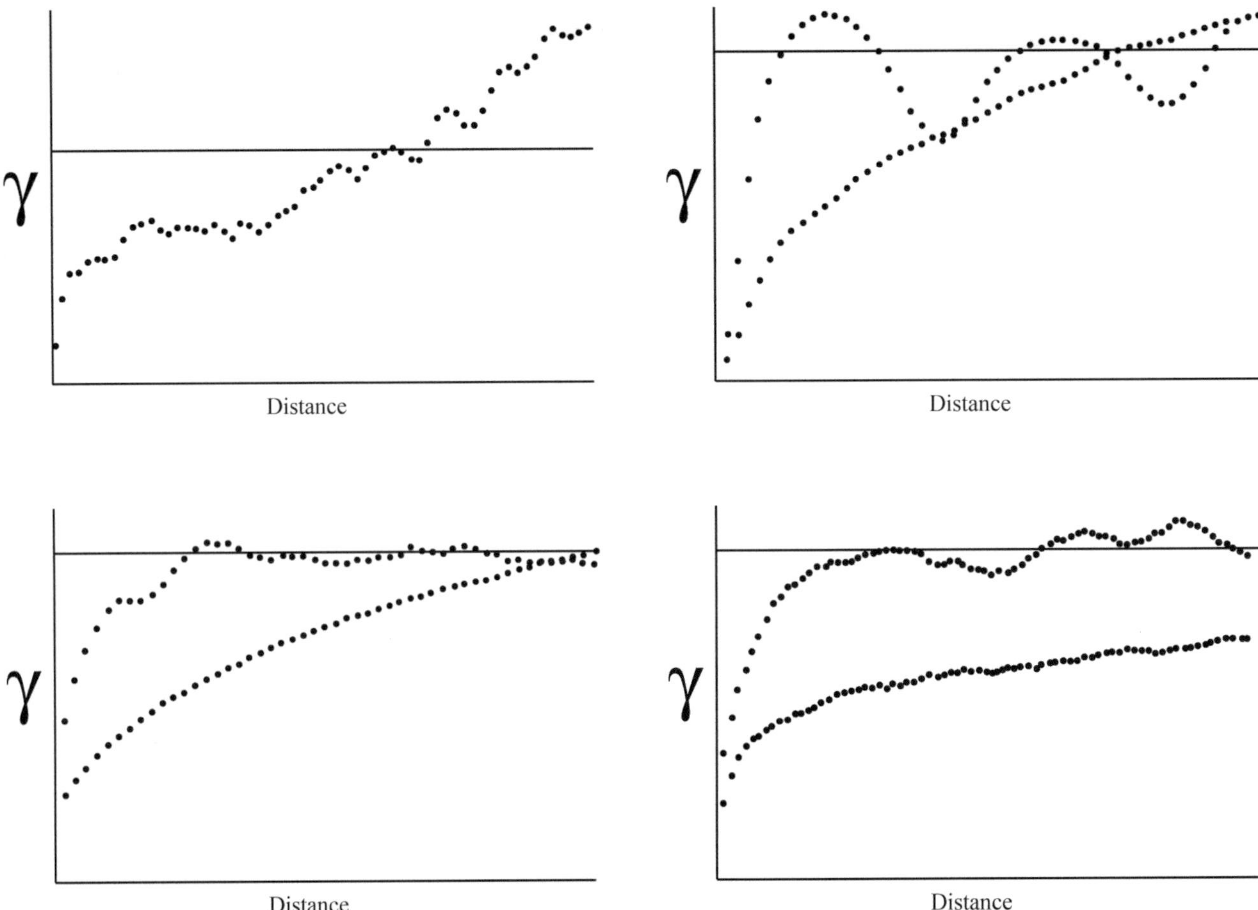

Fig. 6.5 Common variogram shapes (variograms are standardized, but with no specific distance scale). The figures illustrate trends (*top left*), cyclicity (*top right*), geometric anisotropy (*bottom left*), and zonal anisotropy (*bottom right*)

Geometric anisotropy is when the variance (sill) is reached at different distances (lags) for different directions. Zonal anisotropy is when for any distance considered in the variogram calculation the variogram never reaches the expected sill variance. Zonal anisotropies can also be considered geometric anisotropies if it assumed that the same sill variance is reached at distances much larger than used to calculate the variogram.

6.3 Modeling 3-D Variograms

The experimental variogram points are not used directly in subsequent calculations; rather, a parametric function is adjusted to those points to obtain a three-dimensional model (Armstrong 1984). The two most important reasons for modeling variograms are:

1. Most subsequent geostatistical calculations, including estimation and simulation methods, require a variogram or covariance value for all possible distances and directions. Since only specific distances and directions are used to obtain the variogram points, these need to be interpreted and interpolated into a $\gamma(\mathbf{h})$ function for all $\mathbf{h}$ values. The modeled $\gamma(\mathbf{h})$ function should carry all the geological information derived from the experimental model, including anisotropies, trends, nugget effects, etc. A smooth interpreted function also allows filtering artifacts of data spacing, location, and sampling practices.
2. The covariance values $C(\mathbf{h})$ derived from $\gamma(\mathbf{h})$ using Eq. (6.3) must have a mathematical property called *positive definiteness*. A positive definite model ensures that the kriging equations used can be solved, that this solution is unique, and that the kriging variance is positive. A positive definite function is a valid measure of distance.

The positive definite condition is required because the kriging estimates are weighted linear combinations of samples:

$$z^*(\boldsymbol{u}) = \sum_{\alpha=1}^{n} \lambda_\alpha z(\boldsymbol{u}_\alpha)$$

The variance of the estimates $Z^*(\mathbf{u})$ must, by definition, be positive. It can be shown (Journel and Huijbregts 1978) that

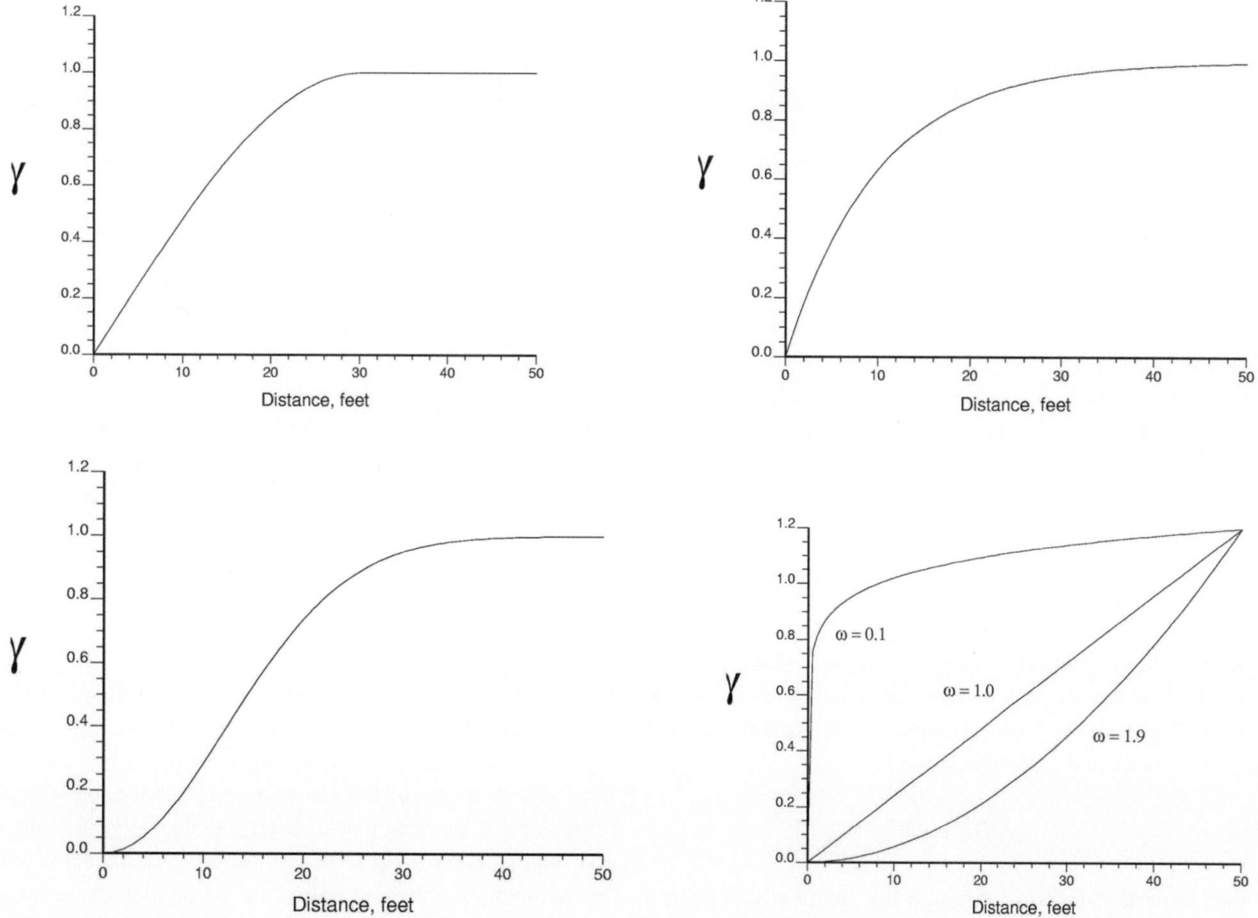

Fig. 6.6 Spherical model (*top left*); exponential model (*top right*); Gaussian model (*bottom left*); and power-law models (*bottom right*)

the variance of a linear combination can be written in terms of the variogram (or covariance) values as:

$$Var\{Z^*(\boldsymbol{u})\} = Var\left\{\sum_{\alpha=1}^{n} \lambda_\alpha Z(\boldsymbol{u}_\alpha)\right\}$$
$$= \sum_{\alpha=1}^{n}\sum_{\beta=1}^{n} \lambda_\alpha \lambda_\beta \gamma(\boldsymbol{u}_\alpha - \boldsymbol{u}_\beta) \geq 0$$

The variance must be non-negative for any choice of locations (**u**) and weights (λ) within the domain. The variogram functions that yield non-negative variance are called semi positive definite. If a covariance is used, then the condition is restricted to positive variance only.

The positive definite condition implies that practitioners typically use specific known positive definite functions; the more commonly used are spherical and exponential, among others described below (Journel and Huijbregts 1978; Cressie 1991; Christakos 1984). Other arbitrary functions can be used, but they must first be proven positive definite (Myers 1991), for which Bochner's theorem (Reed and Simon 1975), based on Fourier transforms, could be used. The additional work and complications arising from attempting to use other functions is rarely worthwhile: traditional parametric models allow achieving a good fit in practice and also allow consideration for the geological information commonly available; and it also allows for a straightforward transfer into existing estimation and simulation codes (Deutsch and Journel 1997).

6.3.1 Commonly Used Variogram Models

Figure 6.6 shows the most common variogram shapes used in mining. These shapes are all parameterized by a scalar vector h and a range parameter. The use of anisotropy with these variogram model shapes is discussed below. The first is the spherical model, for which the spherical covariance, 1—Sph(h), is the volume of two intersecting spheres.

$$Sph(h) = \begin{cases} 1.5(h/a) - 0.5(h/a)^3, h \leq a \\ 1, \text{otherwise} \end{cases}$$

The exponential model is also common, and is similar to the spherical, except that it rises more steeply and reaches the sill asymptotically. The practical range is where the variogram value is 95 % of the sill. Some old definitions of this variogram do not include the "3" and use a range parameter 1/3 of the practical range; modern practice is to consider the practical range:

$$Exp(h) = 1 - \exp(-3h/a)$$

The Gaussian model exhibits a parabolic rather than linear behavior at the origin, which implies more short scale continuity. It is suitable for slowly-varying variables, since the increase in variance is very gradual with distance; examples of such variables are elevations, water-table measurements in hydrogeology, or thicknesses. The practical range is where γ(h) is 95 % of the sill:

$$Gaus(h) = 1 - \exp\left(-3(h/a)^2\right)$$

The power law model is associated to self affine random fractals. The parameter ω of a power model is related to the fractal dimension D. The variogram model is defined by a power $0 < \omega < 2$ and a positive slope, c.

$$\gamma(h) = c \bullet h^{\omega}$$

Other important models, although not frequently used, are the hole effect models: (a) $\gamma(h) = 1 - \frac{\sin r}{r}$ and (b) $\gamma(h) = 1 - \cos h$. The sinusoidal model (a) is valid in three dimensions, with r specified in radians, while model (b) is only valid in one dimension, particularly useful when a strong hole-effect needs to be modeled in a particular direction.

An important notion in variogram modeling is the use of nested structures. The variogram can be fit with a positive sum of valid variogram models—called nested structures. So, for example, the final variogram could be a sum of a spherical variogram explaining part of the variance and an exponential variogram explaining the remaining variance. Both structures should typically have different ranges.

6.3.2 Basic Variogram Modeling Guidelines

Virtually all experimental variograms can be modeled using these various types of models. The specific steps and the order in which they are completed may vary with the software tool used, variogram modeling entails making sometimes subjective decisions on several issues:

Decide on the variable to be modeled. Conventional mineral resource estimation requires that grade variables be modeled. There are estimation methods that also require transformation, including the log-normal (not in common usage), and Gaussian or Indicator-based methods. Most simulation methods require that either the Gaussian or Indicator transforms is used. The transformed variables are generally easier to model, but the characteristics of the continuity models are generally different than the original variables.

Non-transformed variables may require more exploratory analysis and clean-up work using *h*-scatterplots to obtain reasonable models; but the practice of modeling a transformed variable to obtain a variogram model and then back-transforming the variogram mode to obtain a model for the original variable is discouraged. This idea has been applied using Gaussian and logarithm transformations. For a given dataset, it can be shown that the variogram models obtained from the original data often have significant differences with the back-transformed models derived from using the Gaussian or log transforms.

Find a good estimate of the nugget effect. The nugget effect is a variance that results from measurement errors and geological short-scale variance. It can sometimes be obtained from repeat assays of the same sample, subdivided several times. This is done as part of the general quality assurance and quality control (QA/QC) of the sample database, but the number of repeat samples and its spatial distribution is generally insufficient to estimate a nugget effect for the variogram model. More common is to use a direction where there is large amount of data at short spacing, as for example in the down the hole direction of the drill hole. Typically, there is an order of magnitude difference between the lateral drill hole spacing and the sampling frequency down the hole. Since by definition the nugget is the variance at distance zero, then it has to be isotropic, that is, the same in all directions. Therefore, it is licit to estimate it from data in any direction; and closely spaced data will provide the better estimates. The nugget effect obtained from down the hole variograms should always be compared to the sampling variance resulting from the QA/QC program available and heterogeneity or other sampling studies. This comparison should be made always mindful of the stationary domains used in variogram modeling.

Determine the best anisotropy directions. The anisotropy is found by looking at the variogram in multiple directions. These directions can be pre-determined based on geologic knowledge, fixing the directions that may be reasonable candidates for best representing the anisotropy. Sometimes, the software tool used allows specifying multiple directions, with no pre-conceived anisotropy model. If more data is available, a larger number of directions can be run, using tighter tolerances, such that a more precise definition of the anisotropy is possible. But even if the data is sparse such that wide angle and lag tolerances must be used, the anisotropic model is probably more realistic than an isotropic one. Although rare, occasionally geologic phenomena can result in a 3-D isotropic variogram over a relatively small distance scale.

The most common type of anisotropy is geometric anisotropy. This is the case when the directional variograms present the same level of variance (sill) in all directions, but the ranges are different. The variogram model is an ellipsoid characterized by the three principal directions (axes) with three different ranges. All other directional variograms can be derived from this ellipsoid. A linear transformation of the coordinates is sufficient to obtain an effective lag distance h. This transformation involves a rotation to make the ellipsoid axes coincident with the main coordinate axes; and a translation, specified as an affinity matrix, to obtain the equivalent effective ranges. This transformation allows inferring the variogram value for any direction and any distance:

$$h = \sqrt{\left(\frac{h_x}{a_x}\right)^2 + \left(\frac{h_y}{a_y}\right)^2 + \left(\frac{h_z}{a_z}\right)^2}$$

This would be applied for each structure separately. The range of each structure would be one. Common geostatistical software requires the user to specify the orientation of the ellipsoid (three angles) and the three ranges; there is no need to explicitly calculate the scaled h distances.

Zonal anisotropy cannot generally be modeled using a simple coordinate transformation; in this case, one option is to add an additional structure in the specific direction where the zonal component appears. But it is more common to consider it a special case of geometric anisotropy, where the sill is reached asymptotically at large distances. The zonal anisotropy thus appears as a very large range parameter in one or more of the principal directions.

Define the sill at which the variogram reaches the zero correlation distance. There is often confusion about the correct variance to use for variogram interpretation. It is important to have the variance σ^2 to correctly interpret positive and negative correlation, as well as confirm trends in the data. Some authors have discussed the issue about which is the correct variance to use as sill variance for variogram interpretation (Gringarten and Deutsch 1999; Journel and Huijbregts 1978, p. 67; and Barnes 1991).

There are three issues that may affect the decision about which is the correct variance to use: (1) the dispersion variance, which accounts for the difference between our finite domain and the infinite stationary variance; (2) declustering weights, which account for the fact that our summary statistics are not representative of the entire domain, and (3) outlier values, which can cause erratic and unstable estimates of the variance. The main point being made by Journel and Huijbregts (1978), and also Barnes (1991), is that the sample variance $\hat{\sigma}^2$ is *not* an estimator of the stationary variance σ^2; rather, it is an estimate of the dispersion variance of samples of point support within the stationary domain A ($D^2(\bullet,A)$). Only when the domain A approaches an infinite domain does the sample variance $\hat{\sigma}^2$ approach the stationary variance σ^2. However, the data used to estimate the variogram represent the area of interest A and not an infinite domain. Thus, the point where $Y(\boldsymbol{u})$ and $Y(\boldsymbol{u}+\boldsymbol{h})$ are uncorrelated is the dispersion variance $D^2(\bullet,A)$. Thus, the sample variance should be used as the sill of the sample semivariogram understanding that it is, in fact, a dispersion variance.

The second issue relates to the use of the naïve sample variance or the sample variance accounting for declustering weights. The use of declustering weights is important, but they are not used in the calculation of the variogram. There are more pairs in areas of greater sampling density, and therefore the variogram at shorter distances will reflect more the local variance of the clustered data. Also, the sill is reached at the naïve sample variance. Omre (1984) proposed incroporating declustering weights into the variogram calculation; however, they provide no better variogram and are difficult to implement in practice. As a result, declustering weights are not used in variogram calculation, interpretation, and modeling. The sill should be taken as the naive equally weighted variance.

The third issue that must be addressed is the influence of outlier sample values. It is well known that the variance, being a squared statistic, is sensitive to outlier values. For this reason, the sample variance may be unreliable. This is not a problem with correlograms, or transformed data: the Gaussian and indicator transforms remove the sensitivity to outlier data values. While some outliers may be removed (or capped) in traditional variogram modeling, the correct variance for variogram interpretation is the naïve equal-weighted variance. Outliers influence more than just the variogram sill: as shown by Rossi and Parker (1993), the nugget effect and short scale continuity modeled are also affected, and can have a significant impact on the final variogram model. The higher the variability, the more significant the impact of outliers is. Although we still recommend keeping all the data for variogram modeling, there may be cases where the outliers are so extreme and with little or no spatial influence that it may be better to remove or reduce (cap) its value.

Define how to deal with the trends. If the variable being modeled shows a systematic trend, that is, the data is nonstationary over the study domain, it is generally better to remove it before further geostatistical analysis. Trends in the data can be identified from the experimental variogram,

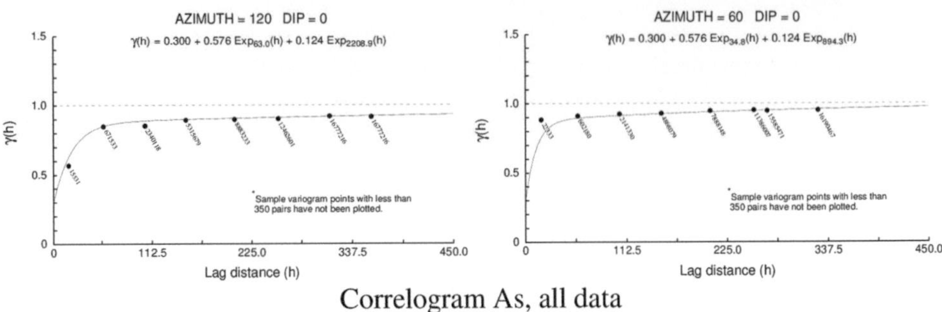

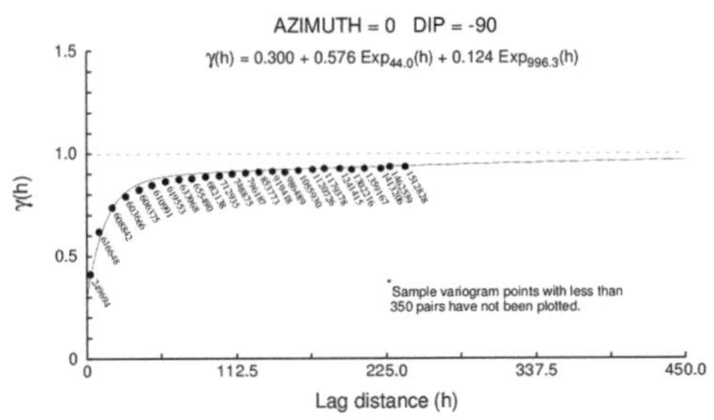

Fig. 6.7 Example of three directions, experimental correlograms with their fit, Arsenic data

which keeps increasing above the theoretical sill. A "power" or "fractal" variogram model could be fit to the experimental variogram; however, these models do not have a sill value (it is infinite), and thus they have no covariance counterpart, which is a problem for most geostatistical applications.

If the trend is removed, variogram analysis and all subsequent estimations or simulations are performed on the residuals. The trend is added back to estimated or simulated values at the end of the study. Although there are difficulties in defining a robust trend model and removing the its deterministic portion from the data, the only practical option is to model trends deterministically. Removing a trend by estimation from the data themselves can introduce a bias; however, this bias may be less significant than the errors introduced by leaving the trend alone. It is often suggested (Journel and Huijbregts 1978 and others) that the variogram should be computed in directions and/or areas where the trend does not appear as significant. Directly considering the residuals in variogram calculation can lead to erroneously high variability.

Fit the variogram models, and decide on the number and type of nested structures. Experimental variograms show different behavior at different distances *h*. Other than the discontinuity at the origin (the nugget effect), the variogram may also present a long range structure superimposed on a short distance structure. These different spatial continuity structures are a reflection of different geologic controls; for example, it is common for more than one geologic factor to influence mineral deposition. Normally, in precious and base metal deposits the higher grade mineralization is controlled by fractures or veins, which tend to have a clear preferential orientation, with a strong anisotropy and short range. These will evidence themselves as a distinct the short-scale continuity and anisotropy, considering also that most of the clustered samples in these types of deposits sample the higher grade mineralization. A larger intrusive body or the geometry of the host rock may be responsible for a larger-scale anisotropic feature.

The use of nested structures provides enough flexibility to model most combinations of geologic controls. However, there is no gain in over-modeling and over-fitting. It is extremely rare that more than three nested structures would be necessary; only in those cases where, for example, a zonal anisotropy or a hole-effect are modeled as an additional structure in a specific direction.

In summary, good practice is to pick a single *isotropic* nugget effect, choosing the same number of variogram structures for all directions based on the most complex direction; zonal anisotropies are modeled as geometric anisotropies, with one direction having a very long, unrealistic range to account for the lower variance; the same sill parameter is used for all variogram structures in all directions, and allowing a different range parameter in each direction. An interesting possibility afforded by at least one semi-automatic fitting program is that different nested structures may present different anisotropies. While care must be taken not to over-fit the models, these can lead to reasonable and defensible models, particularly if there is sufficient data and geologic knowledge to back them up.

Figure 6.7 show as an example of an experimental correlogram and its model for three main directions of anisotropy. The correlograms are of Arsenic values, and the graph shows the fit, the number of pairs used to calculate each variogram point, and the model fitted, with the ranges corresponding to each direction. Two exponential structures were used to fit the models, with significant anisotropy in SE vs. NE directions, as well as in the vertical direction.

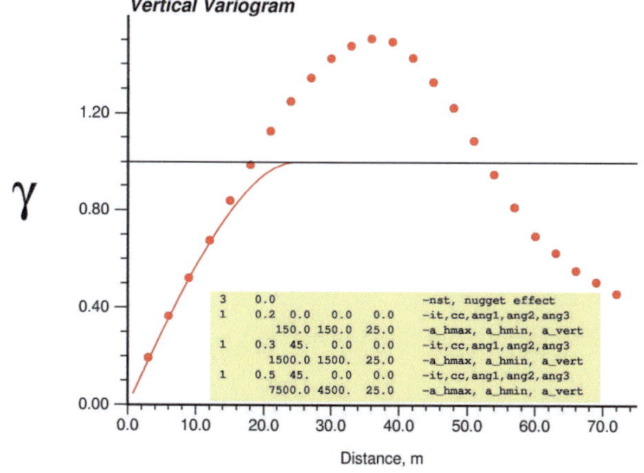

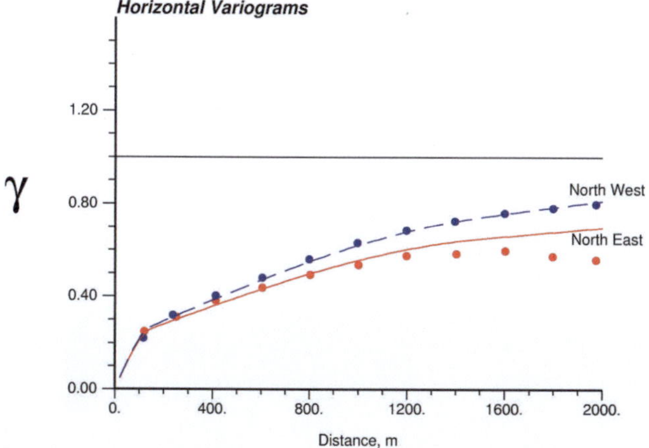

Fig. 6.8 Example directional variograms and fitted models from a Northern Alberta Oil Sands mine

Figure 6.8 shows vertical and horizontal standardized variograms of the mass fraction of bitumen from an oil sands depsit in Northern Alberta. The area of interest is approximately 5 × 5 km, and about 100 m thick. The vertical variogram at the top of the figure shows a relatively short variogram range of 25 m, a trend above the sill and then a drop below the sill. The trend/cyclic behavior is a result of the bitumen grade being lower at the top and the bottom of the formation. The horizontal variograms are nearly isotropic; however, the North East directional variogram shows greater continuity at large scale than the North West directional variogram. The variogram model parameters are shown on the vertical variogram plot. No nugget effect and three spherical structures were used to fi the variogram. The structures explain 20, 30 and 50 % of the variance, respectively. The vertical range for all structures was fixed at 25 m. The horizontal ranges are isotropic at 150 m and 1,500 m for the first two structures and anisotropic at 7,500 and 4,500 for the North East and North West directions for the last structure. The fit to the points is considered good despite the fact that the vertical trend is not fully represented in the vertical variogram.

Figure 6.9 shows another example of a Cu correlogram, taken from the El Pachón Project, in the Argentinean Central Andes, currently owned by Xstrata Copper. It shows three main directions of anisotropy, a variogram map showing in plan view the NW-SE main directions, and the summary model below. The experimental correlograms were fitted with three spherical structures, with the model been almost twice as continuous in the main direction of continuity compared to the other two.

Regardless of its final form, the model should always be checked by viewing directional variograms in several intermediate directions; that is, the 3-D model obtained based on a few directions should represent accurately all possible directions.

6.3.3 Goodness of Variogram Fit and Cross Validation

A good deal of research has been dedicated to fitting the variogram models and measuring their quality. Different models can fit the same experimental data, so it is natural to ask which model is better. The tendency is to search for objective (statistical) measures of goodness to judge a model that is to a large extent subjective.

One common statistical measure is weighted least-squares, whereby the distance from each modeled point to the corresponding experimental variogram value is measured and minimized. This is commonly used in software that provides an automatic or semi-automatic fit. The weights can be proportional to the number of pairs for each lag, the number of informed values for each direction, or a scheme that assigns more weight to the first few lags of the model (Cressie 1985; Isaaks 1999).

While it may be useful to have a measure of closeness between the model and the experimental data, there is no guarantee that the lowest sum of squared differences would yield a better model. A good fit of important features of variogram models such as the nugget effect, anisotropy, and short-scale structure may not, in fact, lead to minimum squared difference. A good practice is to fit the nugget effect with a down the hole variogram, then impose that nugget effect on the variogram model in other directions. This will generally make the least-squares measure worse, not better.

Cross validation is sometimes performed to compare alternative variogram models. The comparison is done based on the results of the final objective, which most often is some kind of estimation. The exercise consists of estimating locations with known sample values, and comparing the estimated and the sampled value; the alternative variogram models would result in different sets of estimates; the better variogram model would be the one that yields a lower average error. The are two ways that this exercise can be

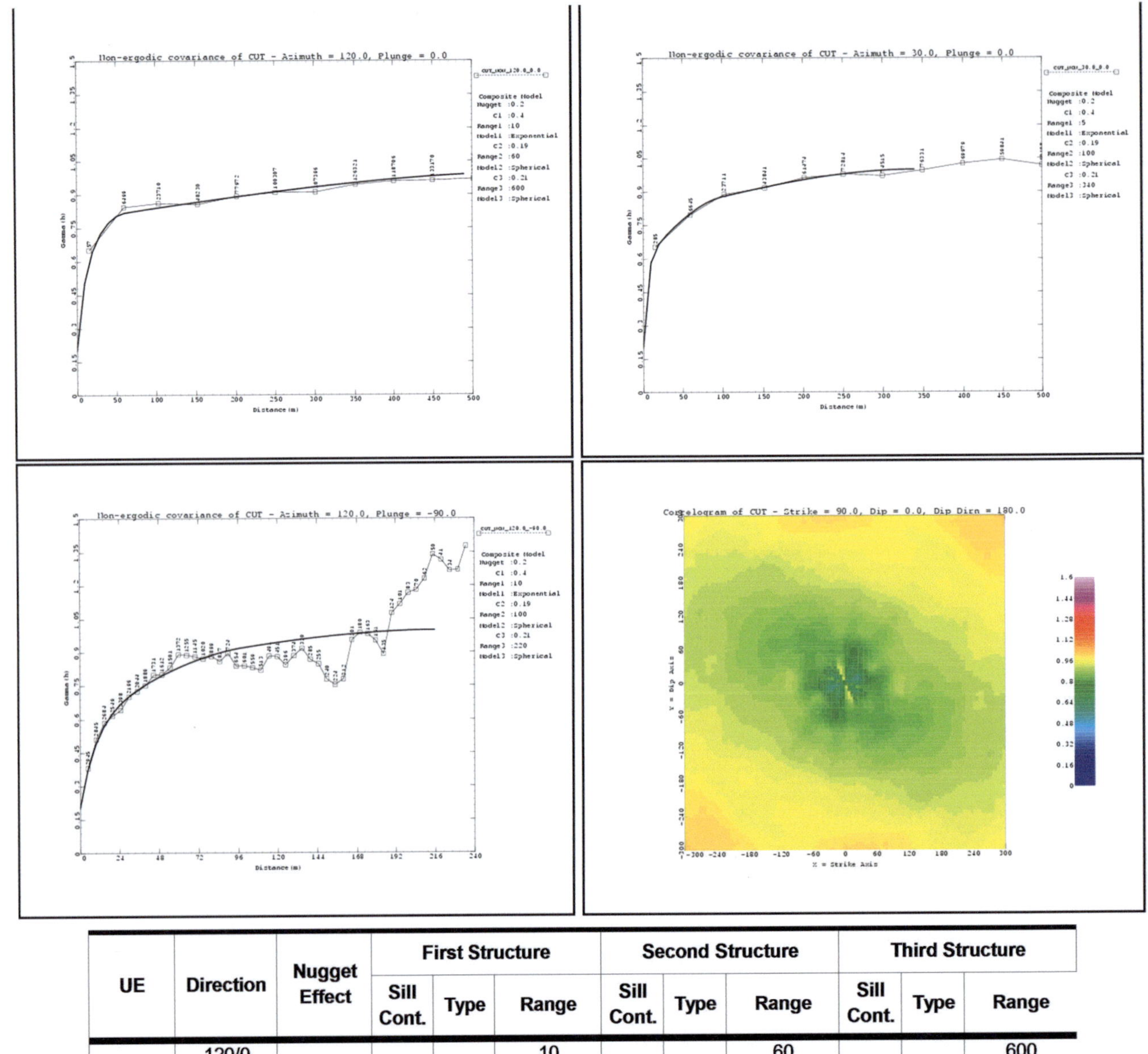

UE	Direction	Nugget Effect	First Structure			Second Structure			Third Structure		
			Sill Cont.	Type	Range	Sill Cont.	Type	Range	Sill Cont.	Type	Range
	120/0				10			60			600
51	30/0	0.2	0.4	exp	5	0.19	sph	100	0.21	sph	340
	120/-90				10			100			220

Fig. 6.9 Cu correlogram for a domain, El Pachón Project, Argentina. Courtesy of Xstrata Copper

performed: (1) a spatial, but otherwise classical, leave-one-out re-estimation, whereby one sample at a time is removed from the data set and re-estimated from the remaining data; to avoid undue influence of the nearest datum in a drill hole, one or more of the closest samples are typically not used in the re-estimation; and (2) a subset of the data (say, 40 or 50 % of the total) is removed completely from the data set, and re-estimated using the remaining data.

There have been a number of objections to the leave-one-out cross validation option (Clark 1986; Davis 1987; Journel 1987; Isaaks and Srivastava 1989; Solow 1990; Goovaerts 1997):

- The method is generally not sensitive enough to detect minor differences from one variogram model to the next. Other kriging parameters such as search strategy, number of data used in the re-estimation, the use or not of octant or quadrant searches, and so on, are generally more consequential than the variogram models themselves.
- The analysis is performed on samples or composites, when in fact we are interested in a different volume support (blocks). This does not allow for any definite conclusion about the final run, since the samples may not be representative of the domain. Even if the variogram model

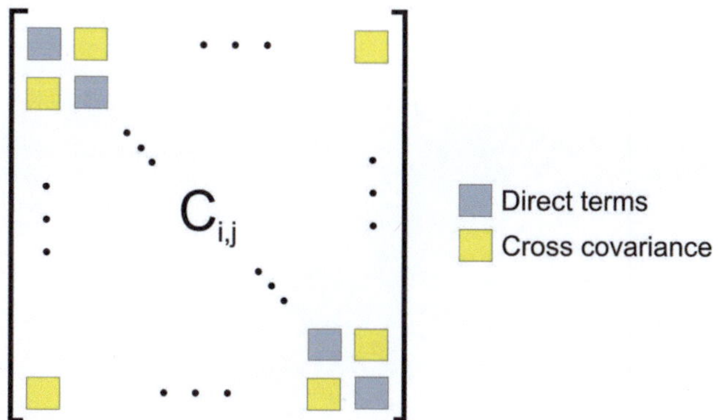

Fig. 6.10 Schematic illustration of a Multivariate Covariance Matrix. The diagonal terms are direct covariances, and the cross-covariances populating the rest of the matrix

performs poorly in the re-estimation, it will not necessarily perform poorly in the final estimation run.

- The sill of the variogram cannot be cross-validated from the re-estimation.
- Variogram values smaller than the minimum distance between samples cannot be validated, such as the nugget effect and the behavior of the model near the origin.

The second option also shares some of the drawbacks mentioned, including the fact that we are re-estimating samples; in addition, there has to be a good number of samples within the domain to be able to split the data in test and a "ground-truth" subsets and carry out the exercise.

It is difficult to define a useful goodness of fit test for a variogram model. The user's experience, subjective geologic information, and consideration of the objectives of the study must be considered to result in a robust variogram model. All subjective decisions must be clearly documented, and is generally preferable to skirting responsibility by relying on blind tests or automatic variogram model fitting software.

6.4 Multivariate Case

Most deposits have more than one variable of interest, and the inference and prediction of one variable can be improved using information from a second, or secondary, variable. Moreover, it is important to respect the relationships between the variables when creating models of more than one variable. The different variables may be of economic value, contaminants, density, or processing characteristics such as hardness. Consider an expanded notation to deal with k = 1,…,K variables.

$$\{Z_k(\mathbf{u}), \mathbf{u} \text{ in } \mathrm{A}, k = 1, \ldots, K\}$$

Quantifying the spatial structure of all *K* variables requires developing the direct variogram models as discussed above, and also the cross-variograms that describes the correlation of one variable to the other. The schematic illustration in Fig. 6.10 below shows the K by K matrix of bivariate relations that must be inferred. There are K direct relationships—one for each variable—and K(K − 1) cross relationships. The cross relationships are almost always taken to be symmetric, that is, the relationship between i and j is the same as j and i. There are some interesting circumstances (referred to as the *lag effect* in some literature) when this is not true.

While calculating the number of variograms is not a major computational effort, the problem lies in the fact that the variograms cannot be modeled independently from each other.

Similar to the case of a single variable, there are a number of permissible models that can be used to model cross correlations (Journel and Huijbregts 1978; Goovaerts 1997). The requirement that must be met is that the variance of each variable is non-negative, and that the matrix of variogram models must be mathematically valid. The linear model of co-regionalization (LMC) is the result of a specific set of correlated variables, and was first proposed by Journel and Huijbregts (1978, p. 172). Other types of co-regionalizations can be assumed, see for example Zhu and Journel (1993), or the Markov models mentioned elsewhere in this book.

The direct and cross variograms between variables k and k' where k, k' = 1,…,K are defined as follows:

$$2\gamma_{k,k'}(\mathbf{h}) = E\{[Z_k(\mathbf{u}) - Z_k(\mathbf{u}+\mathbf{h})] \\ [Z_{k'}(\mathbf{u}) - Z_{k'}(\mathbf{u}+\mathbf{h})]\},\ k,k' = 1,\ldots,K$$

The calculation principles explained above can be applied to the full set of K(K + 1)/2 direct and cross variograms; however, note that the data must be equally sampled, that is, data for variable k and k' must be available at the same data locations. In presence of unequally sampled data, it is necessary to directly compute cross covariances and convert them to variograms for fitting. The covariance would be calculated directly as

$$C_{k,k'}(\mathbf{h}) = E\{Z_k(\mathbf{u})Z_{k'}(\mathbf{u}+\mathbf{h})\} - m_k m_{k'},\ k,k' = 1,\ldots,K$$

The relationship between direct variograms and direct covariances was given above. In case of cross variograms and covariances, the collocated covariance is required to convert the two:

$$\gamma_{k,k'}(\mathbf{h}) = C_{k,k'}(0) - C_{k,k'}(\mathbf{h}),\ k,k' = 1,\ldots,K$$

The collocated covariance between a variable and itself is the variance of that variable: $C_{k,k}(0) = \sigma_k^2$. The collocated cross covariances can be calculated directly when the data are equally sampled. They must be estimated when the data are not equally sampled.

The interpretation of cross variograms is similar to that of direct variograms; however, the sill of a cross variogram is the collocated covariance $C_{k,k'}(0)$, which could be negative. Thus, cross variograms may be both positive and negative; they are the product of differences and not squared differences. The example shown below is for the case of negatively correlated variables; thus when the spatial covariance $C_{k,k'}(\mathbf{h})$ becomes zero, the cross variogram is at the collocated covariance $C_{k,k'}(0)$, which is negative.

In presence of multiple variables, the set of $K(K+1)/2$ direct and cross variograms must be calculated, interpreted, checked for geological reasonableness, then fitted by the linear model of coregionalization (LMC). The LMC implicitly assumes that each variable is a linear combination of common underlying random variables with zero mean. This leads us to model all direct and cross variograms from the same pool of $j=0,\ldots,nst$ nested structures denoted with an upper case $\Gamma^i(\mathbf{h})$, where, by convention, $i=0$ corresponds to the nugget effect:

$$\gamma_{k,k'}(\mathbf{h}) = \sum_{i=0}^{nst} b^i_{k,k'} \bullet \Gamma^i(\mathbf{h}), \quad k,k' = 1,\ldots,K$$

The b coefficients are adjusted to fit the experimental variograms just like the variance contribution parameters are adjusted to fit the variograms of single variables. The anisotropy and range parameters are also adjusted in the specification of each constituent nested structure: $\Gamma^i(\mathbf{h})$. It will be necessary to use negative b coefficients for cross variograms between variables that are negatively correlated. The b coefficients can be adjusted as necessary to achieve a good fit, but the resulting set of $K(K+1)/2$ direct and cross variograms must be jointly positive definite. This is achieved by ensuring that each of the $i=0,\ldots,nst$ K by K matrices of b coefficients is positive definite. There are a number of software programs to ensure this including spreadsheet plugins.

The following oil sands example illustrates a simple application of a cross-variogram. Bitumen content and fines content are two critical factors affecting recovery in the oil sands. Evaluating their spatial variability is of some importance for process control in the extraction plant. Consider the normal score transforms of the two variables, shown in Fig. 6.11:

A model of co-regionalization can be derived to account for the cross correlation. Figure 6.12 shows the cross fines/bitumen variogram.

The first step in fitting is to choose the pool of nested structures $\Gamma^i, i=1,\ldots,n_{st}$. Each nested structure is defined by its type and ranges. It is chosen so that all of the deemed significant features on the experimental variograms can be modeled. The variograms may have different precision in different directions, but it is important to look at all direct

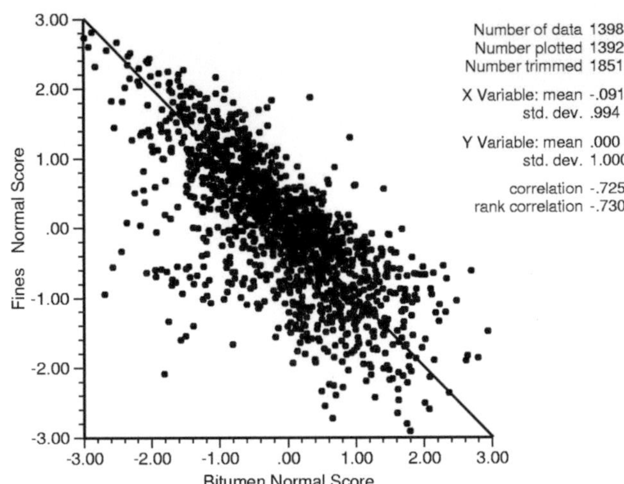

Fig. 6.11 Scatterplot of Gaussian bitumen and fines variables

and cross variograms together. A structure can exist in the direct variogram and not exist in the cross variogram, but any structure that occurs in the cross variogram must be in the direct variograms. The variograms for normal score transforms of % bitumen (Z) and % fines (Y) are shown:

$$\gamma_Z(\mathbf{h}) = 0.3 + 0.3\cdot\Gamma^1(\mathbf{h}) + 0.25\cdot\Gamma^2(\mathbf{h}) + 0.15\cdot\Gamma^3(\mathbf{h})$$
$$\gamma_{Z,Y}(\mathbf{h}) = -0.25 - 0.1\cdot\Gamma^1(\mathbf{h}) - 0.25\cdot\Gamma^2(\mathbf{h}) - 0.1\cdot\Gamma^3(\mathbf{h})$$
$$\gamma_Y(\mathbf{h}) = 0.4 + 0.2\cdot\Gamma^1(\mathbf{h}) + 0.25\cdot\Gamma^2(\mathbf{h}) + 0.15\cdot\Gamma^3(\mathbf{h})$$

where $\Gamma^1(\mathrm{h})$ is spherical with range 200 m, $\Gamma^2(\mathrm{h})$ is spherical with range 1,000 m and $\Gamma^3(\mathrm{h})$ is spherical with range 5,000 m. This is a licit model of co-regionalization since

$$0.3\cdot 0.4 \geq (-0.25)^2, 0.3\cdot 0.2 \geq (-0.1)^2,$$
$$0.25\cdot 0.25 \geq (-0.25)^2, \text{and } 0.15\cdot 0.15 \geq (-0.1)^2$$

The LMC can be applied to any number of variables, and in all cases each matrix of Γ coefficients should be positive definite. For practical reasons, normally no more than 3 or 4 variables are considered simultaneously; otherwise, fewer variables are considered or principal components of the original variables are modeled instead.

6.5 Summary of Minimum, Good and Best Practice

At a minimum, the variogram analysis performed should include models for each variable within each estimation domain defined, an assessment of nugget effects and anisotropies encountered in each case, and a detailed discussion on their geological background. The documentation of the work should be detailed, highlighting the approximations used, the

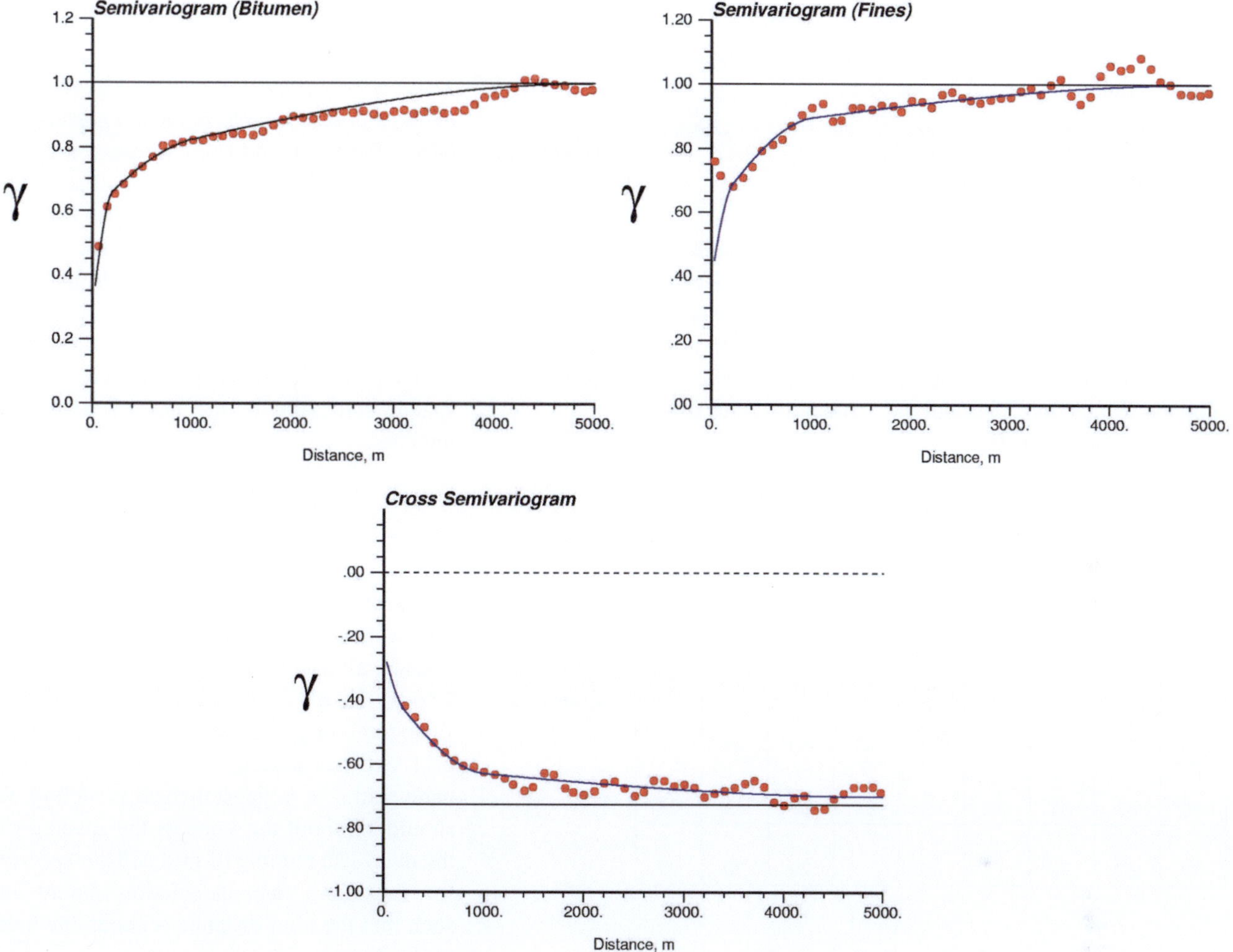

Fig. 6.12 Example direct and cross variograms for an oil sands deposit in the most continuous direction. The direct variograms are shown at the *top* and the cross variogram is shown *below*

data limitations encountered, and the possible artifacts that result from the data. The process of definition of the model should also be explained in detail.

In addition, good practice requires a more detailed exploratory variogram analysis upfront. All relevant data in each estimation domain should be analyzed and used if possible. The definition of relative nugget effects, the models chosen, the anisotropies encountered and how they were modeled, as well as a discussion on the alternatives should be made. In addition, it requires detailed understanding of the uncertainty sources for the resulting models. Alternative models should be tried for the most important estimation domain, and sufficient checks made to have a high degree of confidence on the final model. The description of the variogram models in terms of its geologic significance should be detailed and unambiguous, and all aspects of the model should be demonstrated in terms of geologic factors, and not data artifacts. All the work should be well documented and presented, detailing the quality control procedures in place.

Best practice includes, in addition to the aspects discussed above, the use of all possible geologic, grade, and other ancillary data to obtain the variogram models. Exhaustive exploratory variogram analysis should be done, leaving no parameter or aspect of the model without analysis and justification, and different issues such as data qualities and different data supports analyzed and justified accordingly. All potential sources of variogram model uncertainty should be quantified, and discussions on possible and relevant cross validation and re-sampling should be presented. The models should be presented both mathematically as well as graphically.

6.6 Exercises

The objective of this exercise is to calculate, interpret, and model variograms. Some specific (geo)statistical software may be required. The functionality may be available in different public domain or commercial software. Please acquire the required software before beginning the exercise. The data files are available for download from the author's website—a search engine will reveal the location.

6.6.1 Part One: Hand Calculations

The following data set is taken from *Practical Geostatistics* 2000 (Clark and Harper). You are being asked to review this data and calculate some variograms. The empty boxes have no data. Consider the data to be spaced on a square grid of side a. Consider the horizontal direction (across the page) to be the X direction. Consider the vertical direction (up the page) to be the Y direction.

44		40	42	40	39	37	36	
42		43	42	39	39	41	40	38
37	37	37	35	38	37	37	33	34
35	38		35	37	38	36	35	
36	35	36	35	34	33	32	29	28
38	37	35		30		29	30	32

Question 1: Contour the data, show some basic statistics and comment.

Question 2: Calculate the semivariogram for distances of a and $2a$ in the X direction, distances of a and $2a$ in the Y direction, distances of $\sqrt{2}a$ and $2\sqrt{2}a$ in the 45° direction and distances of $\sqrt{2}a$ and $2\sqrt{2}a$ in the −45° direction. Tabulate and plot your results. Comment.

6.6.2 Part Two: Small Set of Data

Question 1: Consider the thickness in the data file `red.dat`. Normal score transform the thickness and perform calculations with the normal score transform of thickness. Assemble all of the data pairs in Excel (there should be 67×67=4,489 pairs, but only (4489−67)/2=2211 unique pairings). The columns of interest are the distance between the pairs, the values of thickness for both points and the squared difference between the values. Sort the data by the distance. Remove the first 67 rows because the distance and the squared difference should be zero.

HINT: *There is no easting to calculate the distance between points, just elevation and northing.*

Question 2: Group the values into sets (these sets could also be considered lags or bins) of, say, 200 pairs. Calculate the average distance, covariance between the pairs of values, correlation coefficient, and variogram for each bin. Plot the results.

Question 3: Perform reasonable sensitivity studies with different number of lags. Comment on the spatial continuity of the normal scores of thickness.

6.6.3 Part Three: Large Set of Data

Consider the normal scores of Cu data in the `largedata.dat` data file. In practice, the data would be separated by rock type, but consider all data for this exercise.

Question 1: Choose an areal and vertical grid size for variogram map calculation. Note that this grid size has nothing to do with the ultimate grid size in geostatistical modeling. It should be about the same as the spacing of the data. The number of grid cells to specify for variogram map calculation should be such that the total distance is about one half of the domain size.

Calculate the variogram map (variogram volume in 3-D). Plot the horizontal slice through the center of the volume (the example above came from a vein-type gold deposit with 140 intersections). Try a smaller cell size and a larger cell size to ensure that the results are stable with respect to the grid size. Comment on any areal anisotropy and the azimuth directions for subsequent directional variogram calculations. Plot XZ and YZ sections through the variogram map. The only sections through a variogram map that make sense are those through the origin.

Question 2: Given the data spacing in the `largedata.dat` dataset and your work with variogram maps in Question 1, discuss the selection of variogram parameters such as the angle tolerances, lag spacing, and lag tolerance. Establish three directional directional variograms. Experiment with different parameters and establish the stability of your calculated experimental variograms.

Question 3: Fit the directional variograms with a licit variogram model and comment on your results. Comment on aspects of the variogram that are uncertain.

6.6.4 Part Four: Cross Variograms

Question 1: Extend *Part Three* to direct and cross variograms of the normal scores of Cu and Mo, three directional variograms of the normal scores of Mo, and three directional cross variograms between the normal scores transforms of Cu and Mu. Plot the correct sill on the cross variograms and comment on the results.

Question 2: The most difficult aspect of using multiple variables and cross variograms is fitting a model of coregionalization. The only practical model of coregionalization is the linear model of coregionalization (LMC). Recall the LMC and the constraints that it imposes on variogram modeling.

Question 3: Fit an LMC to the variograms you calculated in Question One. Document the procedure you followed and show the final nine experimental variograms with the fitted model.

6.6.5 Part Five: Indicator Variograms for Continuous Data

Consider the same 3-D data as in the previous two parts for indicator variograms.

Question 1: Find or recalculate/remodel the normal scores variogram of Cu. Consider directional variograms.

Question 2: Use the `bigaus` program from GSLIB to calculate the 0.1, 0.25 and 0.9 quantile indicator variograms using the normal scores variogram model. The 0.1 and 0.9 quantile variograms will be identical since the Gaussian distribution is symmetric.

Question 3: Calculate the experimental indicator variograms corresponding to the 0.1, 0.25 and 0.9 quantiles and plot with the results of Question 2. Comment on any differences. Pay particular attention to the 0.1 and 0.9 quantile indicator variograms and any *asymmetry* in the experimental variograms.

Question 4: Three indicator variograms are suitable for checking bivariate Gaussianity; however, nine indicator variograms are more reasonable to discretize the range of variability found in typical continuous variable distributions. Calculate the nine indicator variograms corresponding to the deciles of the Au grade distribution. These variograms should be fit with smoothly varying parameters. Fit the variograms with, say, two spherical variogram structures and plot the parameters as a function of threshold.

References

Armstrong M (1984) Improving the estimation and modeling of the variogram. In: Verly G et al (ed) Geostatistics for natural resources characterization, part I, Reidel, Dordrecht, pp 1–19

Babakhani M, Deutsch CV (2012) Standardized pairwise relative variogram as a robust estimator of spatial structure. Unpublished CCG Research Paper 2012-310

Barnes, RJ (1991) The variogram sill and the sample variance. Math Geol 23(4):673–678

Christakos G (1984). On the problem of permissible covariance and variogram models. Water Resour Res 20(2):251

Clark I (1986) The art of cross-validation in geostatistical applications. In: Proceeding of the 19th APCOM, pp 211–220

Clark I, Harper WV (2000) Practical geostatistics. Ecosse North America, Columbus, p 340

Cressie N (1985) Fitting variogram models by weighted least squares. Math Geol 17(5):563–586

Cressie N (1991) Statistics for spatial data. Wiley, New York, p 900. Reprinted 1993

David M (1977) Geostatistical ore reserve estimation. Elsevier, Amsterdam

Davis BM (1987) Uses and abuses of cross-validation in geostatistics. Math Geol 17(5):563–586

Deutsch CV, Journel AG (1997) GSLIB: geostatistical software library and user's guide, 2nd ed. Oxford University Press, New York, p 369

Goovaerts P (1997) Geostatistics for natural resources evaluation. Oxford University Press, New York, p 483

Gringarten E, Deutsch CV (1999) Methodology for variogram interpretation and modelling for improved reservoir characterization. In: Paper SPE 56654, presented at the SPE annual technical conference and exhibition held in Houston, 3–6 October

Isaaks EH (1999) SAGE2001 User's Manual, Software license and documentation. http://www.isaaks.com

Isaaks E, Srivastava RM (1988) Spatial continuity measures for probabilistic and deterministic geostatistics. Math Geol 20(4):313–341

Isaaks EH, Srivastava RM (1989). An introduction to applied geostatistics. Oxford University Press, New York, p 561.

Journel A (1988) New distance measures: the route towards truly non-Gaussian geostatistics. Math Geol 20(4):459–475

Journel AG (1987) Geostatistics for the environmental sciences: An introduction. U.S. Environmental Protection Agency, Environmental Monitoring Systems Laboratory

Journel AG, Huijbregts ChJ (1978) Mining geostatistics. Academic, New York

Myers DE (1991) Pseudo-cross variograms, positive definiteness and cokriging. Math Geol 23:805–816

Omre H (1984) The variogram and its estimation. In: Verly G, David M, Journel AG, Marechal A (eds) Geostatistics for natural resource characterization: NATO ASI series C, v. 122—Part I. Reidel Publication. Co., Dordrecht, pp 107–125

Reed M, Simon B (1975) Methods of modern mathematical physics, vol II. Academic Press, New York

Rossi ME, Parker HM (1993) Estimating recoverable reserves: is it hopeless? Presented at the Forum 'Geostatistics for the Next Century', Montreal, 3–5 June

Solow AR (1990) Geostatistical cross-validation: a cautionary note. Math Geol 22:637–639

Srivastava RM (1987) A non-ergodic framework for variogram and covariance functions. Unpublished Master of Science Thesis, Stanford University, Stanford, CA

Srivastava RM, Parker HM (1988) Robust measures of spatial continuity. In: Amstrong M (ed) Geostatistics. Reidel, Dordrecht, pp 295–308

Zhu H, Journel A (1993) Formatting and integrating soft data: stochastic imagining via the Markov-Bayes algorithm. In: Soares A (ed) Geostatistics-Troia. Kluwer, Dordrecht, pp 1–12

Mining Dilution

7

Abstract

Dilution is a critical issue that affects many aspects in mining. It is generally due to the geometric characteristics of the ore body, the mining operation, the characteristics of geologic contacts, and the limitations of the mining equipment to recover material to the desired boundaries or contacts. There are three types of dilution that need to be considered at the time of mineral resource estimation. The dilution due to geologic contacts and the dilution due to the mixing of material types within a block are best tackled by geologists and resource estimators at the time of modeling. Operational dilution is generally planned for by the mining engineer at the time of developing a mine plan, but it also occurs unexpectedly, and is called unplanned dilution.

7.1 Recoverable Versus In-Situ Resources

The objective of the resource model is to predict the tonnage and grade that the beneficiation plant will receive at specified time intervals. This is true at all times in a mining operation: at the initial evaluation of the project, as part of pre-feasibility and feasibility studies, and in the context of long-term and short-term resource models in operating mines. The procedures for estimating and managing dilution need to be updated regularly to capture all the new information and experience collected as the deposit is being mined. A model that attempts to satisfy this requirement is called a "recoverable model" (David 1977; Journel and Huijbregts 1978; Rossi and Parker 1993).

A recoverable resource model is an estimate of the tonnage and grade of economic material above certain cutoffs, but can also include other geo-metallurgical and geo-mechanical characteristics that affect mill performance. Revenue is a function of grades, product prices, metallurgical recoveries, and operating costs such as mining, metallurgical, and general and administration (G&A) costs:

$$\text{Revenue} = \text{Price} * \text{Recovery} * \text{Grade(s)} - (\text{Mining Cost} + \text{Metallurgical Costs} + \text{G \& A Costs}) \quad (7.1)$$

The grade for which revenue is nil is called the *break even (or economic) cutoff grade*. Depending on which costs are considered, different types of cutoffs are used. At the break-even point, Revenue in Eq. 7.1 is zero, and the corresponding economic cutoff grade is:

$$\text{Economic Cutof Grade} = \frac{\text{Mining} + \text{Metallurgical} + \text{G \&A Costs}}{\text{Price} * \text{Recovery}} \quad (7.2)$$

Costs are usually expressed on a per unit basis, such as dollars per ton. The units used in the calculation have to be consistent, which often requires conversion factors.

Another important cutoff in an open pit mining operation is the *marginal cutoff*, similar to the economic cutoff, except that the mining cost is not considered. This is to fairly valuate the rock when mining is progressing and the material has to be mined. The only decision is where to send it, the mill, a stockpile, or the waste dump. The mining cost must be spent and is considered a sunk cost. The marginal cutoff is used for example in grade control, as discussed in Chap. 13.

Cutoff calculations become complex if there are several metals to consider, each with different metallurgical recoveries and costs. Also, there may be different mining costs associated with sending material to the mill, as opposed to the waste dumps or stockpiles. In the case of stockpiles, re-handling costs should also be considered. Finally, G&A costs are a mixture of costs, not all of them directly related to the operation. Mining companies have different policies for which

M. E. Rossi, C. V. Deutsch, *Mineral Resource Estimation,* DOI 10.1007/978-1-4020-5717-5_7,

costs to include in these calculations on a project by project basis. For example, the company's headquarters corporate overhead may or may not be included. Each block must be valued separately considering all of the revenues and costs, and then blocks with positive total revenue are considered ore.

In what follows "cutoff" implies the economic cutoff described by Eq. 7.1 above, unless otherwise defined.

At a very early stage of the project the main concern is to determine if the deposit contains enough mineralization to warrant further study and investment, that is, very little may be known about the potential of the deposit to become an operating mine. Technical details and specifications for mine planning and metallurgy are required to estimate tonnages and grades delivered to the mill. In this case, since the proportion of the mineralization that would be recovered is unknown, it is preferable to estimate "in-situ" resources.

Accounting for mine and mill considerations at the time of estimating resources is not yet universally accepted. The sources of dilution and ore loss are well known, but not easily quantifiable. Some practitioners prefer to calculate a model of mineralization without engineering constraints. Dilution then has to be added to the block model by the mine planning engineer, usually using global factors. In general, all resource models should be recoverable.

The differentiation between a recoverable *resource* model and a *reserve* model is based on the wording of the different Resource Classification Systems currently in use (see Sect. 12.3). The term "reserves" is used for material that has been reasonably proven to be minable with an economic benefit. This implies that a well-defined mine plan is in place, that metallurgical studies have proven that the ore is amenable to beneficiation, that there is a viable market for the product, and that there are no legal or environmental impediments for mine development. In addition, a reserve model may include some additional operational dilution not explicitly included in the recoverable resource model.

The available drill hole information has a much smaller volume and scale than mine planning volumes and ore/waste selection. Drill holes are a few centimeters in diameter, and each sample typically represents between 10 and 50 kg of material. In contrast, a very selective open pit mine would consider mining units of 5 × 5 × 5 m (approximately 325 metric tons assuming a 2.65 t/m^3 density), while the larger, massive deposits plan on units that are as large as 25 × 25 × 15 m (approximately 25,000 metric tons). Some underground mines can be more selective, but the volume of the planning unit is still orders of magnitude larger than the drill hole.

The volume of extraction is represented with a "Selective Mining Unit", or SMU. The SMU is defined as the smallest volume that the operation can recover, and depends on the mining method, the equipment size, the data available at the time of selection and the selectivity characteristics of the operation. For convenience, it is generally represented as a rectangular block, even though mines never extract ore and waste as perfect parallelepipeds.

Fig. 7.1 Bucyrus SME 60 Shovel used at the large tonnage Escondida Cu Mine, Northern Chile (photo courtesy of BHP Billiton). Benches are 15 m high

For open pit mines, the vertical dimension of the SMU is the bench height, although occasionally some mines operate on double- or half-bench heights. The lateral dimensions represent the minimum width of the extraction equipment, with consideration to digging depth, the material's angle of repose, the equipment's maneuverability, and the available information to support estimates of the grade at short distances. If it is a massive electric shovel, see Fig. 7.1, with a nominal loading capacity of 90,000 tons of material per day, the minimum width will be about 18–20 m. For such a large operation, the bench height is usually 15 m, and thus the SMU would be 20 × 20 × 15 m.

If the equipment considered is a front-end loader (such as the one shown in Fig. 7.2), the width of the bucket varies between 5.6 and 6.2 m (depending on the model), so it is generally accepted that the minimum width for selectivity will be about 8–10 m. Typical bench height is 10 m, so that a common SMU size for this type of operation could be 10 × 10 × 10 m.

These two examples assume that there would be sufficient grade control sampling and adequate grade control practices to estimate reliable values at the SMU scale mentioned. The SMU size could be bigger with difficult deposits and poor grade control sampling. The ore and waste may be defined by a sharp visual contact. In such cases, the equipment may be able to mine to contacts with only 2 or 3 m of dilution/lost ore.

Underground mining methods vary widely in selectivity. They are often more selective than open pits, but there are significant exceptions, such as mines that use block or

Fig. 7.2 Caterpillar 992 front-end loader used at Cerro Vanguardia's Osvaldo Diez vein. Cerro Vanguardia is a Gold-Silver deposit located in the Patagonia Region of Southern Argentina (photo courtesy of Cerro Vanguardia S.A.). Benches are 5 m high

Fig. 7.3 A hand-specimen approximately 3 inches in size showing typical Porphyry Cu mineralization (Chrysocolla in Type-D veinlet, courtesy of BHP Billiton)

sublevel caving methods. In a traditional cut-and-fill operation, with 5 m lifts, the SMU depends on the geometry of the orebody, but usually is 5×5×5 m, assuming that the mine can separate ore and waste from the stope.

The definition of an SMU is convenient for block modeling, but does not realistically represent the extraction process: shovels and loaders do not load cubes! Moreover, individual SMUs cannot be selected independently although the concept of an SMU assumes *free selection.* The actual practice of ore and waste selection shows that the SMU concept is a convenient approximation. Mining along boundaries is generally more selective than the nominal SMU size for the mine, and typically an isolated SMU-size pod of waste or mineral will not be mined.

7.2 Types of Dilution and Ore Loss

There are several sources of dilution and ore loss. Dilution and ore loss are always closely linked, and references to dilution include both cases. The main sources of dilution may be classified into three different categories (Rossi 2002):

Internal Dilution or Change of Support is a consequence of predicting resources at a different volume than the original data (Parker 1980). The resource estimate requires a degree of averaging within blocks and is generally modeled using the volume-variance or change of support correction, as discussed in detail in the next section. This mixture of material necessarily includes high and low grade mineralization, which will be more significant if the mineralization is less continuous. Also, the larger the block size considered, the larger the amount of mixing of mineralization or internal dilution.

The photo in Fig. 7.3 is a hand specimen of typical Porphyry Cu mineralization, where, within the solid rock mass, high-grade veinlets of Chrysocolla (Cu mineralization) are seen. If this mineralization was to be sampled on a very fine scale, the dispersion of the Cu grades resulting from the laboratory assays could be represented by a distribution like the one shown in Fig. 7.4, top. If the sample volume taken were to be larger, then there would be more mixing of material in any given sample, thus the higher-grade veinlets being mixed with the lower grade material surrounding them. In this case, a distribution like the one shown in Fig. 7.4 (bottom) may be obtained.

Note that the means of the distributions are the same (grades are mass fractions and they scale up linearly, so that the overall average is maintained), but the standard deviation and coefficient of variation is smaller for the larger volume distribution. Also, the minimum and the maximum of the distribution are closer to the overall mean. There is also a general tendency for the larger-volume distribution to be more symmetric than the original distribution.

Since mineralization is not homogeneous, mixing of different grade material always occurs. This is true for all types of mineralization, and depends on the nature of the geologic events that produced the mineralization. The presence of mineralized veinlets, highly fractured zones or units, and more or less permeable lithologies impact the amount of internal dilution to be expected.

Geologic Contact Dilution is defined as the dilution and ore loss resulting from the extraction of material of different geologic characteristics. This type of dilution can often be accounted for when using sub-cells or partial blocks in the definition of the resource block model (Chap. 3): the grades and other characteristics of each geologic unit that comes

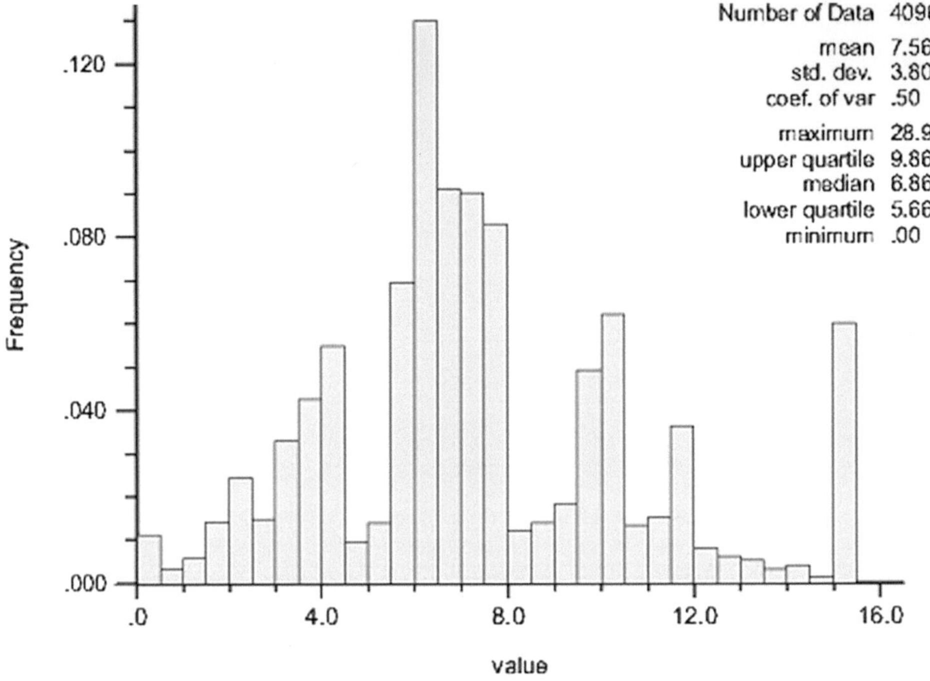

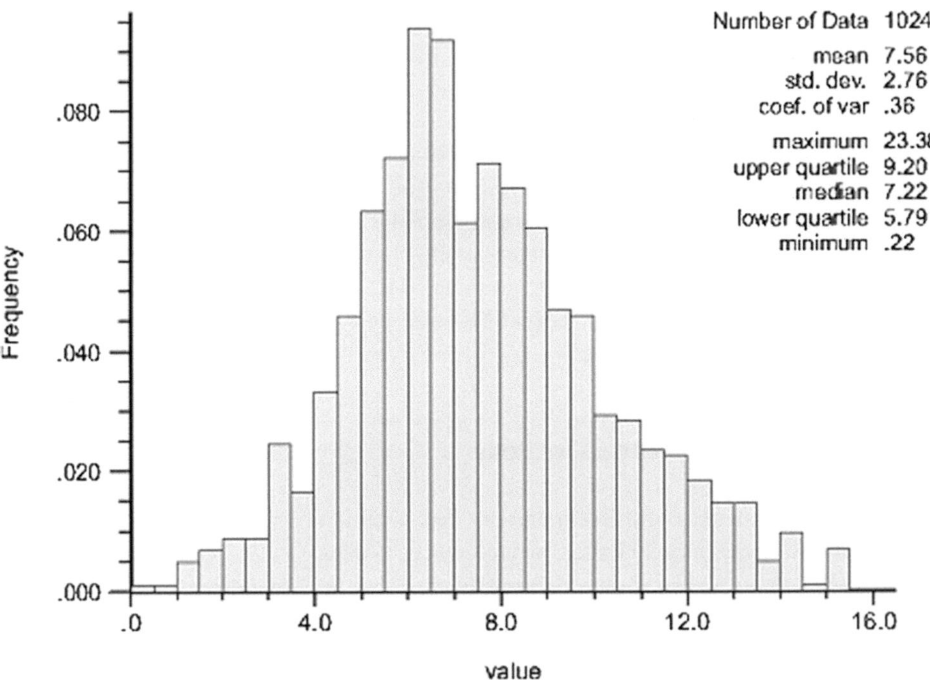

Fig. 7.4 The point distribution above is corrected to a block (SMU) distribution below

into contact within each block can be averaged according to the proportions of each within the mining blocks of the model.

The impact and relative importance of this type of dilution depends on the geometry of the boundaries between geological units and the differences in grade between units. In high tonnage, massive base metals deposits the impact of geologic contact dilution will be small if compared to deposits with complicated geometries, such as vein type or skarn deposits, or a stratigraphically controlled deposit with significant folding and faulting. For a fixed block size, say an SMU, contact dilution can be characterized for individual geologic zones or estimation domains by the ratio of surface contact volume (SCV) to the overall extraction volume (V), SCV/V, as measured by the volume represented by blocks with geologic contacts to the overall volume of the unit.

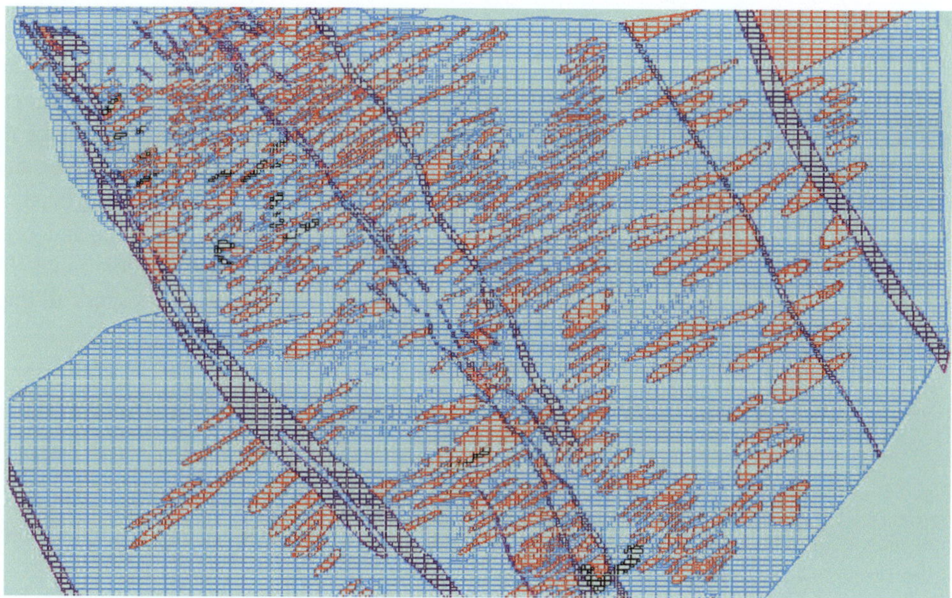

Fig. 7.5 Sectional view of a deposit with a pseudo-stratigraphic control. The lithology units are represented by *red* (volcanic breccias) and *blue* (andesites), with cross cutting dykes (in *purple*). Blocks are 5 × 5 m and can be used for scale; the vertical extension shown is about 800 m. The block model (with sub-cells) is overlaid on the geologic model; supporting drill holes are not shown. Courtesy of Minera Michilla S.A., Chile

This unit-less factor provides an indication of how important contact dilution may be. A ratio of 0.05 or higher generally indicates high contact dilution, and is characteristic of vein-type, skarns, or thin, tabular deposits, while values less than 0.01 correspond to bulk tonnage, massive, or porphyry type deposits.

For massive deposits, contact dilution is generally a local issue, since the bulk of the tonnage will be mined away from contacts, and therefore its importance from a global resource model may be limited. Still, it can impact the positioning of a final pit wall or stope, as well as the corresponding volume of waste that needs to be removed to access the ore (mining strip ratios). It is a very different case for skarn-type and small, narrow tabular or vein-type deposits, where contact dilution may be the most consequential type of dilution.

Figure 7.5 shows a cross section of a lithology model for the Lince-Estefanía Cu deposit, with the corresponding block model with sub-cells overlaid on the view. Notice how the general stratigraphy is crosscut by intrusive dykes. Also notice that, by virtue of the relative high contact surface area to volume ratio, the impact of geologic contact dilution is likely to be significant. The contact dilution can be incorporated into the block model using two alternative but conceptually similar techniques:

1. The sub-cell method, as shown in Fig. 7.5, provides a better definition of the geologic contacts. As discussed in Chap. 3, these sub-cells are then re-blocked to the parent block size of the model to provide the diluted grades and maintaining the proportions of each geologic unit within each block.
2. A direct calculation of the proportion (percentage) of each unit within each block, storing the percentage of each unit within the block.

The average grade of the block is expressed as the proportion-weighted average of the grades of each individual geologic unit within the block:

$$Z_V^* = \sum_{i=1}^{n} p_i \cdot z_i^* \tag{7.3}$$

where z_v represents the block grade average, $\boldsymbol{p_i}$, $\boldsymbol{i=1,\ldots,n}$, represent the percentage of total mass for each of the $\boldsymbol{n}$ geologic units that may be present in the block, and z_i represent the grade of each individual unit within the block.

Another, less desirable option, is to empirically introduce into the block model factors that penalize the grades of blocks at or near contacts, according to pre-specified criteria. This was done, for example, for one of the Escondida Mine's resource models. In this method, if a contact between a high grade and waste geologic zones passes through any given block, the grade of that block is downgraded arbitrarily. The limitations of this procedure are significant, since the factors applied are empirical and global, as opposed to diluting according to the locally estimated grades.

Another method that can be used to estimate dilution and ore loss due to geologic contacts is to draw ore envelopes around the mineralized zones, and then estimate an overbreak, or additional volume for mining. This can be done on sections or benches, and provides an estimate of the total grade and tonnage of material that will be recovered. A similar method is also used by mining engineers to estimate operational dilution. The method is best suited for deposits with well-defined ore zones with hard boundaries, such as vein type or epithermal Au deposits.

Geologic contact dilution is quantified from the geologic model. Thus, the local accuracy of the contact dilution estimate depends on the quality of the geologic model.

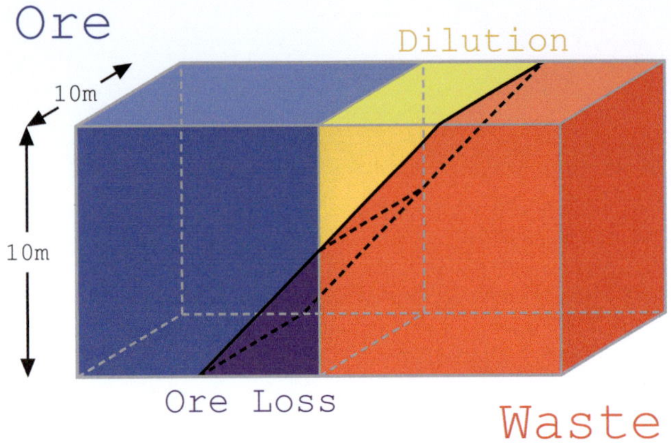

Fig. 7.6 Schematic of operational mining dilution and ore loss. Dilution and ore loss are represented for a bench height of 10 m and an angle of repose of broken ore of 45°. The overall volume of each is 125 m^3 if a 10 × 10 × 10 m SMU is assumed

Operational Mining Dilution includes dilution and ore loss that occurs at the time of mining. Mining equipment unavoidably mixes material, because the precision with which the equipment can follow a dig line is limited, even with Global Positioning Systems (GPS). If the ore/waste contacts correspond with the geologic contacts, operational and contact dilution is the same. More commonly, however, the contacts of ore and waste that occur at the time of mining are defined in economic terms, and they do not necessarily follow geologic contact zones.

One possible estimate of this type of dilution can be obtained by simple geometric calculations. Figure 7.6 illustrates the case of an open pit mine, where the dilution and ore loss is incorporated into the resources considering a specific bench height and assuming an angle of repose for the material. The total metal lost depends on the characteristics of the contact, including the grade of ore lost and the grade of the diluting material. A good reference for quantification of dilution for underground deposits from a mine planning perspective can be found in Pakalnis et al. (1995).

Another source of dilution and ore loss is blast heave and movement that shifts the position of the material to be mined and complicates the modeled dig-lines. Significant research has been done in this area (Yang and Kavetsky 1990; Harris 1997; and Zhang 1994), but to date there are few operations that attempt to accurately quantify and account for blast heave.

Ore loss and dilution also occurs when the extracted material is transported to the wrong destination: waste sent to the mill, or ore sent to the dumps. Control equipment such as GPS and Truck Dispatch systems has reduced the frequency of this error, but the destination control problem persists and can be significant.

Sometimes it is important to distinguish between planned and unplanned dilution; there may be some unexpected operational practices in the mine that are increasing dilution. In some, ore losses and dilution are accounted for using factors obtained from some degree of production reconciliation, and applied to the resource model globally.

A well-planned geostatistical conditional simulation study, as discussed in detail in Chap. 10 can be used to help understand dilution and ore loss (Guardiano et al. 1995; GeoSystems International 1999). Such a conditional simulation study can address all three types of dilution.

7.3 Volume-Variance Correction

Internal dilution is sometimes modeled using geostatistical tools for volume-variance correction. The most common distribution shape change methods for volume-variance correction are the Affine Correction, the Indirect Lognormal, and the Discrete Gaussian methods. These methods correct a distribution of a grade attribute sampled at an initial support (often called the point scale distribution) into an SMU block distribution. These analytical methods are fast and generally applicable to small scale changes. Classical references on these methods include Journel and Huijbregts (1978) and Isaaks and Srivastava (1989).

The relationship between volume and variance is shown in Fig. 7.7. The variance decreases as the volume increases due to the averaging out of high and low values. The averaging is affected by the size and shape of the volume, the continuity of the variable, and the averaging process. For most variables in mining, since they average arithmetically, the mean does not change as the volume increases and the variance of the distribution decreases. There are exceptions, however, mostly when considering some geotechnical and metallurgical performance variables.

The point distribution of an attribute will have a larger variance than the block distribution of the same attribute. The corrections described in this section apply to the distribution of samples within a chosen estimation domain. The goal is to take the representative distribution of point scale data and infer a global block or SMU distribution.

The traditional variance defined in Chap. 2 is the squared difference of the samples with respect to the overall mean implicitly states the support size (samples). A more general Dispersion Variance is defined as:

$$D^2(v,V) = \sigma^2(v,V) = \frac{1}{n}\sum_{i=1}^{n}(z_{i,v} - m_V)^2 \qquad (7.4)$$

where v represents a smaller support such as the sample size, and V represents a larger block support mean value, such as the stationary population or the SMU-sized block distribution.

The dispersion variance quantifies the reduction in variance for specific increases in volume. The dispersion

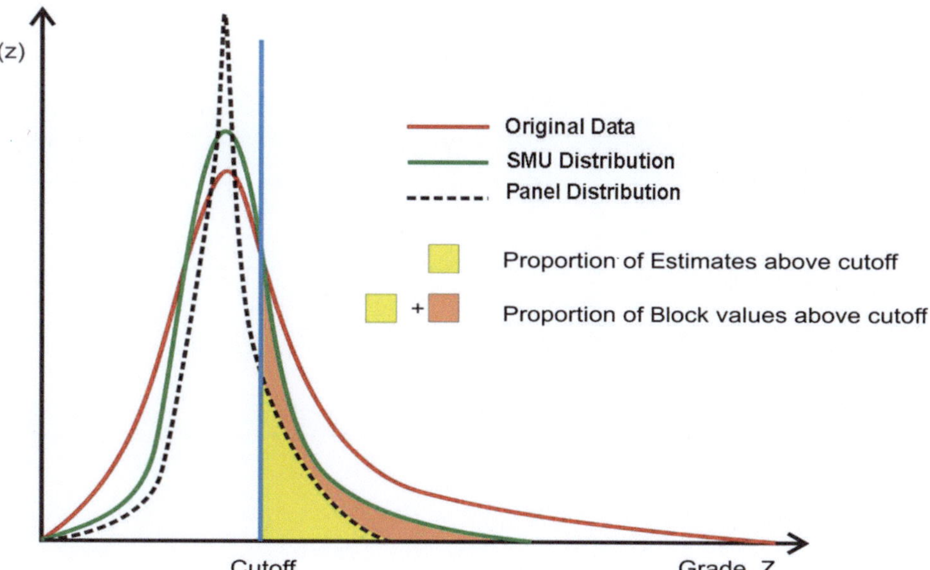

Fig. 7.7 Schematic showing volume-variance relations for original data, SMU-sized distribution, and a larger panel distribution

variance is the same expected squared difference as the variance defined before, except that it is related to specific support sizes for the data and the mean.

The dispersion variance can be expressed as a function of average covariances or variograms, see Isaaks and Srivastava (1989) or Journel and Huijbregts (1978):

$$D^2(v,V) = \overline{C}(v,v) - \overline{C}(V,V) \tag{7.5}$$

where $\overline{C}(v,v)$ and $\overline{C}(V,V)$ are the average covariance values for the samples at smaller sample support v and the SMU support respectively, as defined in Chap. 2. Note that these are spatial averages, and therefore are location-independent.

The additive property of variances leads to the following expression:

$$D^2(v,G) = D^2(v,V) + D^2(V,G), \quad \forall\, v \subset V \subset G \tag{7.6}$$

where v, V, and G represent increasingly larger volumes.

Equation 7.6 states that the variance of samples within a deposit can be found as the sum of the variance of samples within blocks of a certain size plus the variance of those blocks within the deposit. This relationship was found experimentally by D. Krige in the 1950's, and is thus often called Krige's relation.

In Eq. 7.6 two terms are usually known: (1) the variance of the data $D^2(v,G) = \sigma^2$ and (2) the variance within blocks $D^2(v,V)$, which can be estimated from the covariance or variogram model (Eq. 7.1). The variance between blocks (for example, the SMU variance within the Deposit, $D^2(V,G)$) can be obtained.

The variance within blocks ($D^2(v,V)$) is obtained from discretizing the SMU block V using n_v sample points, and calculating the average covariance ($\overline{C}(V,V)$) or variogram value for all possible pairs within the block. The number of discretization points used to estimate $D^2(v,V)$ affects somewhat its final value. As a rule of thumb, it is generally accepted that a $5 \times 5 \times 5$ grid of points within the SMU block is sufficient to obtain a robust estimate of $D^2(v,V)$. Considering too many discretization points could lead to numerical precision problems. One option is to obtain the dispersion variance for several discretization grids. Figure 7.8 shows the resulting dispersion variance for a given variogram model and SMU size for several discretizing grids. Note how the dispersion variance stabilizes after a reasonable number of discretization points have been used.

The dispersion variance is a key parameter needed to predict recoverable resources (recall Sect. 7.1). The volume-variance correction is often characterized by a single parameter, known as the Variance Correction Factor (VCF). The VCF (or more simply, f) is defined as the ratio of the SMU block variance to the original sample variance:

$$\begin{aligned} VCF = f &= \frac{D^2(V,G)}{D^2(v,G)} = \frac{D^2(v,G) - D^2(v,V)}{D^2(v,G)} \\ &= 1 - \frac{D^2(v,V)}{D^2(v,G)} \end{aligned} \tag{7.7}$$

The factor f is a measure of how much the variance of a sample distribution will change, therefore giving an idea of the importance of the volume-variance correction in the estimation of recoverable resources. An f value close to one implies that the variances of samples within the deposit $(D^2(v,G))$ and of SMU blocks $(D^2(v,G))$ within the deposit are fairly similar. This is either because the SMUs are small (small volume, highly selective mine), or the spatial distribution is fairly continuous, that is, there is relatively little mixing of

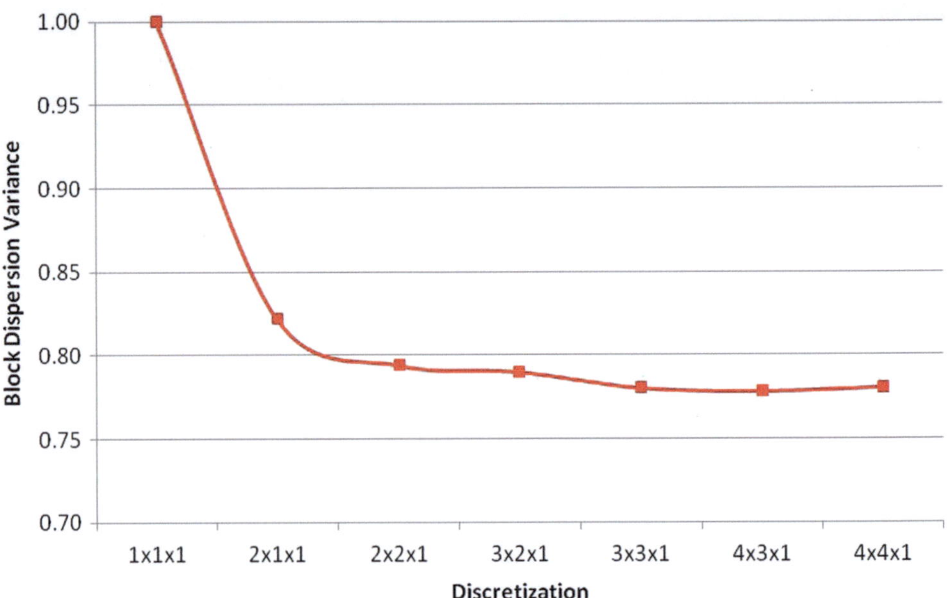

Fig. 7.8 An example of block dispersion variances resulting from different discretization grids. The variogram model and the block size are fixed. The discretization in Z is always 1 because bench height and composite length are the same in this example. Note that a $3 \times 3 \times 1$ grid in this case is sufficient to approximate the block dispersion variance

high and low grades within an SMU. The opposite is true for low f values.

As volume increases from the data support to an SMU support, the mean stays the same and the variance changes by a predictable amount (summarized in the f factor). The shape of the distribution also changes. The influence of the central limit theorem is felt to some extent, since the average of identically distributed values tends to a normal distribution. The grades inside an SMU in fact are not independent; therefore, the distribution of SMU grades does not always approach a normal distribution.

7.3.1 Affine Correction

The affine correction is the simplest method for volume-variance correction. It is based on the concept that the distribution does not change its shape while the variance is reduced, therefore assuming that there is no increase in symmetry of the resulting distribution. Although there is no additional explicit assumption about the point and SMU distributions, the permanence of shape assumption is limiting, since it is known that the distribution shape will change as the variable is averaged within larger volumes. Therefore, in practice, the range of application of this method is limited to small changes in variances, for which changes in distribution shape are small.

The affine correction works by transforming each value of the sample distribution into a different value of the SMU distribution, according to the following relationship:

$$z' = \sqrt{f} \cdot (z - m) + m \tag{7.8}$$

where z is any value of the original distribution, z' is the corresponding value of the SMU distribution, f *is the variance correction factor*, and m is the mean of both sample and SMU distributions.

According to Journel and Huijbregts (1978, p. 471), the affine correction can be applied up to about a correction factor of 30% ($f > 0.7$), although in the experience of these authors this is optimistic. Even for volume-variance corrections much smaller than 30% the affine correction seems to provide the wrong prediction, see Rossi and Parker (1993) and the example below.

7.3.2 Indirect Log-normal Correction

The indirect log-normal correction (ILC) is based on the idea that the change of support is described by two Log-normal distributions that have the same mean, but different variances. This is considered true regardless of the characteristics of the two original distributions (point and SMU support), except that they need be positively skewed.

The quantiles of the original distribution are transformed into the SMU distribution following an exponential equation:

$$q' = aq^b \tag{7.9}$$

with the coefficients a and exponent b given by:

$$a = \frac{m}{\sqrt{f \cdot CV^2 + 1}} \left[\frac{\sqrt{CV^2 + 1}}{m} \right]^b$$

and

$$b = \sqrt{\frac{\ln(f \cdot CV^2 + 1)}{\ln(CV^2 + 1)}}$$

where m is the mean, CV is the coefficient of variation of the point distribution, and f is the variance correction factor (VCF) previously defined.

However, since the distributions will not in general be exactly lognormal, then the transformation of Eq. 7.9 will not result in the same mean for the transformed and untransformed distributions. So, a final step is required to ensure that the original mean is obtained:

$$q'' = \frac{m}{m'} \cdot q' \tag{7.10}$$

After applying Eq. 7.10, the quantiles of the SMU distribution have been rescaled to the correct mean. Interestingly, the differences between the first transformed mean and the rescaled mean can be used as a measure of the dissimilarity between the original distribution and a Log-normal distribution. The final correction may cause the variance to be slightly different than the target variance.

7.3.3 Other Permanence of Distribution Models

As a generalization of the previous methods, the same principle can be applied to other distributions, most practically to those that are characterized by two parameters, such as the Gaussian, Lognormal, and even Gamma distributions.

Under the assumption that a sample distribution can be approximated by a multivariate Gaussian distribution, then the resulting block distribution will also be multi-Gaussian, with the same mean and corrected variance, as described before.

Similarly, the sample distribution can be assumed to be multi-Lognormal, in which case the resulting SMU distribution is also assumed to be multi-Lognormal (although, as in the case of the affine correction, this is an assumption known to be incorrect), with the same mean and corrected variance.

As these methods have had little use in practice, the reader is referred to Journel and Huijbregts (1978, pp. 468–469) for the specific formulae and further details on the limitations of these methods.

7.3.4 Discrete Gaussian Method

The permanence of distribution assumption is a limitation because most real-life mining distributions cannot be easily fitted with a two-parameter distribution (Gaussian or Lognormal). They have multiple modes and mixtures of populations that can only be overcome by using a method that makes no such assumption. The discrete Gaussian model (DGM) has been proposed as a more robust method to obtain the volume-variance correction.

The key idea of the DGM is that the distributions for different supports will be Gaussian after transformation to Gaussian units. The transformation to Gaussian units is achieved in two steps: (1) a normal scores transformation like that described in Chap. 2, then (2) fitting the relationship between the original grades and the normal scores transform with a series of Hermite polynomials. These polynomials are orthogonal, which is important because the variance of the original grades is then a simple summation of the squares of the coefficients. A change to the variance is achieved by scaling the coefficients of the Hermite polynomials by a change of support coefficient related to the factor f. As expected, the corrected distribution gradually becomes more Gaussian in shape as the scale increases.

The fitting of Hermite polynomials and the details of the mathematics are embedded in widely used computer programs and documented in references such as Armstrong and Matheron (1986); Rivoirard (1994) or Machuca-Mory et al. (2007). An overview will be presented here. An anamorphosis function needs to be fit to the sample data. The anamorphosis function is defined by a Hermite polynomial expansion fit to the data. Hermite polynomials are related to the Gaussian distribution and are defined by Rodrigues' formula (Abramovitz and Stegun 1964, p. 773). The anamorphosis function is equivalent to the normal score transformation in that it provides a mapping of the point variable Z to the Gaussian variable Y and vice-versa:

$$z(\mathbf{u}) = \Phi(y(\mathbf{u})) \approx \sum_{p=0}^{\infty} \Phi_p H_p(y(\mathbf{u}))$$

where Φ_p is the coefficient of each polynomial term, and $H_p(y(\mathbf{u}))$ is the Hermite polynomial value. This fitting can be thought of as a polynomial fit to the Q-Q plot between the original grades and the normal scores.

The anamorphosis function is fit by calculating the value of the Φ coefficients of the Hermite polynomials. The first coefficient is simply the mean of the Z samples:

$$\Phi_0 = E\{\Phi(Y(\boldsymbol{u}))\} = E\{Z(\boldsymbol{u})\}$$

Higher order coefficients are found with the following approximation:

$$\begin{aligned}\Phi_p &= E\{Z(\mathbf{u}) \cdot H_p(Y(\mathbf{u}))\} \\ &= \int \Phi(y(\mathbf{u})) \cdot H_p(y(\mathbf{u})) \cdot g(y(\mathbf{u})) \cdot dy(\mathbf{u}) \\ &\approx \sum_{\alpha=2}^{n} (z(u_{\alpha-1}) - z(u_\alpha)) \cdot \frac{1}{\sqrt{p}} H_{p-1}(y(\mathbf{u}_\alpha)) \cdot g(y(\mathbf{u}_\alpha))\end{aligned}$$

where $g(y(\mathbf{u}_\alpha))$ is the probability value y corresponding to a standard Gaussian distribution. Since the polynomials are orthogonal and thus there is no correlation between them, the variance of the Z samples can be identified to:

$$Var\{\Phi(Y(\boldsymbol{u}))\} = Var\{Z(\boldsymbol{u})\}$$

$$\approx \sum_{p=1}^{n}\sum_{q=1}^{n} \phi_p \phi_q \operatorname{cov}\{H_p(Y(\boldsymbol{u})), H_q(Y(\boldsymbol{u}))\} = \sum_{p=1}^{n} \phi_p^2$$

The modeled anamorphosis function can be checked against the original data by comparing the distributions resulting from the samples to the distribution from the anamorphosis. The distributions should be identical, although in practice extreme values can be difficult to model.

Then, the sample histogram at the SMU block support is obtained using the bi-Gaussian assumption. To correct the sample distribution to a predicted-SMU distribution the anamorphosis function is modified by adding a change of support coefficient r:

$$Z(v) = \Phi(y_v(v)) \approx \sum_{p=0}^{\infty} r^p \cdot \Phi_p H_p(Y(\boldsymbol{v}))$$

The calculation of r requires the dispersion variance of the SMU-sized blocks, in obtained from the variogram model derived from samples values (Chap. 7). The anamorphosis function corresponding to the SMU support v assumes that the distribution of $[Y(\mathbf{u}), Y(\mathbf{v})]$ is bi-Gaussian, and is found with:

$$\sigma_v^2 = \sigma_u^2 - \bar{\gamma}_{v,v} \approx \sum_{p=1}^{n}\sum_{q=1}^{n} r^p \phi_p r^p \phi_q \operatorname{cov}$$

$$\{H_p(Y(\mathbf{u})), H_q(Y(\mathbf{u}))\} = \sum_{p=1}^{n} r^{2p} \phi_p^2$$

from which the r coefficient can be obtained. The distribution of grades representing SMU volumes is easily determined with the obtained r coefficient, the fitted coefficients and the Hermite polynomials. Although apparently complex, the procedure is automated and widely available in different programs.

The DGM is deemed to be more robust than the affine or indirect lognormal correction because the normal scores transform is general, and no additional assumptions are necessary for the original or the SMU distributions.

7.3.5 Non-Traditional Volume-Variance Correction Methods

There are other methods used for volume-variance correction, some of them empirical. These range from adjusting the kriging plan used to estimate the blocks to get the predicted dispersion variance, to the use of probabilistic estimation techniques (Chap. 9), to the application of conditional simulations (Chap. 10).

7.3.6 Restricting the Kriging Plan

The concept is based on tuning the kriging plans to control smoothing to match the resulting block distribution to the expected SMU distribution as closely as possible.

This method was proposed originally by Parker and is discussed in Rossi and Parker (1993) and Rossi et al. (1993). It utilizes the notion that the smoothing property of kriging (see Chap. 8, and Journel and Huijbregts 1978, pp. 450–452) can be controlled to obtain an estimated block distribution that closely matches the predicted SMU distribution. Certain parameters of the kriging plan, such as search neighborhoods, minimum and maximum number of samples and drill holes, the use or not of octant searches, etc. can impact the degree of smoothing of the resulting block distribution.

Restricting the kriging plan has the advantage of being simple, although rarely the kriged block distribution will match exactly the desired SMU distribution. More commonly, the matching is achieved for certain cutoffs of interest along the grade-tonnage curve. It is local in the sense that the method is estimating individual block grades, which combine to form a distribution similar to the desired SMU distribution.

One of the disadvantages of the method, as pointed out by Journel and Kyriakidis (2004), is that it is specific to each mineral deposit, and cannot be formulated in general terms. Also, the increased restrictions on the Kriging plans result in higher variance of the resulting block distribution, typically at the expense of higher conditional bias. The spatial distribution of estimates is still smooth, that is, the variogram of the estimates will show a significantly lower nugget effect and continuous behavior at the nugget effect.

It is important to note that the requirement of conditional unbiasedness of the kriged block model is incompatible with the requirement of predicting tons and grade received at a future date by the processing plant, see for example Isaaks and Davis (1999) and Isaaks (2004). This has been empirically verified in practice. Still, too much conditional bias in the output kriged model can lead to significant prediction biases that should be avoided.

The SMU estimates at this time are interim estimates awaiting much more data from blast hole sampling or infill drilling. At the time of final estimation for grade control, care should be taken to avoid conditional bias. It is often more important at the prefeasibility and feasibility stage of resource estimation to get predictions that reasonably reflect the recoverable resource that will ultimately be obtained.

7.3.7 Probabilistic Estimation Methods

Several probabilistic estimation methods, described in detail in Chap. 9, can be used to incorporate the volume-variance effect into the resource estimation process.

One option is to modify the point probability distributions resulting from the multiple indicator kriging (MIK) technique into block probability distributions using either an affine, ILC, or DGM correction. A variant of is the procedure has been used by Newmont Gold at its Gold Quarry mine in Nevada (Hoerger 1992), which, appears to work reasonably well when there is sufficient production data for a correct calibration.

A different option within the application of MIK is to apply the volume-variance correction to a cumulative probability distribution, at the composite scale, resulting from MIK. The compositing refers here to simply averaging the MIK probability distribution values to larger panels. A discussion of this method can be found in Chap. 9 and in Journel and Kyriakidis (2004).

Methods used to estimate distributions that are based on the Gaussian or Lognormal assumptions are also applied to incorporate the volume-variance effect into the resource estimation model. The available options include Multi-Gaussian Kriging (Verly 1984), Disjunctive Kriging (Matheron 1976) and its derivative, Uniform Conditioning (Roth and Deraisme 2000), and the Lognormal Shortcut methods (David 1977). The change of support models afforded by these methods is generally robust, as long as the corresponding underlying Gaussian or Log-normal assumptions are reasonable.

The volume-variance correction methods described share in the same limitations: they do not account for other types of dilution and the information effect. They assume that every block can be selected individually and independently from any other (free selection), and that the selection itself is made based on a known true grade (perfect selection).

7.3.8 Common Applications of Volume-Variance Correction Methods

The methods for volume-variance correction described are applied to ore resource modeling in several manners. The traditional application has been the correction of the global resource model to match the predicted grade-tonnage curve according to the volume-variance effect predicted (David 1977; Journel and Huijbregts 1978). This application is now less common for multiple reasons:

a. The volume-variance correction performed in such a way is a global correction, and therefore of little practical use, except for the overall assessment of resources from a deposit; the mineralization's internal dilution should be somehow incorporated into the resource block model based on more local corrections, so that downstream work, such as mine planning, takes its effect into account.
b. Forcing the overall resources to match the volume-variance corrected distribution implies ignoring all other dilution sources described above. Therefore, the reported overall resources are known to be wrong, since they are based on the incorporation of a single source of dilution. The resource model should incorporate more dilution than predicted by volume-variance correction to include geologic contact dilution, the information effect, and planned operational dilution.

Another application is correcting the drill hole data such that an estimate of the expected SMU distribution is obtained prior to estimating the resources. This provides a target distribution against which the resource model can be compared.

The example shown in Fig. 7.9 corresponds to the Cerro Vanguardia operation, which mines gold and silver vein deposits in the Patagonia Region in Southern Argentina. Figure 7.9 shows the distributions of the 2 m composites used for estimation, as well as the DG-predicted and the Affine-predicted SMU distributions. Note that in this case, the SMU is a $5\times10\times5$ m cube, to account for the open pit mining method currently used. The example shown is from the Osvaldo Diez vein, one of more than 40 Au-Ag bearing veins identified in the district, and the source of most of the mine's production through the late 1990's and early 2000's. It is instructive to note several points:

- The graph in Fig. 7.9 shows the Au cutoff grades applied to the distribution on the X axis, the left Y-axis shows the predicted proportion of tonnage above the corresponding cutoff, while the right Y-axis shows the corresponding grade above cutoff.
- The grade-tonnage curves allow an immediate analysis for the cutoffs of interest, and how the distributions change for different grade ranges.
- The volume-variance correction factor is estimated at 28 %, implying that there is a very significant change in variance from the original 2 m composite to the $5\times10\times5$ m SMU distributions.
- The Affine correction is not the appropriate method to use in this case. It is presented here to highlight the differences in the resulting distributions. Among other reasons, the artificial minimum generated by the Affine correction is quite high, and, although not shown here, the DG model was proven by production data to be more robust.
- The difference between the tonnage and grades for any given cutoff between the SMU distributions and the composites distribution is an indication of the how severe the predicted volume-variance correction is.

In the literature there are several other detailed examples and comparisons of the different volume-variance corrections, see for example Verly (2000) and Rossi and Parker (1993).

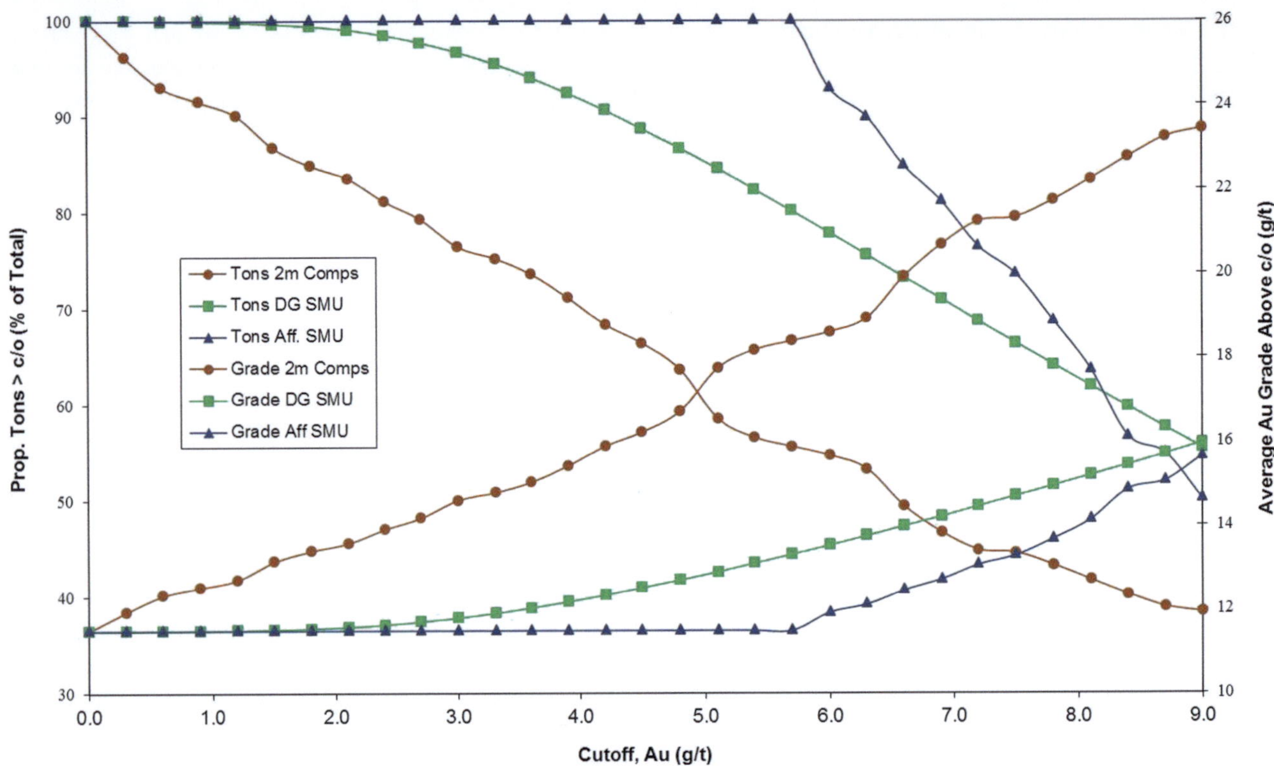

Fig. 7.9 Grade-Tonnage Curves for the Osvaldo Diez Vein, Cerro Vanguardia Mine, Argentina. There is a high volume-variance effect. The 2 m composites distribution is shown along with the DG-predicted and affine-predicted SMU distributions

The volume-variance correction of drill hole information for each estimation domain can also provide a target global distribution of blocks (SMUs), grade-tonnage curves that can be used to calibrate and/or check the grade-tonnage curves resulting from the resource block model, and in particular for specific cutoffs. The comparison between the actual versus target distributions can also be done through distribution parameters, such as the Coefficient of Variation (CV), a robust measure of variability.

Figure 7.10 shows a comparison of the grade-tonnage curves of the DGM-predicted SMU and the estimated block model grades for the high enrichment units of the Escondida Norte Porphyry Copper deposit. Note that for most cutoff grades the estimated grades of the block model are slightly smoother than the corresponding DG-predicted SMU distribution. The conclusion from Fig. 7.10 is that the estimated resource model is incorporating additional dilution, besides the internal dilution represented by the DG model. In this case, the SMU size is $20 \times 20 \times 15$ m, 15 m composites were used to estimate the block model, and the cutoffs of interest are in the range of 0.3 to 0.7 % Cu.

Another application of the volume-variance correction is to help define the selectivity of the mine. This can be approximated by quantifying the impact that different mining equipment used in the operation has on dilution, and based on changes in the volume of the SMU. Most commonly, operations study the impact of changes in bench heights. However, there are limitations to the use of volume-variance methods to predict optimal bench heights, because of the free and perfect selection assumptions.

7.4 Information Effect

The Information Effect describes the fact that, at the time of mining, the information used to decide which portion of the deposit is ore and which is waste is based on more information than that available when obtaining a resource model.

Ore/waste selection is described in more detail in Chap. 13. Although more data is available, the ore/waste selection is always made with an estimate and not the true grades. This is imperfect selection in the sense that an estimation error is always present. Additionally, the selection process is not free, meaning that each SMU is not selected as ore or waste independently of other SMUs in the vicinity. There may be other geometrical and mining constrains that restrict the accessibility of each SMU. All these approximations and sources of error are implicit in the Information Effect.

The problem of selection can be mathematically described by the following recovery equations:

$$i_v(\mathbf{u}; z_c) = \begin{cases} 1 \ if & z_v(\mathbf{u}) \geq z_c \\ 0 \ if & z_v(\mathbf{u}) < z_c \end{cases}$$

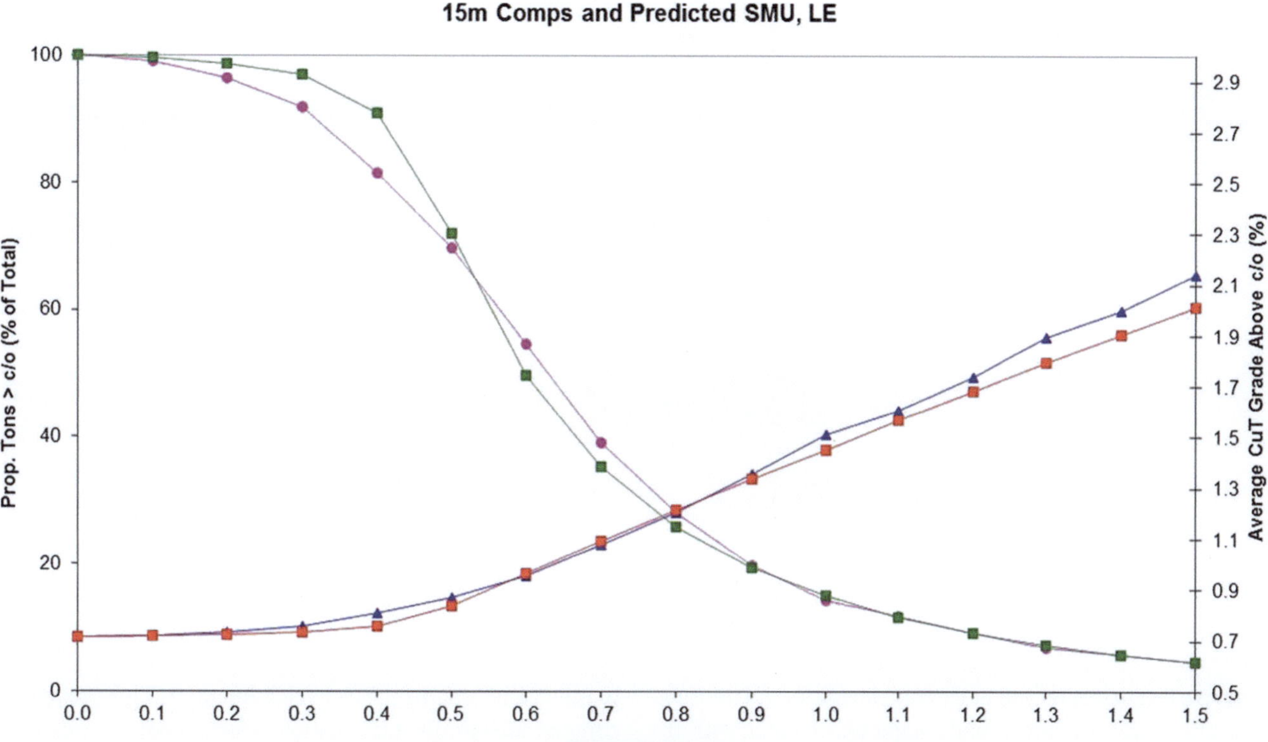

Fig. 7.10 Grade-Tonnage curves of the high secondary enrichment units of the Escondida Norte Porphyry Cu deposit

where $i_v(\mathbf{u}; z_c)$ represents an indicator of perfect selection for the SMU v and z_c is the cutoff grade. If the value of the SMU z_v is higher than the cutoff, then the SMU is recovered $\left(i_v^p(\mathbf{u}; z_c) = 1\right)$. The total tonnage, quantity of metal, and grade thus recovered for any panel or region V is

$$t_v(z_c) = \sum_{j=1}^{N} i_v(\mathbf{u}_j; z_c),\ v \in [1, N];\ \mathrm{x_j} \in V \quad (7.11)$$

$$q_v(z_c) = \sum_{j=1}^{N} i_v(\mathbf{u}_j; z_c) \cdot z_v(\mathbf{u}), v \in [1, N];\ \mathbf{u}_j \in V \quad (7.12)$$

$$m_v(z_c) = \frac{q_v(z_c)}{t_v(z_c)} \quad (7.13)$$

For simplicity, the density (tonnage factor) in the above equations is assumed to be 1.0. Equations 7.11–7.13 assume perfect selection, that is, knowledge of the true SMU value. However, in reality, only an estimate of that true value is available.

Graphically, the ore/waste selection problem can be represented by a scatter plot of the unknown true SMU values vs. the estimated SMU values shown in Fig. 7.11. Consider, for example, a $z_c = 2.0$ cutoff; there are four possible outcomes:

a. The SMU is estimated to be ore and is recovered as such; in this case, no error (or misclassification) is made (Quadrant I).
b. The SMU is estimated to be waste, and is recovered as such; as before, no error (or misclassification) is made (Quadrant IV).
c. The SMU is estimated to be ore, and is in fact waste; in this case, dilution is sent to the processing plant (Quadrant II).
d. The SMU is estimated to be waste, and is in fact ore (Quadrant III); in this case, ore loss occurs as economic material is being discarded.

The imperfect selection described is a major component of the information effect. The economic performance of any operating mine is impacted by this unavoidable selection error. Commonly, little attention is paid to optimizing that selection, relative to its economic impact.

The simple scenario shown in Fig. 7.11 becomes more complicated if there are several destinations for the ore, such as crushed ore to the mill, crushed ore to the leach pad, and Run-of-Mine ore to a different leach pad. In this case, there are four possible destinations including waste. Optimal procedures for ore/waste selection are discussed in Chap. 13.

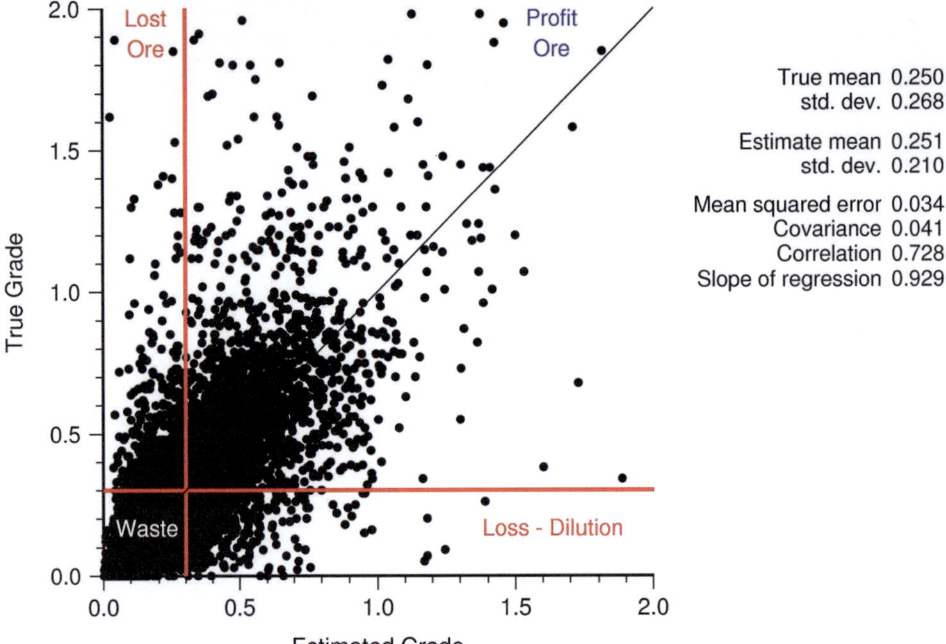

Fig. 7.11 Scatter Plot of Hypothetical True vs. Estimated SMU values. The $Z_c=0.3$ cutoff value defines 4 quadrants in the graph, two if which correspond to misclassification. (SMUs represented by dots)

Imperfect selection and other components of the information effect are difficult to understand and predict with the often-used empirical models. The better alternative is to use geostatistical conditional simulations (Chap. 10), which allows the reproduction, based on simulated data, of the entire process of blast hole sampling and ore/waste selection, as discussed in Chaps. 10 and 13, and exemplified in Chap. 14. This approach has been used successfully in recent years in practice (Guardiano et al. 1995; Badenhorst and Rossi 2012) and Journel and Kyriakidis (2004).

Variants of the probabilistic estimation methods discussed above in the context of volume-variance correction (based on Gaussian or Log-normal assumptions) can be modified to incorporate the information effect. One such method is advocated by Roth and Deraisme (2000), and is based on a Bi-Gaussian assumption between the true, unknown SMU value, and its estimate. The Uniform Conditioning method (as well as others) can be applied to incorporate a correction to the predicted SMU grades and tonnages above cutoff.

Besides the more complete and complex conditional simulation approach, there are several ad-hoc methods that deal with the information effect. One such method, commonly used, is to conservatively bias the ore resource model (similar to what is shown in Fig. 7.10) to compensate for the information effect and future losses. This entails purposefully introducing a certain degree of dilution in the resource model. As all empirical methods, it can only be successfully applied if there is sufficient knowledge about the deposit and valid production data to adequately calibrate the amount of additional dilution incorporated into the model.

A conceptually similar method consists in defining an SMU larger than the SMU that the operation can realistically mine, and assume perfect selection on it. This procedure compensates for the information effect and the fact that the theoretical SMU can never be selected (extracted) perfectly, without any further ore loss and dilution. The impact of assuming a larger-than-expected SMU can be quantified in terms of additional dilution incorporated into the model.

These empirical methods are subjective, and rely heavily on assumptions that cannot be easily verified or quantified. As such, they should be considered only approximations to the incorporation of the information effect into the resource model.

The amount of additional data available at the time of ore/waste selection is significantly more than that available at the time of developing a resource model for a feasibility study. Therefore, predictions about mineable tonnage and grades for economic cutoffs can be much different and improved at the time of selection, if only because of the massive amount of information available.

7.5 Summary of Minimum, Good and Best Practices

The minimum practice in modeling resources requires the following:

a. All models should have an assessment of the global internal dilution by estimation domains. This assessment should be used to quantify the impact of internal dilution, and compare it with the dilution introduced into the block model due to the smoothing property of Kriging.
b. The geologic contact dilution should also be included through geometric considerations if deemed important enough, or discussed in the documentation of the model if considered negligible. The methods used could include

the use of factors to penalize block values along contacts. A more direct approach is preferred, estimating the grade of each geologic unit within the block and then obtaining the average block grade using Eq. 7.1.

c. The information effect is usually handled with factors, sometimes calibrated to production figures, and often applied by mining engineers to the ore resource model at the time of developing the mine plan. In any case, the block model documentation should clearly state its limitations in terms of dilution, and to what extent it can be considered "recoverable".
d. If an indirect or empirical method has been used to incorporate additional dilution into the model to compensate for planned and unplanned operational dilution, such as using a larger SMU size, this should be clearly stated in the documentation.

In addition to the above, good practice requires:

a. A more specific method to include internal dilution into the resource model. This can be done through any of the methods mentioned in Sect. 7.3, and in all cases should include a fair assessment of the uncertainties and tradeoffs involved.
b. Geologic contact dilution should be explicitly incorporated into the block model, and a statement about the uncertainty of the position of the contacts should be included. The information effect should be dealt with using at least a reasonable empirical approximation, or a modification of the estimation method.
c. All the work should be well documented and clearly presented, detailing the checks performed and the quality control procedures in place.

Best practice consists of using uncertainty models to deal with all three types of dilution described: block averaging, geologic model uncertainty, and operational dilution. The full conditional simulation study would:

1. Incorporate the uncertainty of the geologic model, thus implicitly considering geologic dilution.
2. The internal dilution is more accurately incorporated by direct block simulation or simply by averaging the simulated values into the SMU size.
3. The simulation model should also incorporate operational dilution and the information effect by simulating the complete mining process.

Thus, most of the possible sources of dilution and ore loss are modeled simultaneously. In such case, it is not necessary to apply any of the volume-variance correction methods, unless it is done as checks on simulation models, for example. The work is only completed when, as always, a very thorough validation and checking of the models is completed and documented. Preferably, the simulations models should be validated against production, or at least alternative models, and through thorough statistical and graphical checking, see Chap 11.

7.6 Exercises

The objective of this exercise is to review change of support calculations. Some specific (geo)statistical software may be required. The functionality may be available in different public domain or commercial software. Please acquire the required software before beginning the exercise. The data files are available for download from the author's website—a search engine will reveal the location.

7.6.1 Part One: Assemble Variograms and Review Theory

You will use the Cu variable from the `largedata.dat` dataset. The key parameter in all scaling is the variogram; however, the normal scores transforms of grades do not average linearly and we cannot use the normal scores variograms for scaling. The variograms of the Cu grades directly are required. Of course, the direct grade variogram should be similar to the normal scores variogram.

Question 1: Compute and fit a 3-D Cu variogram (like that modeled in Chap. 6). Comment on the "stationarity" of the variogram model, that is, does it flatten off at the variance of Cu grades?

Question 2: Write a short review of the key theoretical results needed for variogram scaling: (1) the definition of the average variogram/ average covariance, (2) the definition of the dispersion variance and the link to the average variogram, (3) krige's relation or the additivity of variance, and (4) the scaling of variogram sill parameters.

Question 3: Derive the volume scaling law of the nugget effect, that is, demonstrate that the following relation is exact: $C_V = |v|/|V|\ C_v$. Where C_V and C_v are the nugget effects at scales V and v, respectively.

7.6.2 Part Two: Average Variogram Calculation

Average variogram or "gammabar" values tell us the variance at any scale. The discretization required for stable numerical integration is a consideration. Average variogram values can be calculated between two disjoint volumes V and v'; however, classic histogram and variogram scaling requires the average variogram to be calculated for $V=v'$, that is, for the same volume and itself. This brings up the zero effect as another complicating factor.

Question 1: Consider your reference Cu variogram model and a 10 m cubed block size for a number of sensitivity studies. Create a plot with the average variogram versus discretization level (starting with $1 \times 1 \times 1$ and going to $20 \times 20 \times 20$). Plot two lines—one with the zero values for coincident discretization points and another for this corrected.

Question 2: Calculate the average variogram for regular cubic block sizes from 1 through 20 m with the zero effect correctly handled. Comment on your choice of discretization level. Plot and tabulate (1) the average variogram versus block size, and (2) the block variance versus block size.

7.6.3 Part Three: Change of Shape Models

The global mean does not change with scale. The variance changes in a predictable manner; however, the shape change is not precisely known.

Question 1: Consider cubic block sizes of 5, 10, and 20 m. Calculate the scaled distributions using the (1) affine, (2) indirect lognormal, and (3) discrete Gaussian models. Plot the original Cu histogram and all of the scaled histograms. Comment on the results.

Question 2: Attempt to quantify the importance of the shape change by plotting grade tonnage curves at the 10 m scale. Discuss the different models and explain where you require such a model.

References

Abramovitz M, Stegun I (1964) Handbook of mathematical functions. Dover Publication, New York, p 1046 (9th print)

Armstrong M, Matheron G (1986) Disjunctive kriging revisited (Parts I and II). Math Geol 18(8):711–742

Badenhorst C, Rossi M, (2012) Measuring the impact of the change of support and information effect at Olympic Dam. In: Proceedings of the IX international geostatistics congress, Oslo, June, Springer, pp 345–357

David M (1977) Geostatistical ore reserve estimation. Elsevier, Amsterdam

GeoSystems International Inc. (1999) Conditional simulation study for the michilla mine. Unpublished Internal Report, Minera Michilla S.A.

Guardiano FB, Parker HM, Isaaks EH (1995) Prediction of recoverable reserves using conditional simulation: a case study for the fort knox gold project, Alaska. Unpublished Technical Report, Mineral Resource Development, Inc.

Harris GW (1997) Measurement of blast induced rock movement in surface mines using magnetic geophysics. Unpublished M.S. Thesis, Department of Mining Engineering, University of Nevada Reno

Hoerger S (1992) Implementation of indicator Kriging at Newmont Gold Company, In: Kim YC (ed) Proceedings of the 23rd international APCOM symposium, published by the Society of Mining, Metallurgy, and Exploration, Inc., Tucson, April 7–11, pp 205–213

Isaaks EH (2004) The kriging oxymoron: a conditionally unbiased and accurate predictor, 2nd ed. In: Proceedings of geostatistics banff 2004. Springer, 2005, 1:363–374

Isaaks EH, Davis B (1999) The kriging oxymoron, Presented at the 1999 Society of Mining Engineers Annual Convention, Denver

Isaaks EH, Srivastava RM (1989) An introduction to applied geostatistics. Oxford University Press, New York, p 561

Journel AG, Huijbregts ChJ (1978) Mining geostatistics. Academic Press, New York

Journel AG, Kyriakidis P (2004) Evaluation of mineral reserves: a simulation approach. Oxford University Press, New York

Machuca-Mory D, Babak O, Deutsch CV (2007) Flexible change of support model suitable for a wide range of mineralization styles, Transactions, Society of Mining Engineering, SME

Matheron G (1976) A simple substitute for conditional expectation: the disjunctive kriging, In: Guarascio M, David M, Huijbregts C (eds) Advanced geostatistics in the mining industry. Reidel, Dordrecht, pp 221–236

Pakalnis R, Poulin R, Vongpaisal S (1995) Quantifying dilution for underground mine operations. Annual meeting of the Canadian Institute of Mining, Metallurgy and Petroleum, Halifax, May 14–18, 1995

Parker HM (1980) The volume-variance relationship: A useful tool for mine planning. In: Geostatistics. McGraw-Hill, pp 61–91

Rivoirard J (1994) Introduction to disjunctive kriging and non-linear geostatistics. Claredon Press, Oxford, p. 190

Rossi ME (2002) Recursos Geológicos o Reservas Mineras? In: Proceedings from the Sextas Jornadas Argentinas de Ingeniería de Minas, San Juan, Argentina, Mayo 30–Junio 1

Rossi ME, Parker HM (1993) Estimating recoverable reserves: is it hopeless? Presented at the Forum 'Geostatistics for the Next Century', Montreal, June 3–5

Rossi ME, Parker HM, Roditis YS (1993) Evaluation of existing geostatistical models and new approaches in estimating recoverable reserves, XXIV APCOM'93, Montreal, October 31–November 3

Roth C, Deraisme J (2000). The information effect and estimating recoverable reserves, In: Kleingeld WJ, Krige DG (eds) Proceedings of the sixth international geostatistics congress, Cape Town, April, pp 776–787

Verly G (1984) Estimation of spatial point and block distributions: The multiGaussian model. Ph.D. Dissertation, Department of Applied Earth Sciences, Stanford University

Verly G (2000) Accounting for mining dilution and misclassification in resource block modeling. In: Kleingeld WJ, Krige DG (eds) Proceedings of the sixth international geostatistics congress, Cape Town, April, pp 788–797

Yang RL, Kavetsky A (1990) A three dimensional model of muckpile formation and grade boundary movement in open pit blasting. Int J Min Geol Eng 8:13–34 (Chapman y Hall, London)

Zhang S (1994) Rock movement due to blasting and its impact on ore grade control in Nevada open pit gold mines. Unpublished M.S. Thesis, Department of Mining Engineering, University of Nevada Reno

Recoverable Resources: Estimation

8

Abstract

The prediction of the tonnages and grade of ore recoverable with a particular mining plan is a central problem in mineral resource estimation. The conventional approach to this problem is to estimate the mineral grade for volumes relevant to the mining plan and base the recoverable resource calculations on those estimates. Details of that approach are presented in this Chapter.

8.1 Goals and Purpose of Estimation

In general, less than one billionth of the volume of a deposit is sampled. The grades and other attributes must be estimated in the unsampled region. Geological variability makes this estimation difficult. There are a number of estimation schemes designed for different goals. One goal could be simplicity and reproducibility—polygonal areas of influence or inverse distance may be suitable. Another goal could be to reveal large-scale geologic trends—block kriging, splines or inverse distance could be appropriate. Avoiding preventable errors is critical (Dominy et al. 2002).

The most important goal, however, is to predict future grade and tonnage of material that may be mined. There are two situations of importance to us: (1) *interim* estimation with widely spaced data when additional information will be available before final decision making, for example, long term estimates in an open pit context before grade-control samples, and (2) *final* estimation for the purposes of selecting ore and waste. For example, at the time of grade control in an open pit or when estimating stope grades in an underground mine with limited future flexibility. The local precision of estimates in the first case is not the top priority; the emphasis is on the accuracy of the estimated global recoverable reserves. The global results are not the emphasis in the second case; local precision and exactitude matters.

The goal in the first situation of interim estimation is to have the tonnages and grade of ore predicted accurately over a reasonably large production area or time period. The other important factor is that additional information will become available in the future. The goal of final estimation is to calculate estimates that are correct in expected value, that is, the true value recovered will be equal to the estimates in expected value. There may be additional goals such as correct classification (see Chap. 13).

The recommended approach to estimation is different for interim and final estimates. The rationale behind the difference is partially explained with the concept of conditional bias.

8.1.1 Conditional Bias

Conditional bias occurs when the expected value of the true grade (Z_V) conditional on the estimated grade ($Z^*_V=z$) is not equal to the estimated grade (McLennan and Deutsch 2004):

$$E\{Z_V \mid Z_V^* = z\} \neq z$$

where the V is symbolic of some volume of estimation, for example, a selective mining unit (SMU).

Conditional bias is almost always present due to the smoothing effect of all linear estimation procedures, including kriging, in the presence of sample data that are *widely* spaced. The true grade is typically less than the estimated grade when the estimated grade is high and the true grade is typically greater than the estimated grade when the estimated grade is low. It is interesting to note that simple kriging creates smooth estimates and has no conditional bias. Common practice, however, is to use ordinary kriging and consider limiting the search to adapt to local departures from stationarity; ordinary kriging will always have some conditional bias.

M. E. Rossi, C. V. Deutsch, *Mineral Resource Estimation,* DOI 10.1007/978-1-4020-5717-5_8,

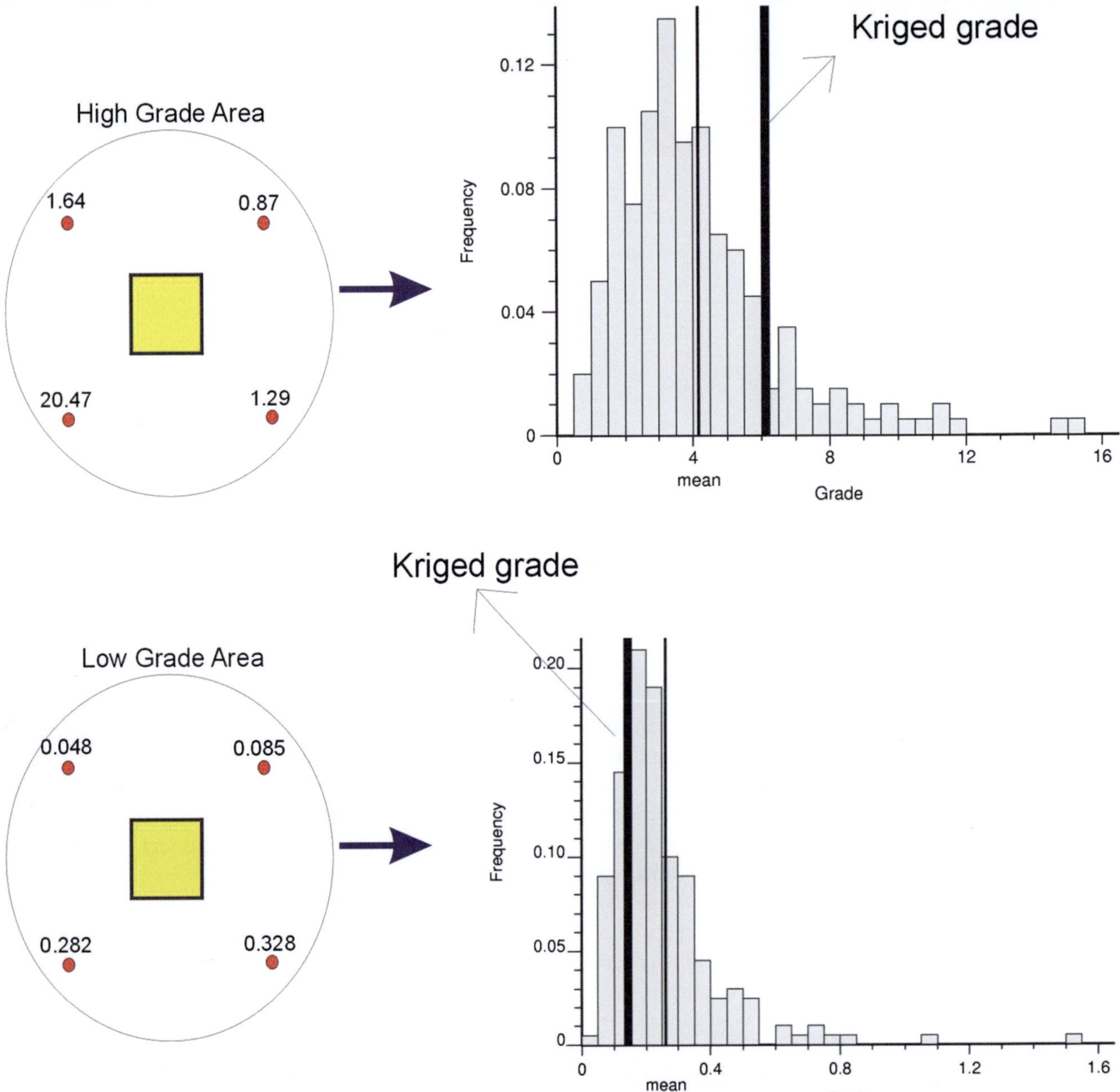

Fig. 8.1 The sketches to the left show the data configuration and a central block being estimated (10 m on a side) in a high and low-grade case. The histograms show the distributions of true grades conditional to the 20 m spaced sample data. The heavy vertical line is the kriged grade and the light vertical line is the mean true grade. The kriged grade is too high in the high-grade case and too low in the low-grade case

Figure 8.1 shows a small example of conditional bias. Consider mineralization with a lognormal histogram with a mean of 1.0 and a variance of 4.0. The variogram has a relative nugget effect of 20% and an isotropic range of 40 m. Consider four samples on a regular 20 m grid to estimate a central 10 m square block in a high and low-grade case. This is a favorable estimation scheme since the nugget effect is relatively low and the range is large relative to the sample spacing. Nevertheless, there is conditional bias, that is, the kriged grade is too high in the high-grade case and too low in the low-grade case.

Conditional bias is largely (but not completely) removed by using many samples in the estimation. This may be a good idea when making *final* estimates, because it would not be reasonable to accept estimates that are known to be wrong in expected value. Large search routines retaining many data are implemented to minimize conditional bias and provide the best estimate. However, in mining, final estimates are obtained using closely spaced data, which means that there is seldom a good application for large searches and many samples. Typical blast hole spacing in an open pit can vary from 5×5 to 10×10 m patterns. The price of large searches and many samples is that block estimates are very smooth; the farther apart the drill holes, the greater the smoothing. In some special circumstances, the price to pay for smoothing may be unavoidable if making *final* estimates, but it will generally be unacceptable for *interim* estimates.

One approach for interim estimates is to modify the estimation procedure to match the smoothing to the ultimate block distribution. The block distribution is predicted analytically, see Chap. 7. The search in estimation is reduced in size to increase the variability of the estimates, using an iterative approach until the adequate amount of smoothing is achieved. These estimates will have conditional bias, but that is not a concern since decisions are being made on the accuracy of the overall grade distribution, and not on block by block estimates.

Although there are many papers on conditional bias, the subject continues to be poorly understood and controversial in the geostatistical community. Authors like Krige (1994, 1996, 1999) claim that resource estimates should have no conditional bias at all. Sinclair and Blackwell (2002) argue that conditional bias contributes to the discrepancies between resource models and production. Guertin (1984) and Pan (1998) proposed different types of corrections for conditional bias. Isaaks (2004) argues that a conditionally unbiased estimate that is also an accurate recoverable resource predictor, except for one theoretical case that in practice is never found, is an oxymoron.

These authors have found that smoothing in mineral resource (interim) estimates in long term open pit mine planning is less acceptable than conditional bias, and in fact that some conditional bias is necessary for the resource estimates to better predict the mined tonnages and grades. These authors have also found that for final estimates using tightly spaced data, large search radii and many samples do not improve the estimates. The practitioner should understand the purpose of the estimates and strive to manage and understand the consequences of smoothing and/or conditional bias.

There are alternate methods to avoid smoothing. Chapter 9 presents methods based on probabilistic estimation that avoids considering a single kriged estimate as the block grade that will be encountered. Chapter 10 presents methods based on simulation, which by construction do not smooth. In Chap. 13 it will be argued that, for ore/waste selection and grade control, simulation-based methods are preferable to any form of kriging.

8.1.2 Volume Support of Estimation

In some cases we are interested in point estimation, that is, estimation at the scale of the data. The smoothing of most estimation from widely spaced drill holes implies that the variability of the estimates is not the same as the variability of the data. However, in mining, the overriding interest is in estimating a certain selective mining unit (SMU), a volume of material of a specific size that characterizes mining selectivity. The definition of the selective mining unit (SMU) volume size is *the minimum volume of material on which ore and waste can be separated, which is a function of mining method and selectivity*. This size is related to the ability of the equipment to select material; but it is also based on the data available for ore/waste classification (blast holes and/or dedicated grade control drilling), the procedures used to translate that data to mineable dig limits, and the efficiency with which the mining equipment excavates those dig limits.

Several sources of dilution must be accounted for, including internal dilution due to grade variability within the SMU, external dilution resulting from geological and geometric contacts, and planned and unplanned operational dilution. Dilution and estimation domains definition (Chap. 4) are the two most important factors for accurately estimating recoverable resources.

Recoverable resources imply that we are interested in evaluating a truncated statistic of the overall grade distribution. The classical formulae are found after defining an economic cutoff for any set of SMU estimates (Journel and Huijbregts 1978, p. 480). The tonnage is simply the sum of all unit tonnages (or area of the histogram) that are above that threshold:

$$T(z_0) = T_0 \cdot [1 - F_Z(z_0)]$$
$$= T_0 \cdot \int_{z_c}^{+\infty} f_Z(z)dz = T_0 \cdot \frac{1}{N_A} \sum_{i=1}^{N_A} t_i(\mathbf{u}_i; z_c)$$

where T_0 is the total in-situ tonnage at cutoff 0 and z_c is the grade cutoff applied.

The quantity of metal is calculated as the summation of the quantity of metal of each individual unit:

$$Q(z_0) = T_0 \cdot \int_{z_c}^{+\infty} z \cdot f_Z(z)dz = T_0 \cdot \frac{1}{N_A} \sum_{i=1}^{N_A} z \cdot t_i(\mathbf{u}_i; z_c)$$

where z is the grade of the unit. Finally, the average grade of the recovered material is:

$$m(z_0) = \frac{Q(z_0)}{T(z_0)}$$

8.1.3 Global and Local Estimation

The estimation methods mentioned here yield local estimates, in the sense that the estimated values are specific to a location within the deposit, and are derived from nearby samples. A global estimate is an estimate for an entire domain or

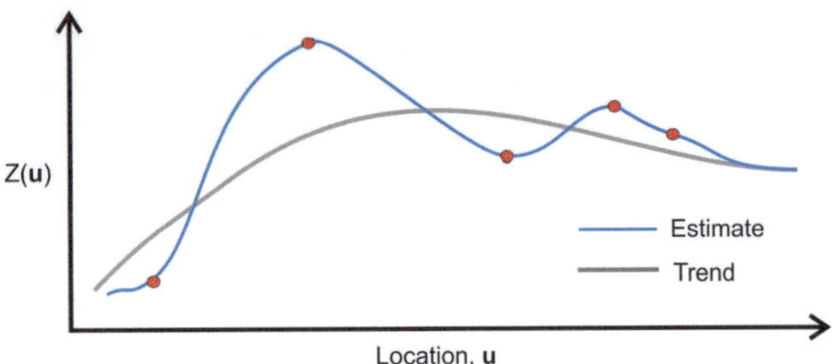

Fig. 8.2 A sketch showing estimation in one dimension

deposit, such as those discussed in Chap. 2, where methods were proposed to remove bias resulting from clustering.

8.1.4 Weighted Linear Estimation

Estimates are often made as weighted linear estimators. A common approach is to estimate the values as deviations from a mean or trend surface, see Fig. 8.2. The estimate reverts to the mean some distance away from the data, see the far right edge. The deviations from the mean surface are estimated at unsampled locations with some method of interpolation. The most common interpolation scheme is a weighted linear estimate:

$$z^{*}(\mathbf{u}_0) - m(\mathbf{u}_0) = \sum_{i=1}^{n} \lambda_i \cdot [z(\mathbf{u}_i) - m(\mathbf{u}_i)]$$

where the *denotes an estimate, $\mathbf{u}_0$ denotes the unsampled location being estimated, $z(\cdot)$ denotes the variables value, $m(\cdot)$ denotes the mean or trend value and $i=1,\ldots, n$ is the index of data values.

Estimation then becomes an exercise in determining the weights λ_i using certain criteria. Factors considered when assigning weights may include the closeness to the location being estimated; the redundancy between data values; the anisotropic continuity (preferential direction); and the magnitude of continuity/variability.

8.1.5 Traditional Estimation Methods

Simple (traditional) estimation techniques can be used to assign values to blocks. Polygonal methods and Inverse Distance (ID) methods are often applied at early stages of a mining project or for checking. These methods are not particularly accurate, but can provide an order-of-magnitude resource estimate. They can also be used to check the results of more sophisticated geostatistical estimation methods.

According to Popoff (1966) polygonal methods have been used since the early 1900s. Variants of the polygonal method include the sectional estimation method (Stone and Dunn 1996), the classic polygonal method, and the computerized nearest-neighbor (NN) method (Sinclair and Blackwell 2002).

8.1.6 Classic Polygonal Method

The polygonal estimate is also based on assigning areas of influence around drill hole intercepts. The drawing of polygons around drill hole data is based on the perpendicular bisectors between a sample and all its neighbors, see Fig. 8.3. The perpendicular bisector of a line segment is a line for which points are at the same distance from either side of the line segment. This concept can be extended to three-dimensions, although polygons are typically drawn in two-dimensions and the volume of influence is defined perpendicular to the polygonal plane.

The polygon of influence is such that each point within the polygon is closest to the central sample than any other sample. Special care must be taken with samples on the outer edges. These samples are not completely surrounded by other samples, so bounding the polygon is important. There are several alternatives, including the use of geologic boundary (if available), or, more commonly, a fixed maximum distance from the sample. In any event, the closing of the polygons on the outer edges may be arbitrary and can have a significant impact on the final results.

The polygonal method corresponds with the intuitive idea that the amount of information provided by each sample is proportional to its area (or volume) of influence. In this sense, the method has found modern application as a spatial declustering tool, calculating weights to avoid biased statistics based on spatial drill hole data aggregation (Chap. 2). One example is that it can be used to provide a deposit-wide estimate of the average grade (Isaaks and Srivastava 1989).

8.1.7 Nearest-Neighbor Method

This is a variant of the polygonal method, but in this case the grades or attributes are assigned directly to a block model. This is the most common computerized polygonal estimation method and it has evolved into two common uses.

The first and more traditional use is the calculation of mineral resources. A grid of blocks is assigned grades by the closest drill hole data sample or composite. The method does not average values from different samples. The original variance of the data is maintained. There is no smoothing and grades from one block to the next change abruptly producing artificial discontinuities.

The nearest-neighbor method is not as good as Inverse Distance methods and Kriging at estimating grades. From theory and practice (see for example Knudsen et al. 1978; Baafi and Kim 1982; Readdy et al. 1982; Knudsen 1990),

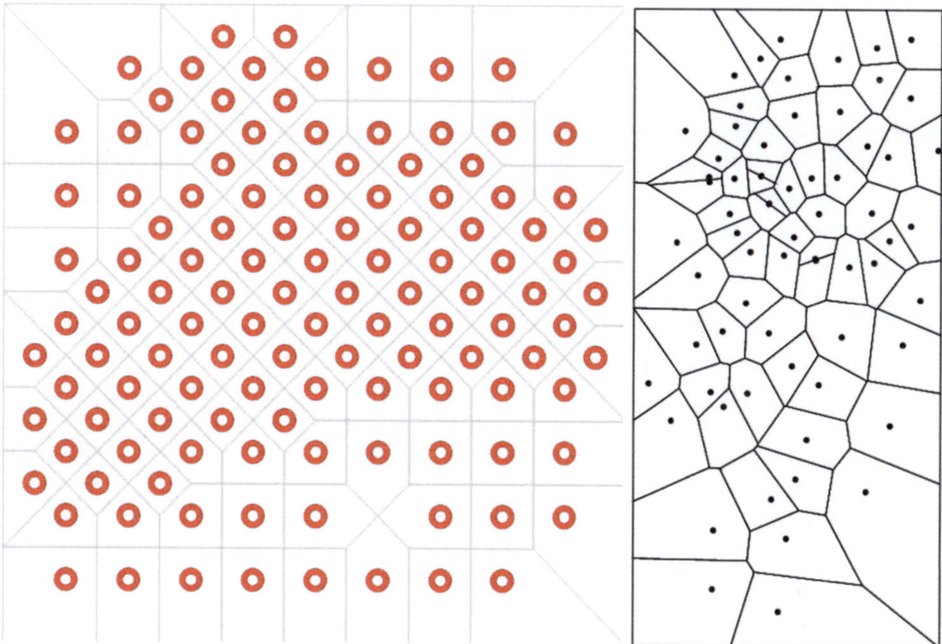

Fig. 8.3 Two schematic examples of the polygons of influence method; no distance units are given

it is known that the errors are larger than for other methods. For many deposits with positively skewed distributions, large errors in individual blocks lead to a tendency to overestimate the average grade and underestimate tonnage of the resources above cutoff. The NN method is mostly used as a checking tool (Chap. 11).

The second application is to decluster grades, assuming that the block grade distribution is a fair representation of the declustered drill hole data. While it is in concept equivalent to a polygonal declustering technique, it is much easier to implement, since most geologic and mining software packages incorporate the algorithm. It can also be implemented as an Inverse Distance method with specific parameters, see below.

8.1.8 Inverse Distance Weighting

Inverse Distance methods are a family of weighted average methods. They result in estimates that are smoothed versions of the original data. Inverse distance methods are based on calculating weights for the samples based on the distrance from the samples to the point or block of interest. The linear estimator is written as:

$$z^* = \frac{\sum_{i=1}^{N} w_i \cdot z_i}{\sum_{i=1}^{N} w_i}$$

where w_i are the weights assigned to each composite data and z_i is the corresponding composite value, for all composites ($i=1,\dots, N$) used in the estimation, and z^* is the estimated value.

The calculation of the weights w_i is based on inverse of the distance between the composite and the point being estimated. This is written as:

$$w_i = \frac{1}{c + d_i^{\omega}}$$

where d_i is the distance between the composite and the point being estimated, ω is the exponent and c is a constant to avoid over weighting very close data. The weights are standardized to sum to 1 to ensure a globally unbiased estimate.

Two generalizations are possible. One is to modify the exponent in Eq. 3.10. The most common exponents used are $\omega=2$ (Inverse Distance Squared, IDS,) and $\omega=3$ (Inverse Distance Cubed, IDC,). IDS is used with smoothly varying attributes, such as topographic surfaces, thickness of geologic units including coal beds, some stratabound deposits, and interpolation of in-situ bulk density values. Larger exponents, such as $\omega=3$ (IDC), are used when large weights are desired for the closest composites. This is applicable when the variable being estimated is more erratic and the current data spacing is large relative to the data that will ultimately be available for decision making, as for example with open pit gold grade distributions. The extreme case is to increase the exponent so that only the closest composite receives any weight at all, which is equivalent to a nearest neighbor estimation.

The opposite extreme is when the exponent is 0, which amounts to an equally weighted moving average, as described in Chap. 2. Isaaks and Srivastava (1989, pp. 257–259) demonstrate the impact of the exponent on the weights assigned to each composite.

The second generalization is based on using distances calculated with anisotropy, that is, preferred directions of continuity. The anisotropy can be introduced by re-scaling the directional distances appropriately, as has been used by some major gold companies, also observed in the Western Australia goldfields. Overall, the application of ID method has been steadily decreasing through the years in favor of geostatistical methods.

Kriging is a method that allows calculating weights that are optimal according to the least-squares, or minimum expected error variance, criteria. Although estimation schemes sometimes provide a measure of how good the estimates are, there is no good measure of uncertainty attached to the estimates. Probabilistic estimation (Chap. 9) or simulation (Chap. 10) is needed for this purpose.

8.2 Kriging Estimators

The basis for the kriging framework is to calculate the weights that minimize an expected error variance. There are many flavors of kriging, but the basic forms differ mostly on the assumptions they make regarding the local or stationary domain mean. This is expressed as conditions on the set of weights. Linear kriging has been presented in several classic references, such as Journel and Huijbregts (1978); Isaaks and Srivastava (1989); Deustch and Journel (1997); and Chilès and Delfiner (2011).

The more common types of kriging are:

- Simple kriging (SK): minimizes the error variance with no constraints on the weights. The mean is a known constant (inferred from the available samples) for the entire domain.
- Ordinary kriging (OK): the local mean is implicitly re-estimated as a constant within each search neighborhood. OK is a common technique used to obtain interim estimates.
- Kriging with a trend model or universal kriging (KT or UK): this method estimates residuals from a specified location-dependent mean $m(\mathbf{u})$. The location-dependent mean could be a specified constant (local-varying mean), or a deterministic trend typically specified as a function of the coordinates. This method is also called non-stationary kriging because of the location-dependent mean.
- Kriging with an external drift: in this variant, the trend model is scaled from a secondary variable.
- Factorial kriging: the RF model $Z(\boldsymbol{u})$ is split into independent components (factors), which are then independently estimated.
- Non-linear kriging, including Gaussian-based (disjunctive kriging, uniform conditioning, multi-Gaussian), indicator kriging (median, multiple, probability), and lognormal kriging. These are discussed in Chap. 9.

The choice of method depends on the geologic setting, the amount of information available, and the characteristics of the RF model envisioned. The most common estimation method is OK, although the different variants that model trends have become more popular in more recent years.

8.2.1 Simple Kriging

The purpose of kriging is to determine a set of optimal weights that minimize the expected error variance. Consider a linear estimator:

$$Z^*(\mathbf{u}) = \sum_{i=1}^{n} \lambda_i \cdot [z(\mathbf{u}_i) - m] + m$$
$$= \sum_{i=1}^{n} \lambda_i \cdot z(\mathbf{u}_i) + \left[1 - \sum_{i=1}^{n} \lambda_i(\mathbf{u}_i)\right] \cdot m$$

where $z(\mathbf{u}_i)$ are the data values and $Z^*(\mathbf{u})$ is the estimate. The constant mean m is assumed known and stationary (location-independent). In this case, the SK estimator is unbiased by definition, and the estimation is performed in effect on the residuals data values. The known mean m is subtracted from the data values and then added back after the residuals have been estimated. The estimation error is then expressed as a linear combination of the residuals $Y^*_{SK}(\mathbf{u}) - Y(\mathbf{u})$.

The error variance is defined as

$$E\{[Y^*(\mathbf{u}) - Y(\mathbf{u})]^2\} = E\{[Y^*(\mathbf{u})]^2\} - 2 \cdot E\{Y^*(\mathbf{u}) \cdot Y(\mathbf{u})\} + E\{[Y(\mathbf{u})]^2\}$$

and can be expressed as a linear combination of covariance values of the residuals:

$$\sum_{i=1}^{n}\sum_{j=1}^{n} \lambda_i \lambda_j E\{Y(\mathbf{u}_i) \cdot Y(\mathbf{u}_j)\}$$
$$-2 \cdot \sum_{i=1}^{n} \lambda_i E\{Y(\mathbf{u}) \cdot Y(\mathbf{u}_i)\} + C(0) = \sum_{i=1}^{n}\sum_{j=1}^{n} \lambda_i \lambda_j C(\mathbf{u}_i, \mathbf{u}_j)$$
$$-2 \cdot \sum_{i=1}^{n} \lambda_i C(\mathbf{u}, \mathbf{u}_i) + C(0)$$

It can be seen that the error variance is written in terms of the (1) weights used for the estimate (the λ values), (2) the variance ($C(0)$), (3) the covariance between the data locations and the location, ($C(\mathbf{u}, \mathbf{u}_i)$) and (4) the covariance between all pairs of data ($C(\mathbf{u}_i, \mathbf{u}_j)$). The covariance is required because the estimate is linear and the estimation variance is a quadratic form. The required covariance values are calculated from the variogram model.

The optimal weights λ_i, $i = 1, \ldots, n$ are determined by taking partial derivatives of the error variance with respect to the weights and setting them to zero:

$$\frac{\partial \sigma_E^2}{\partial \lambda_\iota} = 2 \cdot \sum_{j=1}^{n} \lambda_\beta C(\mathbf{u}_\iota, \mathbf{u}_j) - 2 \cdot C(\mathbf{u}, \mathbf{u}_\iota) = 0, \quad \iota = 1, \ldots, n$$

which results in a system of n equations with n unknown weights, known as the simple kriging (SK) or normal equations system (Luenberger 1969):

$$\sum_{j=1}^{n} \lambda_j C(\mathbf{u}_i, \mathbf{u}_j) = C(\mathbf{u}, \mathbf{u}_i), \quad i = 1, \ldots, n$$

The minimized estimation variance also known as the kriging variance is:

$$\sigma_E^2 = C(0) - \sum_{i=1}^{n} \lambda_i C(\mathbf{u}, \mathbf{u}_i)$$

A more general form of the simple kriging estimator that accounts for the support or volume of the location being estimated and the samples used in the estimation, recall Chap. 7, is written below. In the more general case, an estimate of a particular block size; that is, an estimate of $Z_V(\mathbf{u})$ with $z_v(\mathbf{u}_i)$, $i = 1, \ldots, n$ is being made:

$$z_k^*(\mathbf{u}) - m(\mathbf{u}) = \sum_{i=1}^{n} \lambda_i \bullet [z_v(\mathbf{u}_i) - m(\mathbf{u}_i)]$$

$$\sum_{j=1}^{n} \lambda_j \bar{C}(v_i, v_j) = \bar{C}(v_i, V), \ \forall\, i = 1, \ldots, n$$

with

$$\sigma_K^2 = \bar{C}(V, V) - \sum_{i=1}^{n} \lambda_i \bar{C}(v_i, V)$$

The solution exists and is unique if the matrix $\left[\bar{C}(v_i, v_j)\right]$ is positive definite (Chap. 6). Also note that the existence of duplicate points will result in a singular covariance matrix, since the distance will be zero. Some basic properties of the simple kriging estimator are:

- SK is unbiased by definition, since the stationary mean is assumed known. In practice, it is inferred from the average value of the samples within the stationary domain. Because stationary domains are seldom found in mining applications, the SK method is used little in mineral resource estimation. It is commonly used, however, to obtain conditional simulations (Chap. 10).
- Kriging in all its forms is, by construction, a minimum error variance estimator. No other set of weights will provide a lower estimation variance than kriging. While least-squares optimization is generally thought of as a valuable property of kriging, there are cases where minimum variance may be the wrong optimization criteria to use (Srivastava 1987).
- Kriging is also an exact interpolator, meaning that, at known locations, the estimate is the sampled value. If the location $\mathbf{u}$ to be estimated coincides with a datum location $\mathbf{u}_0$, the normal system returns the datum value for the estimate. Kriging honors the (hard) data values at their locations.
- The kriging variance depends only on the covariance values, not on the actual sample values. The kriging variance can be known before any estimation is performed.
- The kriging weights are, like the variance, non-data dependent. Therefore, the same covariance values and data configuration will result in the same kriging weights, regardless of the values of the individual samples used to estimate the unknown location $\mathbf{u}$.
- Kriging takes into account the geometry of the volume being estimated through the consideration of the volumes V. It is well understood in practice that estimating a larger volume V is easier, that is, the estimation variance will be smaller.
- The distance of the information to the location being estimated is taken into account as a structural distance; in the case of Inverse Distance methods, the distance used is Euclidean, always the same regardless of geology, depositional environment, or variable being estimated. Kriging is an improvement because it considers a distance that is specific to the geologic environment. The structural continuity of the variable being considered is modeled, including its anisotropy and other features that result from measuring data spatial correlation.
- The configuration of the data is quantified through the term $\left[\bar{C}(v_i, v_j)\right]$, which accounts for redundancy and thus data clustering.
- The smoothing effect of kriging can be forecast. Since the estimation variance, the covariance values among data, and the covariance values between the samples and the volume V being estimated can be pre-calculated, it is possible to obtain theoretical distributions for different volume supports (SMU, Chap. 7).

8.2.2 Ordinary Kriging

Ordinary kriging is based on the same minimum error variance linear estimate at a location where the true value is unknown. But contrary to SK, OK does not make any prior assumptions about the mean. By requiring global unbiasedness, Ordinary kriging (OK) constrains the sum of the weights to be 1.0, and as a result the mean does not need to be known. We assume the unknown mean for the volume being estimated constant:

$$[z^*(\mathbf{u}) - m] = \sum_{i=1}^{n} \lambda_i \bullet [z(\mathbf{u}_i) - m]$$

$$z^*(\mathbf{u}) = \sum_{i=1}^{n} \lambda_i z(\mathbf{u}_i) + \left[1 - \sum_{i=1}^{n} \lambda_i\right] \cdot m$$

The condition $\sum_{i=1}^{n} \lambda_i = 1$ is the unbiasedness condition when the mean m is not known. This is the essence of ordinary kriging: the estimation variance is minimized under the condition that the sum of the weights is 1.0. It can be shown that ordinary kriging amounts to re-estimating, at each new location **u**, the mean m as used in the SK expression. Since OK is most often applied within moving search neighborhoods, i.e., using different data sets for different locations **u**, the implicit re-estimated mean denoted $m^*(\mathbf{u})$ depends on the location **u**. Thus the OK estimator is a type of SK, where the constant mean value m is replaced by a location-dependent estimate $m^*(\mathbf{u})$.

Ordinary kriging is a non-stationary algorithm. It corresponds to a non-stationary RF model with varying mean but stationary covariance. This ability to rescale locally the RF model $Z(\mathbf{u})$ to a different mean value $m^*(\mathbf{u})$ explains the robustness of the OK algorithm. Ordinary kriging has been and is likely to remain the anchor algorithm of geostatistics.

The OK system is also a system of normal equations, but with an additional constraint: the sum of weights equal to 1. The Lagrange formalism is again used to obtain the optimal weights and derive the OK system of equations. Using the more general notation to take into account the different support of the samples and the blocks being estimated, the derivation of the OK system is

$$Q(\lambda_i, i=1,\ldots,n,\mu) = \sigma_E^2 + 2\mu\left[\sum_{j=1}^{n} \lambda_j - 1\right] \rightarrow \text{minimum}$$

Taking the partial derivative with respect to the weights and the Lagrange multiplier,

$$\frac{\partial Q}{\partial \lambda_\alpha} = -2 \cdot \overline{C}(V, v_\alpha) + 2 \cdot \sum_{\beta=1}^{n} \lambda_\beta C(v_\alpha, v_\beta) + 2 \cdot \mu = 0, \quad \forall \alpha = 1,\ldots,n$$

$$\frac{\partial Q}{\partial \mu} = \sum_{j=1}^{n} \lambda_j - 1 = 0$$

with μ being the Lagrange parameter introduced due to the constraint that the weights sum up to 1. The resulting OK system and the corresponding OK variance are

$$\begin{cases} \sum_{j=1}^{n} \lambda_j C(v_i, v_j) + \mu = \overline{C}(V, v_i), \forall i = 1,\ldots,n \\ \sum_{j=1}^{n} \lambda_j = 1 \end{cases}$$

$$\sigma_K^2 = \overline{C}(V,V) - \mu - \sum_{\alpha=1}^{n} \lambda_\alpha \overline{C}(V, v_\alpha)$$

8.2.3 Kriging with a Trend

The term universal kriging has been traditionally used to denote what is, in fact, kriging with a prior trend model. The terminology, kriging with a trend model (KT) is more appropriate since the underlying RF model is considered to be the sum of a trend component plus a residual:

$$Z(\mathbf{u}) = m(\mathbf{u}) + R(\mathbf{u})$$

The trend component defined as $m(\mathbf{u}) = E\{Z(\mathbf{u})\}$, is usually modeled as a smoothly varying deterministic function of the coordinates vector **u** whose unknown parameters are fitted from the data:

$$m(\mathbf{u}) = \sum_{l=0}^{L} a_l f_l(\mathbf{u})$$

where $m(\mathbf{u})$ is the local mean, a_l, $l = 0 \ldots L$ are unknown coefficients of the trend model, and $f_l(\mathbf{u})$ are low order monomials of the coordinates. The trend value $m(\mathbf{u})$ is itself unknown since the parameters a_l are unknown.

The residual component $R(\mathbf{u})$ is usually modeled as a stationary RF with zero mean and covariance $C_R(\mathbf{h})$.

The Kriging with the trend model (KT) system is also a system of constrained normal equations. The KT estimator is written as

$$Z_{KT}^*(\mathbf{u}) = \sum_{i=1}^{n} \lambda_i^{(KT)}(\mathbf{u}) Z(\mathbf{u}_i)$$

and the KT system is

$$\begin{cases} \sum_{j=1}^{n} \lambda_j^{(KT)}(\mathbf{u}) C_R(\mathbf{u}_j - \mathbf{u}_i) + \sum_{k=0}^{K} \mu_k(\mathbf{u}) f_k(\mathbf{u}_i) = C_R(\mathbf{u} - \mathbf{u}_i), \quad i = 1,\ldots,n \\ \sum_{j=1}^{n} \lambda_j^{(KT)}(\mathbf{u}) f_k(\mathbf{u}_j) = f_k(\mathbf{u}), \quad k = 0,\ldots,K \end{cases}$$

where the $\lambda_j^{(KT)}(\mathbf{u})$'s are the KT weights and the $\mu_k(\mathbf{u})$'s are the $(K+1)$ Lagrange parameters associated with the $(K+1)$ constraints on the weights.

Ideally, the functions $f_k(\mathbf{u})$ that define the trend should be specified by the physics of the problem. For example, if a periodic component is known to contribute to the spatial or temporal variability of $z(\mathbf{u})$, a sine function $f_k(\mathbf{u})$ with specific period and phase.could be considered; the amplitude of the periodic component, i.e., the parameter a_k, would then implicitly be estimated from the z data through the KT system.

In the absence of any information about the shape of the trend, the split of the z data into trend and residual components is somewhat arbitrary. What is regarded as stochastic fluctuations $R(\mathbf{u})$ at large scale may later be modeled as a

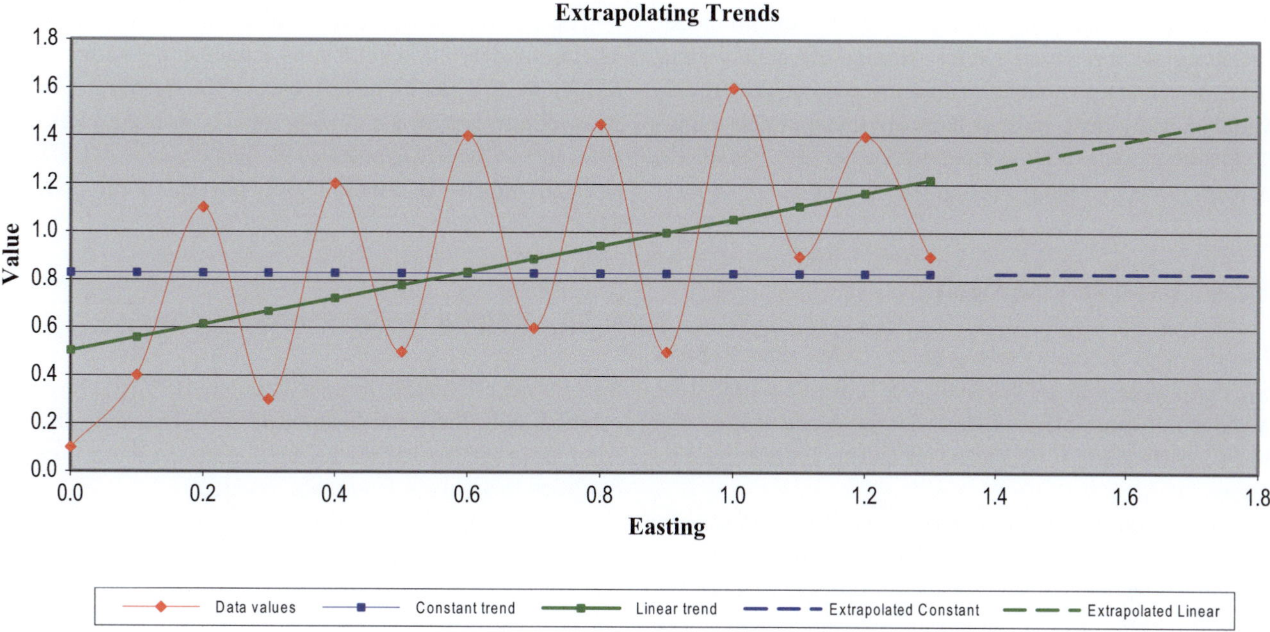

Fig. 8.4 Sketch illustrating the differences between linear and constant trend extrapolations. Differences are quickly amplified for locations away from actual data

trend if additional data allow focusing on the smaller-scale variability. In the absence of physical interpretations, the trend is usually modeled as a low-order (≤2) polynomial of the coordinates **u**, e.g., with $\mathbf{u}=(x, y)$:

- a linear trend in 1D: $m(\mathbf{u})=a_0+a_1x$
- a linear trend in 2D limited to the 45° direction: $m(\mathbf{u})=a_0+a_1\,(x+y)$
- a quadratic trend in 2D: $m(\mathbf{u})=a_0+a_1\,x+a_2\,y+a_3x^2+a_4\,y^2+a_5\,x\,y$

By convention, $f_0(\boldsymbol{u})=1$, for all **u**. Hence the case $K=0$ corresponds to ordinary kriging with a constant but unknown mean: $m(\boldsymbol{u})=a_0$.

Trend models using higher-order polynomials ($n>2$) or arbitrary non-monotonic functions of the coordinates **u** are better replaced by a random function component with a large-range variogram.

When only z data are available, the residual covariance $C_R(\mathbf{h})$ is inferred from linear combinations of z data that filter the trend $m(\mathbf{u})$, as proposed by Delfiner (1976). For example, differences of order 1 such as $[z(\mathbf{u}+\mathbf{h})-z(\mathbf{u})]$ would filter any trend of order zero $m(\mathbf{u})=a_0$; differences of order 2 such as $[z(\mathbf{u}+2\mathbf{h})-2\,z(\mathbf{u}+\mathbf{h})+z(\mathbf{u})]$ would filter any trend of order 1 such as $m(\mathbf{u})=a_0+a_1\,\mathbf{u}$.

However, in most practical situations it is possible to locate subareas or directions along which the trend can be ignored, in which case $Z(\mathbf{u})\approx R(\mathbf{u})$, and the residual covariance can be directly inferred from the local z data.

When the trend functions $f_k(\boldsymbol{u})$ are not based on physical considerations, as is often the case in practice, and in interpolation conditions, it can be shown that the choice of the specific functions $f_k(\boldsymbol{u})$ does not change the estimated values $z_{KT}{}^*(\boldsymbol{u})$ or $m_{KT}{}^*(\boldsymbol{u})$. When working with moving neighborhoods the important aspect is the residual covariance $C_R(\boldsymbol{h})$, not the choice of the trend model.

The traditional notation for the trend does not reflect the general practice of kriging with moving data neighborhoods. Because the data used for estimation change from one location **u** to another, the resulting implicit estimates of the parameters a_l's are different. Hence the following notation for the trend is more appropriate:

$$m(\mathbf{u}) = \sum_{k=0}^{K} a_k(\mathbf{u}) f_k(\mathbf{u})$$

The trend model, however, is important in extrapolation conditions, i.e., when the data locations $\mathbf{u}_\alpha$ do not surround within the covariance range the location **u** being estimated. Extrapolating a constant yields significantly different results for either $z_{KT}{}^*(\boldsymbol{u})$ or $m_{KT}{}^*(\boldsymbol{u})$ than extrapolating either a line or a parabola (non-constant trend), see Fig. 8.4. However, estimates based on extrapolation in mineral resource estimates are generally not acceptable per current Reporting Standards (Chap 12). At most, a limited portion of resources estimated in extrapolation conditions would be inferred, which are deemed unreliable and not adequately known to use in engineering studies and economic evaluations.

The practitioner is warned against overzealous modeling of the trend and the unnecessary usage of universal kriging (KT) or intrinsic random functions of order k (IRF-k, Matheron 1973). In most interpolation situations the simpler and well-proven OK algorithm within moving search neighborhoods will suffice (Journel and Rossi 1989). In extrapolation situations, almost by definition, the sample z data alone cannot justify the trend model chosen (Deutsch 2002).

There are several variants of kriging with a trend model, including random trend model, kriging the actual trend, using a secondary variable to impose a trend on the primary variable, and others. Most of these are found in non-mining applications since they are more difficult to justify given the amount of data normally encountered in mining.

8.2.4 Local Varying Mean

Kriging with a locally varying mean (LVM) is a variant of SK that works with the residuals, but is different in that the mean is not constant everywhere. In this sense, it is similar to OK, particularly if the mean is constant within certain regions.

The general approach is to model the trend so that at every location the value $m(\mathbf{u})$ is known and possibly different. As before, a covariance model for the residuals is needed. The influence of the local mean depends on the amount of primary data available in the neighborhood. If there are a large number of nearby samples, then the influence of the mean is mitigated; in areas where the primary data is scarce, the mean has a large influence. LVM is appropriate for modeling geological trends and smooth secondary data, and is more commonly applied in simulations (Chap. 10).

8.2.5 Random Trend Model

A similar model to KT results from interpreting the trend as a random component. The random trend is denoted $M(\mathbf{u})$, and is added to a residual $R(\mathbf{u})$ independent from it:

$$Z(\mathbf{u}) = M(\mathbf{u}) + R(\mathbf{u}), \text{ with } E\{Z(\mathbf{u})\} = E\{M(\mathbf{u})\}$$

Prior data that allows describing the trend are assumed available. For example, they can be prior information $m(\mathbf{u})$ about the local z data. These trend data allows inference of the M-covariance $C_M(\mathbf{h})$, and the corresponding residual data can be used to infer the covariance of the residuals $R(\mathbf{u})$. Based on the independence assumption, the z data covariance is then

$$C_Z(\mathbf{h}) = C_M(\mathbf{h}) + C_R(\mathbf{h})$$

Kriging is then performed using the z data and the covariance model $C_Z(\mathbf{h})$. The resulting kriging estimates and variances depending on the $E\{M(\mathbf{u})\}$ and $C_M(\mathbf{h})$. The variance of $M(\mathbf{u})$ can be made non-stationary and used to measure the reliability of the prior guess $m(\mathbf{u})$. However, in this case the M-covariance is not anymore stationary and its inference may become problematic.

The random trend model is equivalent to the Bayesian kriging model, but is simpler to implement. The weakness of both models lies in the inference of the statistics of $M(\mathbf{u})$, whether interpreted as a random trend or as prior guess on z data, and on the key hypothesis of independence of the $M(\mathbf{u})$ and $R(\mathbf{u})$ values. As usual, the only physical reality is z, not M or R.

8.2.6 Kriging the Trend and Filtering

Rather than estimating the sum $Z(\mathbf{u}) = m(\mathbf{u}) + R(\mathbf{u})$ one could estimate only the trend component $m(\mathbf{u})$. Starting directly from the original z data the KT system shown above is easily modified to yield a KT estimate for $m(\mathbf{u})$:

$$m^*_{KT}(\mathbf{u}) = \sum_{i=1}^{n} \lambda_i^{(m)}(\mathbf{u}) Z(\mathbf{u}_i)$$

and the KT system:

$$\begin{cases} \sum_{j=1}^{n} \lambda_j^{(m)}(\mathbf{u}) C_R(\mathbf{u}_j - \mathbf{u}_i) + \sum_{k=0}^{K} \mu_k^{m}(\mathbf{u}) f_k(\mathbf{u}_i) = 0, \quad i = 1,\dots,n \\ \sum_{j=1}^{n} \lambda_j^{(m)}(\mathbf{u}) f_k(\mathbf{u}_j) = f_k(\mathbf{u}), \quad k = 0,\dots,K \end{cases}$$

where the $\lambda_j^{(m)}$'s are the KT weights and the μ_k^{m}'s are Lagrange parameters. Note that this system differs from the KT system of the variable $Z(\mathbf{u})$.

This algorithm identifies the least-squares fit of the trend model when the residual model $R(\mathbf{u})$ is assumed to have no correlation: $C_R(\mathbf{h}) = 0$ for all $\mathbf{h} \neq 0$.

The direct KT estimation of the trend component can also be interpreted as a low-pass filter that removes the random (high-frequency) component $R(\mathbf{u})$. The same principle underlies the algorithm of factorial kriging and that of the Wiener-Kalman filter (Kalman 1960).

Factorial kriging is a technique that aims to either extract features for separate analysis or filter features from spatial data. The technique was originally proposed by Matheron in the early days of geostatistics (1971), and takes its name in relation to factor analysis (Journel and Huijbregts 1978; Goovaerts 1997). Factorial kriging is of greatest interest to geophysicists and those concerned with image analysis.

8.2.7 Kriging with an External Drift

Kriging with an external drift is a particular case of KT above. It considers a single trend function $f_I(u)$ defined at each location from some external, secondary variable.

The trend model is limited to two terms $m(\mathbf{u}) = a_0 + a_1 f_1(\mathbf{u})$, with the term $f_1(\mathbf{u})$ set equal to the secondary variable.

Let y(**u**) be the secondary variable; the trend model is then:

$$E\{Z(\mathbf{u})\}=m(\mathbf{u})=a_0+a_1 1\, y(\mathbf{u})$$

$y(\boldsymbol{u})$ is assumed to reflect the spatial trends of the z variability up to a linear rescaling of units, corresponding to the two parameters a_0 and a_1.

The estimate of the z variable and the corresponding system of equations are identical to the KT estimate and system with K=1, and $f_1(\mathbf{u})=y(\mathbf{u})$, i.e.:

$$Z^*_{KT}(\mathbf{u})=\sum_{i=1}^{n}\lambda_i^{(KT)}(\mathbf{u})Z(\mathbf{u}_i), \text{ and}$$

$$\begin{cases}\sum_{j=1}^{n}\lambda_j^{(KT)}(\mathbf{u})C_R(\mathbf{u}_j-\mathbf{u}_i)+\mu_0(\mathbf{u})+\mu_1(\mathbf{u})y(\mathbf{u}_i)\\ =C_R(\mathbf{u}-\mathbf{u}_\alpha),\ \mathrm{i}=1,\ldots,n\\ \sum_{j=1}^{n}\lambda_j^{(KT)}(\mathbf{u})=1\\ \sum_{j=1}^{n}\lambda_j^{(KT)}(\mathbf{u})y(\mathbf{u}_j)=y(\mathbf{u})\end{cases}$$

where the $\lambda_\beta^{(KT)}$'s are the kriging (KT) weights and the μ's are Lagrange parameters.

Kriging with an external drift is an efficient algorithm to incorporate a secondary variable in the estimation of the primary variable $z(\mathbf{u})$, and is appropriate for linearly related secondary data. The fundamental relation between the two variables must make physical sense.

In mining, there are few cases where this technique has been applied. There are two reasons for this: (1) primary $z(\mathbf{u})$ data sets in mining tend to be large, and the addition of a linearly-related trend from a secondary variable is difficult to justify; and (2) there are few variables where this linear relationship can be safely assumed. Kriging with an external drift is more common in other applications. For example, if the secondary variable $y(\mathbf{u})$ represents the travel time to a seismic reflective horizon, assuming a constant velocity, the depth $z(\mathbf{u})$ of that horizon should be proportional to the travel time $y(\mathbf{u})$. Hence a relation of this type makes sense.

Two conditions must be met before applying the external drift algorithm: (1) The external variable must vary smoothly in space, otherwise the resulting KT system may be unstable; and (2) the external variable must be known at all locations $\mathbf{u}_0$ of the primary data values and at all locations $\mathbf{u}$ to be estimated.

Note that the residual covariance rather than the covariance of the original variable $Z(\boldsymbol{u})$ must be used in the KT system. Both covariances are equal in areas or along directions where the trend $m(\boldsymbol{u})$ is deemed non-existent. Note also that the cross covariance between variables $Z(\boldsymbol{u})$ and $Y(\boldsymbol{u})$ plays no role in this system; this is different from cokriging. In a sense, the $Z(\boldsymbol{u})$ variable borrows the trend from the $Y(\boldsymbol{u})$ variable. Therefore, the $Z^*(\boldsymbol{u})$ estimates reflect the trends of the $Y(\boldsymbol{u})$ variability, not necessarily the z variability.

8.3 Cokriging

The term kriging is reserved for estimation using data from the same attribute as that being estimated. For example, an unsampled gold grade value $z(\mathbf{u})$ is estimated from neighboring gold grade values.

Cokriging is a similar estimate that uses data defined on different attributes. For example, the gold grade $z(\mathbf{u})$ may be estimated from a combination of gold and copper samples values. There must be a spatial correlation between the primary and secondary variables that can be inferred from available information. As is the case when considering a single variable, there are three basic variants of cokriging: simple cokriging (SCK), ordinary cokriging (OCK), and cokriging with a trend model (CKT). Conceptually these cokriging methods are the same as the ones explained above; however, there is the additional complication of dealing with at least two variables, which is reflected in the heavier notation.

8.3.1 Simple Cokriging

Consider a linear combination of primary and secondary data values:

$$y_0^*(\boldsymbol{u})=\sum_{i=1}^{n_0}\lambda_i^0\cdot y_0(\boldsymbol{u}_i^0)+\sum_{j=1}^{n_l}\lambda_j^l\cdot y_l(\boldsymbol{u}_j^l)$$

The estimation variance may be written as

$$Var\left\{Y_0-Y_0^*\right\}=\rho_{0,0}(0)-2\sum_{i=1}^{n_0}\lambda_i^0\cdot\rho_{0,0}(\boldsymbol{u}_i^0-\boldsymbol{u})$$
$$-2\sum_{j=1}^{n_j}\lambda_j^l\cdot\rho_{1,0}(\boldsymbol{u}_j^l-\boldsymbol{u})+\sum_{i=1}^{n_0}\sum_{j=1}^{n_0}\lambda_i^0\cdot\lambda_j^0\cdot\rho_{0,0}(\boldsymbol{u}_i^0-\boldsymbol{u}_j^0)$$
$$+\sum_{i=1}^{n_l}\sum_{j=1}^{n_l}\lambda_i^l\cdot\lambda_j^l\cdot\rho_{1,1}(\boldsymbol{u}_i^l-\boldsymbol{u}_j^l)+2\sum_{i=1}^{n_0}\sum_{j=1}^{n_l}\lambda_i^0\cdot\lambda_j^l\cdot\rho_{0,1}(\boldsymbol{u}_i^0-\boldsymbol{u}_j^l)$$

Minimizing this estimation variance results in the simple cokriging system of equations:

$$\begin{cases}\sum_{\beta_1=1}^{n_1(u)}\lambda_{\beta_1}(u)C_{ZZ}(u_{\alpha_1}-u_{\beta_1})\\ +\sum_{\beta_2=1}^{n_2(u)}\lambda'_{\beta_2}(u)C_{ZY}(u_{\alpha_1}-u'_{\beta_2})=C_{ZZ}(u_{\alpha_1}-u),\ \alpha_1=1,\ldots,n_1(u)\\ \sum_{\beta_1=1}^{n_1(u)}\lambda_{\beta_1}(u)C_{YZ}(u'_{\alpha_2}-u_{\beta_1})+\sum_{\beta_2=1}^{n_2(u)}\lambda'_{\beta_2}(u)C_{YY}(u'_{\alpha_2}-u'_{\beta_2})\\ =C_{YZ}(u'_{\alpha_2}-u),\quad \alpha_2=1,\ldots,n_2(u)\end{cases}$$

The cokriging estimator and the resulting estimation variance are

$$z^*(u) - m_z = \sum_{\alpha_1=1}^{n_1(u)} \lambda_{\alpha_1}(u)\left[z(u_{\alpha_1}) - m_z\right] + \sum_{\alpha_2=1}^{n_2(u)} \lambda'_{\alpha_2}(u)\left[y(u'_{\alpha_2}) - m_y\right]$$

$$\sigma^2(u) = C_{ZZ}(0) - \sum_{\alpha_1=1}^{n_1(u)} \lambda_{\alpha_1}(u) C_{ZZ}(u_{\alpha_1} - u) - \sum_{\alpha_2=1}^{n_2(u)} \lambda'_{\alpha_2}(u) C_{ZY}(u'_{\alpha_2} - u)$$

The equations for simple cokriging are essentially the same as for simple kriging, but taking into account the direct and cross covariances. As before, the system of equations must lead to a valid result and the cokriging variance has to be positive, which means that the covariance matrix is positive definite. The condition is satisfied when using permissible coregionalization model and no two data values (of the same variable) are collocated.

We often avoid full cokriging because it is tedious to calculate, interpret, and fit the necessary variograms. The linear model of coregionalization (Chap. 6) is restrictive, and there is no real benefit in cases where the same amount of data is present for both primary and secondary variables. We are motivated to consider cokriging when there are many more secondary data than primary data and when a simple trend/local mean model is considered inadequate.

8.3.2 Ordinary Cokriging

In the case of a single secondary variable (Y), the ordinary cokriging estimator of Z(**u**) is written as

$$Z^*_{COK}(\mathbf{u}) = \sum_{\alpha_1=1}^{n_1} \lambda_{\alpha_1}(\mathbf{u}) Z(\mathbf{u}_{\alpha_1}) + \sum_{\alpha_2=1}^{n_2} \lambda'_{\alpha_2}(\mathbf{u}) Y(\mathbf{u}'_{\alpha_2})$$

where the λ_{α_1}'s are the weights applied to the n_1 z samples and the λ'_{α_2}'s are the weights applied to the n_2 y samples.

Cokriging requires a joint model for the matrix of covariance functions including the *Z* covariance $C_Z(\boldsymbol{h})$, the *Y* covariance $C_Y(\boldsymbol{h})$, the cross *Z*-*Y* covariance $C_{ZY}(\boldsymbol{h}) = Cov\{Z(\boldsymbol{u}), Y(\boldsymbol{u}+\boldsymbol{h})\}$, and the cross *Y*-*Z* covariance $C_{YZ}(\boldsymbol{h})$.

More generally, the covariance matrix requires K^2 covariance functions when *K* different variables are considered in a cokriging exercise. The inference becomes extremely demanding in terms of data and the subsequent joint modeling is particularly tedious. Algorithms such as kriging with an external drift and collocated cokriging have been developed to shortcut the tedious inference and modeling process required by cokriging.

Another reason that cokriging is not used extensively in practice is the screen effect of the better correlated data (usually the z samples) over the data less correlated with the z unknown (the y samples). Unless the primary variable, that which is being estimated, is under-sampled with respect to the secondary variable, the weights given to the secondary data tend to be small, and the reduction in estimation variance brought by cokriging is not worth the additional inference and modeling effort.

Other than tedious inference and matrix notations, cokriging is the same as kriging. Cokriging with trend models and a cokriging that filters specific components of the spatial variability of either *Z* or *Y* could be developed. These notation-heavy developments will not be given here, but can be found in Journel and Huijbregts (1978) and Goovaerts (1997).

The three most commonly applied types of cokriging are as follows:

1. **Traditional ordinary cokriging:** the sum of the weights applied to the primary variable is set to one, and the sum of the weights applied to any other variable is set to zero. In the case of two variables, these two conditions are

$$\sum_{\alpha_1} \lambda_{\alpha_1}(\mathbf{u}) = 1 \text{ and } \sum_{\alpha_2} \lambda'_{\alpha_2}(\mathbf{u}) = 0$$

The problem with this traditional formalism is that the second condition tends to limit severely the influence of the secondary variable(s).

2. **Standardized ordinary cokriging:** often, a better approach consists of creating new secondary variables with the same mean as the primary variable. Then all the weights are constrained to sum to one.

In this case the expression could be rewritten as

$$Z^*_{COK}(\mathbf{u}) = \sum_{\alpha_1=1}^{n_1} \lambda_{\alpha_1}(\mathbf{u}) Z(\mathbf{u}_{\alpha_1}) + \sum_{\alpha_2=1}^{n_2} \lambda'_{\alpha_2}(\mathbf{u}) [Y(\mathbf{u}'_{\alpha_2}) + m_Z - m_{.}]$$

with the single condition $\sum_{\alpha_1=1}^{n_1} \lambda_{\alpha_1}(\mathbf{u}) + \sum_{\alpha_2=1}^{n_2} \lambda'_{\alpha_2}(\mathbf{u}) = 1$, where $m_Z = E\{Z(\boldsymbol{u})\}$ and $m_Y = E\{Y(\boldsymbol{u})\}$ are the stationary means of *Z* and *Y* respectively.

3. **Simple cokriging:** there is no constraint on the weights. Just like simple kriging, this version of cokriging requires working on data residuals or, equivalently, on variables whose means have all been standardized to zero. This is the case, for example, when applying simple cokriging in a Gaussian method, such as MG or UC, because the normal score transforms of each variable have a stationary mean of zero.

Except when using traditional ordinary cokriging, covariance measures should be inferred, modeled, and used in the cokriging system rather than variograms or cross variograms.

8.3.3 Collocated Cokriging

Collocated cokriging makes two simplifications (Zhu 1991). The first is that only one secondary variable is considered; the second is that the cross covariance is assumed to be a linear scaling of the variance. The reasoning behind this is that the collocated *y* value is surely more important than the other y values available in the neighborhood and likely screens the influence of multiple secondary data. Under this assumption, the cross variogram is no longer needed, and the ordinary cokriging estimator is re-written as

$$Z^*_{COK}(\mathbf{u}) = \sum_{\alpha_1}^{n_1} \lambda_{\alpha_1}(\mathbf{u}) Z(\mathbf{u}_{\alpha_1}) + \lambda'(\mathbf{u}) Y(\mathbf{u})$$

The corresponding cokriging system requires knowledge of only the Z covariance $C_Z(\mathbf{h})$ and the *Z-Y* cross covariance $C_{ZY}(\mathbf{h})$. A further approximation through the Markov model allows simplifying the latter:

$$C_{ZY}(\mathbf{h}) = \mathrm{B} \cdot C_Z(\mathbf{h}), \text{ for all } \mathbf{h}$$

where $B = \sqrt{\frac{C_Z(0)}{C_Y(0)}} \cdot \rho_{ZY}(0)$; $C_Z(0)$ and $\mathrm{C_Y}(0)$ are the variances of *Z* and *Y*, and $\rho_{ZY}(0)$ is the linear coefficient of correlation of collocated *z-y* data.

If the secondary variable $y(\mathbf{u})$ is densely sampled but not available at all locations being estimated, it may be estimated at those missing locations conditional to the *y* data. Under the collocated model and since the *y* values are only secondary data, the estimation of the missing *y* values should not impact the final estimate of the *Z* variable.

The Markov model is becoming widely used due to its simplicity. It can only be used when collocated secondary data will be used. If the secondary data are smooth then considering *y* values beyond the collocated values should not help.

Retaining only the collocated secondary datum does not affect the estimate (close-by secondary data are typically very similar in values), but it may affect the resulting cokriging estimation variance: that variance is overestimated, sometimes significantly. In an estimation context this is generally not a problem, because kriging variances are rarely used. In a simulation context where the kriging variance defines the spread of the conditional distribution from which simulated values are drawn, this may be a problem.

8.3.4 Collocated Cokriging Using Bayesian Updating

Bayesian updating is a technique closely related to collocated cokriging, but is designed for many secondary variables that are available to predict the primary data. The method can be subdivided into the following steps:

1. Calculate a prior distribution of uncertainty based on spatial information of the same type.
2. Calculate a likelihood distribution of uncertainty based on multivariate information at the location we are predicting.
3. Merge the prior and likelihood distributions into an updated (posterior) distribution.
4. Perform post processing with the updated distribution.

The prior distribution in (1) is calculated as the conditional distribution of each variable at each unsampled location *conditional to the surrounding data of the same type*. This is kriging from the surrounding data.

The likelihood distribution in (2) is calculated as the conditional distribution of each variable at each unsampled location *conditional to other data types at the same location*. This is accomplished using cokriging or perhaps some form multivariate linear regression.

The updated distribution (3) is created by merging the prior with the likelihood distribution. The arithmetic is exactly the same as collocated cokriging. The separate contributions of the secondary data and the data or the same variable are more easily understood compared to collocated cokriging.

Bayesian updating is appealing because it is simple. In cases where multiple secondary variables are available, there are few approaches that can be used with comparable ease. Still, the major steps in Bayesian updating are involved, and include: (1) data assembly and calculation of correlation; (2) calculation of likelihoods using secondary data; (3) calculation of prior probabilities of all variables, combining likelihoods and prior distributions into posterior distribution; (4) cross validation and checking; and (5) summarizing uncertainty and displaying results. The resulting distributions of uncertainty can be used for a qualitative assessment of local uncertainty.

The results of Bayesian updating should be used to complement conventional analysis. They provide a quantification of how the secondary data merge together to predict the primary variable(s). Bayesian updating, as do all other estimation techniques, is as good as and does not go beyond the data that have been input to the algorithm.

8.3.5 Compositional Data Interpolation

Interpolation of compositional data is a multivariate problem. From a geostatistical perspective, cokriging is the typical method. It is an exact linear estimator and under certain conditions is unbiased and provides a minimum estimation variance estimate.

Conditions that ensure these properties are true include: (1) the domain of interpolation, $\Re^D$, is unconstrained; (2) the variables are distributed according to a model that permits valid interpretation of results, for example normal or lognormal distributions; (3) data are second order stationary. The second condition is important when the data distribution is not unconstrained in $\Re^D$, and a transformation is required to do so. For Lognormally distributed data for example are restricted to positive real space and are transformed to follow a normal distribution via natural logarithms. Valid interpretation of results requires a back-transformation process for the expected value and variance of the estimates in transformed space. It must also be true that estimates in both spaces are exact, minimum estimation variance best and unbiased.

For compositional data, cokriging is not as well developed. Most available work applies cokriging to alr transformed compositions (Pawlowsky 1989; Pawlowsky et al. 1995; Martin-Fernandez et al. 2001). Direct cokriging of compositions has also been formulated (Walvoort and de Gruijter 2001), although direct statistical analysis of compositions is not advocated by Aitchison (1986) or Pawlowsky-Glahn and Olea (2004).

The kriging method explained by Walvoort and de Gruijter (2001) is a reformulation of cokriging as a constrained optimization problem. The objective is to minimize the estimation variance while adhering to the following constraints:

- Components of the compositions are positive: $\mathbf{x}^*(\mathbf{u}_0) \geq \mathbf{0}$
- The constant sum property of compositions: $\mathbf{1}^T\mathbf{x}^*(\mathbf{u}_0) = c$
- Ordinary kriging formulation: $\sum \Phi_k = \mathbf{I}$

It is difficult to state if this method provides the best or correct solution; however it is an alternative and has some advantages such as the ability to handle zeros in the data.

8.3.5.1 Additive Logratio Cokriging

This method was derived for alr transformed compositions. It can be applied to compositions following normal, lognormal or additive logistic distributions. The disadvantage of the technique is a lack of an analytical back-transformation of cokriging results, limiting its use to basic interpolation and mapping. Recall the alr transform and its inverse (alr^{-1}) where c is the constant sum constraint of the compositions, $\mathbf{x}(\mathbf{u}) \in \wp^D$ and $\mathbf{y}(\mathbf{u}) \in \Re^d$:

$$\begin{aligned}\text{alr}(\mathbf{x}(\mathbf{u}_k)) &= \mathbf{y}(\mathbf{u}_k)\\ &= \left[\log\left(\frac{x_i(\mathbf{u}_k)}{x_D(\mathbf{u}_k)}\right)\right], i=1,\ldots,d, k=1,\ldots,N\\ \text{alr}^{-1}(\mathbf{y}(\mathbf{u}_k)) &= \mathbf{x}(\mathbf{u}_k)\\ &= \left[\frac{\exp(y_1(\mathbf{u}_k))}{\sum_{i=1}^d \exp(y_i(\mathbf{u}_k))+1},\ldots,\right.\\ &\qquad \left.\frac{\exp(y_d(\mathbf{u}_k))}{\sum_{i=1}^d \exp(y_i(\mathbf{u}_k))+1}, c-\sum_{i=1}^d x_i(\mathbf{u}_k)\right]\end{aligned}$$

If $\mathbf{X}$ follows an additive logistic normal distribution, then $\mathbf{Y}=\text{alr}(\mathbf{X})$ follows a multivariate normal distribution, then the properties of $\mathbf{Y}$ being minimum estimation variance and unbiased are achieved. However, there is no interpretation for the expected value and variance of estimates of the original compositions. No analytical back-transformation process exists to calculate $\text{E}\{\mathbf{x}(\mathbf{u}_0)\}$ from $\text{E}\{\mathbf{y}(\mathbf{u}_0)\}$ and $\text{Var}\{\mathbf{x}(\mathbf{u}_0)\}$ from $\text{Var}\{\mathbf{y}(\mathbf{u})\}$ (Pawlowsky-Glahn and Olea 2004). Approximations are available, but Pawlowsky-Glahn and Olea (2004) identify that it is unverified if optimality conditions hold for back-transformed values in $\wp^D$ since the Euclidean metric does not apply in simplex space. The alr cokriging estimator is

$$\mathbf{x}^*(\mathbf{u}_0) = \text{alr}^{-1}\left(\mathbf{c} + \sum_{k=1}^N \Phi_k \text{alr}\left[\mathbf{x}(\mathbf{u}_k)\right]\right)$$

$$\mathbf{y}^*(\mathbf{u}_0) = \mathbf{c} + \sum_{k=1}^N \Phi_k \mathbf{y}(\mathbf{u}_k)$$

Application of the previous equation will result in an exact interpolation of compositions. The smoothing effect occurs as with kriging applied to unconstrained data. No other assertions can safely be made about results in the space of interpolated compositions.

8.3.6 Grade-Thickness Interpolation

When using service variables, the interpolation of the two variables involved, grade*thickness and thickness, are estimated independently, as long as the thickness and grade variables are not correlated. If they are, then some form of co-estimation (cokriging) is required. These variables can be estimated with any technique, geostatistical or not, but the most commonly used is Ordinary Kriging.

The final estimated grade is then obtained by dividing the two estimated variables:

$$G^*(x) = \frac{Acc^*(x)}{T^*(x)}$$

where $Acc^*(x)$ is the estimate accumulation variable (grade*thickness), $T^*(x)$ is the estimated thickness, and $G^*(x)$ is the final estimated grade. This is uncomplicated.

However, if we are interested in obtaining the estimation variance of $G^*(x)$, then we need to calculate the estimation variance of a quotient. This is less trivial, and the detailed development can be found in Journel and Huijbregts (1978, pp. 424–428). The final calculation will depend on whether grade and thickness are intrinsically coregionalized, and also whether we are calculating the estimation variance block by block, or globally for entire domains.

8.4 Block Kriging

Most variables estimated or simulated in mining average linearly. This allows to use the linear kriging estimator to obtain directly linear averages of the variable $z(\boldsymbol{u})$. Some of the most common examples are kriged block estimates of grades. Section 8.2 above already discussed estimated values on a different support than the original samples.

Consider the estimation of a block average, defined as:

$$z_v(\mathbf{u}) = \frac{1}{|V|}\int_{V(u)} z(\mathbf{u}')\cdot d\mathbf{u}' \simeq \frac{1}{N}\sum_{j=1}^{N} z(\mathbf{u}_j')$$

where $V(\mathbf{u})$ represent the block or panel centered at $\mathbf{u}$, and the $\mathbf{u}_j'$'s are the N points used to discretize the volume V.

Block kriging then amounts to estimating the individual discretization points $z(u_j')$ and then average them to obtain the block value $z_v(u)$. The "block kriging" system applies a different set of covariance values: the right-hand-side point-to-point covariance values are replaced by averages (point-to-block) covariance values of the form:

$$\bar{C}(V(\mathbf{u}),\mathbf{u}_\alpha) = \frac{1}{|v_\alpha|\cdot|v_\beta|}\int_{v(u_\alpha)} d\mathbf{u} \int_{v(u_\beta)} C(\mathbf{u}-\mathbf{u}')d\mathbf{u}' \simeq \frac{1}{N}\sum_{j=1}^{N} C(\mathbf{u}_j' - \mathbf{u}_\alpha)$$

Two major mistakes are sometimes made with respect to block kriging. The first is to estimate using block kriging variables that do not average linearly. Most of these are either geometallurgical or geotechnical variables (Chap. 5), for which a different estimation strategy is needed.

The second is to incorrectly apply non-linear transforms. One common example is log-normal kriging: the average of log-transforms is not the log-transform of the average of the $z(\mathbf{u}')$'s. Thus, the anti-log of a block estimate is not a kriged estimate, and is a biased estimate of the block value $z_V(\mathbf{u})$. Another example is indicator kriging (Chap 9).

8.5 Kriging Plans

The kriging plan mostly determines the quality of the grade estimate. The strategy used in the estimation of the variables has a very significant impact on the final estimates. Often is more significant than the actual kriging method chosen, which is the case for those kriging variants that use neighborhoods for local estimation (Boyle 2010; Rivoirard 1987; Vann et al. 2003).

There are several variables and parameters that constitute the kriging plan. The kriging neighborhood itself is chosen based on practical considerations, and will be different for different geologic and estimation domains.

Some of the factors that influence the decision on neighborhood size and geometry include: (1) The influence of far away data is screened or diminished by the closer data; (2) the variogram values at long distances are derived from the model $\gamma(\mathbf{h})$, not from the data; (3) if the number of data n chosen is large, the covariance matrices used to obtain the kriging weights are very large, and the weights themselves will be small and very similar for most data values; and (4) there are some cases when solving the kriging system may pose computational problems. This may be the case if clustering is present, the variogram model is very continuous, and the nugget effect is very small or 0; or when kriging on a string (Deutsch 1994, 1996).

The maximum search radii should be the limit of reliability and effectiveness of the variogram model, not just its range; recall that zonal anisotropy models can have very long ranges. It should also be related to data density. The distance may be anisotropic, following the anisotropy observed in the variogram model. If each nested structure of the variogram model is allowed to have different anisotropies, then search neighborhoods can be customized for different kriging passes.

Often the neighborhood is split into sectors (quadrants or octants), with only the nearest data being retained within each sector. This reduces the effect of clustered data, aiding the effect obtained from the term $\left[\bar{C}(v_i,v_j)\right]$ in the covariance matrix.

Another important decision is the minimum and maximum number of samples to be used in the estimation. Also, the related decisions regarding the minimum of drill holes and the number of samples per quadrant or octant to use. There is a direct relation between the number of samples used and the conditional bias vs. recoverable resource accuracy debate.

Other implementation decisions include the use of hard and soft boundaries among estimation domains, the use of multiple kriging passes or not, the number of discretization nodes if doing block kriging, and the use of different data sources, including secondary data.

All these parameters can be modified to a certain extent to obtain a resource model that achieves specific goals. Ideally, the process of setting up the kriging plans becomes iterative because some kind of calibration procedure is being used. The type of calibration that can be used depends on whether the mineral deposit being estimated is in production or not.

Comparisons with prior production are best, assuming that the data that reflects production is indeed reliable. These may be blast hole data in open pits; stope definition or grade control data in underground mines; or grades and tonnages fed to the mill. But in all cases, the reliability of production data should be demonstrable, since the resource estimation would be tuned to match production data. Care should be taken when considering the amount and location of the production data, ensuring that it is relevant to the future mining predicted by the resource model.

If no production data is available, then validations and some subjective criteria need to be used to define the kriging plans. The process is still iterative, but issues like reproduction of global declustered means; smoothing and dilution; behavior of alternative models; past resource models; grade profiles across contacts; and so on. These are all part of the resource model validation process discussed in Chap. 11.

8.6 Summary of Minimum, Good and Best Practices

In addition to some general comments on presentation and block model reporting, this section presents the details of what is considered minimum, good, and best practices in obtaining and handling ore resource models.

The estimated values obtained should be thoroughly checked and validated, as detailed in Chap. 11, where a summary of minimum, good, and best practices for model checking and validation is also presented. These checks should result in, among other comments, a statement about whether the model can be considered "recoverable" or fully diluted. Assuming that all checks have shown that the model is acceptable, other aspects need to be considered.

One important consideration is the reporting and transmittal of the resource model to mine planning. It is important to communicate the technical characteristics of the model, the sources and amount of dilution included, and the modeling methodology for all variables.

The minimum practice consists of documented parameters within each estimation domain, including specific kriging plans and using specific variogram models, according to the each domain's geologic and statistical characteristics. Clear documentation and justification of the kriging method chosen should be presented, as well as basic checks, see corresponding section in Chap. 11. The reporting of the model should clearly state the ore resource estimates, globally, by mining phases, benches, or whichever other volume may be appropriate. Grade-tonnage curves should be clear, and should include all relevant economic cutoffs and estimated variables. Subjective expressions of uncertainty are appropriate, as part of the limitations of the model. A report documenting in detail all relevant aspects should be prepared, including recommendations for improvements.

In addition to the above, good practice requires a more detailed justification of the estimates. Typically a degree of calibration is also required, if possible according to past production. At a minimum, the resource model should be able to reproduce past production to within a reasonable tolerance. The rationale for the kriging plan adopted should be clearly stated, as well as the "history" of the different estimation runs and iterations possibly performed. The model should be thoroughly checked for all variables involved, and according to the details described in Chap. 11. Comparisons with prior models and alternative models should also be made, and all differences explained, whether due to incremental data, or methodological differences. Potential risk areas and global uncertainty measures should be included, in addition to the standard resource classification scheme. The documentation trail should be complete and thorough, with assumptions and perceived limitations of the model clearly stated. Recommendations for future work, presented as part of a suggested risk mitigation plan, should be included.

Best practice includes the use of alternate models to check the results of the intended final resource model. All issues relating to the choice of estimation method, the parameters used, and the data selection adopted should be clearly stated and justified. All possible production and calibration data should be used to indicate whether the model is performing as expected, possibly including simulation models to calibrate the recoverable resource model. The validation of the estimates should include a validation of the calibration data. The model should be fully diluted, and should quantitatively describe the amount of the different types of dilution included. Checking and validation, as detailed in Chap. 11, should be fully implemented and documented. Model reporting should be complete, including all zones, domains, variables, and aspects of the model that are considered relevant. The documentation of the model should also be complete, and should include the best available visualization tools for the benefit of the uninitiated. All risk issues should be dealt with in detail, and if possible, quantified. This will generally require performing one or more simulation studies.

8.7 Exercises

The objective of this exercise is to review the theory and practice of kriging. Some specific (geo)statistical software may be required. The functionality may be available in different public domain or commercial software. Please acquire the required software before beginning the exercise. The data files are available for download from the author's website—a search engine will reveal the location.

8.7.1 Part One: Kriging Theory

Question 1: Derive the estimation variance in terms of the covariance. Explain where the assumption of stationarity comes into the derivation.

Question 2: Derive the simple kriging equations by taking the derivative of the estimation variance with respect to the weights. Explain why the kriging weights do not depend on the data values and why this is important. Explain why kriging is unbiased and exact.

Question 3: The covariance between the simple kriging estimate at an unsampled location and the data values used in kriging is correct, that is, the covariance model used in kriging. Demonstrate this result. Demonstrate that this result is **not** true for ordinary or universal kriging. Would the covariance be too high or too low with ordinary kriging.

Question 4: Derive the variance of the kriging estimate $Var\{Y^*(\mathbf{u})\}$ and express the result in terms of the kriging variance. Comment on the importance of this result relative to the smoothness of kriging.

8.7.2 Part Two: Kriging by Hand Question

Consider the configuration to the right. The global mean is 1.3 and the variance is 0.2. Calculate the simple kriging estimate at the unsampled location given that the isotropic covariance function C(h) = exp(−3h/275). Show all steps clearly.

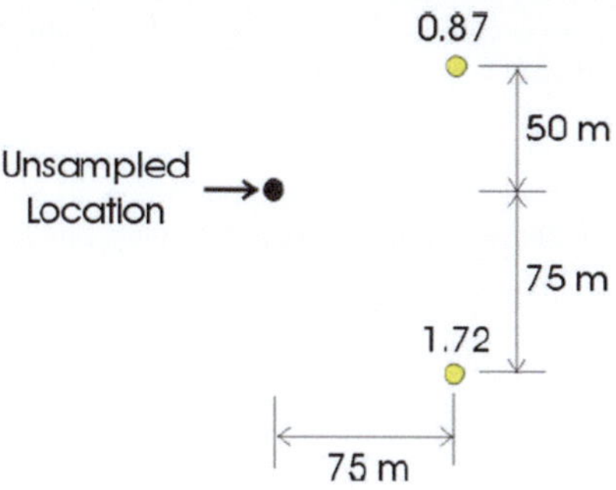

8.7.3 Part Three: Conditional Bias

A serious problem with kriging is conditional bias. Use the `largedata.dat` data to experiment with conditional bias in cross validation mode.

Question 1: Set up to cross validate the `largedata.dat` data. Consider ordinary kriging with a search radius equal to the variogram range. Vary the number of data between 2 and 40 (consider 2, 4, 8, 16, and 40), plot the cross plots of true versus estimate, fit linear regression lines, and plot the slope of the regression line versus the number of data used for kriging. Comment on the results.

Question 2: Plot grade tonnage curves for the distributions of question 1 and see which one most closely matches the grade tonnage curve from volume variance (if those results are available). Comment on the application of this procedure in practice.

8.7.4 Part Four: Kriging a Grid

Question 1: Set up to estimate a reasonable grid using `largedata.dat`. Create a 3-D model with stationary simple kriging with a discretization of $1\times1\times1$ (this most closely approximates what will happen in simulation).

Question 2: Evaluate the results of your kriging: (1) plot some sections or visualize in 3-D checking that the map looks reasonable and that the original data are reproduced at their locations, (2) compare the histogram of the estimates to the original data, and (3) compare the variogram of the estimates to the original data. Many of these evaluations will have to be done on a by-rock-type basis. Comment on the results.

Question 3: Create a model with ordinary kriging and compare the results to the initial simple kriged model. Investigate how the variability of the block estimates depends on the number of data used (particularly for a small number, say, 4).

Question 4: Create a model with block kriging by discretizing the cells by at least 9 points. Keep all other parameters comparable to a run that you have constructed. Comment on the results.

References

Aitchison J (1986) The statistical analysis of compositional data: monographs on statistics and applied probability. Chapman and Hall, London, p 416

Baafi EY, Kim YC (1982) Comparison of different ore reserve estimation methods using conditional simulation. AIME annual meeting, Preprint, pp 82–94

Boyle C (2010) Kriging neighbourhood analysis by slope of regression and weight of mean—evaluation with the Jura data set. Min Technol 119(2):49–58

Chilès JP Delfiner P (2011) Geostatistics: Modeling spatial uncertainty, 2nd ed. Wiley Series in Probability and Statistics, New York, p 695

Delfiner P (1976) Linear estimation of non-stationary spatial phenomena. In: Guarascio M, David M, Huijbregts CJ (eds) Advanced geostatistics in the mining industry. Reidel, Dordrecht, pp 49–68

Deutsch CV (1994) Kriging with strings of data. Math Geol 26(5):623–638 (November)

Deutsch CV (1996) Correcting for negative weights in ordinary kriging. Comp Geosci 22(7):765–773

Deutsch CV (2002) Geostatistical reservoir modeling. Oxford University Press, New York, p 376

Deustch CV, Journel AG (1997) GSLIB: Geostatistical software library and user's guide, 2nd edn. Oxford University Press, New York, p 369

Dominy SC, Noppé MA, Annels AE (2002) Errors and uncertainty in mineral resource and ore reserve estimation: the importance of getting it right. Explor Min Geol 11:77–98

Goovaerts P (1997) Geostatistics for natural resources evaluation. Oxford University Press, New York, p 483

Guertin K (1984) Correcting conditional bias in ore reserve estimation. PhD Thesis, Stanford University

Isaaks EH (2004) The kriging oxymoron: A conditionally unbiased and accurate predictor, 2nd edn. In: Proceedings of geostatistics banff 2004, 1:363–374, Springer 2005

Isaaks EH, Srivastava RM (1989) An introduction to applied geostatistics. Oxford University Press, New York, p 561

Journel AG, Huijbregts ChJ (1978) Mining geostatistics. Academic Press, New York

Journel AG, Rossi ME (1989) When do we need a trend model? Math Geol 22(8):715–738

Kalman RE (1960) A new approach to linear filtering and prediction problems. Trans ASME, Ser D: J Bas Eng 82, 35.45

Knudsen HP (1990) Computerized conventional ore reserve methods. In: Kennedy BA (ed) Surface mining, 2nd ed. SME, Littleton, pp 293–300

Knudsen HP, Kim YC, Mueller E (1978) A comparative study of the geostatistical ore reserve estimation method over the conventional methods. Mining Eng 30:54–58

Krige DG (1994) A basic perspective on the roles of classical statistics, data search routines, conditional biases and information and smoothing effects in ore block valuations. Conference on mining geostatistics, Kruger National Park

Krige DG (1996) A Practical analysis of the effects of spatial structure and data available and used. In: Conditional biases in ordinary kriging. Fifth international geostatistics congress, Wollongong

Krige DG (1999) Conditional bias and uncertainty of estimation in geostatistics. Keynote address for APCOM'99 international symposium, Colorado School of Mines, Golden, October, 1999

Luenberger DL (1969) Optimization by vector space methods. Wiley, New York, p 326

Martin-Fernandez JA, Olea-Meneses RA, Pawlowsky-Glahn V (2001) Criteria to compare estimation methods of regionalized compositions. Math Geol 33(8):889–909

Matheron G (1971) The theory of regionalized variables and its applications. Fasc. 5, Paris School of Mines, Paris, p 212

Matheron G (1973) The intrinsic random functions and their applications. Adv Appl Probab 5:439–468

McLennan JA, Deutsch CV (2004) Conditional non-bias of geostatistical simulation for estimation of recoverable reserves. CIM bulletin, 97

Pan G (1998) Smoothing effect, conditional bias, and recoverable reserves; Can Inst Min Metall Bull 91(1019):81–86

Pawlowsky V (1989) Cokriging of regionalized compositions. Math Geol 21(5):513–521

Pawlowsky V, Olea RA, Davis JC (1995) Estimation of regionalized compositions: A comparison of three methods. Math Geol 27(1):105–127

Pawlowsky-Glahn V, Olea RA (2004) Geostatistical analysis of compositional data. Oxford University Press, New York, p 304

Popoff CC (1966) Computing reserves of mineral deposits: Principles and conventional methods. Information Circular 8283, US Bureau of Mines

Readdy LA, Bolin DS, Mathieson GA (1982) Ore reserve calculation. In: Hustrulid WA (ed) Underground mining methods handbook. AIME, New York, pp 17–38

Rivoirard J (1987) Two key parameters when choosing the kriging neighborhood. Math Geol 19(8):851–856

Sinclair AJ, Blackwell GH (2002) Applied mineral inventory estimation. Cambridge University Press, New York, p 381

Srivastava RM (1987) Minimum variance or maximum profitability. CIMM 80(901):63–68

Stone JG, Dunn PG (1996) Ore reserve estimates in the real world. Society of Economic Geologists, Special publication 3:150

Vann J, Jackson S, Bertoli O (2003) Quantitative kriging neighbourhood analysis for the mining geologist—a description of the method with worked case examples. In: 5th international mining geology conference, Bedigo, 17–19 November, 2003(8):215–223

Walvoort DJJ, de Gruijter JJ (2001) Compositional kriging: a spatial interpolation method for compositional data. Math Geol 33(8):951–966

Zhu H (1991) Modeling mixtures of spatial distributions with integration of soft data. Ph.D. Thesis, Stanford University, Stanford

Recoverable Resources: Probabilistic Estimation

9

Abstract

The block estimates made by conventional inverse distance or kriging techniques have no reliable measure of uncertainty attached to them. The approach presented in this chapter consists of directly predicting the variability/uncertainty in the mining block grades based on a probability distribution model. The limitations and assumptions supporting these models are summarized, as well as some of the most important issues regarding the estimation of point and block distributions.

9.1 Conditional Distributions

Conditional probability functions are alternatives to the estimation of a single point or block grade. The original data are interpreted to provide a conditional distribution function that is updated locally to obtain posterior probability distribution functions at each unsampled point/block location. This function is represented as a cumulative conditional distribution function, or ccdf, and describes the range of possible values that the estimate can take. The ccdf is written as

$$F(z,\mathbf{u}\,|\,(n)) = Prob\{Z(\mathbf{u}) \le z\,|\,(n)\}$$

where "$|(n)$" means conditional to the nearby information used to derive the ccdf. This function contains all information that may be available about the unknown location. Basic distribution parameters that can be extracted are: E-type or average estimates; probabilities of the grade exceeding critical thresholds; probabilities of the grade being within certain thresholds; and so on.

Nonlinear Transforms The first step to understand the methods used to estimate probability functions is to understand nonlinear geostatistics. All nonlinear kriging algorithms are actually linear kriging (SK or OK) applied to specific nonlinear transforms of the original data. The nonlinear transform used specifies the nonlinear kriging algorithm considered. The transforms lead to methods that are classified as parametric or non parametric.

The parametric approach to building probabilistic models was developed early in geostatistics (Matheron 1971, 1973; Marcotte and David 1985). The methods are based on assuming a multivariate or bivariate distribution for the RF model $\{Z(\mathbf{u}), \mathbf{u} \in A\}$. This assumption entails that all ccdfs are fully specified by a limited number of parameters.

The most common parametric approach is the Gaussian-based methods, of which the log-normal transformation can be considered a special case. Since Gaussian distributions are uniquely characterized by its mean and variance, the problem of determining the ccdf model $Prob\{Z(\mathbf{u}) \le z\,|\,(n)\}$ becomes one of estimating the two parameters of the model. Specific examples are multi-Gaussian kriging; Disjunctive Kriging; Uniform Conditioning; and Lognormal kriging. The limitation of the parametric approach lies as much in establishing the appropriateness of the model, as in estimating its parameters.

Non-parametric transformations derive in methods that do not make strong univariate assumptions about the distribution; rather, they directly estimate a number of probabilities of the $Prob\{Z(\mathbf{u}) \le z\,|\,(n)\}$ function, and then interpolate these to obtain the full ccdf. All variants of indicator kriging fall into this category, including probability kriging.

The non-parametric methods rely heavily on the quantity and quality of data available. The modeling process is more time consuming because it requires inference of more spatial continuity models. There must be sufficient information within the stationary domain to implement a robust and reliable non-parametric estimation.

M. E. Rossi, C. V. Deutsch, *Mineral Resource Estimation,* DOI 10.1007/978-1-4020-5717-5_9,

9.2 Gaussian-Based Kriging Methods

The popularity of Gaussian methods comes from their simplicity, the properties of the multivariate Gaussian distribution, and also from the fact that they can produce acceptable estimates. While some characteristics of the Gaussian distribution imply significant drawbacks, its simplicity and ease of use make Gaussian-based probabilistic estimation and simulation highly popular. Some of the convenient properties of the multivariate Gaussian RF were discussed in Chap. 2.

Gaussian methods are maximum entropy methods in the sense that, for a given mean and variance, the distribution will tend to produce estimates that are as disorganized as possible. The implication is that in cases where the connectivity of extreme values is high (low entropy), Gaussian methods will not reproduce that connectivity. The geology of most deposits is characterized by structuring, and this is reflected in typical geologist's interpretations, geologic models, and the theories about the genesis of deposits.

The maximum entropy or destructuring effect of the multi-Gaussian RF can be better understood analyzing their indicator variograms (Journel and Posa 1990). It can be shown that for $y_p \to 0$ or $y_p \to 1$, $y_p^2 \to \infty$ and the two-point cdf tends to be independent. Therefore, the indicator variogram tends to its theoretical sill, $\sigma^2 = \mathrm{p} \cdot (1 - \mathrm{p})$ where p is the CDF value for the indicator transform.

Another important property of Gaussian methods is homoscedasticity, which means that the conditional variance $\sigma_K^2(\mathbf{u})$ does not depend on the actual data values. This is an unusual property in the statistical world, since no other RF has it. In practice, it is known that many variables are heteroscedastic, evidencing the proportional effect discussed before. The dependency of the variance to the data values is removed when transforming the data to the Gaussian distribution.

If developing conditional simulations (Chap. 10), the back-transformation is delayed until the simulation is complete. However, the back-transformation is necessary when estimating a conditional distribution for $Z(\mathbf{u})$ (Verly 1984; Journel 1980); the proportional effect and other characteristics may not be reproduced adequately in the estimates. The back-transform is quite sensitive to the variance of the conditional distribution in Gaussian units, $\sigma_y^2(\mathbf{u})$, making the Z-estimate potentially unstable. There are procedures that can be applied to dampen this effect (Journel 1980; Parker et al. 1979), but they have limitations.

9.2.1 Multi-Gaussian Kriging

The multivariate Gaussian RF model is the most widely used of the parametric models. Multi-Gaussian kriging is straightforward. If the multivariate Gaussian assumption is made, the simple kriging estimate $y_{SK}^*(\mathbf{u})$ and simple kriging variance $\sigma_{SK}^2(\mathbf{u})$ of the normal score data are the parameters of the conditional Gaussian distribution frequency (Journel and Huijbregts 1978, p. 566).

The covariance modeled from the sample normal score covariance must be used in the MG kriging system, but otherwise is the same system of equations described in Chap. 8. With both parameters estimated, the full ccdf is then modeled using:

$$[G(\mathbf{u}; y|(n))]_{SK}^* = G\left(\frac{y - y_{SK}^*(\mathbf{u})}{\sigma_{SK}(\mathbf{u})}\right)$$

The MG SK estimate requires strict stationarity, that is, a known mean at each location over the entire domain and all possible sub-domains.

MG can also be implemented using OK and KT as options for quasi or non-stationary domains. This is done by simply using the OK or KT algorithms (in any of its variants) on the normal score data. However, the kriging variances of the OK or KT systems are no longer the correct variances for the estimated Gaussian distribution. Only the SK variance is theoretically correct (Journel 1980). All other variances, because they result from constrained systems, will tend to inflate the variance of the estimated Gaussian distribution. This inflated variance could induce a bias in the back transformed estimates. Therefore, the OK and KT variances must be corrected to account for this difference.

An alternative for KT is to de-trend the original z data first, and then normal score transform the residuals. The residuals are assumed strictly stationary, with a mean of 0, and therefore MG with SK can be more robustly applied. However, the MG estimated distribution is still dependent on the effectiveness of the de-trending process.

9.2.2 Uniform Conditioning

Uniform Conditioning (UC) is a Gaussian method used to estimate the recoverable resources of a mining (SMU) panel. The method requires a panel estimate and a change of support model. The panel grade is assumed to be known, then the distribution of SMUs within that panel can be established from a bivariate Gaussian model. The name uniform conditioning arises because of the assumption that the estimates of the recoverable resources are conditioned to the same data configuration for every panel grade.

The idea is to estimate a panel much larger than the SMUs. The size of the panel is based on fitting a reasonable number of SMUs within it, and is generally irrespective of drill hole spacing; however, often the actual drill hole spacing or

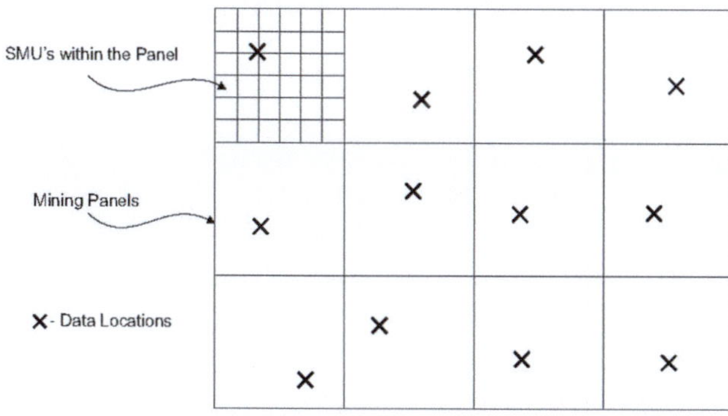

Fig. 9.1 Schematic showing the setting for UC. Notice that there are three support sizes of interest: the data, the SMUs and the panels

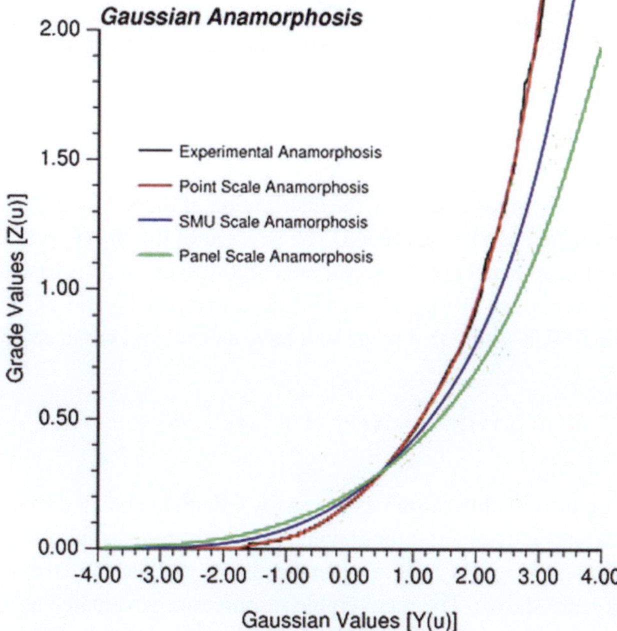

Fig. 9.2 Graphical representation of anamorphosis and the transformation of grades

slightly larger is used. The panel grades should be estimated with little uncertainty. Figure 9.1 shows a schematic of the scales that uniform conditioning considers.

The earliest reference to UC is Matheron (1974). Some further references for UC are Remacre (1987), Guibal and Remacre (1984), Rivoirard (1994), Vann and Guibal (1998), Chilès and Delfiner (2011), and Roth and Deraisme (2000). In general, there are few published detailed references on the theory and practice of UC. The method has gained some popularity due to its simplicity and availability in commercial software. The following description is based on the more recent compilation by Neufeld (2005).

There are 6 basic steps required to complete UC: (1) estimate the panel grades; (2) fit the Discrete Gaussian Model (DGM) to the data; (3) determine the change of support coefficients for the SMU and panel sized blocks; (4) transform the $Z^*(V)$ panel estimates to the Gaussian variable $Y^*(V)$ using the panel anamorphosis function; (5) transform the $Z_c(v)$ cutoff grades to $Y_c(v)$ using the SMU anamorphosis function; and (6) calculate the proportion and quantity of metal above each cutoff.

The first step in UC is to estimate the panel grades, most commonly using ordinary kriging. Uniform conditioning relies on a robust estimate of the panel grade. Also, the panel grade estimates have to be uniformly robust, that is, with nearly the same amount of data. Large panels are used because the block kriged panel grades are more robust for larger panels. Also, in UC the SMUs are seen as discretizing the larger panel, and there should be a sufficient number of them. Ordinary kriging in the original grades units is the most popular option used, although the panel grade can also be estimated directly in Gaussian units, see, for example, Guibal (1987). The size of the panel and the SMUs must first be chosen to complete the OK of the panel. Also, the grade directional variograms and corresponding models $\gamma_z(\boldsymbol{h})$ must be obtained. Although not required, it is common practice to set the discretization of the block (panel) kriging to the SMU resolution within the panel.

The second step is to fit a DGM model to the point scale data, which is used to perform the change of support from sample support to the larger panels and also to the SMU-size blocks. The DGM model was described in Chap. 7, and can also be found in several references, for example Journel and Huijbregts (1978, p 472). Figure 9.2 shows a graphical representation of Gaussian anamorphosis.

The DGM model is applied to get the anamorphosis function for the SMU sized blocks. The same procedure is repeated to calculate the change of support for the panel sized blocks, using the r' coefficient corresponding to the larger panel size V. The dispersion variance calculated from the variogram models for the panels provides the theoretical variance to be used to obtain r'.

The panel variance is found as before solving for r' in the following equality:

$$\sigma_V^2 = \sum_{p=1}^{n} (r')^{2p} \phi_p^2$$

It can be shown that the correlation coefficient of the bi-Gaussian distribution $[Y(\boldsymbol{v}), Y(\boldsymbol{V})]$ is:

$$\rho(v, V) = \frac{r'}{r}$$

Figure 9.3 shows the anamorphosis transformation for both the panel and the SMU grades.

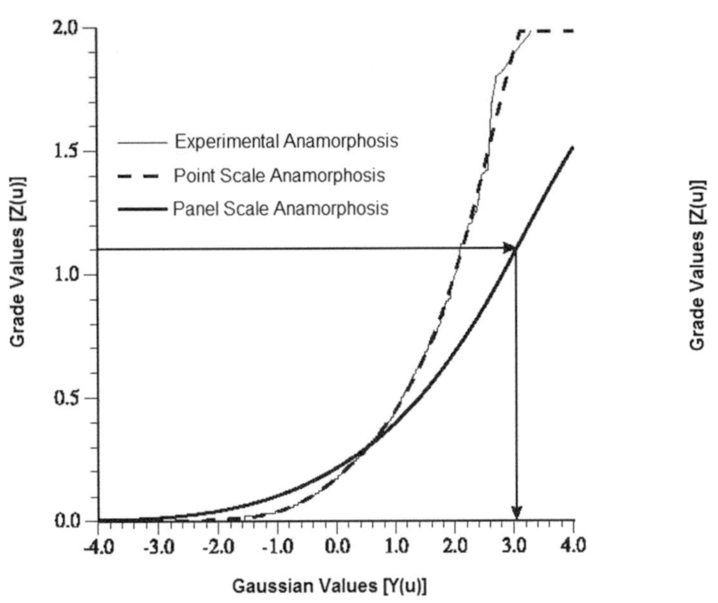

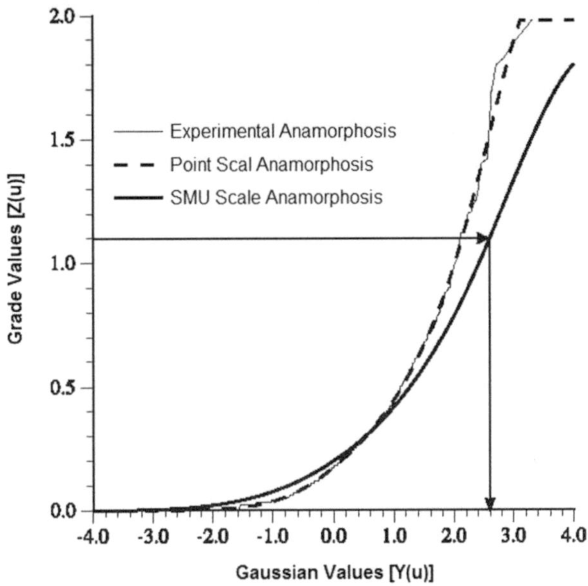

Fig. 9.3 Transformation of the panel grades (*left*) and the SMU grades (*right*) to Gaussian units

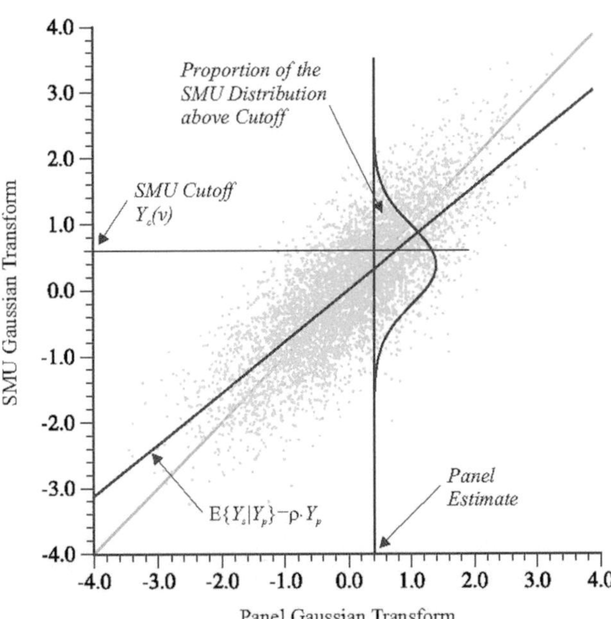

Fig. 9.4 Transformation of the panel grades (*left*) and the SMU grades (*right*) to Gaussian units. This figure is adapted from the Isatis User's Manual

Step 4 is to transform the estimated panel grades to Gaussian units using the panel anamorphosis equation.

Step 5 is to transform the cutoff grades to Gaussian values using the modeled SMU anamorphosis. Under the bivariate normal assumption, knowing the Gaussian panel grade allows us to calculate the mean and variance of the conditional SMU distribution. For this transformation the SMU anamorphosis is used.

Finally, the proportion and the grade above cutoff for the conditional SMU distribution can be calculated. Given that the panel grade is known, the distribution of the SMUs within that panel can be calculated. The average of the SMUs is the estimated panel grade; the variance in Gaussian units is based on the change of support coefficients. For a panel grade, $y(V)$, the SMUs within that panel will have a mean and variance of

$$E\{y(v)\} = \frac{r'}{r} \cdot y(V); \quad Var\{y(v)\} = 1 - \left(\frac{r'}{r}\right)^2$$

The recoverable resources are easily calculated using the bivariate Gaussian assumption and the anamorphosis function. Consider Fig. 9.4, where a panel estimate and SMU distribution are shown. The recoverable resources are evaluated as a truncated statistic, the SMU cutoff $Y_c(v)$.

The tonnage (or proportion) above cutoff can be calculated as:

$$P(z_c) = P\left[z(v) \geq z_c \mid z(V)\right] = P\left[y(v) \geq y_c \mid y(V)\right] =$$

$$1 - G\left(\frac{y_c - \left(\frac{r'}{r}\right) y(V)}{\sqrt{1 - \left(\frac{r'}{r}\right)^2}}\right)$$

The quantity of metal and the mean grade can be calculated in two different ways, which give very similar results: (1) by integrating the conditional distribution above cutoff (Rivoirard 1994):

$$Q(z_c) = \int_{y_c}^{\infty} \Phi_v(y(v)) \cdot g(Y(v) \mid Y(V)) \cdot d(y(v))$$

or (2) by using the fitted Hermite polynomials, see Remacre 1987.

The final grade above cutoff is simply $M(z_c) = \frac{Q(z_c)}{P(z_c)}$. In practice, and to avoid spurious results, $P(z_c)$ is only considered when it is greater than 1 %.

UC is based on a couple of important assumptions: (1) that the Gaussian transformed point data, SMU and panels are all bivariate normal; and (2) that the change of support model for the SMUs can be extended to the panels, an assumption common to other change of support models.

UC has at least two important limitations. First, it does not provide any information regarding where the high or low grade SMUs are within the panel, which is convenient for most mine planning. This is not a theoretical limitation, since it is part of the underlying premise of UC: panel grades can be predicted reliably, but SMU grades cannot. But it is a major practical limitation, since mine planners require the location of the SMUs in order to calculate recoverable reserves. This is the most significant unresolved practical issue.

The second significant limitation is that panels with the same estimate will have the same grade and proportion curves irrespective of the surrounding data. Surrounding drill hole samples are used to estimate the panel grade, but they are not used for determining the SMU distribution. For example, consider two panel estimates: one panel is in a homogenous zone where the surrounding samples have all the same values, and the other where the surrounding samples result in the same estimated panel average but are highly variable. When the panel estimates are the same the estimated recoverable reserves are the same. This limitation implies that UC is more sensitive to departures from stationarity compared to other methods, despite using OK to estimate panel grades.

Therefore, UC is only recommended at an early stage of project development. UC is useful when data are uniformly sparse and the SMU blocks cannot be reliably estimated. Other estimation methods that directly estimate smaller blocks would not be considered reliable with widely spaced exploration drilling. When there is sufficient infill drilling or for an operational mine, it is likely that other methods will result in better local estimates.

9.2.3 Disjunctive Kriging

Disjunctive kriging (DK) was introduced by Matheron (1974, 1976). It is a method that also relies on a bi-Gaussian assumption and uses Hermite polynomials to transform the original data into additive functions. The purpose is to estimate recoverable grades and tonnages above cutoffs for any size block.

The method is based in decomposing the variable into a sum of uncorrelated orthogonal factors, for which the optimal solution is found with a simple kriging (SK) of each component.

It is a mathematically involved method that will not be described here in detail. Some additional references are Armstrong and Matheron (1986), Chilès and Delfiner (2011), and Rivoirard (1994). Also, a readable summary is presented in Kumar (2010). There have been only a few applications in the mining industry, the most significant drawback being its theoretical complexity. Also, since DK depends heavily on stationarity because it requires the theoretically correct SK, the estimates are very sensitive to the stationarity assumption. Experience has shown that the estimated values tend to be extremely smooth, with little variability around the mean.

9.2.4 Checking the Multivariate Gaussian Assumption

The normal score transform (or its equivalent anamorphosis) ensures that the one-point distribution is Gaussian. This is a necessary but not sufficient condition to prove that a RF model is multivariate Gaussian. Theoretically, the normality of the two-point, three-point, and in general n-point distributions should be checked (Verly 1984).

While two-point statistics can be inferred from the data, three-point and higher order statistics are much more difficult to obtain. The corresponding analytical expressions exist, but sparse data and irregular grids do not allow inference from sampled values. Also, the practical significance of the multivariate Gaussian assumption is limited, since the vast majority of the MG applications are bivariate. In practice only the two-point distribution is checked for bi-normality: if the bi-Gaussian assumption holds, then the multi-Gaussian formalism is adopted.

Checking the Gaussianity of two-point distributions implies checking that the experimental cdf values of any set of sample pairs separated by any vector $\boldsymbol{h}$ match the theoretical Gaussian distribution. In practice, the comparison is made for corresponding p-quantile values, $y_p = y_p'$, such that the two-point Gaussian distribution becomes:

$$G(\mathbf{h}; y_p) = Prob\{Y(\mathbf{u}) \le y_p, Y(\mathbf{u}+\mathbf{h}) \le y_p\}$$
$$= p^2 + \frac{1}{2\pi} \int_0^{\arcsin C_Y(h)} \exp\left[\frac{-y_p^2}{1+\sin\theta}\right] d\theta \quad (9.1)$$

The steps required to complete the actual checking process are:

1. The variogram $\gamma_Y(\mathbf{h})$ of the normal score data is computed and modeled, and from this the covariance model $C_Y(\mathbf{h})$ is obtained.

2. Recall that the two-point Gaussian cdf $G(\mathbf{h}; y_p)$ is the non-centered indicator covariance for the threshold y_p (Chap. 6). Then, if expression 9.1 is evaluated for a series of p-quantile values y_p, the corresponding indicator variogram is derived with:

$$\gamma_I(\mathbf{h}; y_p) = p - G(\mathbf{h}; y_p)$$

In this case, the indicator function is defined as $I(\mathbf{u}; y_p) = 1$ if $Y(\mathbf{u}) \le y_p$, 0 otherwise.

3. The experimental indicator variograms of the normal score data $\hat{\gamma}_I(\mathbf{h}; y_p)$ are obtained for the same p-quantiles.
4. The indicator variograms obtained experimentally $\hat{\gamma}_I(\mathbf{h}; y_p)$ and the theoretical Gaussian-deduced $\gamma_I(\mathbf{h}; y_p)$ are compared graphically. Based on the quality of the comparison, the bi-Gaussian assumption may be rejected or accepted.

An additional check is to verify that the pattern of indicator spatial correlation is symmetric with respect to the p-quantiles, that is $\gamma_I(\mathbf{h}; y_p) = \gamma_I(\mathbf{h}; y_{p'})$, where $p' = 1 - p$. The experimental indicator variograms of the normal score data for p and p' should match well. This checking is not common because it is difficult to assess if the differences are significant.

9.3 Lognormal Kriging

A particular case is found if the stationary multivariate RF is assumed lognormal. In this case, the RF $Y(\mathbf{u}) = \ln Z(\mathbf{u})$ has a multivariate Gaussian distribution with mean m', covariance $C'(\mathbf{h})$, and variance $\sigma'^2 = C'(0)$. The relationship between the arithmetic and logarithmic moments is (Journel and Huijbregts 1978, p. 570):

$$m = e^{m' + \sigma'^2/2}$$

and

$$C(h) = m^2\left[e^{C'(h)} - 1\right] \Rightarrow \sigma^2 = m^2\left[e^{\sigma'^2} - 1\right]$$

In the method implementation the original data is log transformed: $y(\mathbf{u}) = \ln z(\mathbf{u})$. The $z(\mathbf{u})$ variable has to be strictly positive. Simple or ordinary kriging of the log data yields an estimate $y^*(\mathbf{u})$ for $\ln z(\mathbf{u})$. Unfortunately, a good estimate of $\ln z(\mathbf{u})$ is not necessarily a good estimate of $z(\mathbf{u})$; in particular the antilog back-transform $e^{y^*(u)}$ is a biased estimator of $Z(\mathbf{u})$ as can be derived from the relationships above. The unbiased back-transform of the simple lognormal kriging estimate $y^*(\mathbf{u})$ is actually:

$$z^*(u) = e^{[y^*(u) + \sigma_{SK}^2(u)/2]}$$

where $\sigma_{SK}^2(\mathbf{u})$ is the simple lognormal kriging variance. In practice, the theoretically non-biased estimate $z^*(\mathbf{u})$ often differs from the expected value m. This is due to the exponentiation involved in that back-transform. It is particularly problematic since it significantly amplifies any error in the estimation of the lognormal estimate $y^*(\mathbf{u})$ or its SK variance $\sigma_{SK}^2(\mathbf{u})$.

Another problem commonly encountered is that the estimated values increase as the kriging variance increases. This is a serious issue because higher kriging variances correspond to sparsely populated areas, and thus it can result in overestimation of the estimated values. Among others, this problem was critical for several gold mines in northern Nevada (USA) in the 1980s, because the predicted grades increased away from the main mineralized zones. A similar artificial grade trend can be produced by this method in many other types of deposits, such as porphyry copper deposits.

This extreme sensitivity to the back-transform explains why lognormal kriging is not used any more. The method has been largely replaced by other Gaussian approaches or the indicator kriging approach. While there are exceptions, the use of lognormal kriging is mostly confined to South Africa, where the method was originally developed (Sichel 1952, 1966; Krige 1951).

9.4 Indicator Kriging

The indicator kriging-based estimation methods are non-parametric in the sense that they do not make any prior assumption about the distribution being estimated. The objective of the method is not to estimate parameters of an assumed distribution, but directly estimate the distribution itself (Journel 1983).

Consider the binary transform of the original $Z(\mathbf{u})$ variable defined as:

$$I(\mathbf{u}; z_k) = \begin{cases} 1, & \text{if } Z(u) \le z_k \\ 0, & \text{otherwise} \end{cases} \tag{9.2}$$

The indicator formalism consists of discretizing the continuous variable z with a series of K threshold values z_k, $k = 1, \ldots, K$. The experimental cdf of the n samples within the stationary domain are considered a prior distribution, which can be obtained through an equal-weighted average:

$$F(\mathbf{u}; z_k) = Prob\{Z(\mathbf{u}) \le z_k\} = \frac{1}{n}\sum_{\alpha=1}^{n} i(\mathbf{u}_\alpha; z_k)$$

This is the proportion of the samples $z(\mathbf{u}_\alpha)$ below the cutoff z_k. In this prior cumulative distribution frequency, the samples could be weighted to account for spatial clustering. The indicator RV $I(\mathbf{u};z_k)$ has only two possible outcomes, 1 or 0. Thus, by definition of the expected value:

$$E\{I(\mathbf{u};z_k)\} = 1\cdot Prob\{I(\mathbf{u};z_k)=1\} + 0\cdot Prob\{I(\mathbf{u};z_k)=0\}$$
$$= E\{I(\mathbf{u};z_k)\} = 1\cdot Prob\{I(\mathbf{u};z_k)=1\} = Prob\{Z(\mathbf{u})\le z_k\}$$

These relations still hold for conditional expectations, such that:

$$E\{I(\mathbf{u};z_k)|(n)\} = Prob\{Z(\mathbf{u})\le z_k|(n)\} = F(\mathbf{u};z_k|(n))$$

The practical consequence is that a conditional cdf can be built by assembling K indicator kriging estimates. This ccdf represents a probabilistic model for the uncertainty about the unsampled value $z(\mathbf{u})$.

$$[i(\mathbf{u};z_k)]^* = E\{I(\mathbf{u};z_k)|(n)\}^* = Prob^*\{Z(\mathbf{u})\le z_k|(n)\}$$

which can be obtained with a weighted linear average. The optimal weights are given by a kriging system on the indicator data:

$$E\{I(\mathbf{u};z_k)\,|\,(n)\}^* = [i(\mathbf{u};z_k)]^* = \sum_{\alpha=1}^{n} \lambda_\alpha(\mathbf{u};z_k) i(\mathbf{u}_\alpha;z_k)$$

Since several thresholds k are used, this is usually called multiple indicator kriging (MIK). The weights and the ccdf $F(\mathbf{u};z_k)$ are dependent on both the location and a number of thresholds z_k, $k=1,\ldots,K$. Thus, there is one indicator variogram $\gamma_I(\mathbf{u};z_k)$ and one kriging system per threshold. While the inference is more time-consuming, the flexibility is greater, and no prior assumption of any type of distribution is made.

Another advantage of indicator methods is that no back-transformation is required, since working with indicator variables directly yields a ccdf model for the RV $Z(\mathbf{u})$. Another important aspect of the IK method is that it can be applied equally to continuous or categorical variables. In what follows, references made to continuous variables also apply to categorical ones.

There are challenges with IK including (1) inference of the distribution details, particularly above the highest threshold used in the kriging, (2) the greater effort required to infer all of the required parameters such as the variograms, (3) the inevitable multivariate Gaussian flavor in high order distributions (>2) because of averaging and (4) the practical use of either probabilities or a smooth estimator.

9.4.1 Data Integration

The indicator formalism allows for a more straightforward integration of different data types. There are four types of data that can be used in indicator coding. The first is local hard indicator data $i(\mathbf{u}_\alpha;z)$ originating from local hard data $z(\mathbf{u}_\alpha)$, as in Eq. 9.2.

The second type is a local hard indicator data $j(\mathbf{u}_\alpha;z)$ originating from ancillary information that provides hard inequality constraints on the local value $z(\mathbf{u}_\alpha)$. If $z(\mathbf{u}_\alpha)\in[a_\alpha,b_\alpha]$ then:

$$j(\mathbf{u}_\alpha;z) = \begin{cases} 0 \text{ if } z\le a_\alpha \\ \text{missing if } z(\mathbf{u}_\alpha)\in[a_\alpha,b_\alpha] \\ 1 \text{ if } z>b_\alpha \end{cases}$$

A third type is the coding of soft indicator data $y(\mathbf{u}_\alpha;z)$ originating from ancillary information providing prior probabilities about the value $z(\mathbf{u}_\alpha)$:

$$y(\mathbf{u}_\alpha;z) = Prob\{Z(\mathbf{u}_\alpha)\le z|\text{local information}\}\in[0,1]$$

The fourth type of coding is the global prior information common to all locations $\boldsymbol{u}$ within the stationary area:

$$F(z) = Prob\{Z(\mathbf{u}_\alpha)\le z\},\ \forall\,\mathbf{u}\in A$$

This global prior is different than the local prior histogram of samples used in the indicator kriging system. Figure 9.5 shows a graphical representation of all four.

9.4.2 Simple and Ordinary IK with Prior Means

In Simple Indicator Kriging, the expected value of the indicator transform for each category is assumed known and constant throughout the study area. The linear estimate is then a linear combination of the n nearby indicator RVs and the CDF value.

$$i^*(\mathbf{u};z_k) - F(z_k) = \sum_{\alpha=1}^{n}\lambda_\alpha[i(\mathbf{u}_\alpha;z_k) - F(z_k)] =$$

$$i^*(\mathbf{u};z_k) = \sum_{\alpha=1}^{n}\lambda_\alpha\cdot i(\mathbf{u}_\alpha;z_k) + F(z_k)\left[1-\sum_{\alpha=1}^{n}\lambda_\alpha\right]$$

The SIK system of equations is then:

$$\sum_{\beta=1}^{n}\lambda_\beta C_I(\mathbf{u}_\alpha-\mathbf{u}_\beta;z_k) = C_I(\mathbf{u}_\alpha-\mathbf{u};z_k);\quad \alpha=1,\ldots,n$$

where $C_I(\mathbf{h};z_k)=Cov\{I(\mathbf{u};z_k), I(\mathbf{u}+\mathbf{h};z_k)\}$ is the indicator covariance at cutoff z.

If K cutoff values are retained, simple IK requires K indicator covariances in addition to the K cdf values.

There are cases when the prior means can be considered non-stationary and can be inferred from a secondary variable. In such cases one can use the general non-stationary

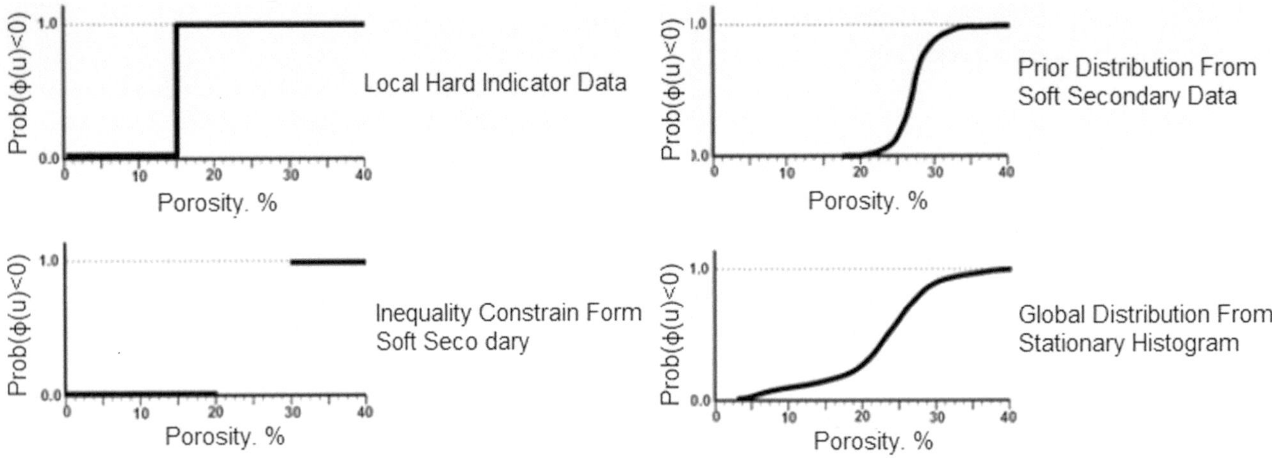

Fig. 9.5 Data types used in indicator coding

expression of simple kriging. For example, if $I(\mathbf{u};s_k)$ represents the indicator of presence/absence of category k at location $\mathbf{u}$, and $p(\mathbf{u};s_k)$ is a prior probability of presence of category k at $\mathbf{u}$, the updated probability is given by simple IK:

$$[Prob\{\mathbf{u} \in k|(n)\}]^*_{IK} = p(\mathbf{u};s_k)\cdot[i(\mathbf{u}_\alpha;s_k) - p(\mathbf{u};s_k)]$$

where $p(\mathbf{u};s_k)$ is the prior indicator mean at location $\mathbf{u}$. The SK system remains as above assuming a stationary residual indicator covariance model (Goovaerts 1997; Deutsch 2002).

For ordinary indicator kriging, the expected value of the indicator transform for each category is assumed unknown but constant within a local neighborhood.

$$i^*(\mathbf{u};z_k) = \sum_{\alpha=1}^{n} \lambda_\alpha \cdot i(\mathbf{u}_\alpha;z_k) + F(z_k)\left[1 - \sum_{\alpha=1}^{n} \lambda_\alpha\right]$$

subject to $\sum_{\alpha=1}^{n} \lambda_\alpha = 1$. The resulting Ordinary IK system of equations is

$$\begin{cases} \sum_{\beta=1}^{n} \lambda_\beta C_I(\mathbf{u}_\alpha - \mathbf{u}_\beta; z_k) + \mu = C_I(\mathbf{u}_\alpha - \mathbf{u}; z_k); \quad \alpha = 1,\ldots,n \\ \sum_{\beta=1}^{n} \lambda_\beta = 1 \end{cases}$$

Ordinary indicator kriging is common because it is more robust with respect to departures from stationarity and, unlike multi-Gaussian kriging, there is no theoretical requirement that simple kriging be used.

9.4.3 Median Indicator Kriging

One of the considerations used in indicator kriging to choose the K cutoff values z_k is that the corresponding indicator covariances $C_I(\mathbf{h};z_k)$ could be significantly different from one another. Occasionally, however, sample indicator variograms appear proportional or are similar to each other. The corresponding continuous RF model $Z(\mathbf{u})$ is the so-called mosaic model:

$$\rho_Z(\mathbf{h}) = \rho_I(\mathbf{h};z_k) = \rho_I(\mathbf{h};z_k;z_{k'}), \ \forall z_k, z_{k'}$$

where $\rho_Z(\mathbf{h})$ and $\rho_I(\mathbf{h};z_k;z_{k'})$ are the correlograms and indicator cross correlograms of the continuous RF $Z(\mathbf{u})$.

The single correlogram function is better estimated either directly from the sample z correlogram or from the sample indicator correlogram at the median cutoff $z_k = M$, i.e., $F(M) = 0.5$. Indeed, at the median cutoff, the indicator data are evenly distributed as 0 and 1 values with, by definition, no outlier values.

Indicator kriging under the mosaic model is called median indicator kriging. It is a particularly simple and fast procedure since it calls for a single easy-to-infer median indicator variogram that is used for all K cutoffs. Moreover, if the indicator data configuration is the same for all cutoffs, one single IK system needs to be solved with the resulting weights being used for all cutoffs. For example, in the case of simple IK,

$$[i(\mathbf{u};z_k)]^*_{SK} = \sum_{\alpha=1}^{n} \lambda_\alpha(\mathbf{u}) i(\mathbf{u}_\alpha;z_k) + \left[1 - \sum_{\alpha=1}^{n} \lambda_\alpha(\mathbf{u})\right] F(z_k)$$

where the $\lambda_\alpha(\mathbf{u})$'s are the SK weights common to all cutoffs z_k and are given by the single SK system:

$$\sum_{\beta=1}^{n} \lambda_\beta(\mathbf{u}) C(\mathbf{u}_\beta - \mathbf{u}_\alpha) = C(\mathbf{u} - \mathbf{u}_\alpha), \quad \alpha = 1,\ldots,n$$

The covariance $C(\mathbf{h})$ is modeled from either the z sample covariance or, better, the sample median indicator covariance. Note that the weights $\lambda_\alpha(\mathbf{u})$ are also the SK weights of the simple kriging estimate of $z(\mathbf{u})$ using the $z(\mathbf{u}_\alpha)$ data.

9.4.4 Using Inequality Data

Normally the indicator data $i(\mathbf{u}_\alpha;z)$ originate from data $z(\mathbf{u}_\alpha)$ that are deemed perfectly known; thus the indicator data $i(\mathbf{u}_\alpha;z)$ are hard in the sense that they are valued either 0 or 1 and are available at any cutoff value z.

There are applications where some of the z information takes the form of inequalities such as:

$$z(\mathbf{u}_\alpha) \in [a_\alpha, b_\alpha]$$

or $z(\mathbf{u}_\alpha) \leq b_\alpha$ equivalent to $z(\mathbf{u}_\alpha) \in (-\infty, b_\alpha)$, or $z(\boldsymbol{u}_\alpha) > a_\alpha$ equivalent to $z(\mathbf{u}_\alpha) \in (a_\alpha, +\infty)$. The indicator data corresponding to the constraint interval are available only outside that interval:

$$i(\mathbf{u}_\alpha; z) = \begin{cases} 0 \text{ for } z \leq a_\alpha \\ \text{undefined for } z \in (a_\alpha, b_\alpha] \\ 1 \text{ for } z > b_\alpha \end{cases}$$

The use of inequality data does not pose any complication. The undefined or missing indicator data in the interval $(a_\alpha, b_\alpha]$ are ignored, and the IK algorithm applies identically. The constraint interval information is honored by the resulting ccdf.

The IK solution is particularly fast if median IK is used. However, the data configuration may change if constraint intervals of type are considered; in such case, one may have to solve a different IK system for each cutoff.

9.4.5 Using Soft Data

Hard indicator data and inequality data are treated similarly—except for missing values at some thresholds. Soft indicator data, however, are truly different data types and a form of cokriging should be adopted to combine them in the estimate. The Markov-Bayes formalism is commonly considered (Zhu 1991; Zhu and Journel 1992).

9.4.6 Exactitude Property of IK

Assume the location $\mathbf{u}$ to be estimated coincides with a datum location $\mathbf{u}_\alpha$, whether a hard datum or a constraint interval of type. Then, the exactitude of kriging (simple, ordinary, or median IK) entails that the ccdf returned is either a zero variance cdf identifying the class which the datum value $z(\mathbf{u}_\alpha)$ belongs to; or a ccdf honoring the constraint interval up to the cutoff interval amplitude:

$$\begin{aligned} [i(\mathbf{u}_\alpha; z_k)]^* &= Prob^*\{Z(\mathbf{u}_\alpha) \leq z_k | (n)\} \\ &= \begin{cases} 0, \text{ if } z_k \leq a_\alpha \\ 1, \text{ if } z_k > b_\alpha \end{cases}, \text{ if } z(\mathbf{u}_\alpha) \in (a_\alpha, b_\alpha] \end{aligned}$$

Because the ccdf returned by IK honors both hard z data and constraint intervals, the corresponding expected value-type (E-type) estimate also honors that information. More precisely, at a datum location $\mathbf{u}_\alpha$, $[z(\mathbf{u}_\alpha)]_E^* = z(\mathbf{u}_\alpha)$, if the z datum is hard; and $\left[z_k(\mathbf{u}_\alpha)\right]_E^* \in (a_\alpha, b_\alpha]$ if the information available at $\mathbf{u}_\alpha$ is the constraint interval $z_k(\mathbf{u}_\alpha) \in (a_\alpha, b_\alpha]$. In practice, the exactitude of the E-type estimate is limited by the finite discretization into K cutoff values z_k. For example, in the case of a hard z datum, the estimate is: $\left[z(\mathbf{u}_\alpha)\right]_E^* \in (z_{k-1}, z_k]$ with z_k being the upper bound of the interval containing the datum value $z(\mathbf{u}_\alpha)$.

9.4.7 Change of Support with IK

The indicator variable $I(\mathbf{u};z)$ results from a non-linear transform of the original $z(\mathbf{u})$ samples; therefore, the block indicator variable $I_V(\mathbf{u};z)$ is not a linear average of point indicators $I(\mathbf{u};z)$. Thus, the averaging of discretization points within a block, as done for linear variables, does not result in the estimated block indicator:

$$i_V^*(\mathbf{u}; z) \neq \frac{1}{|V|} \int_{V(u)} i(\mathbf{u}'; z) \cdot du'$$

Or, equivalently:

$$\left[F_V(\mathbf{u}; z | (n))\right]^* \neq \frac{1}{N} \sum_{\alpha=1}^{N} \left[F(\mathbf{u}'_\alpha; z | (n))\right]^*$$

The ccdf that results from averaging the proportions of point values within the block $V(\mathbf{u})$ is called a "composite" ccdf, $[F_N(\mathbf{u}; z|(n))]^*$ (Goovaerts 1997), and is an estimate of the proportion of point values within $V(\mathbf{u})$ that do not exceed the threshold z. This is quite different than what a true block ccdf would give, which is by definition the probability that the average value is no greater than the threshold z.

Practitioners have attempted to correct the IK-derived point ccdf within the block using a change of support method, the most common being the affine correction discussed in Chap. 7. The process is simply to correct the IK-estimated point ccdf to represent a "block" ccdf. All z_k thresholds are corrected without changing any of the estimated indicator values $i(\boldsymbol{u}; z_k)$:

$$z_{V,k}(\mathbf{u})^* = \sqrt{f}\left[z_k - m(\mathbf{u})^*\right] + m(\mathbf{u})^*$$

where f is the classical variance reduction factor (Isaaks and Srivastava 1989). The dispersion variance is derived from the z value variogram models.

In one published case (Hoerger 1992), the affine correction is applied in the log-space under the permanence of distribution assumption. This change of support model is difficult to validate. The method is more useful if there is production or closely spaced data that can be used to validate the results.

Often IK is not used to provide the entire block ccdf $[F_V(\mathbf{u}; z|(n))]^*$, but simply the e-type average of the block. This is calculated from the point e-type averages within the block. In this case, the process is similar to linear kriging, and the practitioner must decide whether the extra work required in defining and solving the k-indicator kriging systems is worthwhile.

The results from an e-type ordinary IK can be quite different from a linear ordinary kriging. The geology of the deposit and the characteristics of the spatial distribution of the z-variable typically explain the differences. The criteria used to decide whether an e-type IK estimate is worthwhile are: (1) there is sufficient number of samples available for all grade ranges defined by the k-thresholds; (2) the z-variable shows a highly variable distribution, typically characterized by a coefficient of variation above 1.0 and often above 2.0; and (3) the indicator variogram models and other statistics suggest overprinting of mineralization styles. A good reference on the application of Indicator Kriging is Zhang (1994).

9.5 The Practice of Indicator Kriging

There are several steps required to complete an indicator kriging estimate as described in the literature (Journel 1983; Deustch and Lewis 1992), whether the final objective is a point ccdf, an e-type point or block estimate, or a composite ccdf.

Step 1: Obtaining an unbiased (declustered) global histogram The available samples over the domain A may not be representative of the domain due to spatial clustering. The idea of IK can be seen as a modification of the prior cdf derived from the sample for the entire domain to a posterior ccdf specific to each location. The more representative prior cdf used should result in a better, more representative local ccdf.

Step 2: Choosing the K threshold values Indicator kriging begins with establishing a series of thresholds. The number of thresholds chosen is a trade-off between the amount of data available, and the resolution sought in the ccdf model $F(\mathbf{u}; z_k)$. Also, the indicator variograms should be sufficiently different in spatial structure.

While there should be a sufficient number of thresholds to obtain good resolution, too many cutoffs induce more order relation problems. Criteria used to define the thresholds include: (1) the classes should define approximately equal-quantity of metal, and not equal amplitude; (2) additional thresholds are usually placed around z-values that are consequential to the project, such as economic cutoff grades; however, the economic cutoff grades themselves need not used; (3) each class should have a sufficient number of data for robust interpolation. It is common to use between 8 and 15 thresholds, although there are exceptions.

Step 3: Modeling the indicator variograms There are K indicator variograms to be modeled, one for each threshold used. These variograms tend to have reasonably well-defined structures and are easy to interpret, since there are no outliers. There should not be any proportionality between them; if there is, the proportional variogram models will result in the same kriging weights, and thus one of the thresholds can be dropped.

Indicator variograms standardize all points and models to a unit variance of $p \cdot (1 - p)$, where p is the proportion or average value at that threshold. It is common practice to work on standardized indicator variograms, where the sill of each variogram is divided by their respective variance $p_k \cdot (1 - p_k)$. In this case, the model is interpreted to the theoretical sill of 1.0, making the joint modeling of the indicator variograms easier and more consistent.

It is advisable also to use the same type of structure for the models, since they all relate to the same RF. They can be spherical, exponential, or any other licit model; however, the contribution of each structure will vary from one threshold to the next, usually modeled as a smooth transition.

The changes in terms of variance contribution and anisotropy ranges and directions should be smooth from one indicator to the next. It is good practice to plot the relative nugget effects, short scale ranges, and the anisotropy parameters for all thresholds to check the transitions from one model to the next.

For positively skewed distributions, higher thresholds will show less continuity, so a de-structuring effect is expected and can be quantified.

While smooth transitions from one model to the next are desirable because they minimize order relation problems, different patterns of anisotropy can and should be expected. This is particularly true if multiple populations co-exist, or an explainable change in geologic control is evident. For example, low values in many mineral deposits tend to be more disseminated into the host rock; while higher grade mineralization tends to be more spatially restricted and located in favorable structures, veins, or higher porosity areas of the host rock. In such a case, the thresholds for lower grades will be more isotropic, while the anisotropy ratios will increase for higher grade thresholds.

Step 4: Kriging plans The strategy for implementing the estimation follows the same general criteria discussed in Chap. 8. However, there are specific recommendations for implementing IK.

The same search neighborhood and the same number of data should be used for all K thresholds. Even if the indicator

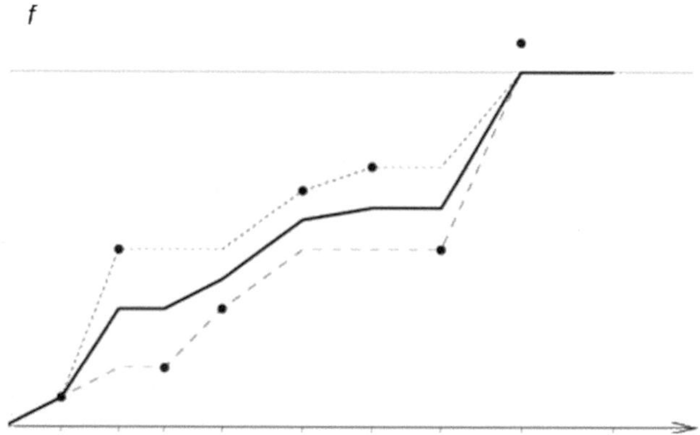

Fig. 9.6 Order relation problems and their correction. The dots are the ccdf values returned by IK. The corrected ccdf is obtained by averaging the forward and downward corrections

variograms show different anisotropies due to the mixing of geologic controls, the search neighborhood should be the same. If quadrant or octant searches are used, the conditions applied to those also must be the same for all thresholds. Changing the search neighborhood between thresholds changes the data used and causes the estimated probabilities to vary significantly in a non-physical manner; there are more order relation deviations, see below.

It is reasonable to orientate the samples search according to the interpreted anisotropy model. However, this should only be done if no significant changes in anisotropy direction is observed for all K variogram models; and in any case, the search neighborhood should be made more isotropic than what the variogram models suggest. This is to ensure that all directions away from the point being estimated are represented in the sample pool.

Step 5: Correcting for order relation deviations It is necessary to ensure that the IK-estimated ccdf at each location **u** respects the axioms of a cdf:

$$F(\mathbf{u};z\,|\,(n)) \geq F(\mathbf{u};z'\,|\,(n)), \quad \forall\; z > z'$$

and

$$F(\mathbf{u};z\,|\,(n)) \in [0,1]$$

Since the K thresholds are estimated independently of each other, the estimated ccdf values may not satisfy these order relations. While the number of deviations may be large, perhaps ½ of the total ccdfs, the absolute value of those deviations should not be large. Journel (1987) recommends checking the implementation of IK if the deviations are greater than 0.01; in practice a limit of 0.05 is more reasonable.

There are several sources of order relation problems. Most commonly, they are due to inconsistent variogram models and the kriging implementation strategy. Also, negative indicator kriging weights and lack of data in some classes increase order relation violations. Although the order relation problems can be minimized, they cannot be completely avoided.

Correcting for order relations in the case of categorical variables is simpler compared to continuous variables. If the estimated probability of a category s_k is outside the licit bounds, then the solution is to reset the estimated value $F^*(\mathbf{u};s_k|(n))$ to the nearest bound, 0 or 1. This resetting corresponds exactly to the solution provided by quadratic programming.

The other constraint is more difficult to resolve because it involves K separate krigings. One solution consists of kriging only $(K-1)$ probabilities leaving aside one category s_{ko}, chosen with a large enough prior probability p_{ko}, so that:

$$F^*(\mathbf{u};s_{k_0}\,|\,(n)) = 1 - \sum_{k \neq k_o} F^*(\mathbf{u};s_k\,|\,(n)) \in [0,1]$$

Another solution, which should be applied after the estimated distribution is corrected (if necessary) to the interval [0,1], is to re-standardize each estimated probability $F^*(\mathbf{u};s_k|(n))$ by the sum $\sum_k F^*(\mathbf{u};s_k\,|\,(n)) < 1$.

Correcting for order relations of continuous variable ccdfs is more delicate because of the ordering of the cumulative indicators. Figure 9.6 shows an example with order relation problems.

The following correction algorithm implemented in GSLib (Deutsch and Journel 1997) considers the average of an upward and downward correction:

1. Upward correction resulting in the upper line in Fig. 9.6 showing order relations problems:
   - Start with the lowest cutoff z_1.
   - If the IK-returned ccdf value $F(\mathbf{u};\, z_1|(n))$ is not within [0,1], reset it to the closest bound.
   - Proceed to the next cutoff z_2. If the IK-returned ccdf value $F(\mathbf{u};\, z_2|(n))$ is not within $[F(\mathbf{u};\, z_1|(n)),1]$, reset it to the closest bound.
   - Loop through all remaining cutoffs z_k, $k=3, \ldots, K$.
2. Downward correction resulting in the lower line in Fig. 9.6 showing order relations problems:
   - Start with the largest cutoff z_K.
   - If the IK-returned ccdf value $F(\mathbf{u}_\alpha;\, z_K|(n))$ is not within [0,1], reset it to the closest bound.
   - Proceed to the next lower cutoff z_{k-1}. If the IK-returned ccdf value $F(\mathbf{u}_\alpha;\, z_{k-1}|(n))$ is not within $[F(\mathbf{u}_\alpha;\, z_K|(n)),1]$, reset it to the closest bound.
   - Loop downward through all remaining cutoffs z_k, $k=K-2, \ldots, 1$.
3. Average the two sets of corrected ccdfs resulting in the thick middle line in Fig. 9.6.

Practice has shown that the majority of order relation problems are due to a lack of data; more precisely, to cases when IK is attempted at a cutoff z_k which is the upper bound of

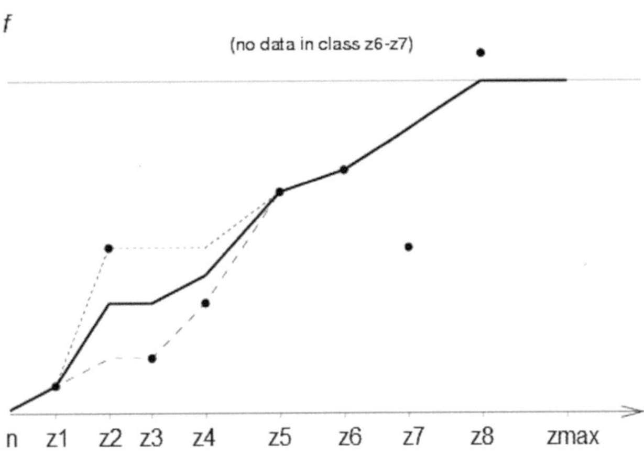

Fig. 9.7 Order relation problems and their correction ignoring the class $(z_6, z_7]$ that did not contain any z data. The ccdf value $F(\mathbf{u}_\alpha; z_7|(n))$ is ignored and the correction is applied to the remaining cutoff values

a class $[z_{k-1}, z_k]$ that contains no z data. In such case the indicator data set is the same for both cutoffs z_{k-1} and z_k and yet the corresponding indicator variogram models are different; consequently, the resulting ccdf values will likely be different with a good chance for order relation problems.

A solution would be to retain only those prior cutoff values z_k such that the class $(z_{k-1}, z_k]$ has at least one datum as in Fig. 9.7:

Order relation problems represent the most severe drawback of the indicator approach. They are the price to pay for trying to reproduce (even approximately) more than a single sample covariance.

Step 6: Interpolation between the *K* values of the ccdf Once the K ccdf values have been corrected, it is necessary to interpolate values between thresholds and extrapolate beyond the first and last threshold to obtain a complete distribution. A more complete description of the models commonly implemented can be found in Deustch and Journel (1997).

Some implementations of IK assumes a particular distribution model for the interval between two consecutive thresholds such as the linear model, which assumes a uniform distribution and is generally accepted between thresholds; the power model is commonly used for extrapolating the lower tail, between 0 and threshold z_1, and sometimes the upper tail, between threshold z_k and 1; and the hyperbolic model, which is most used to control the extrapolation of the upper tail. In practice, it is better to use the non-parametric global distribution shape adapted to the IK-estimated ccdf values.

Step 7: Using the IK model and calculating recoverable resources The ccdf distribution can be used to provide any statistic of interest at location **u**. The ccdf is a measure of uncertainty, from which probability intervals can be derived.

A special case is the probability of exceeding certain thresholds, such as the economic cutoffs. Naturally, the e-type mean and other truncated statistics can also be obtained.

In mining and other earth science applications it is critical to know the probability of the block grade exceeding a cutoff grade:

$$z_V(\mathbf{u}; z_c) = Prob\{Z_V(\mathbf{u}) \geq z_c \,|\, (n)\} = 1 - F_V(\mathbf{u}; z_c \,|\, (n))$$

which results in the probability of the estimated block grade being recoverable.

The uncertainty about any value $z(\mathbf{u})$ can be derived from probability intervals, such as:

$$Prob\{Z_V(\mathbf{u}) \in [a,b] \,|\, (n)\} = F_V(\mathbf{u}; b \,|\, (n)) - F_V(\mathbf{u}; a \,|\, (n))$$

where a and b are the thresholds that define the interval of interest. For example, there is frequently more than one economic cutoff. Aside from the higher grade mill cutoff, there may be a lower grade cutoff used to determine the material that is to be processed using a lower cost option, such a leaching, or that may be simply stockpiled for later processing. To adequately engineer the stockpile or heap leach it is important to know the probability of the block falling within the two economic cutoffs.

Probably the most common application of IK, however, is an e-type estimate. The e-type estimate can be written:

$$m(\mathbf{u})^* = \sum_{k=1}^{K+1} m_k \left[i(\mathbf{u}; z_k) - i(\mathbf{u}; z_{k-1}) \right]$$

where $i(\mathbf{u}; z_k)$ is the kriged indicator value for threshold k, $i(\mathbf{u}; z_0) = 0$, $i(\mathbf{u}; z_{K+1}) = 1$, and m_k is the (declustered) class mean, that is, the mean of the data falling in the interval $[z_{k-1}, z_k]$. Note that if only the e-type is required, then no correction for change of support is necessary, since $m_V(\mathbf{u})^* = m(\mathbf{u})^*$ for variables that upscale linearly.

9.6 Indicator Cokriging

The IK estimators discussed above independently krige each threshold and thus they do not make full use of the information contained in the series of indicators. The full information contained in the cross indicators can be accounted for by using a co-kriging estimator across defined thresholds. Note that this co-IK is not between multiple data types, but between the same data type considering all possible thresholds.

The co-indicator kriging (co-IK) estimate is then defined as:

$$[i(\mathbf{u}; z_{k_0})]^*_{coIK} = \sum_{k=1}^{K} \sum_{\alpha=1}^{n} \lambda_{\alpha,k}(\mathbf{u}; z_{k_0}) \cdot i(\mathbf{u}; z_k)$$

The corresponding co-kriging system would call for a matrix of K^2 direct and cross indicator covariances of the type:

$$C_I(\mathbf{h}; z_k, z_{k'}) = Cov\{I(\mathbf{u}; z_k), I(\mathbf{u}+\mathbf{h}; z_{k'})\}$$

The direct inference and joint modeling of the K^2 covariances is not practical for large K. The shape of the cross variograms is quite smooth relative to the direct indicator variograms making it impossible to fit the set of covariances/variograms with the linear model of coregionalization. Also, the kriging matrices that need to be inverted would be significantly larger. There has been however some solutions proposed. Suro-Pérez and Journel (1991) proposed to reduce the co-IK system by working on linear transforms of the indicator variables, which are less cross-correlated, such as indicator principal components. Another solution calls for a prior bivariate distribution model.

A prior bivariate distribution model amounts to forfeiting actual data-based inference of some or all of the indicator (cross-) covariances. Most commonly, the bivariate Gaussian model after normal scores transform of the original variable is adopted: $Z(\mathbf{u}) \rightarrow Y(\mathbf{u}) = \varphi(Z(\mathbf{u}))$. A slight generalization of the bivariate Gaussian model is offered by the (bivariate) isofactorial models used in disjunctive kriging (DK). The generalization is obtained by a nonlinear rescaling of the original variable $Z(\mathbf{u})$ by transforms $\psi(\cdot)$ different from the normal score transform $\varphi(\cdot)$. In either case, all indicator (cross-) covariances are then determined from the Gaussian Y covariance. At this point, it makes more practical sense to simply rely on the complete multivariate Gaussian model.

More importantly, general experience indicates that co-IK improves little from IK (Goovaerts 1994). Such is the case if primary and secondary variables are equally sampled, as happens with indicator data defined at various cutoff values. In addition, when working with continuous variables, the corresponding cumulative indicator data do carry substantial information from one cutoff to the next one; in which case, the loss of information associated with using IK instead of co-IK is not as large as it appears. Finally, co-IK will generally create more order relation problems, which require additional manipulation of the estimated cdf.

9.7 Probability Kriging

An alternative to indicator co-kriging is to use not only the transformed indicators, but also their standardized rank transforms, which are distributed in [0,1]. The idea is that greater resolution could be achieved near data locations.

The so-called probability kriging (PK) estimate (Sullivan 1984), actually a ccdf estimate for $Z(\boldsymbol{u})$, is written in its simple kriging form:

$$\left[i(\mathbf{u}; z_k\right]^*_{PK} - F(z_k) = \sum_{\alpha=1}^{n} \lambda_\alpha(\mathbf{u}; z_k) \cdot \left[i(\mathbf{u}_\alpha; z_k) - F(z_k)\right] + \sum_{\alpha=1}^{n} \nu_\alpha(\mathbf{u}; z_k) \cdot \left[p(\mathbf{u}_\alpha) - 0.5\right]$$

where $p(u_\alpha) = F(z(u_\alpha)) \in [0,1]$ is the uniform (cdf) transform of the datum value $z(\boldsymbol{u}_\alpha)$, the expected value of which is 0.5; $F(z) = Prob\{Z(u) \leq z\}$ is the stationary cdf of $Z(\boldsymbol{u})$. The co-kriging weights $\lambda_\alpha(\mathbf{u}; z_k)$ and $\nu_\alpha(\mathbf{u}; z_k)$ correspond to the indicator and the uniform data, respectively.

The corresponding simple PK system requires the inference and modeling of (2K+1) (cross-) covariances: K indicator covariances, K indicator-uniform cross covariances, and the single uniform transform covariance. That modeling is still demanding and represents a drawback in practice. For this reason, there have been few practical applications of PK.

9.8 Summary of Minimum, Good and Best Practices

This section presents the details of what is considered minimum, good, and best practices in probabilistic estimation, as well as discussing specific presentation and reporting issues. The block model obtained should be thoroughly checked and validated, as detailed in Chap. 11, according to the criteria suggested for minimum, good, and best practices. As before, these validations should result in, among other comments, a statement about whether the model can be considered "recoverable" or fully diluted.

In the case of having an estimated distribution of possible values for each block, it is even more important to consider the reporting and transmittal of the resource model to downstream clients. One possibility is to obtain the e-type average for the block, and present it; however, having developed a range of possible values, other information, such as probabilities of exceeding important cutoffs, should be provided. In this sense, it is important to consider that probabilities can be seen as proportions of blocks, a concept that may make the understanding and manipulation of this information easier for downstream clients.

The minimum practice in estimating distributions should include making the estimation implementation specific to each estimation domain. All other criteria given in Sect. 8.4 should apply here, including documentation and justification of the Kriging method chosen, the reporting of the ore resources, which should be based on E-type estimates, the use of grade-tonnage curves, and the full documentation of the more relevant aspects. Subjective expressions of uncertainty, assumptions and limitations, and recommendations for improvements should be included.

In addition, good practice requires a more detailed justification of the model. The main differentiation includes the use of calibrations, and comparisons with alternative and prior models. A comparison with past production should be done, if available, and the model should be fully diluted. As before, significant emphasis in describing uncertainty and potential risk areas of the model should be discussed in detail, and a risk mitigation plan should be suggested.

Best practice consists of using alternate models to check the results of the intended final ore resource model. All relevant production and calibration data should be used to indicate whether the model is performing as expected, possibly including simulation models to calibrate the recoverable resource model. The model should quantitatively describe the amount of the different types of dilution included. All other tasks related to checking and validation, model presentation, reporting, and visualization should also be completed. Again, all risk issues should be dealt with in detail, and if possible, quantified, which requires complementing the estimation process with specific simulation studies.

9.9 Exercises

The objective of this exercise is to review indicator kriging (IK) and multivariate Gaussian kriging (MG) for uncertainty assessment. Some specific (geo)statistical software may be required. The functionality may be available in different public domain or commercial software. Please acquire the required software before beginning the exercise. The data files are available for download from the author's website—a search engine will reveal the location.

9.9.1 Part One: Indicator Kriging

Question 1: Consider the indicator variograms from Part 5 of Chap 6 using the `largedata.dat` data file. Setup indicator kriging to estimate with the nine deciles as threshold values. You may want to choose a smaller area near the centre of the dataset. Run indicator kriging and create a map at the median threshold for checking.

Question 2: Post process the indicator kriging results to calculate the local average (etype estimate) and the local conditional variance. Map the results. The local average should look like a kriged map. The conditional variance should account for the proportional effect, that is, higher grade areas should have higher conditional variance when the drillhole spacing is uniform.

9.9.2 Part Two: MG Kriging for Uncertainty

The objective of this exercise is to become familiar with how kriging can be used to get uncertainty without resorting to simulation.

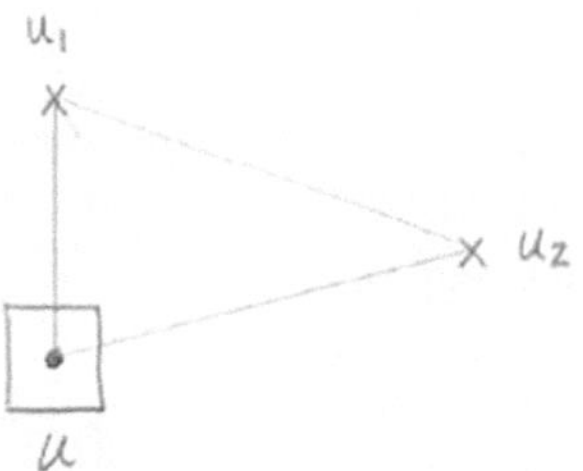

Copper grades in a particular domain were found to fit an exponential distribution with a mean m of 1%. Consider a particular location **u** informed by two nearby samples at locations **u1** and **u2** such that $|\mathbf{u}\text{-}\mathbf{u1}|=20$ m, $|\mathbf{u}\text{-}\mathbf{u2}|=37$ m, and $|\mathbf{u1}\text{-}\mathbf{u2}|=38$ m. We are interested in the uncertainty in the copper grade at the location **u** and in the uncertainty in the copper grade of a 10 m cubed block centered at **u**. The copper grade at **u1** is 2.5% and the grade at **u2** is the mean value of 1%.

Question 1: Provide a detailed description of how you would go about characterizing the uncertainty about the unsampled value $z(\mathbf{u})$ and calculating a best estimate and 90% probability interval. You are to adopt a multivariate Gaussian model. State all the steps and approximations, comment extensively, and use figures/sketches where appropriate.

The multivariate distribution of the stationary copper grade random function is assumed multinormal after appropriate normal score transform. Variogram analysis was performed on the standard normal transform of the copper grades. This resulted in an isotropic spherical model with a nugget effect of 10% and a range of 100 m.

Question 2: Write the equations for the transform to and from Gaussian units. In general, the relationship must be fit, but it is possible to write equations in this case. Transform the data values to Gaussian units. You can do this question on a piece of paper with the help of a calculator.

Question 3: Establish the parameters of the analytical conditional distribution in Gaussian space. This requires the solution of two equations with two unknowns. Once again, this can be done on paper with the help of a calculator.

Question 4: Back transform 99 evenly spaced percentiles to establish the conditional distribution in the units of copper grade. You will probably want to use Excel or a short program for this question. Calculate the mean grade and the 90% probability interval.

Question 5: Correct the conditional distribution of copper grades to represent a 10 m cubed block grade. Assume that the stationary *Z*-copper grade variogram is the same shape as the normal scores variogram. Calculate the mean grade and the 90 % probability interval.

References

Armstrong M, Matheron G (1986) Disjunctive kriging revisited (Parts I and II). Math Geol 18(8):711–742

Chilès JP, Delfiner P (2011) Geostatistics: Modeling spatial uncertainty, 2nd edn. Wiley, New York, p 695

Deustch CV, Journel AG (1997) GSLIB: Geostatistical software library and user's guide. 2nd edn. Oxford University Press, New York, p 369

Deustch CV, Lewis RE (1992) Advances in the practical implementation of indicator geostatistics. In: Kim YC (ed) Proceeding of the 23rd international APCOM symposium, Society of Mining, Metallurgy, and Exploration. 7–11 April, Tucson, p 169–179

Deutsch CV (2002) Geostatistical reservoir modeling. Oxford University Press, New York, p 376

Goovaerts P (1994) Comparative performance of indicator algorithms for modeling conditional probability distribution functions. Math Geol 26(3):389–411

Goovaerts P (1997) Geostatistics for natural resources evaluation. Oxford University Press, New York, p 483

Guibal D (1987) Recoverable reserves estimation at an Australian gold project, Geostatistical case studies, vol 1. In: Matheron G, Armstrong M (eds) Reidel, Dordrecht, p 149–168

Guibal D, Remacre A (1984). Local estimation of the recoverable reserves: Comparing Various methods with the reality on a porphyry copper deposit. In: Verly G et al. (eds) Geostatistics for natural resources characterization, vol 1. Reidel, Dordrecht, p 435–448

Hoerger S (1992) Implementation of indicator kriging at Newmont Gold Company. In: Kim YC (ed) Proceeding of the 23rd international APCOM symposium. Society of Mining, Metallurgy, and Exploration. 7–11 April, Tucson, p 205–213

Isaaks EH, Srivastava RM (1989) An introduction to applied geostatistics. Oxford University Press, New York, p 561

Journel AG (1980) The lognormal approach to predicting local distributions of selective mining unit grades. Math Geol 12(4):285–303

Journel AG (1983) Non–parametric estimation of spatial distributions. Math Geol 15(3):445–468

Journel AG (1987) Geostatistics for the environmental sciences: An introduction. U.S. Environmental Protection Agency, Environmental Monitoring Systems Laboratory

Journel AG, Huijbregts ChJ (1978) Mining geostatistics. Academic, New York

Journel AG, Posa D (1990) Characteristic behavior and order relations for indicator variograms. Math Geol 22(8):1011–1028

Krige DG (1951) A statistical approach to some basic mine valuation problems on the Witwatersrand. J Chem Metall Min Soc S Afr 52:119–139

Kumar A (2010) Guide to recoverable reserves with disjunctive kriging. Guidebook Series, Vol 9. Centre for Computational Geostatistics, University of Alberta, Edmonton

Marcotte D, David M (1985) The bi-Gaussian approach: A simple method for recovery estimation. Math Geol 17(6):625–644

Matheron G (1971) The theory of regionalized variables and its applications. Fasc. 5, Paris School of Mines, p 212

Matheron G (1973) The intrinsic random functions and their applications. Adv Appl Prob 5:439–468

Matheron G (1974) Les Fonctions de transfert des petits panneaux. Centre de Géostatistique, Fontainebleau, France, No. N–395

Matheron G (1976) A simple substitute for conditional expectation: the disjunctive kriging. In: Guarascio M., David M, Huijbregts C (eds) Advanced geostatistics in the mining industry. Reidel, Dordrecht, p 221–236

Neufeld C (2005). Guide to recoverable reserves with uniform conditioning. Guidebook series, Vol 4. Centre for Computational Geostatistics, Edmonton

Parker H, Journel A, Dixon W (1979) The use of conditional lognormal probability distribution for the estimation of open-pit ore reserves in stratabound uranium deposits, a case study. In: Proceedings of the 16th APCOM, p 133–148

Remacre AZ (1987) Conditioning by The panel grade for recovery estimation of non-homogeneous ore bodies. In: G. Matheron G, Armstrong M (eds) Geostatistical case studies. Reidel, Dordrecht, p 135–148

Rivoirard J (1994) Introduction to disjunctive kriging and non-linear geostatistics. Claredon Press, Oxford, p 190

Roth C, Deraisme J (2000) The information effect and estimating recoverable reserves. In: Kleingeld WJ, Krige DG (eds) Proceedings of the sixth international geostatistics congress. Cape Town, April, p 776–787

Sichel HS (1952) New methods in the statistical evaluation of mine sampling data. Trans Inst Min Metall Lond 61:261

Sichel HS (1966) The estimation of means and associated confidence limits for small samples from lognormal populations. In: Proceedings of the 1966 symposium South African Institute of Mining and Metallurgy

Sullivan JA (1984) Non parametric estimation of spatial distributions. PhD Thesis, Stanford University, p 367

Suro-Pérez V, Journel AG (1991) Indicator principal component kriging. Math Geol 23(5):759–788

Vann J, Guibal D (1998) Beyond ordinary kriging—An overview of non-linear estimation. In: Beyond ordinary kriging: Non-linear geostatistical methods in practice. The Geostatistical Association of Australasia, Perth, 30 October, p 6–25

Verly G (1984) Estimation of spatial point and block distributions: The multigaussian model. PhD Dissertation, Department of Applied Earth Sciences, Stanford University

Zhang S (1994) Multimetal recoverable reserve estimation and its impact on the Cove ultimate pit design Min Eng July 1998:73–79

Zhu H (1991) Modeling mixtures of spatial distributions with integration of soft data. PhD Thesis, Stanford University, Stanford

Zhu H, Journel A (1992) Formatting and integrating soft data: Stochastic imagining via the Markov-Bayes algorithm. In: Soares A (ed) Geostatistics-Troia. Kluwer, Dordrecht, p 1–12

10 Recoverable Resources: Simulation

Abstract

Local uncertainty estimates do not account for the variability from one location to another. The idea of simulation is to assess the joint uncertainty between multiple realizations allowing a more complete representation of block uncertainty and the uncertainty between multiple block locations. The tools described in this Chapter allow transferring uncertainty of the resource estimates into risk in downstream studies. These studies are mine design, mine planning, or operational optimization studies; the risk assessment is achieved after applying transfer functions to the conditional simulation models.

10.1 Simulation versus Estimation

Simulated models provide the same information that an estimated block model does, but, in addition, it also provides a joint model of uncertainty. A "complete" resource model should not only include an estimated grade, or even an estimated distribution, but also a more detailed assessment of uncertainty and the consequences of that uncertainty (Dimitrakopoulos 1997).

Estimation provides a value that is, on average, as close as possible to the actual (unknown) value, based on some definition of goodness or quality. It is unbiased, has minimal quadratic error, uses linear combinations of the available data, and has an unavoidable smoothing effect. Simulations reproduce the original variability observed in the data and allow an assessment of uncertainty. This implies that the extreme values of the original distribution are preserved, see Fig. 10.1. The uncertainty model also provide the tools for risk analysis when applying to it a transfer function

Estimation honors local data, is locally more accurate, and has a smoothing effect appropriate for visualizing trends, but is inappropriate for simulating extreme values and provides no assessment of local uncertainty. Simulation also honors the local data, but additionally reproduces the histogram, honors spatial variability, and is able to provide an assessment of uncertainty.

Geostatistical conditional simulations have become popular as tools that provide models of uncertainty at different stages of a mining project. They have been used as grade control tools in daily operations (Rossi 1999), to analyze risk related to resource classifications (Rossi and Camacho 2001), to assess the uncertainty of minable reserves at the project's feasibility stage (Guardiano et al. 1995; Glacken 1996; Van Brunt and Rossi 1999; Journel and Kyriakidis 2004; Leuangthong et al. 2006; Badenhorst and Rossi 2012), and to assess mineralization potential in certain settings. Other applications include assessment of recoverable reserves and drill hole spacing optimization studies.

Geostatistical conditional simulations are used to build models that reproduce the full histogram and modeled measures of spatial continuity of the original, conditioning data. They honor the characteristics of the spatial variable of interest as represented by the conditioning data.

The simulation model should correctly represent the proportion of high and low values, the mean, the variance, and other univariate statistical characteristics of the data, as represented by the histogram. It should also correctly reproduce the spatial continuity of the variable, including the connectivity of low and high contaminant zones, anisotropies, relative nugget effect, and other characteristics of the variogram model.

Conditional simulations are built on fine grids, fine enough to provide a sufficient number of nodes within the block size of interest. The vertical resolution of the grid should be a function of the support data, for example the size of the mining bench, if modeling a variable mined by open pit. Larger grid sizes may be used sometimes because of the amount of computer time and hard disk space involved.

M. E. Rossi, C. V. Deutsch, *Mineral Resource Estimation,* DOI 10.1007/978-1-4020-5717-5_10,

In building a conditional simulation model, many of the conditions and requirements of linear and non-linear estimations apply, most importantly regarding stationarity decisions. Shifts in geologic settings require the separation of the data into different populations. Detailed knowledge of the behavior of extreme and outlier values in the sampled population is required. Issues such as limiting the maximum simulated grade should be carefully considered.

The simulation method itself should be decided based on the type of deposit, the Random Function model chosen, the quantity and quality of available samples, the possibility of using "soft" or fuzzy information, and the desired output. All these are subjective decisions. These and other implementation parameters, along with the chosen algorithm and simulated domain, have a bearing on the output simulations and the uncertainty model.

Conditional simulation methods can be grouped in similar manner as estimation methods were in Chaps. 8 and 9. There are simulation methods for continuous and discrete, or categorical, variables; there are Gaussian and indicator-based approaches, such as Sequential Gaussian (Isaaks 1990) and Sequential Indicator (Alabert 1987a). The latter is more complicated, based on multiple indicator kriging techniques, and requires definition of several indicator cutoffs. The former is simpler and quicker, although more restrictive in its basic assumptions. As with any estimation exercise, variogram models are necessary.

There are other types of simulations, including object-based simulation methods, and sequential annealing, a particular case of optimization. Also, there are several types of multivariate simulations.

A conditional simulation model results in a set of grades or realizations for each node. These realizations, all equiprobable by construction, describe the model of uncertainty for each block, i.e., provide the cumulative conditional distribution function (ccdf) for that node. Preferably, a large number of simulations are needed to describe the ccdf better. However, a smaller number is generally used due to practical limitations. It has been these authors' experience that between 20 and 50 simulations are generally sufficient to characterize the range of possible values for the simulated values

Uncertainty is not a property of the physical attribute being modeled, but rather of the Random Function (RF) model developed. The RF model includes the stationarity decisions made; the simulation algorithms chosen; and the implementation parameters used. Therefore, (a) the uncertainty model that can be derived from conditional simulations is subjective and only relevant to the underlying RF model; and (b) applications or risk assessments that can be derived from that uncertainty model are only useful and "realistic" if they are relevant to the problem of interest.

A common example is the assessment of uncertainty of a block model, used to define resources and reserves of a deposit. The conditional simulation should be constructed using the same underlying RF model used in the construction of the block model, if it is to describe resource uncertainty. Likewise, the same geologic model and estimation domains used to constrain the block model have to be used to constrain the simulation model. Otherwise, the uncertainty model will not fully relate to the resource model.

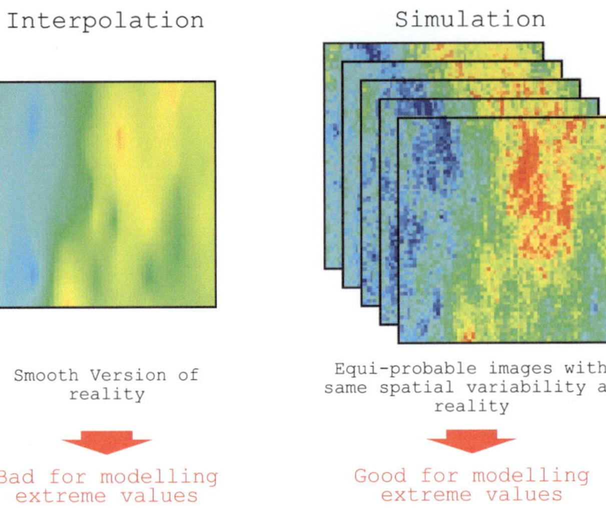

Fig. 10.1 Comparison of estimated and simulated models

10.2 Continuous Variables: Gaussian-Based Simulation

Gaussian simulations are most common in mining. Among these, sequential Gaussian simulation (SGS) is the most frequently used method, although there are several others that are been used as well.

Gaussian methods are maximum entropy methods, in the sense that for a specific covariance model, they provide the most "disorganized" spatial arrangement possible. While the covariance model controls the degree of mixing of high and lows, there are some highly structured spatial distributions that may be more difficult to reproduce with a Gaussian simulation (Fig. 10.2). This results in a model that potentially understates the continuity of the distribution's extreme values. In practice, however, this effect can be controlled through the definition of stationary geologic domains and to some extent the covariance model. Gaussian simulations are most popular because of their convenient properties and easier implementation; but also because they result, for most cases, in reasonable representations of spatial distributions.

The Turning Bands (TB) method was the first simulation algorithm developed (Matheron 1973; Journel 1974), as the simulation of a general trend plus a random error. It was the only method available for several years, although never applied in industry on a large scale. New methods, including

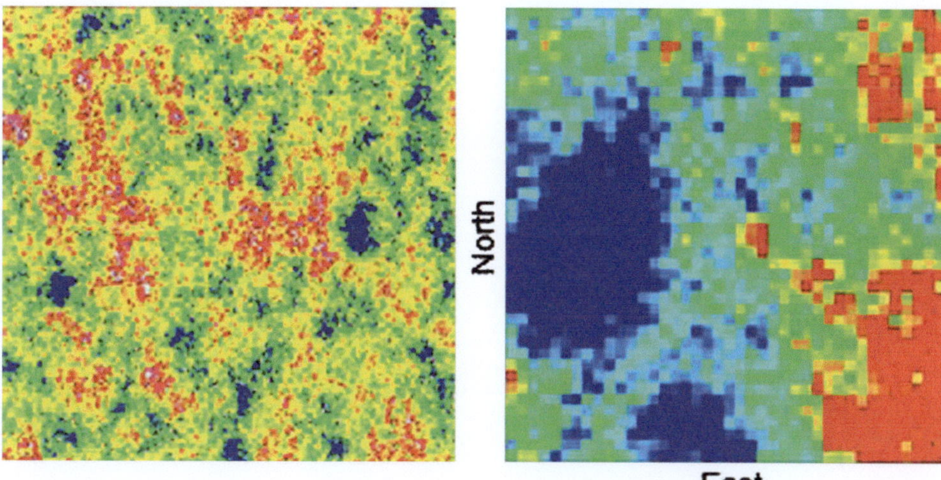

Fig. 10.2 Comparison of high (*left*) and low (*right*) spatial entropy distributions

indicator-based simulations, were developed during the second half of the 1980's. These methods were tested in several mining scenarios and applications, but because of hardware limitations and other practical reasons not fully implemented within industry.

It was not until 1991 that the first, full scale implementation and application was completed within the mining industry. Two sequential Gaussian simulations (SGS) were developed by H.M. Parker and E.H. Isaaks (personal communications). The conditional simulation study was developed to support a Feasibility Study for the Lihir Au project in Papua New Guinea, at the time owned by Kennecott Mining Corporation. Soon after that, the first conditional simulation-based grade control method was implemented by N. Schofield (at the time with FSS International Consultants) for the Marvel Loch mine in Western Australia, and followed a few months later with the implementation of a similar grade control method at the San Cristóbal mine in Northern Chile (Aguilar and Rossi 1996).

As computer hardware capabilities improved drastically throughout the 1990's, more and more simulation studies were tested and published, although full, industry-scale implementation remained relatively scarce. Gradually the number of implementations grew, and by the second half of the following decade, several mining companies began to use conditional simulations routinely, with large companies like BHP-Billiton adding these methods to their internal good practices guidelines for project evaluation.

10.2.1 Sequential Gaussian Simulation

The sequential Gaussian simulation algorithm (Isaaks 1990) is based on a multiGaussian RF model assumption. It is the simulation version of the MG algorithm, and it benefits from all the convenient properties that the multi-Gaussian RF offers: all conditional distributions are Gaussian, and simple kriging is the only method that yields (exactly) the estimated Gaussian mean and variance.

The sequential Gaussian simulation entails the following steps:

1. Complete a full exploratory data analysis of the original data, including variography and domain definition.
2. After defining the domains, analyze whether the data needs to be de-trended, that is, whether the simulation should be done on the residuals.
3. Apply the normal scores transformation to the original data to obtain the corresponding Gaussian distribution.
4. Obtain the Gaussian variogram models for the transformed variable.
5. Define a random path through each domain to be simulated. The path for the simulation is defined randomly to avoid artifacts.
6. Estimate through simple kriging the conditional distribution for each node to be simulated in the Gaussian space. The estimated simple kriging value $Y^*(\mathbf{u})$ is the mean of the conditional distribution, and its variance the simple kriging variance, $\sigma_{sk}(\mathbf{u})$. If the simulation is done on residuals after de-trending, then the Gaussian mean of the conditional distribution is 0.
7. Draw randomly from the conditional distribution to obtain a simulated value for the node, $Y_s(\mathbf{u})$.
8. Incorporate the simulated value $Y_s(\mathbf{u})$ as conditioning data for nodes simulated later. This is necessary to ensure variogram reproduction.
9. Repeat and continue the process until all nodes and all domains have been simulated.
10. Verify that the univariate distribution (histogram) of the simulated values is Gaussian; also check that the simulated values reproduce well the Gaussian variogram model used in the simulation.
11. Back-transform the Gaussian simulated values to the original variable space.
12. Add back the trend if the simulation was performed on residuals.
13. Verify that the histogram of the back-transformed data is similar to the original distribution; also verify that the

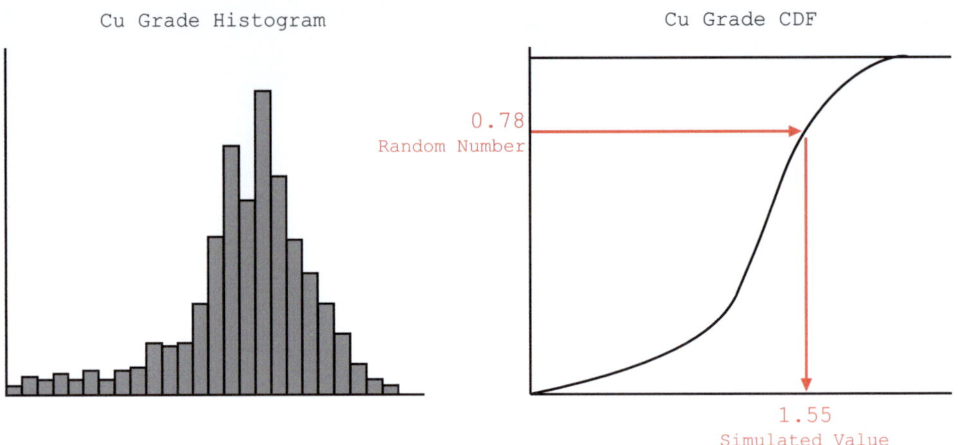

Fig. 10.3 Schematic of a point being drawn using Monte Carlo Simulation

variogram obtained from the simulated values is similar to the variogram of the original values.

14. Verify that the model presents a reasonable spatial distribution and that no other errors or omissions have been made in simulating each stationary domain.

The price to pay for using SGS is that the values will show less connectivity than the original data. This is due to the maximum entropy property of the Gaussian distribution. The significance of this issue is generally small, depending on the overall spatial distribution and the definition of stationary domains used. Conditional simulations are much more sensitive to departures from strict stationarity compared to estimation methods, and so domain definition is key to obtaining representative simulation models.

A Monte Carlo simulation (MCS) technique is used to draw a simulated value from the estimated conditional distribution at each node, see Step 7 above. A random number between 0 and 1 is generated, and the simulated value is obtained by reading the associated quantile from the estimated cumulative distribution. Figure 10.3 illustrates this with an example for a Cu grade distribution.

10.2.1.1 Practical Considerations in the Implementation of SGS

The transformation of the data to a standard normal distribution using the normal scores (NS) transform is a graphical one-to-one (rank preserving) transform and was discussed in Chap. 2.

Declustering weights are recommended for performing the transformation. Another potentially significant issue is despiking. Despiking is the term used to describe the process of removing the ties that the original variable may present, and may be significant since the NS transform does not allow for ties. The method chosen for resolving the ties can be consequential, particularly for distributions that have a very significant percentage of ties. This is common, for example, in data from epithermal Au deposits, where usually most sampled values are below detection limit, perhaps up to 70 or 80 %. In these types of distributions, a few high values contain most of the metal in the drill hole database. Random despiking is fast and usually does not create artifacts in the later back-transformed distribution. A general rule-of-thumb is that, for distributions with more than 50 % percent of tied values, the local-average despiking method (Verly 1984) may be safer and worth the extra effort it requires. In areas where there may be too few data, a global transformation table can be used.

The search path needs to be random to avoid artifacts in the simulated model. It is also important to ensure that each cell is visited once and only once. Nodes that already have a value are skipped, and the original preserved in the simulation.

The data can be assigned to grid nodes, which significantly speeds up the simulation. This is because searching for previously simulated nodes and original data is accomplished in one step and based on a regular grid. However, there is a price to pay, since assigning the closest of several possible data to a node will lose some information. This loss of information is dependent on the data density and the node spacing. The node spacing in the vertical direction should be the same as the original composite length, which ensures that most of the drill hole data would be used, the possible exception being inclined or sub-horizontal drill holes, as well as twinned, or closely drilled holes.

The number of data to be used in the simulation (original composites and previously simulated nodes) can be a consequential decision. More data gives a more accurate estimate of the conditional mean and variance and results in better reproduction of the variogram model, but will take longer to run. Also, fewer data can provide a more robust model with respect to departures from strict stationarity, better reproducing the local variability.

Simple kriging should be used to estimate the Gaussian mean and variance. In certain circumstances, practitioners use ordinary kriging instead, generally with the intention of avoiding the consequences of departures from stationarity.

The price to pay for this choice is that the variance of the Gaussian simulated values is inflated, and that the OK estimator does no longer provide the exact Gaussian mean, only an approximation. While these authors generally discourage simulating using OK, if there is a significant number of original data, such as blast holes, the variance inflation issue may be less significant.

As with all other simulation techniques, the most critical and often time-consuming step is to check the results. Honoring the declustered histograms of the Gaussian values and of the original distribution is a first check. The reproduction of the variogram model for both the Gaussian and the original data should be checked as well. The simulated spatial distribution should show the characteristics and trends of the original data, adequately reproducing trends and local means and variances.

If there is reliable production information, the grade-tonnage curves derived from the simulated models should reproduce well the actual values from production. Likewise, if grade control data exists and is not used in the simulation, it can be used to validate the conditional simulation models.

It is good practice to implement the simulation initially on a small area, fine-tuning the simulation parameters that result in a better reproduction of histograms and variogram models, as well as production data if available. After the implementation details are defined, then the full, multiple simulation runs are completed.

10.2.2 Turning Bands

The Turning Bands (TB) method was the first 3-D geostatistical simulation method, originated by Matheron (1973) and developed by Journel (1974). Although the sequential simulation method has been popular for many years, turning bands simulation is still used. The turning bands method generates a 3-D simulation results from several independent 1-D simulations along lines that can be rotated in 3-D space. This unique way of simulating provides 3-D unconditional realizations.

After transforming the original data to Gaussian values, TB consists of two main steps: (1) develop an unconditional simulation in Gaussian units, with the experimental histogram reproduced by transformation and the covariance or variogram from the data being also reproduced; and (2) condition the turning bands simulation through a post-processing using kriging (Journel and Huijbregts 1978). This method exactly honors the conditioning data and also preserves the variability of the unconditional simulation realizations. This method has been adapted to conditioning multiple-points structures (Ren et al. 2004; Ren 2005).

The initial step in Turning Bands is to obtain an unconditional simulation based on the covariance models derived from the existing data. Initially, N lines are defined in the three-dimensional space: D_i, $i=1, N$. On each line, a one-dimensional RF $Y(\mathbf{u}_{Di})$, $i=1, N$ is defined; these N RFs are independent from each other. For each line, there is also a 3-D Random Function: $Z_i(\mathbf{u})$, $i=1,\dots N$.

At first, the line D_i is simulated at each point $\mathbf{u}_{Di}$ on the line. Moving averages is typically used for the initial 1-D simulation. Second, the simulated value $y(\mathbf{u}_{Di})$ at $\mathbf{u}_{Di}$ is assigned to all the points inside the slice or band perpendicular to line D_i at $\mathbf{u}_{Di}$:

$$Z_i(\boldsymbol{u}) = y(\boldsymbol{u}_{Di})$$

Then, at each point $\mathbf{u}$ in the 3-D space, sum all the values from the N slices or bands to obtain a realization for this point:

$$Z_s(x) = \frac{1}{\sqrt{N}} \sum_{i=1}^{N} Z_i(x)$$

The unconditional simulation is generated after obtaining the values for all the points in the 3-D model.

Data conditioning of the TB simulations is done by developing two kriging runs: first, the conditioning data is kriged to obtain $y_{kc}(\mathbf{u})$, which reproduces data trends. The second step is to krige with the unconditional simulated values at these conditioning data locations to obtain $y_{ku}(\mathbf{u})$. Then, the conditional simulation values $y_{cs}(\mathbf{u})$ are calculated as the unconditionally simulated values adjusted by the difference between both kriged values:

$$y_{cs}(\mathrm{u}) = y_{uc}(\mathrm{u}) + \left[y_{kc}(\mathrm{u}) - y_{ku}(\mathrm{u})\right]$$

The simulation is performed in the Gaussian space to ensure histogram reproduction, while covariance reproduction is ensured by the use of the unconditional simulation.

The conditional simulation procedure can be understood using a 1-D example, see Fig. 10.4. At each real data location, the unconditional simulated value is taken out, and the conditioning datum is put in. Near the data location, the kriging estimator smoothes the change between the sample data and the unconditional simulated values outside the range of kriged values. Therefore, the conditional simulated values at these N data locations will be the data values. Beyond the range of correlation, the conditional simulated values will be the unconditional simulated values.

There are some inherent limitations of the turning bands algorithm: (1) the post processing for conditioning is cumbersome; (2) the 1-D covariances must be worked out separately for each nested structure and covariance shape because they are different in 2-D and 3-D; and (3) only isotropic covariances can be used; anisotropy is introduced by geometric transformation.

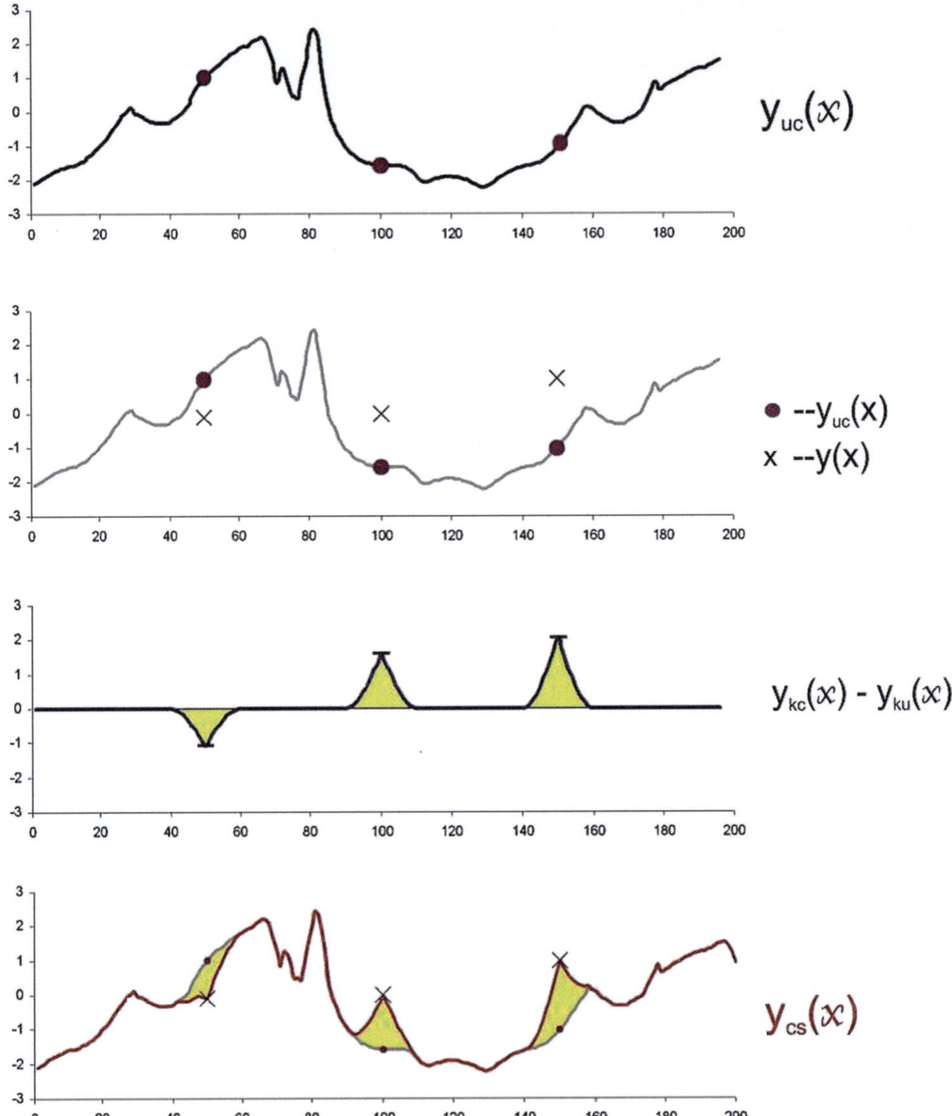

Fig. 10.4 1-D example of conditioning by kriging

In earlier implementations of TB and because of hardware limitations, the position of the original *N* lines was evident in the resulting simulated image. The solution to avoid artifacts is to use a very large number *N* of lines, which is currently more practical than in years past. Artifacts, particularly if the method is poorly implemented, can be a significant disadvantage.

10.2.3 LU Decomposition

When the total number of conditioning data *plus* the number of nodes to be simulated is small (fewer than a few hundred) and a large number of realizations is requested, simulation through LU decomposition of the covariance matrix provides the fastest solution (Luster 1985; Alabert 1987b).

Let $Y(\mathbf{u})$ be the stationary Gaussian RF model with covariance $C_Y(\mathbf{u})$. Let $\mathbf{u}_\alpha$, $\alpha=1, \ldots, n$, be the locations of the conditioning data and $\mathbf{u}'_i$, $i=1, \ldots, N$, be the N nodes to be simulated. The large covariance matrix $(n+N)\cdot(n+N)$ is partitioned into the data-to-data covariance matrix, the node-to-node covariance matrix, and the two node-to-data covariance matrices:

$$\mathbf{C}_{(n+N)(n+N)} = \begin{bmatrix} \left[C_Y\left(\mathbf{u}_\alpha - \mathbf{u}_\beta\right)\right]_{n\cdot n} & \left[C_Y\left(\mathbf{u}_\alpha - \mathbf{u}'_j\right)\right]_{n\cdot N} \\ \left[C_Y\left(\mathbf{u}'_i - \mathbf{u}_\beta\right)\right]_{N\cdot n} & \left[C_Y\left(\mathbf{u}'_i - \mathbf{u}'_j\right)\right]_{N\cdot N} \end{bmatrix} = \mathbf{L}\cdot\mathbf{U}$$

The large matrix $\mathbf{C}$ is decomposed into the product of a lower and an upper triangular matrix, $\mathbf{C}=\mathbf{L}.\mathbf{U}$. A conditional realization $\{y^{(l)}(\mathbf{u}'_i), i=1, \ldots, N\}$ is obtained by multiplication of $\mathbf{L}$ by a column matrix $\omega_{(N+n)1}^{(l)}$ of normal deviates:

$$y^{(l)} = \begin{bmatrix} \left[y(\mathbf{u}_\alpha)\right]_{n\cdot 1} \\ \left[y^{(l)}(\mathbf{u}'_i)\right]_{N\cdot 1} \end{bmatrix} = \mathbf{L}\cdot\omega^{(l)} = \begin{bmatrix} \mathbf{L}_{11} & 0 \\ \mathbf{L}_{21} & \mathbf{L}_{22} \end{bmatrix} \cdot \begin{bmatrix} \omega_1 \\ \omega_2^{(l)} \end{bmatrix}$$

where $[y(\mathbf{u}_\alpha)]_{n1}$ is the column matrix of the n normal score conditioning data and $[y^{(l)}(\mathbf{u}'_i)]_{N1}$ is the column matrix of the N conditionally simulated y values.

Identification of the conditioning data is written as $\mathbf{L}_{11}\ \omega_1 = [y(\mathbf{u}_\alpha)]$; thus matrix ω_1 is set at:

$$\omega_{1=}\left[\omega_1\right]_{n\cdot 1} = \mathbf{L}_{11}^{-1}\cdot[y(\mathbf{u}_\alpha)]$$

The column vector $\omega_2^{(l)} = [\omega_2^{(l)}]_{N1}$ is a vector of N independent standard normal deviates.

Additional realizations, l=1, …, L, are obtained at very little additional cost by drawing a new set of normal deviates $\omega_2^{(l)}$, then by applying the matrix multiplication. The major cost and memory requirement is in the upfront LU decomposition of the large matrix **C** and in the identification of the weight matrix ω_1.

The LU decomposition algorithm requires that all nodes and data locations be considered simultaneously in a single covariance matrix **C**. The current practical limit of the number (n+N) is no greater than a few hundred.

Implementation variants have been considered, attempting to relax the previous size limitation by considering overlapping neighborhoods of data locations. Unfortunately, artifact discontinuities appear if the correlation between all simulated nodes is not fully accounted for.

The LU decomposition algorithm is particularly appropriate when a large number of realizations is needed over a small volume or block (n+N is small). A typical application is the evaluation of block ccdf's. Any block *V* can be discretized into *N* points. The normal score values at these *N* points are simulated *L* times ($l=1,\ldots,L$) through the LU decomposition algorithm and back-transformed into simulated point *z* values: $\{z^{(l)}(\mathbf{u}'_i),\ i=1,\ldots,N; \mathbf{u}'_i\ in\, V \in A\},\ l=1,\ldots,L$. Each set of *N* simulated point values can then be averaged to yield a simulated block value. The distribution of the *L* simulated block values $z_v^{(l)}$, $l=1,\ldots,L$, provides a numerical approximation of the probability distribution (ccdf) of the block average, conditional to the data retained.

10.2.4 Direct Sequential Simulation

Direct Sequential Simulation (DSS, Soares 2001) is based on the idea that a non-Gaussian distribution could be considered in the sequential path, as long as this distribution has the same mean and variance of the Gaussian distribution it replaces, and therefore it is seen as an extension of the more established Gaussian simulation paradigm (Journel 1994). In essence, direct sequential simulation is the same as sequential Gaussian simulation, but without the normal score transform step.

The reasons to be interested in DSS include: (a) the reproduction of the variogram in original units; (b) dealing with variables that do not average linearly after the normal score transform; in this case, Gaussian techniques are inappropriate for data at different scales; and (c) the maximum entropy characteristic of the Gaussian distribution.

Consider simple kriging at node $\mathbf{u}=\mathbf{u}_l$ with N data values $z(\mathbf{u}_a)$, $a=1,\ldots,N$

$$z^*(\mathbf{u}) = \sum_{\alpha=1}^{N}\lambda_\alpha z(\mathbf{u}_\alpha)$$

$$\sigma_{SK}^2(\mathbf{u}) = 1 - \sum_{\alpha=1}^{N}\lambda_\alpha\rho(\mathbf{u}-\mathbf{u}_\alpha)$$

$$\sum_{\beta=1}^{N}\lambda_\alpha\rho(\mathbf{u}_\beta-\mathbf{u}_\alpha) = \rho(\mathbf{u}-\mathbf{u}_\alpha),\quad \alpha=1,\ldots,N$$

A random variable $z_s(\mathbf{u})$ is drawn from the univariate probability distribution function (pdf) $f(\mathbf{u}, z|(N))$

$$Z_S(\mathbf{u}) = Z^*(\mathbf{u}) + R_S(\mathbf{u})$$

With the residuals $R_s(\mathbf{u})$ drawn from a pdf $f_l(r)$, with mean 0 and variance $\sigma^2_{SK}(\mathbf{u})$. The critical point is the independence of $Z^*(\mathbf{u})$ and $R_S(\mathbf{u})$, linked to the homoscedastic property of the Gaussian variance $\sigma^2_{SK}(\mathbf{u})$.

Now consider the next node $\mathbf{u}'=\mathbf{u}_l+1$. *Simple Kriging* using N+1 data, including the previously simulated value $z_s(\mathbf{u})$, is written as:

$$z^*(\mathbf{u}') = \sum_{\alpha=1}^{N}\lambda_\alpha(\mathbf{u}')z(\mathbf{u}_\alpha) + \lambda_{N+1}(\mathbf{u}')z_s(\mathbf{u})$$

$$\sigma_{SK}^2(\mathbf{u}') = 1 - \sum_{\alpha=1}^{N}\lambda_\alpha(\mathbf{u}')\rho(\mathbf{u}'-\mathbf{u}_\alpha) - \lambda_{N+1}(\mathbf{u}')\rho(\mathbf{u}'-\mathbf{u})$$

$$\sum_{\beta=1}^{N}\lambda_\beta(\mathbf{u}')\rho(\mathbf{u}_\beta-\mathbf{u}_\alpha) + \lambda_{N+1}(\mathbf{u}')\rho(\mathbf{u}-\mathbf{u}_\alpha) = \rho(\mathbf{u}'-\mathbf{u}_\alpha),\quad \alpha=1,\ldots,N$$

$$\sum_{\beta=1}^{N}\lambda_\beta(\mathbf{u}')\rho(\mathbf{u}_\beta-\mathbf{u}) + \lambda_{N+1}(\mathbf{u}') = \rho(\mathbf{u}'-\mathbf{u})$$

A simulated value can then be drawn from this distribution:

$$Z_S(\mathbf{u}') = Z^*(\mathbf{u}') + R_S(\mathbf{u}')$$

The two kriged values clearly depend on one another and the kriged value at the second location depends on the first random value.

It can be shown that the covariance between the two values is correct. This is the well-established theory of sequential simulation: it is unbiased, the variance is correct, and the covariance between all simulated values is correct. However, there are concerns with DSS: (a) it simply cannot avoid the influence of Gaussianity; (b) the shape of the *R*-values distribution required to preserve the original histogram; (c) the proportional effect or heteroscedasticity of kriging variances, and (d) the multi-scale multivariate data. The shape of the resulting histogram is difficult to control, and multivariate spatial features are typically very similar to SGS (Deutsch 2002).

While any shape distribution can be used and can be changed locally, the resulting histogram will be subject to three influences: the histogram of original data, the chosen shape of the random distributions, and the Gaussian distribution that results from the averaging of random components, i.e., the Central Limit Theorem.

Correction schemes proposed to obtain the correct local histograms include post-processing of realizations (Journel and Xu 1994; Caers 2000); selective sampling; and establishing a consistent set of distributions (Oz et al. 2003). The latter is based on using the link between Gaussian and real data units to build the shape of the conditional distributions, and is the recommended option to solve DSS's issues with histogram reproduction. This is because, while the link to the Gaussian model gives the ccdf its expected shape, the results are not Gaussian. Also, there is no post-processing or ad-hoc correction, and the block data are reproduced exactly.

Another significant issue is that of proportional effect. For most variables, the mean and its variance are correlated. Simulations that use a transformation, such as indicator simulations and Gaussian techniques are insensitive to the proportional effect. This is because the transform effectively removes the proportional effect, although the data in the original space do show a proportional effect.

In DSS, the kriging variance provides the variance of the local ccdf. This variance depends only on the data configuration and is independent of the data values, which is incorrect most of the times when dealing with data in original units. The kriging variance is not a measure of local variability; it only works well after a Gaussian transformation. However, the central idea of DSS is to avoid that Gaussian transformation.

The best approach if using DDS is to: (a) use a standardized variogram; (b) calculate the standardized kriging variance; and (c) rescale that variance to a local measure of variability. This requires two additional steps: fitting the proportional effect and calculating the local mean at each location.

10.2.5 Direct Block Simulation

Direct block simulation is another simulation option that attempts to simplify the simulation process by directly working at a support other than the original nodes or composites. Journel and Huijbregts (1978) originally proposed a direct block simulation based on separate simulation and conditioning steps. The method is based on using a global change of support (Chap. 7), which is based on a permanence of distribution technique, to correct the point support data to block support. Then the conditioning happens at the block support level.

A different approach has been proposed by Gómez-Hernández (1992) in the context of simulating hydraulic conductivity fields. The idea is that, if the block statistics are known and the point and block distributions are assumed jointly Gaussian, then a joint sequential Gaussian simulation provides direct block simulated values conditioned to the original point support data. The inference of the block covariances can be done using a global change of support using a permanence of distribution assumption, or, as developed by Gómez-Hernández, from a training image and the assumption of a point and block univariate lognormal distribution, that is, the sequential Gaussian simulation performed on the log of the original data.

Marcotte (1993) provided another alternative, based on using Disjunctive Kriging (DK) to obtain the local block cumulative distribution frequencies. The simulated values are drawn from this local block cdf. This method has the substantial disadvantage of using DK, a cumbersome and theoretically difficult method to implement, which requires also a strong prior assumption about the distribution of block grades, but at the same time offers the flexibility of integrating data types with different support, including drill hole data, bulk samples, and mined out stopes or areas.

Godoy (2002) developed an alternative version, the Direct Block Simulation (DBSim) algorithm. The DBSim algorithm is an adaption of the "classical" SGS method.

The main difference between a traditional SGS and the DBSim methods is that the DBSim simulation is done originally on nodes, then immediately re-blocked to the specified block (SMU) size, and the block data (through its discretization points) is used to condition further nodes and blocks in the sequential random path used. DBSim works on block centroids, retaining in memory only the previously simulated block values, while SGS first obtains the full set of nodes on the random path. In the case of SGS, re-blocking from node support to the SMU block size is a separate, independent step.

The implementation of direct block simulations has, as with direct sequential simulation, pitfalls that must be avoided. But in cases where the size of the simulated model is significant, or even impractical if performed at a node scale, it may be an alternative worth investigating.

10.2.6 Probability Field Simulation

The key idea of probability field (P-field) simulation is to perform the simulation in two separate steps (Froidevaux 1992). In the first step the local distributions of uncertainty are constructed. This is done using only the original data so that it can be done only once instead of repeatedly for each realization. The second step is to draw from those distributions using correlated probability values instead of the random numbers used in traditional Monte Carlo simulation.

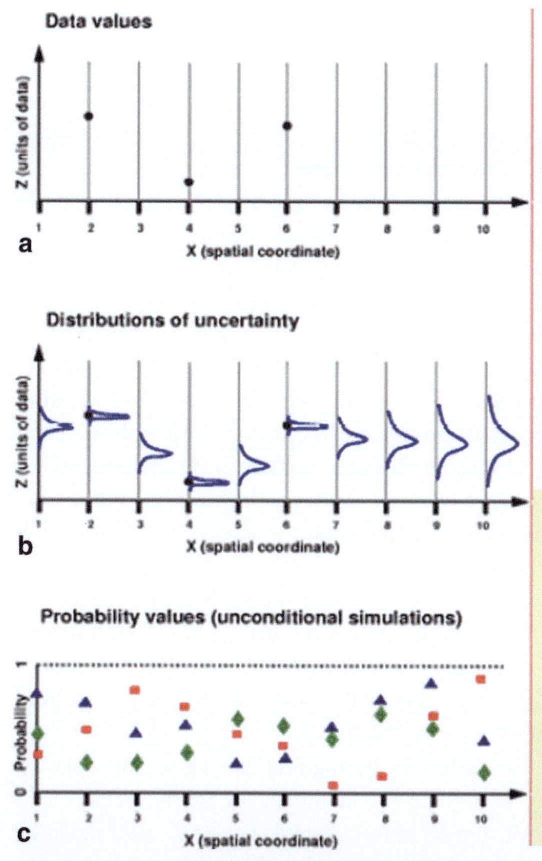

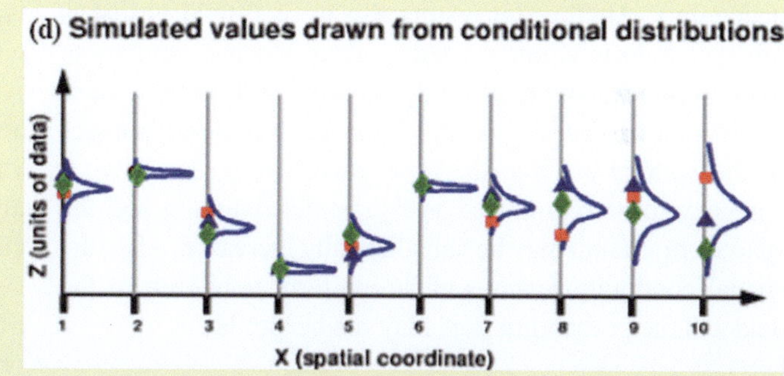

Fig. 10.5 Step required obtaining a P-field simulation

The typical sequential simulation approach is based on drawing from the local conditional distributions random probabilities and then adding each simulated value to the pool of conditioning data. P-field, by dissociating the two steps, starts with the premise that the local cdf's are known, and they become input parameters. Probability values are used to draw from the local cdf's, constituting a probability field, and which are interpreted as outcomes of a RF with uniform distribution and a known covariance function.

Assume that a simulated value has been obtained for each location **u** within the area of interest. The simulated value $z_s(\boldsymbol{u})$ corresponds to a specific $p(\boldsymbol{u})$ probability of the local cdf:

$$F(u, z_s(u)) = p(u)$$

The local probabilities $p(\boldsymbol{u})$ are interpreted as outcomes of the RF $P(\boldsymbol{u})$. If the local cdf's are available and the univariate and bivariate statistics of $P(\boldsymbol{u})$ can be inferred, then the P-field simulation can be completed in a straightforward manner.

The local cdf's can be derived using hard data, soft, exhaustive data, or empirically using a geologist's subjective opinion as to likely range of values for the local cdf's. Perhaps more importantly, the inference of the probability field parameters presents bigger challenges. Froidevaux (1992) recommends starting with two intuitive basic assumptions: (a) the probability field $P(\boldsymbol{u})$ follows a Uniform Distribution, as one would expect; and (b) the covariance of the probability field $P(\boldsymbol{u})$ and the uniform transform of the variable $U(Z(\boldsymbol{u}))$ are the same, such that:

$$C_P(h) = C_U(h), \quad \text{with} \quad U(u) = F(Z(u))$$

In essence, it is assumed that the two-point continuity of the uniform transform of the original variable is similar to that of the probability field. These features include relative nugget effect, anisotropy ratios, and direction and range of maximum continuity. This is a rather strong assumption, and difficult to verify. Typically, the more hard data exists, the less similar would $C_P(h)$ and $C_U(h)$ be.

The implementation of P-field follows these basic steps, see Fig. 10.5 for a schematic explanation:

1. After defining the grid spacing and nodes to be simulated, the local cdf's of the variable being simulated must be obtained. These cdf's can be derived from hard, local data through a non-linear estimation method (Chap. 9), with or without secondary information, and can even be defined empirically. The local cdf's are independent of the simulation processes.

2. Calculate and model the variogram of the uniform transform of the original variable, *U(Z(u)*. Assume that the probability field *P(u)* follows a uniform distribution, and that $C_P(h) \approx C_U(h)$.
3. Generate a non-conditional simulation of *P(u)* honoring the uniform distribution and the covariance $C_P(h)$.
4. At each node, draw a simulated value $z_s(u)$ from the local cdf *F(u, z)* using the probability value *p(u)*: $z_s(u) = F^{-1}(u, p(u))$
5. Repeat the above two steps until a sufficient number of realizations have been obtained.

The main advantage of P-field simulation is its speed, and that the distributions of uncertainty can be constructed to honor all data and checked before any realizations are drawn. Also, the simulations are consistent with the distributions of uncertainty. Another interesting aspect of P-field is that is fairly easy to integrate secondary data into the simulation, without increasing significantly the time and effort it takes to obtain a realization. Some of the potential disadvantages of the method are that the local conditioning data have a tendency to be reproduced as a local discontinuity and the spatial correlation may be unrealistically increased. Also, the spatial continuity features of the uniform transform of the data and the probability field may not be similar.

10.3 Continuous Variables: Indicator-Based Simulation

Sequential Indicator Simulation (SIS) is in essence the same as the sequential Gaussian simulation, except that instead of simulating a Gaussian variable, the indicator transform of the original variable is simulated. Aside from providing a method that is not dependent on Gaussian assumptions, SIS does not require any back-transformation, drawing directly the simulated value in the original space from the local indicator kriging-derived conditional distributions. Details of the indicator formalism were presented in Chap. 9.

Recall that the average of the indicator transform is the global proportion of the stationary domain. The variogram of an indicator variable measures spatial correlation, as opposed to variability:

$$\gamma_I(h) = \frac{1}{2} E\{[I(u;k) - I(u+h;k)]^2\}$$

Inference of cumulative indicator variograms is easier than class indicators because more conditioning data is always used, and cumulative indicators carry information across cutoffs.

The ccdf is calculated using a linear combination of the nearby indicator data:

$$p_k^*(u) = \sum_{\alpha=1}^{n} \lambda_\alpha(u) \cdot I(u_\alpha;k) + [1 - \sum_{\alpha=1}^{n} \lambda_\alpha(u)] \cdot m_k$$

The weights (λ_α(u), α=1,...,*n*) are determined by kriging. Simple kriging is the exact solution to the least-squares optimal estimate. If local departure from stationarity is a concern, an estimate of the local mean can be used, but usually with an inflated variance as cost.

As before, a random number is drawn to determine which class *k* to assign to the node. Since the conditional probabilities were estimated by kriging with a given variogram, the original histogram and variograms of the data will be reproduced by the simulated values.

The steps in SIS are identical to those used for SGS. All the relevant decisions regarding implementation are the same, and the processes used to check the models are also the same. However, in practice, SIS of a continuous variable has proven to be difficult to calibrate. Variance inflation is common, as is the difficulties in controlling the tails of the distribution, particularly the highs on a positively skewed distribution.

SIS is mostly used with variables that exhibit high variability, as for example epithermal gold grades. Dealing with variables characterized by a high coefficient of variation is always difficult, and in these cases SIS can present significant challenges.

While the indicator variograms are reproduced as expected (up to ergodic fluctuations), the reproduction of the original *z* variogram is not assured. For this, a full indicator co-kriging should be performed, which in practice is never done. It can be shown that the SIS simulation reproduces the madogram, $2\gamma_M(h) = E\{|Z(u) - Z(u+h)|\}$, since this function is the integral of all indicator variograms (Alabert 1987a; Goovaerts 1997). But the practical consequence of this is minor, since there is no particular reason to prefer the *z* variogram to the *z* madogram reproduction.

The reward for those who take on the challenges of using SIS on continuous variable is a simulation model that does not rely on Gaussian assumption, and thus do not have that underlying maximum entropy property. SIS of a continuous variable when it is necessary to reproduce well the connectivity of extreme values, and thus Gaussian features may be a concern.

10.4 Simulated Annealing

Simulated annealing is a minimization/maximization technique that has attracted significant attention in recent years. It is based on a thermodynamics analogy, specifically with the way liquids freeze and crystallize, or metals cool and anneal. The basic principles are that: (a) molecules move freely at high temperature; (b) thermal mobility is lost if cooled slowly; (c) atoms can often line up over a distance billions their size in all directions; and (d) such crystal structures are minimum energy states. This state can always be found if cooled slowly to allow time for redistribution of the atoms.

The annealing method mimicks nature's minimization algorithm. It is different than conventional minimization algo-

Fig. 10.6 Annealing in a physical process: grains of brass tend to re-organize as the metal cools

rithms, because these go for nearby, perhaps short-term solutions, and stop. That is, they go immediately downhill as far as possible in any step. Annealing may not go immediately downhill (Fig. 10.6).

The Boltzmann probability distribution expresses the idea that a system in thermal equilibrium has its energy probabilistically distributed among all different energy states: even at low temperature (T), there is a small chance of being in a high energy (E) state:

$$P(E) \approx e^{\frac{-E}{kT}}$$

Therefore, there is a chance to get out of a local energy minimum in favor of finding a better, more global, one. Boltzmann's constant (k) is a constant of nature that relates temperature to energy.

In 1953, Metropolis and coworkers first incorporated these principles into numerical calculations. The idea was to provide a succession of options; a simulated thermodynamic system will change configuration from energy E1 to energy E2 with a probability given by (Fig. 10.7):

$$P(E) = \begin{cases} e^{\frac{-(E_2 - E_1)}{kT}} & \text{if } E_2 > E_1 \\ 1, \text{ otherwise} \end{cases}$$

This general scheme of always taking a downhill step and *sometimes* taking an uphill step has come to be known as the Metropolis algorithm.

In order to implement this idea, a few key elements are needed: (a) The possible system configurations should be described; (b) A random generator of changes is required to present options to the system; (c) An objective function O (analogous to thermodynamic energy) is defined; the goal of the procedure is to minimize this function; (d) A control parameter T (analog of temperature) and (e) an annealing schedule which tells how T is lowered from high to low values. For example, the schedule defines how many random changes are necessary before is a downward step in T is taken, and how large is that step.

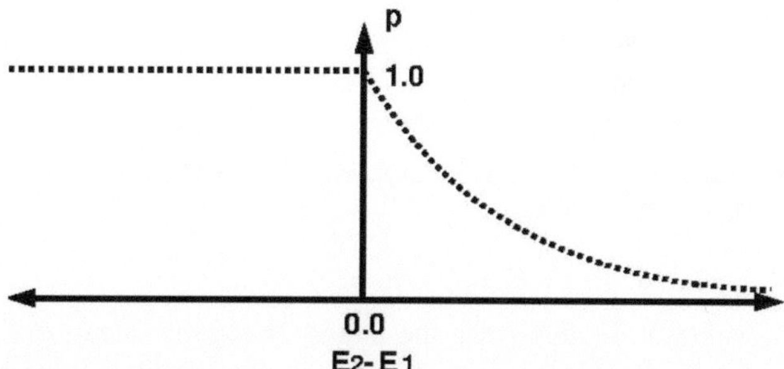

Fig. 10.7 Probability function of a change in energy state according to Metropolis' algorithm

Consider the traveling salesman problem. The salesman visit N cities with given positions (x_i, y_i) returning finally to city of origin. The objective is to visit each city only once and to make the route as short as possible.

This problem is known as an NP-complete problem, whose computation time for an exact solution increases with N as $e^{(c*N)}$. Thankfully, the simulated annealing method limits its calculations to a small power of N. To solve this problem, the annealing procedure is implemented as follows:

1. By numbering the cities from i = 1,…,N, the initial configuration is defined. Any other configuration is the permutation of the numbers 1,…,N.
2. A rearrangement is the swapping of the order of two cities. There are more and less efficient procedures to rearrange the order of the cities.
3. The Objective Function is defined as total distance traveled:

$$E = \sum_{i=1}^{N} \sqrt{(x_i - x_{i+1})^2 + (y_i - y_{i+1})^2}$$

with the last city point $N+1$ identified to the first city, i.e., back to the original starting place.

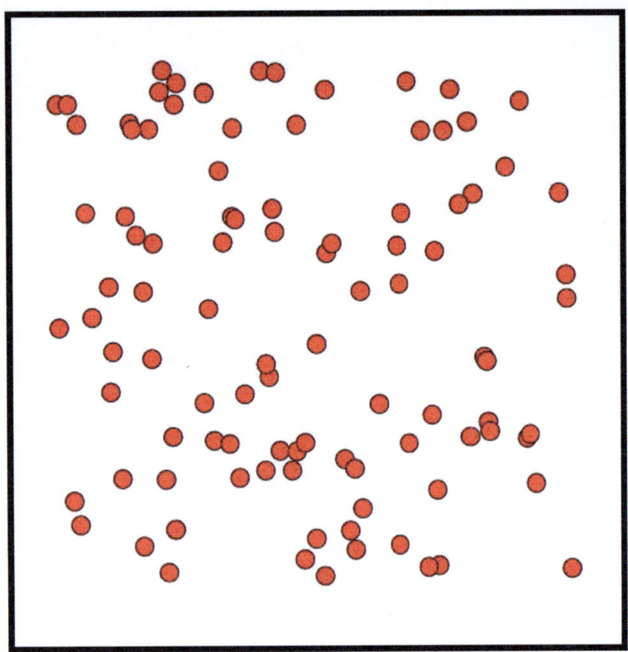

Fig. 10.8 The traveling salesman needs to visit 100 cities within a 1,200 by 1,200 miles area

4. The annealing schedule is defined by choosing a starting value of T larger than the largest change in energy ΔE normally encountered. The procedure is complete when nothing is happening, i.e., there is no improvement on the Objective Function value (Fig. 10.8).

Consider a N=100 city problem within a 1,200 by 1,200 miles area:

Initially, choose randomly multiple configurations (Fig. 10.9) to assess what the non-optimal results would be. A histogram of the total distance traveled of those multiple configurations, 1,000 random paths in this case, is shown in Fig. 10.10.

The problem is solved with a near-optimal solution (Fig. 10.11) that results in 8,136 miles total distance traveled. It generally takes only a few minutes on a capable computer, and the coding is straightforward.

Other characteristics of the sequential annealing solution is that it is straightforward to add components to objective function; it often requires some trial-and-error to get the annealing schedule right; but in the end, provides a solution to an otherwise intractable problem.

The extension to geostatistical mining problems is straightforward. First, an objective function must be formulated which may include with many different components. For example, local drill hole data with geologic information and assays; variogram or other two-point spatial correlation measures; multiple-point spatial connectivity, if available; vertical and areal trends; correlation (collocated or spatial) with secondary data; and historical data.

The objective function generally takes the form of a weighted sum of multiple components. The basic approach is the same:

1. Establish an initial guess;
2. Calculate the initial objective function;
3. Propose a change;
4. Update the objective function;
5. Decide whether to keep the change or not; this is the SA decision rule;
6. Go back and propose a new change;
7. Stop when objective function is low enough. Generally, this means that the original data is matched within acceptable tolerances.

Simulated annealing provides a flexible optimization procedure that has multiple applications and has not been fully exploited within the mining industry. Partly this may be because the parameters need to be set up carefully to avoid artifacts, such as the points where conditioning data exist, and edge effects.

The annealing schedule may be difficult to establish, and it generally requires some experience with the method to avoid either lowering the temperature too slowly, with slow

Fig. 10.9 Two initial configurations chosen randomly

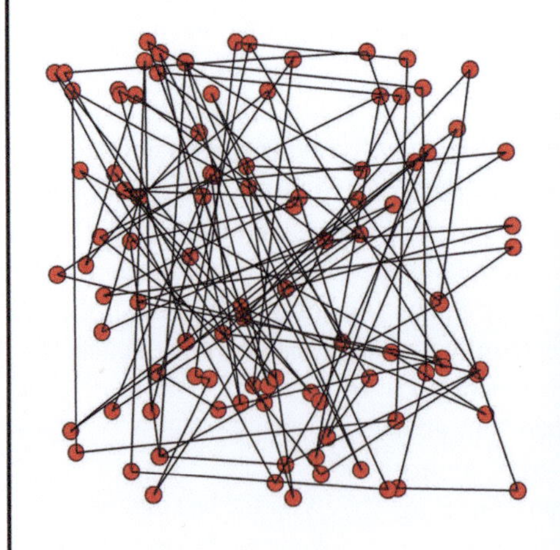

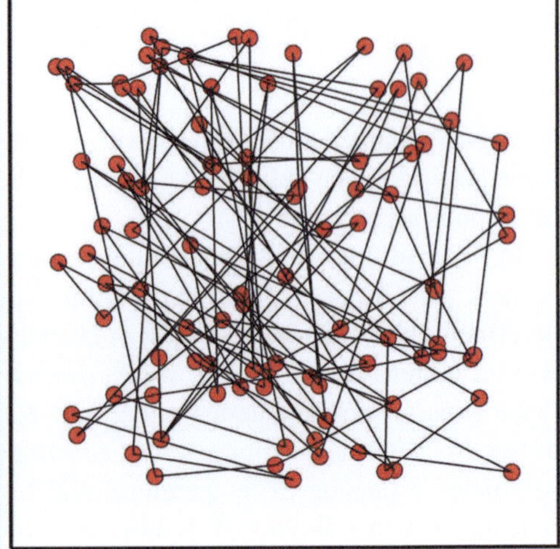

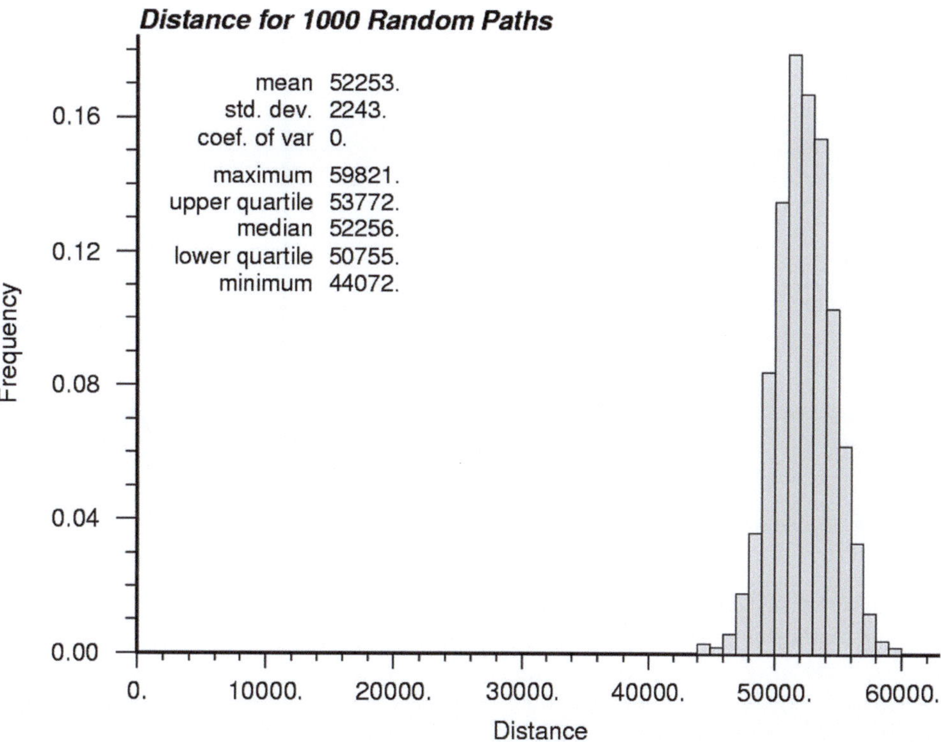

Fig. 10.10 Histogram of 1000 total distances traveled corresponding to random paths. The average is 52,253 total miles traveled

solution times, or too fast, with perhaps the algorithm finding only a suboptimal solution. Rigorous mathematically based schedules that guarantee convergence are possible (Aarts and Korst 1989; Geman and Geman 1984); but they are impractical due to their slow speed.

10.5 Simulating Categorical Variables

Simulating categorical variables aims at providing spatial models for discrete classes. Geologic models provide classic examples of such classes. Consider K mutually exclusive categories s_k, $k=1,\ldots,K$ within a stationary domain. These classes are exhaustive, that is, any location **u** has one and only one of these K categories assigned to it.

10.5.1 SIS For Discrete Variables

If an indicator for the presence $(i(\mathbf{u};\ s_k)=1)$ or absence $(i(\mathbf{u};\ s_k)=0)$ of any class k is defined for each location **u**, then the simple kriging estimate of the indicators provides a probability of s_k occurring at that location:

$$Prob^*\left\{I(u;s_k)=1\,|\,(n)\right\}=p_k+\sum_{\alpha=1}^{n}\lambda_\alpha\left[I(u_\alpha;s_k)-p_k\right]$$

where p_k is the expected value of class k, inferred for example from the declustered data for the entire domain. The weights λ_α are solved using the indicator covariance for s_k in a simple kriging system. It may be wise to choose using local averages, instead of the global mean for the entire domain, to obtain a proportion more representative of the neighborhood of location **u**.

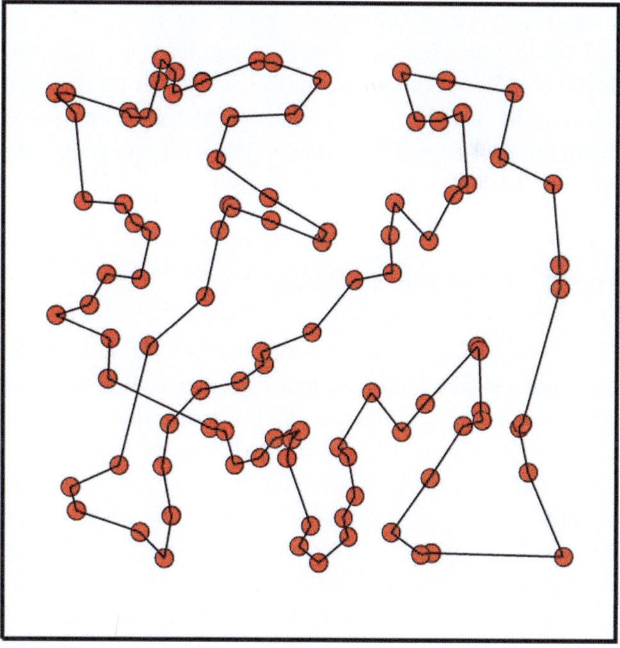

Fig. 10.11 Near optimal solution to the traveling salesman problem. In this example, travel has been reduced to a total of 8,136 miles

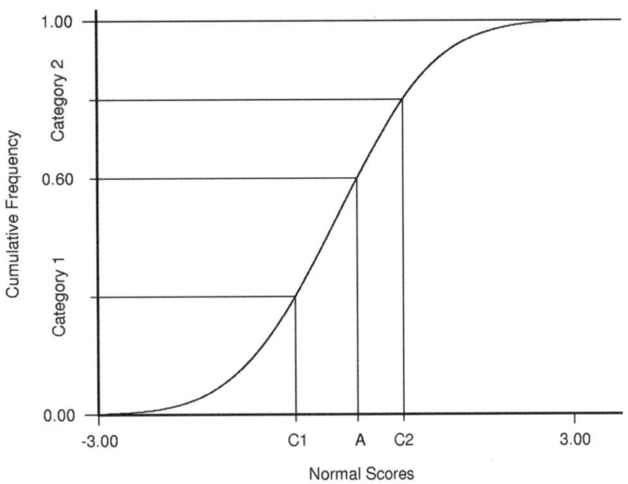

Fig. 10.12 Truncated Gaussian simulation

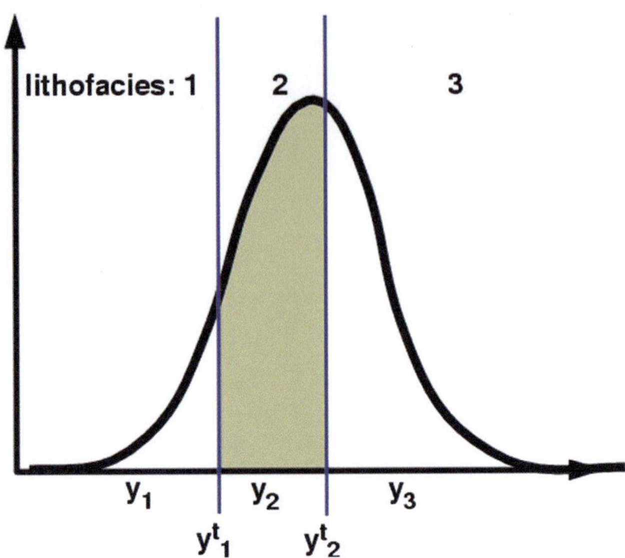

Fig. 10.13 Consequence of ordering, implicit in the truncated Gaussian technique

The sequential indicator simulation (SIS) algorithm is identical for continuous or categorical variables. The key difference is, in the latter case, that the *K* categories can be defined in any order; the local ccdf obtained from indicator kriging provides a cdf-type ordered set of the probability interval [0,1] discretized in *K* intervals. The simulated category at location **u** is defined by the interval in which the random number *p* falls, which is drawn from the uniform distribution [0,1].

Note that the ordering of the K categories is arbitrary and does not affect the simulated model. The ordering does not affect which category is drawn at location **u** or the spatial distribution of categories (Alabert 1987a) because the random number *p* has a uniform distribution.

10.5.2 Truncated Gaussian

The truncated Gaussian technique simulates a continuous standard Gaussian field and truncates it at a series of thresholds to get a categorical variable realization. Only one (Gaussian) variogram model can be specified in this technique. Continuous Gaussian realizations are generated and truncated with the proportions of the different rock types. The conditioning data are coded as the normal score value at the centroids of each category (or rock type) (Fig. 10.12).

An important feature of truncated Gaussian simulation is the ordering of the resultant probability density function (pdf) models. The codes used to characterize the individual classes are generated from an underlying continuous variable. Therefore, normally class 2 will occur between classes 1 and 3. Only rarely would code 1 be next to code 3 (Fig. 10.13). There are specific applications where this could be an advantage, such as in simulating a sedimentary sequence. However, the most common applications in mining are the simulation of lithology; mineralization type; and alteration. In these cases, rarely ordering is part of the natural process being simulated.

10.5.3 Truncated PluriGaussian

A variant of truncated Gaussian is the truncated pluri-Gaussian method. This method uses multiple Gaussian variables, which allow for using different variograms, each with its own spatial variability model, including different relative nugget effects, anisotropies, ranges, and other variogram parameters.

For practical reasons, the number of Gaussian functions is usually kept to 2, one category being the complement of the other. A non-conditional simulation of the indicator variable *i(**u**)* can be obtained by truncating a simulation of the standard Gaussian RF *Y(**u**)*:

$$i^{(l)}(u) = \begin{cases} =1, \text{ if } y^{(l)}(u) \le y_p \\ =0, \quad \text{otherwise} \end{cases}$$

with $y_p = G^{-1}(p)$ being the standard normal p quantile, and at the same time the desired proportion of indicators, $E\{I(u)\} = p$.

The Gaussian RF model is fully characterized by its covariance $C_Y(h)$, and there is direct relationship between $C_Y(h)$ and the indicator covariance after truncation at the *p* quantile. Thus, by inversion of the indicator variograms, the Gaussian variograms are obtained (Fig. 10.14).

Multiple truncations of the same Gaussian realization at different thresholds would result in multiple categorical indicators each with the right proportions (marginal probabilities). But the indicator covariances of the additional categories would be controlled by the Gaussian RF, because the covariance $C_Y(h)$ is the single parameter that defines the Gaussian RF, and thus can only be used to define one indicator covariance. There is a significant drawback if more

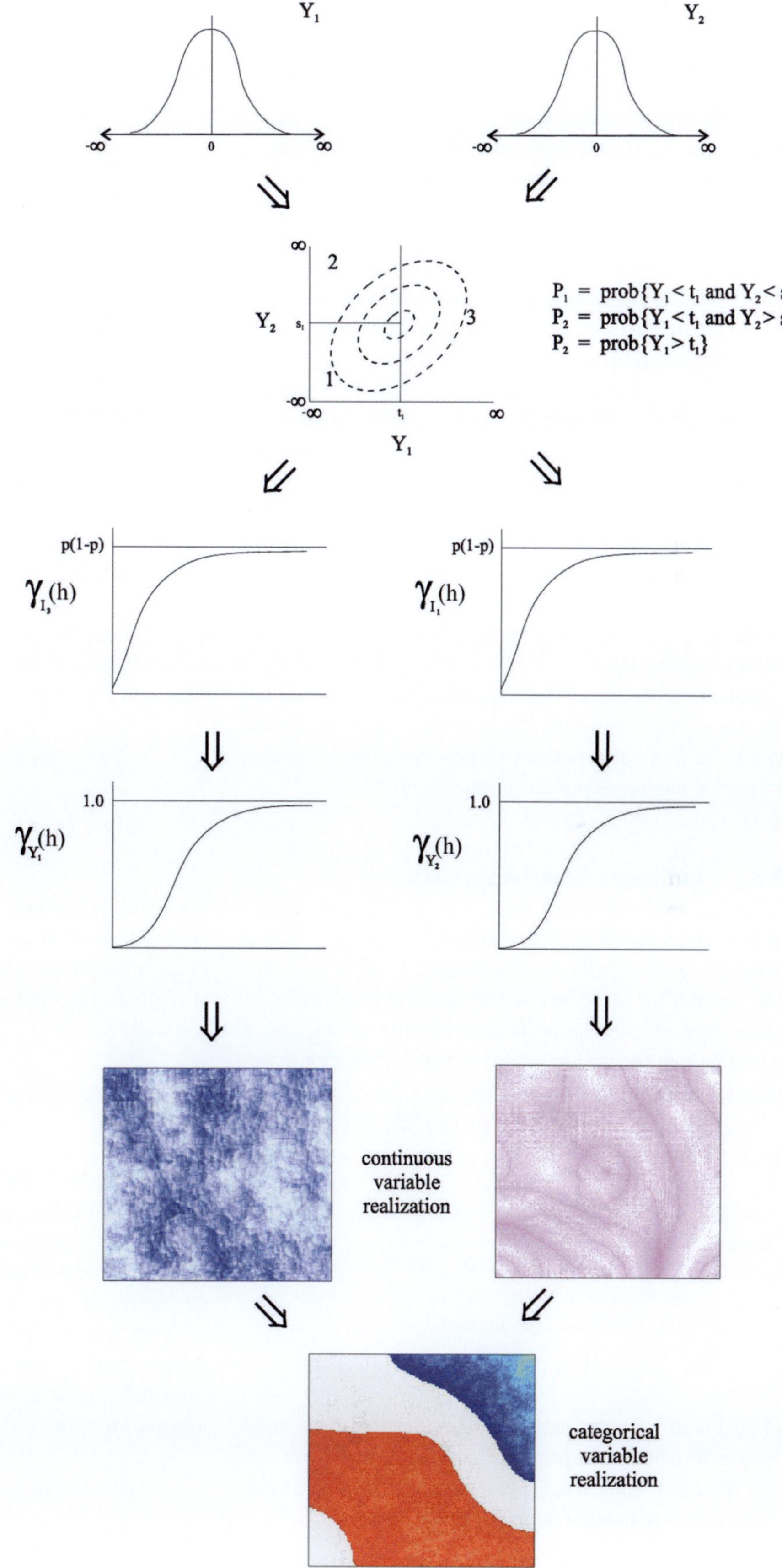

Fig. 10.14 Schematic illustration of the implementation of the truncated pluri-Gaussian method

than two categories are defined, because the additional indicator covariances would not reproduce the correct spatial continuity.

As with the truncated Gaussian method, the order and spatial sequence of the categories is fixed. While this may be reasonable in sedimentary stratigraphic environments, may be unreasonable in a more general setting.

10.6 Co-Simulation: Using Secondary Information and Joint Conditional Simulations

There are two distinct ways in which multiple variables can be accounted for. The first is considering secondary information by conditioning the primary variable to both primary and secondary information. Secondary information can refer to the same primary variable, but presented in a different format, or more generally any other variable to which the primary variable is correlated. This is different than jointly simulating the primary and secondary information, which will be discussed later.

Gaussian techniques (Wackernagel 2003) are commonly used due to their simplicity, but often indicator-based methods may be preferable because of how easily the secondary information can be incorporated into the process.

10.6.1 Indicator-Based Approach

A major advantage of the indicator kriging approach to generating posterior conditional distributions (ccdf's) is its ability to account for secondary or soft data. As long as the soft data can be coded into prior local probability values, indicator kriging can be used to integrate that information into a posterior probability value.

The prior secondary information can take one of the following forms:

- Local hard indicator data $i(\mathbf{u}_\alpha; z)$ originating from local hard data $z(\mathbf{u}_\alpha)$:

$$i(\mathbf{u}_\alpha; z) = 1, \text{ if } z(\mathbf{u}_\alpha) \le z, = 0 \quad \text{if not}$$
$$\text{or } i(\mathbf{u}_\alpha; s_k) = 1, \text{ if } \mathbf{u}_\alpha \in \text{ category } s_k, = 0 \quad \text{if not}$$

- Local hard indicator data $j(\mathbf{u}_\alpha; z)$ originating from ancillary information that provides hard inequality constraints on the local value $z(\mathbf{u}_\alpha)$. For example, if $z(\mathbf{u}_\alpha)$ can only take values within the interval $(a_\alpha, b_\alpha]$, then:

$$j(\mathbf{u}_\alpha; z) = \begin{cases} 0, \text{ if } z \le a_\alpha \\ \text{undefined (missing), if } z \in (a_\alpha, b_\alpha] \\ 1, \text{ if } z > b_\alpha \end{cases}$$

- Local soft indicator data $y(\mathbf{u}_\alpha; z)$ originating from ancillary information providing prior probabilities about the value $z(\mathbf{u}_\alpha)$:

$$y(\mathbf{u}_\alpha; z) = Prob\{Z(\mathbf{u}_\alpha) \le z | \text{local information}\} \in [0,1]$$

- Global prior information common to all locations $\mathbf{u}$ within the stationary area A:

$$F(z) = Prob\{Z(\mathbf{u}) \le z\}, \forall \mathbf{u} \in A$$

At any location $\mathbf{u}$ in A, prior information about the value $z(\mathbf{u})$ is characterized by any one of the four previous types of prior information. The process of building the ccdf with indicator kriging consists of a Bayesian updating of the local prior into a posterior cdf:

$$[\text{Prob}\{Z(\mathbf{u}) \le z | (n+n')\}]^*_{IK} =$$
$$\lambda_0(\mathbf{u})F(z) + \sum_{\alpha=1}^{n} \lambda_\alpha(\mathbf{u}; z) - \sum_{\alpha'=1}^{n'} v_{\alpha'}(\mathbf{u}; z) y(\mathbf{u}'_\alpha; z)$$

The $\lambda_\alpha(\mathbf{u}; z)$'s are the weights attached to the n neighboring hard indicator data, the $v_\alpha'(\mathbf{u}; z)$'s are the weights attached to the n neighboring soft indicator data, and λ_0 is the weight attributed to the global prior cdf. To ensure unbiasedness, λ_0 is usually set to:

$$\lambda_0(\mathbf{u}) = 1 - \sum_{\alpha=1}^{n} \lambda_\alpha(\mathbf{u}; z) - \sum_{\alpha'=1}^{n'} v_{\alpha'}(\mathbf{u}; z)$$

The ccdf model is thus an indicator co-kriging that pools information of different types: the hard i and j indicator data and the soft y prior probabilities. When the soft information is not present or is ignored ($n'=0$), the expression reverts to the known IK expression.

If the spatial distribution of the soft y data is modeled by the covariance $C_I(\mathbf{h}; z)$ of the hard indicator data, then there is no updating of the prior probability values $y(\mathbf{u}'_\alpha; z)$ at their locations $\mathbf{u}'_\alpha$, i.e.,

$$[\Pr ob\{Z(\mathbf{u}'_\alpha) \le z | (n+n')\}]^*_{IK} \equiv y(\mathbf{u}'_\alpha; z), \forall z$$

Most often, the soft z data originate from information related to, but different from, the hard data $z(\mathbf{u}_\alpha)$. In this case, the

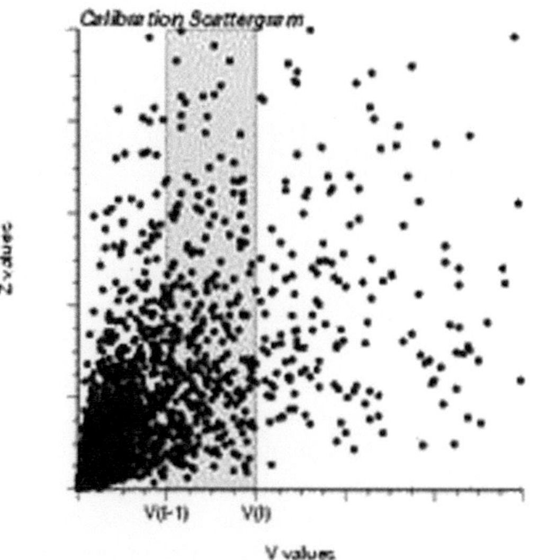

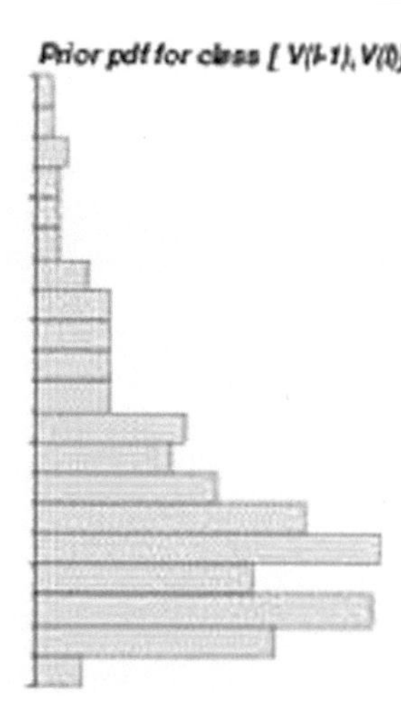

Fig. 10.15 Inference of the soft prior probabilities from a calibration scattergram. The prior z probability pdf at a location $\mathbf{u}'_\alpha$ where the secondary variable is $v(\mathbf{u}'_\alpha)$ in $(v_{l-1}, v_l]$ is identified to the calibration conditional pdf, shown in the right of the figure

soft y indicator spatial distribution is likely different from that of the hard i indicator data:

$$C_Y(\mathbf{h};z) \neq C_{IY}(\mathbf{h};z) \neq C_I(\mathbf{h};z)$$

Then the indicator cokriging amounts to a full updating of all prior cdf's that are not already hard.

At the location of a constraint interval $j(\mathbf{u}_\alpha; z)$, indicator kriging or cokriging amounts to in-filling the interval $(a_\alpha, b_\alpha]$ with spatially interpolated ccdf values. Thus if simulation is performed at that location, a z attribute value would be drawn necessarily from within the interval.

10.6.2 Markov-Bayes Model

With enough data one could infer directly and model the matrix of covariance functions (one for each cutoff z): $[C_Y(\mathbf{h}; z) \neq C_{IY}(\mathbf{h}; z) \neq C_I(\mathbf{h}; z)]$. An alternative to this tedious exercise is provided by the Markov-Bayes model, whereby:

$$\begin{aligned} C_{IY}(\mathbf{h};z) &= B(z)C_I(\mathbf{h};z), \forall h \\ C_Y(\mathbf{h};z) &= B^2(z)C_I(\mathbf{h};z), \forall h > 0 \\ &= |B(z)|\, C_I(\mathbf{h};z), h = 0 \end{aligned}$$

The coefficients B(z) are obtained from calibration of the soft y data to the hard z data:

$$B(z) = m^{(1)}(z) - m^{(0)}(z) \in [-1, +1]$$

with:

$$\begin{aligned} m^{(1)}(z) &= E\{Y(\mathbf{u};z) \mid I(\mathbf{u};z) = 1\} \\ m^{(0)}(z) &= E\{Y(\mathbf{u};z) \mid I(\mathbf{u};z) = 0\} \end{aligned}$$

Consider a calibration data set $\{y(\mathbf{u}_\alpha; z), i(\mathbf{u}_\alpha; z), \alpha = 1,\ldots,n\}$ where the soft probabilities $y(\mathbf{u}_\alpha; z)$ valued in [0,1] are compared to the actual hard values $i(\mathbf{u}_\alpha; z)$ valued 0 or 1. $m^{(1)}(z)$ is the mean of the y values corresponding to i = 1; the best situation is when $m^{(1)}(z) = 1$, that is, when all y values exactly predict the outcome i = 1. Similarly, $m^{(0)}(z)$ is the mean of the y values corresponding to i = 0, best being when $m^{(0)}(z) = 0$.

The parameter B(z) measures how well the soft y data separate the two actual cases i = 1 and i = 0. The best case is when $B(z) = \pm 1$, and the worst case is when B(z) = 0; that is, $m^{(1)}(z) = m^{(0)}(z)$.

The case $B(z) = -1$ corresponds to soft data predictably wrong and is best handled by correcting the wrong probabilities $y(\mathbf{u}_\alpha; z)$ into $1 - y(\mathbf{u}_\alpha; z)$.

When B(z) = 1, the soft prior probability data $y(\mathbf{u}'_\alpha; z)$ are treated as hard indicator data and are not updated. Conversely, when B(z) = 0, the soft data $y(\mathbf{u}'_\alpha; z)$ are ignored; i.e., their weights become zero.

Since the Y covariance model generally presents a strong nugget effect, the Markov model implies that the y data have little redundancy with one another. The undesired effect of this is that too much weight is given to clustered, mutually redundant y data. In practice, only the closest y datum is retained, which leads to using the collocated correlation, i.e., the soft autocovariance at distance 0, $C_Y(h = \mathbf{0}; z)$.

10.6.3 Soft Data Calibration

Consider the case of a primary continuous variable $z(\mathbf{u})$ informed by a related secondary variable $v(\mathbf{u})$. The series of hard indicator data valued 0 or 1, $i(\mathbf{u}_\alpha; z_k)$, k = 1,…,K, are derived from each hard datum value $z(\mathbf{u}_\alpha)$.

The soft indicator data, $y(\mathbf{u}'_\alpha, z_k)$ in [0,1], k = 1,…,K, corresponding to the secondary variable value $v(\mathbf{u}'_\alpha)$, can be obtained from a calibration scattergram of z values versus collocated v values (Fig. 10.15):

The range of v values is discretized into L classes $(v_{l-1}, v_l]$, $l=1,\ldots,L$. For class $(v_{l-1}, v_l]$, the y prior probability cdf can be modeled from the cumulative histogram of primary data values $z(\mathbf{u}'_\alpha)$ such that the collocated secondary data values $v(\mathbf{u}'_\alpha)$ fall into class $(v_{l-1}, v_l]$:

$$y(\mathbf{u}';z) = \Pr ob\{Z(\mathbf{u}'_\alpha) \le z \mid v(\mathbf{u}'_\alpha) \in (v_{l-1}, v_l]\} \qquad 10.15$$

Note that the secondary variable $v(\mathbf{u})$ need not be continuous. The classes can be in fact categories of v values; for example, if the information v relates to different lithologies or mineralization types.

The calibration scattergram that provides the prior y probability values may be borrowed from a different and better sampled field. That calibration scattergram may be based on data other than those used to calibrate the covariance parameters B(z).

10.6.4 Gaussian Cosimulation

If a Gaussian method is used, the data must be transformed to a standard normal distribution. If two variables are considered, the cross correlation between Y_i and Y_j should show a bivariate normal distribution, i.e., an elliptical probability contours along a line through the origin. For third and higher orders, a distribution of k Gaussian variables should show probability contours following a hyper-ellipsoid in k-dimensions.

The first possible approach is a direct co-simulation, as proposed by Verly (1993), and is similar to that for conventional simulation, see Fig. 10.16. It begins with establishing a random path through all the grid nodes. At each grid node the nearby data and previously simulated grid nodes are found, a conditional distribution is constructed by cokriging, and a simulated value is drawn from this distribution. The simulated value is added to the conditioning data, and the process is repeated until all nodes are simulated. The final steps are to back-transform the simulated values and to check the results.

The next approach for a joint simulation is to define a hierarchy of variables. In this case, the variables are not simulated simultaneously, but in order according to a pre-defined hierarchy and conditionally to the previously simulated variables (Almeida and Journel 1994). This idea allows for the implementation of a full cokriging, a collocated cokriging approximation, or a further approximation based on the Markov-Bayes model.

The third approach is to apply a transformation that would make the correlated variables independent (Luster 1985). To obtain the non-correlated variables, there are several options. One is to decompose the original variables $Z_i(\mathbf{u})$ into orthogonal factors, that is, obtain the principal components of the original Z-variable correlation matrix at $|h|=0$ (Luster 1985). The significant assumption here is that the orthogonality of the principal components at lag 0 extends to all possible lags.

Alternative options is to use either the super-secondary variable (Babak and Deutsch 2009), or a stepwise conditional transformation (Leuangthong and Deutsch 2003). The choice between these two transformations depends on the shape of the cross plot between the two variables. If there is a non-linear or constrained relationship between the variables, stepwise conditioning provides a more flexible, albeit data-hungry, method; if the cross plots show mostly a relationship correctly and fully characterized by its linear correlation coefficient, then a simpler super-secondary variable approach may be preferred (Fig. 10.17).

10.6.5 Stepwise Conditional Transform

The stepwise conditional transform (SCT) was introduced by M. Rosenblatt in 1952 and was re-introduced by Leuangthong and Deutsch in 2003 for use in geostatistics. The motivation for it is that it produced independent model variables, thereby avoiding cosimulation.

In a bivariate case, the normal transform of the second variable is conditional to the probability class of the first or primary variable. Extending to a k-variate case, the kth variable is conditionally transformed based on $(k-1)$ first variables.

The following is an example of a bivariate case where Zn is the primary variable and Pb is the secondary. Zn is transformed to a Gaussian distribution using the Normal Scores procedure as before. Pb is transformed as shown in Fig. 10.18. The stepwise conditional transform results in a non-correlated variable at $\mathbf{h}=0$, which indicates that Cosimulation is not required.

The major incentive to use SCT in practice is that it is robust when dealing with complex multivariate distributions. This means that SCT is able to solve issues of non-conformity to multi-Gaussian assumptions, as well as significantly simplify the joint simulation of multiple variables. The major disadvantage is that it is data-intensive, while care must be used not to define classes with less than 50 data values (Leuangthong and Deutsch 2003), which could make the resulting conditional distributions not representative. Another possible limitation is that SCT may produce artifacts in the transformation of secondary variables. This should thoroughly checked for before proceeding with any subsequent simulation.

The more important implementation aspects of SCT are:

1. The number of classes; it is recommended than no less than 10 are used. More classes result in a correlation that

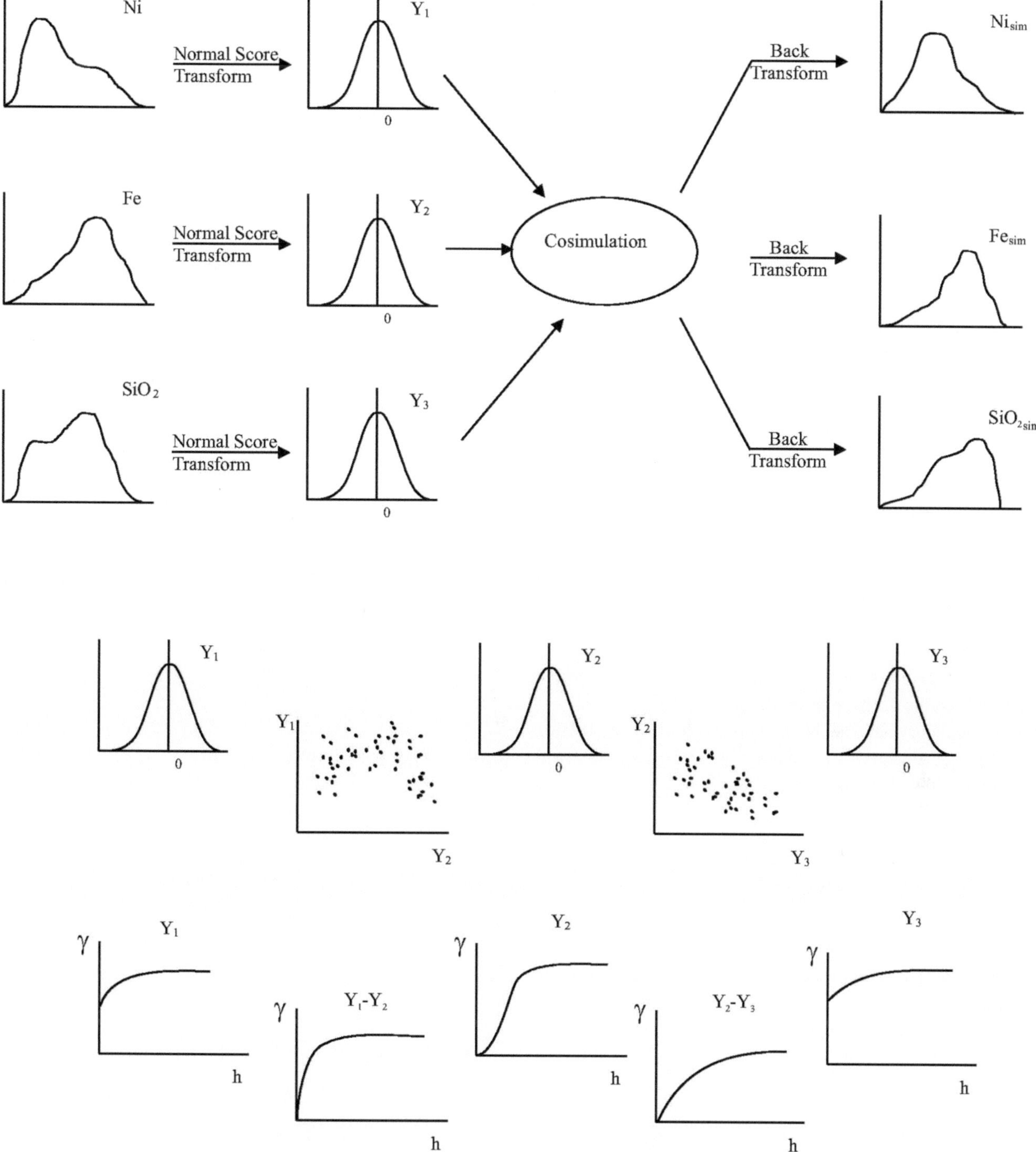

Fig. 10.16 Co-simulation using Gaussian variables

approaches zero, but care should be taken not to produce classes with few data in the transformed secondary conditional distributions.

2. Another aspect to consider is that SCT is sensitive to outliers. Extensive exploratory data analysis is recommended, and data cleaning or capping may be necessary.
3. The order in which the transform is performed is consequential. The choice of primary variable should be made based on which one has the most data; if both variables have similar number of samples then choose as primary the most continuous variable.

The stepwise conditional transformation ensures that the transformed variables, taken together, are multivariate Gaussian with zero correlation. Thus, conventional Gaussian simulation techniques can be applied with no requirement for cokriging or to fit a model of coregionalization. The correla-

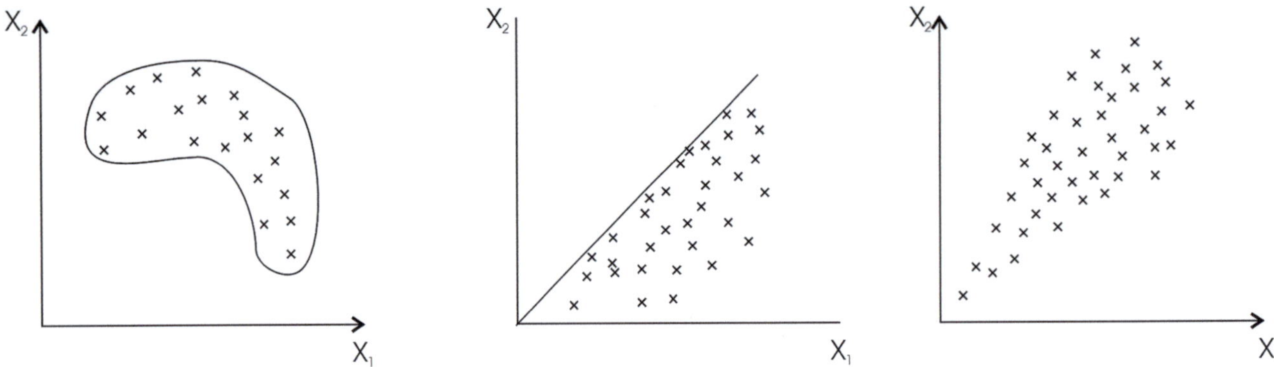

Fig. 10.17 Non-linear, constraint, and hetersoscedastic features

tion between the variables is accounted for in the forward and back transformations.

10.6.6 Super-Secondary Variables

Under a multivariate Gaussian model, it can be shown that multiple secondary data can be co-simulated by merging it into a single secondary variable, such that a standard co-simulation procedure can be applied (Babak and Deutsch 2009). The setting that is most practical is when collocated co-simulation is deemed appropriate.

All secondary data can be merged as a linear combination into a single secondary variable that can be used in the conventional collocated cokriging.

$$y_{\substack{\text{super}\\\text{secondary}}}(\mathbf{u}) = \frac{\sum_{i=1}^{n_{\text{sec}}} c_i \cdot y_{s,i}}{\rho_{\substack{\text{super}\\\text{secondary}}}}$$

The c_i weights are calculated from the well known cokriging equations:

$$\sum_{j=1}^{n_{\text{sec}}} c_j \cdot \rho_{i,j} = \rho_{i,0}, \quad i = 1,\ldots,n_{\text{sec}}$$

The left hand side $\rho_{i,j}$ values represent the redundancy between the secondary data. The right hand side $\rho_{i,0}$ values represent the relationship between each secondary data and the primary variable being predicted. The correlation coefficient of the super secondary value with the primary variable being estimated is based on the cokriging variance:

$$\rho_{\substack{\text{super}\\\text{secondary}}} = \sqrt{\sum_{i=1}^{n_{\text{sec}}} c_i \cdot \rho_{i,0}}$$

The expression inside the square root is one minus the estimation variance, which would is precisely the correlation coef-

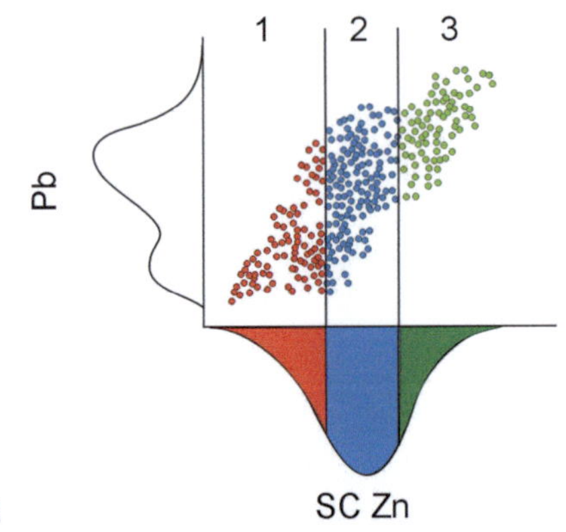

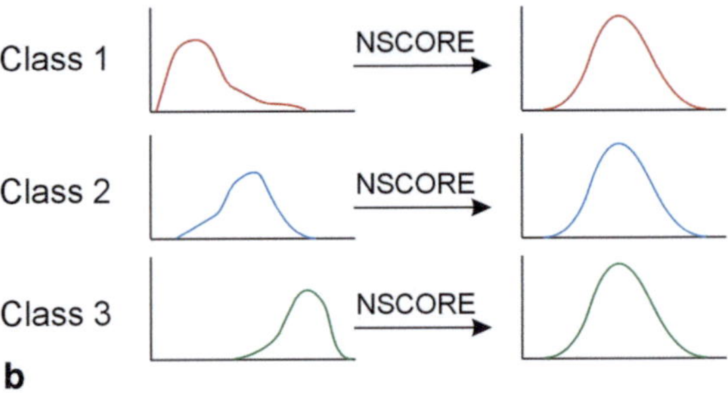

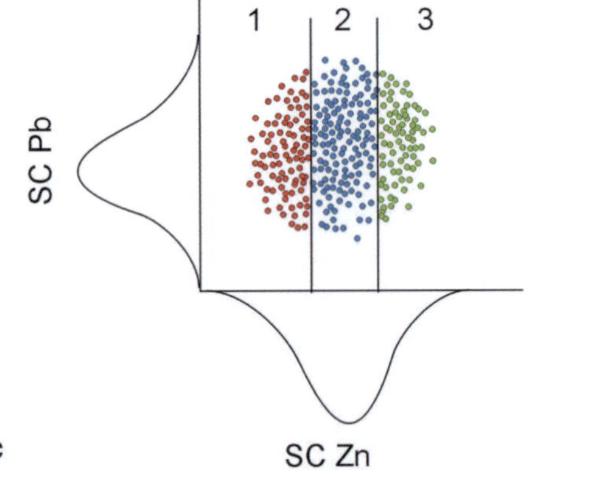

Fig. 10.18 Stepwise conditional transform of Pb using Zn as a primary variable

ficient if one data is being used. Recall that $\sigma_k^2 = 1 - \rho_{i,0} \cdot \rho_{i,0}$ in presence of one data, thus $\rho_{i,0} = \sqrt{1-\sigma_k^2}$ given that the estimation variance is known, as is the case here.

The single super secondary variable is used with the primary data in the well known collocated cokriging equations:

$$y^* = \sum_{i=1}^{n} a_i \cdot y_i + c \cdot y_{\substack{\text{super}\\\text{secondary}}}$$

$$\sigma_K^2 = 1 - \sum_{i=1}^{n} a_i \cdot \rho_{i,0} - c \cdot \rho_{\substack{\text{super}\\\text{secondary}}}$$

10.6.7 Simulation Using Compositional Kriging

Compositional kriging is done without any transform of compositions from the simplex to real space. Solving the compositional kriging system of equations gives a vector that adheres to the constraints of compositions. These do not necessarily form the vector of expected values $\mathbf{m}^*$ and we are not working with multivariate normal data or distributions. A covariance matrix can be calculated. Unfortunately, the space of compositions is the simplex and there is no definition of covariance and cross-covariance in that space (Pawlowsky-Glahn and Olea 2004).

Defining the multivariate distribution from which a composition is simulated using this method is an outstanding problem. We know two constraints on the distribution: $\mathbf{x} \geq 0; \sum_i x_i = c$, but the shape of multivariate conditional distributions is also needed. It is also not clear if $\mathbf{m}^*$ (the estimated averages) and $\mathbf{S}^*$ (the estimation variance) are correctly parameterizing these distributions.

Simulation using alr Cokriging Under certain assumptions, simulation using alr cokriging can be accomplished. If the distribution of compositions is assumed additive logistic normal, then the alr transformed data is multivariate normal distributed. At location $\mathbf{u}_M+1$, kriging gives a multivariate conditional distribution with mean and covariance parameters given by:

$$\boldsymbol{\mu}_y^* = \mathbf{y}^*\left(\mathbf{u}_{M+1}\right) = \mathbf{c} + \sum_{k=1}^{N} \boldsymbol{\Phi}_k \operatorname{alr}\left[\mathbf{x}\left(\mathbf{u}_k\right)\right] = \mathbf{c} + \sum_{k=1}^{N} \boldsymbol{\Phi}_k \mathbf{y}\left(\mathbf{u}_k\right)$$

$$\boldsymbol{\Sigma}_y^* = \operatorname{Cov}\left\{\mathbf{y}^*(\mathbf{u}_{M+1})\right\} = \sum_{k=1}^{M}\sum_{l=1}^{M} \boldsymbol{\Phi}_k \mathbf{C}\left(\mathbf{u}_l - \mathbf{u}_k\right) \boldsymbol{\Phi}_l^T$$

Despite there being no analytical back transform of $\mathbf{m}^*$ and $\mathbf{S}^*$, a vector, $\mathbf{y}^s(\mathbf{u}_M+1)$, can still be simulated from the multivariate conditional distribution. The inverse alr transform can be applied to recover a simulated composition $\mathbf{x}^s(\mathbf{u}_M+1)$.

$$x_k^s(\mathbf{u}_{M+1}) = \operatorname{alr}^{-1}\left(y_k^s(\mathbf{u}_{M+1})\right) = \frac{\exp\left(y_k^s(\mathbf{u}_{M+1})\right)}{\sum_{i=1}^{d} \exp\left(y_i^s(\mathbf{u}_{M+1})\right) + 1}, k = 1,\ldots,d$$

$$x_D^s(\mathbf{u}_{M+1}) = c - \sum_{i=1}^{d} x_i^s(\mathbf{u}_{M+1}) \tag{0.1}$$

while a simulated model can be obtained, there are a number of unresolved and unknown issues of this method. To begin with, kriging of alr transformed data does not necessarily result in an optimal solution of the original compositional data in the simplex (Chap. 8). Also, the covariance between known compositions and simulated compositions is not necessarily correct, and the covariance between two simulated compositions is not necessarily correct. And it is not known whether the global statistics such as the variogram, mean, variance and other moments are reproduced in the simplex.

10.7 Post Processing Simulated Realizations

Conditional simulation generates many possible outcomes, all with the correct variability. Dealing with the multiple realizations has proven a difficult practical issue, typically underestimated by geostatistical researchers.

Simulation corrects for the smoothness of kriging and is theoretically free of conditional bias. Multiple realizations allow for uncertainty assessment, so conditional simulation techniques are required to assess joint uncertainty. Uncertainty visualization is one aspect that needs to be considered (Caers 2011).

The following discusses in general terms some of the most common CS applications in mining, including assessment of point scale uncertainty, change of support, uncertainty around existing mine plans and schedules, optimization studies of various types, and recoverable reserves calculation (Dimitrakopoulos and Ramazan 2008; Dowd 1994).

The multiple realizations represent at each location the most information that can be gathered. The full range of possible values, described as a conditional cumulative distribution frequency (ccdf) is a model of local uncertainty.

Simulation allows for a better understanding of the volume variance relations. The basic idea is to simply simulate values at a tight grid spacing, and then average up to the relevant block size. This "brute force" approach is best to observe volume-variance relations, since all other change of support methods require assumptions that are difficult to verify, or may be theoretical approximations.

More importantly, CS allows mimicking the mining process itself, either through blast hole drilling and sampling, or through production drilling in underground mines. Mining is done to irregular polygonal boundaries. Isolated blocks cannot be freely extracted and thus assuming free selection is optimistic. The simulation of the grade control process, feasible if using CS models from a pre-feasibility stage onwards, allows for an early understanding of the information effect, selectivity, dilution and ore loss.

Another useful application of simulation is forecasting metal recovery and other related metallurgical performance variables. This is done by visiting each location over one realization for all variables and applying a transfer function that accounts for the known metallurgical processes and translates grades for multiple metals into a recovery at that location. This is repeated for all realizations to create multiple recovery models. These recovery models can be used to produce distributions of uncertainty for both local and global recovery. This process requires a clear understanding of the metallurgical processes, including the non-linearity of some of the geometallurgical properties involved.

The production of probability maps is easily achievable using the results of simulation. This is done by visiting each location over multiple realizations to determine local distributions of uncertainty. We then calculate the probability to exceed a cutoff grade from this local distribution. This is done at all locations. A map of probabilities can then be plotted. Specific quantiles of interest can be analyzed, such as 10, 50, or 90%. Any areas that are high on the p10 map will surely be high grade. Any areas that are low on the p90 map will surely be low. The same technique can also be used for classification of resources; for example, yearly volumes within X% of predicted value Y% of the time can be called proven.

Simulation models can also be used to simulate stockpiles (see Boucher et al. 2005, among others). A large volume consistent with blast patterns is selected. This volume is applied to a simulated realization to determine the average grade for the stockpile. This is repeated for all realizations to get average grades from each. This information can be used to determine the probability that the volume will satisfy blending criteria or economic cutoff.

The results of simulation are also able to determine global reserves and their uncertainty. The global reserves are calculated using all the relevant metal grades, recovery models, and the economic cutoff. This is done for each realization. The uncertainty in the reserves can now be found.

Another post-processing application of a simulation-generated model is assessing the link between equipment size and mine selectivity, and thus to determine the value/cost of selective mining. This is applicable to both open pit and underground settings; however, stope design and extraction methods are less flexible than the open pit counterpart, and thus this study should be done well in advance of mine development. In open pit, CS models are used to evaluate equipment selectivity in relation to the available sampling,

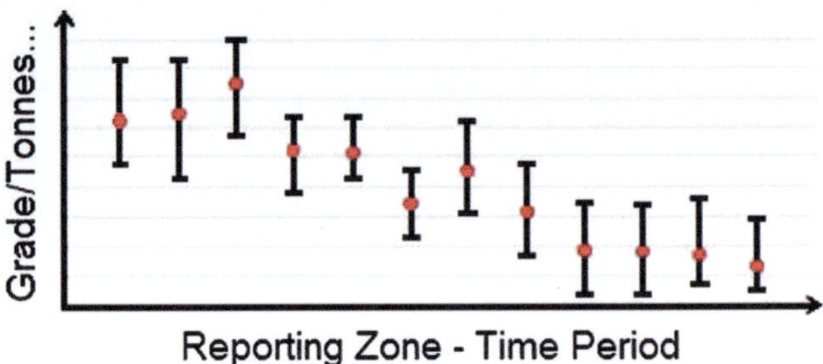

Fig. 10.19 Example of error bounds for a time period

bench height, ore/waste boundary definitions, blast hole spacing, blast heave, and so on. The process always involves one or more transfer functions that allow incorporating relevant economic assessments and cost/benefit curves.

Reporting the results of studies from CS models is usually done by zone (time period, bench, stope, and so on) and reporting common quantiles (P_{10}; P_{50}; P_{90}) or (P_5; P_{50}; P_{95}).

The uncertainty in NPV/ROI type statistics can be shown as error bounds derived from the multiple realizations. In the example in Fig. 10.19, P90 is exceeded 10% of the time, while the value of interest falls below P10 10% of the time.

Usually there is a need to rank the simulated values. By definition, all realizations are equi-probable, but have different impact to the problem being studied. Often, ranking is necessary to limit the number of realizations used in the applications, since considering the entire model of uncertainty may be impractical.

An increasingly popular post processing of uncertainty from uniform conditioning, indicator kriging and simulation is localization (Abzalov 2006; Hardtke et al. 2011). One value is chosen per block in a manner that reproduces the block distribution within larger panels.

The most common criterion used for ranking is total metal content; for example, in pit or stope optimization what matters is the total dollar value of the block, which is in turn a function of the metal content in the block. Occasionally, ranking by total variability (by stationary domain) may be an option, especially if overall local uncertainty measures are desired. The chosen realizations should fairly represent the full space of uncertainty for the variable that is deemed of highest consequence.

10.8 Summary of Minimum, Good and Best Practices

Minimum practice is to thoroughly check and validated the realizations obtained. The details of the validation procedure are given in Chap. 11, but it is a more involved and demanding process than for any estimated block models.

Realizations are more difficult to report and to communicate than any other model, because they contain much more information, and in a format that non-specialists and industry professionals are less familiar with. Visualization tools are useful for qualitative demonstrations of the models, but typically the reporting of the simulation models depends on their objective, and the risk assessment they provide.

The description of joint uncertainty is also a key element that should be considered. Examples are shown of what is meant by joint uncertainty, and its possible applications and consequences on downstream work.

Another aspect that requires some detailed discussion is the issue of ranking the individual simulations. The discussion includes criteria for deciding if there is a need for ranking, the methods that can be used to rank individual simulations, and the tradeoffs that typically are made.

Minimum practice includes simulating specific domains, typically the same estimation (simulation) domains defined through the combination of geology and statistical analyses. A clear statement of objectives, the justification of the simulation method chosen, the related risk assessment of interest, and the implementation-specific parameters should be clearly stated and demonstrated when appropriate. Some of the simulations should be carefully validated to the extent described in detail in Chap. 11, while the overall resulting uncertainty model should be checked against other models, including estimated grade models and known production values, if available. The reporting and visualization of the simulation models are critical, and they should be related always to the original objective of the conditional simulation study. Commonly, only a subset of the simulations obtained is used in the risk assessment study, and so all decisions related to individual simulation ranking should be presented and justified. The simulation work should be documented with some detail, particularly with respect to validation, applications, and perceived model limitations.

Good practice requires, in addition to the above, a more detailed justification of the simulation model. Calibrations on a smaller subset of the overall volume, significantly more iterations and comparisons before an individual simulation is considered acceptable, and much more validation and checking of the resulting uncertainty model than before. Also, the assessment on the geologic model should be done (simulation of categorical variables) and introduced into the overall resource simulation. Comparison with past production should be done in detail whenever possible. The full reporting of all relevant aspects of the models and the corresponding risk assessment is required, as well as full and detailed documentation. It is important to emphasize the correct use of the simulation models, their limitations, and possible future improvements to the work.

Best practice consists of full implementation of the techniques available, and the use of alternative models to verify their relevancy. All potential sources of uncertainty in an ore resource model should be investigated, including the uncertainty of the original drill hole or blast hole data used, the uncertainty of the geologic model and the estimation (simulation) domains defined, etc. The consequences of using alternative implementation parameters should be fully explored and documented. All relevant production and calibration data should be used to indicate whether the simulation model is performing as expected. All possible uncertainty measures in relation to an ore resource model should be quantitatively described and discussed, whether global or local, including internal and geologic contact dilution, the impact of the information effect, etc. Checking and validation should be exhaustive, as well as the model presentation, reporting, and visualization. The full set of individual simulations should be used in the risk assessment study, and it in turn fully described, validated, and documented.

10.9 Exercises

The objective of this exercise is to review a variety of simulation techniques and post processing methods. Some specific (geo)statistical software may be required. The functionality may be available in different public domain or commercial software. Please acquire the required software before beginning the exercise. The data files are available for download from the author's website—a search engine will reveal the location.

10.9.1 Part One: Sequential Indicator Simulation

The objective of this exercise is to construct SIS realizations of a categorical variable. It is common in mining applications to have a deterministic rock type model; however, the boundaries between the rock types are not hard boundaries. We often want to consider fuzzy or soft boundaries. Some topographic data is used for this exercise, but it is adequate to communicate the principles and methodology.

Consider the 2-D `SIC.dat` data that has been provided for your analysis. You will have to perform a cursory data analysis to determine the X/Y limits and the variables present. Choose a reasonable grid size relative to the data spacing. This data was used for an international spatial interpolation contest in 1997. The categorical variable in the data file is of interest here.

Question 1: Plot a 2-D map of the indicator and note the direction of continuity. Calculate, plot and fit indicator variograms in the principal directions.

Question 2: Print the SIS parameter file and comment on the appropriate settings for this exercise. Create two SIS realizations and plot the results. Perform reasonable sensitivity studies on the parameters that you are uncertain of.

Question 3: Create 100 realizations and plot a map of the probability of each category. This map should look like a kriged model.

10.9.2 Part Two: Sequential Gaussian Simulation

Sequential simulation is common because simulation and conditioning to local data is accomplished in one step. Historically, conditional simulation was divided into two steps: unconditional simulation and kriging for conditioning. Matrix methods and moving average methods were used for unconditional simulation, but they were really only applicable for small grids. Turning bands was (and still is) used for unconditional simulation on larger 3-D grids.

A small 2-D example will be considered for testing some alternative Gaussian simulation methods.

Consider a grid of 50 by 50 grid cells each 1 unit square. There are four regularly spaced data (see the picture below). Two data are at the mean, one is a high value and one is low. The variogram is omnidirectional spherical with a range of 10 grid units.

Consider unconditional and conditional simulation of a 50 by 50 domain. Such a small grid permits fast calculations and testing of methods such as LU simulation that requires a small grid size.

Sequential Gaussian simulation (SGS) is popular because of its simplicity and flexibility. SGS draws realizations from the multivariate Gaussian distribution by recursive application of Bayes Law. This part focuses on the application and limitations of SGS.

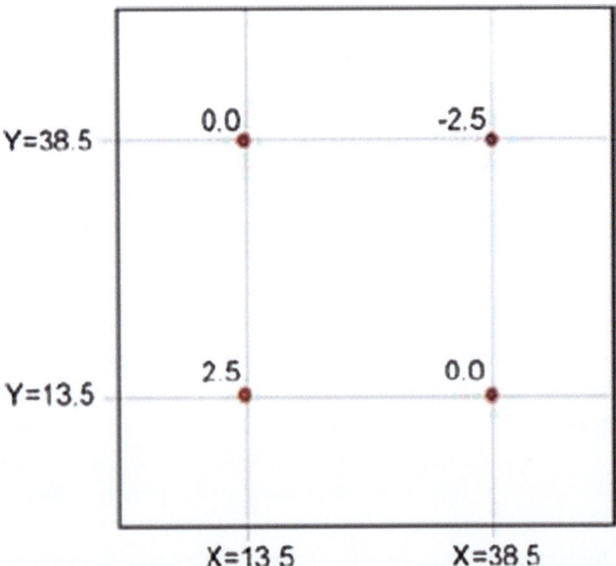

Question 1: Setup a 2-D Gaussian simulation with the conditioning data described above. Run the simulation with a search radius of 10 grid units and 16 previously simulated grid nodes. Use the reference variogram given above. Note the CPU time for 100 realizations. Create four realizations for checking. Plot maps of four realizations. Plot the histogram of the simulated values and compare to the reference distribution. Calculate and plot the variogram with the input reference variogram model.

Question 2: The distinguishing characteristic of sequential simulation is the Markov screening assumption; however, it can cause poor variogram reproduction. Create four realizations with 4, 8, 16, and 32 previously simulated grid nodes and the multiple grid search turned off. Add coordinates and calculate the omnidirectional variogram. Comment on variogram reproduction. Run the "4" and "8" cases with a multiple grid and comment on variogram reproduction.

Simulation is at the scale of the data, that is, composite data of some specified length. The grid specification in simulation programs is not a *block size*, it is the spacing of point scale simulated values. For this reason, we simulate at least 10 point values within the smallest selective mining unit (SMU) size ultimately desired. Geostatistical simulation involves many locations simultaneously. This involved excessive CPU time and storage requirements in the early days of simulation. Fast unconditional simulation techniques were devised and kriging was used as a post processor for conditioning.

Question 3: Give the equation for conditioning unconditional realizations to local data. Clearly explain how this could be implemented in practice.

Question 4: Use turning bands to simulate 4 unconditional realizations of a 50 by 50 by **5** (because the turning bands algorithm is dimension dependent—the tb3d program is for 3-D only) domain (spacing set to 1 in all directions) using the reference variogram given above. Plot the central 2-D slice of the realizations. Check histogram and variogram reproduction of the realizations.

Question 5: Use LU simulation to simulate 4 unconditional realizations of a 50 by 50 by **1** domain (spacing set to 1 in all directions) using the reference variogram given above. Plot the realizations. Check histogram and variogram reproduction.

Question 6: Condition these realizations to the four Gaussian data values given above.

10.9.3 Part Three: Simulation with 3D Data

Theoretically, simulation requires an incremental effort to estimation kriging. In practice, however, the generation of multiple realizations through simulation can amount to a significant increase to computation and time requirements. Consider the 3D data from `largedata.dat`. You may wish to recall your work/results of previous exercises where you created a model of kriged estimates using parameters that were refined after cross validation.

Question 1: Using the same grid definition and similar parameters from kriging (as you determined in Exercise 5), generate 10 realizations of Au. For four realizations (it should not matter which you select), plot the middle slice of the model and compare this to the same slice from the kriged model. Comment on any differences/similarities between (a) the kriged model and a simulated realization, and (b) one realization to another realization.

Question 2: Create a map of local averages (also referred to as the E-Type estimate). For this E-type model, plot the middle slice and compare it to the same slice from the kriged model. Comment on any similarities/differences that you note.

Question 3: Calculate probability maps at 10% and 90% probability. What can you say about the information conveyed in the 10% probability map?

Question 4: Check the proportional effect of the final results.

Question 5: Consider one realization, and scale this realization to an arbitrary volume that consists of 3×3 grid points (that is average a total of 9 simulated values together to obtain a block value).

Question 6: Recalculate measures of uncertainty (etype, local variance, and probability maps) as above.

10.9.4 Part Four: Special Topics in Simulation

The objective of this exercise is experiment with multivariate simulation approaches. Most geostatisticians do not go to the trouble to calculate and fit a full model of coregionalization. It is common to adopt the collocated cokriging shortcut or to adopt a multivariate transformation such as the stepwise transformation. Fitting an LMC is a challenge. Moreover, many software do not permit the full model of coregionalization to be used in cosimulation. Cokriging is more straightforward. The Markov model is used extensively in simulation.

Question 1: Recall the Markov model and the implicit cross variogram that is used if this model is adopted for cosimulation.

Question 2: Consider the Markov assumption with (a) bitumen as the primary variable, and (b) fines as the primary variable. Plot the implicit cross variogram from the Markov model with the experimental cross variograms. Comment on any mismatch.

Question 3: Perform sequential Gaussian simulation for bitumen and cosimulation of fines collocated to bitumen. Simulate one realization and check the crossplot of simulated values. If you have time, run ten realizations and record the correlation coefficient from each set of realizations, and plot a histogram of correlation coefficients. Comment on how this compares to the expected correlation coefficient.

The stepwise conditional transformation is becoming more commonly used to avoid the requirement for cross variograms.

Question 4: Explain the stepwise conditional transformation and perform the transformation twice with the `oilsands-3D.dat`: (a) fines conditional to bitumen, and (b) bitumen conditional to fines. Cross plot the transformed values and confirm that there is no cross correlation.

Question 5: For both cases, calculate the cross variogram between the transformed values in all three directions. Comment on any non-zero correlations that appear.

Question 6: Perform sequential Gaussian simulation for bitumen and fines. Simulate one realization and check the crossplot of simulated values (after back transformation). Comment on how this compares to the original data crossplot. As above, if you have time run ten realizations, record the correlation coefficient from each set of realizations, and plot a histogram of correlation coefficients. Comment on how this compares to (1) the expected correlation coefficient, and (2) the correlation coefficient distribution from Part One.

References

Aarts E, Korst J (1989) Simulated annealing and boltzmann machines. Wiley & Sons, New York

Abzalov M (2006) Localised uniform conditioning: a new approach for direct modeling of small blocks. Math Geol 38(4):393–411

Aguilar CA, Rossi ME (1996) (January) Método para Maximizar Ganancias en San Cristóbal. Minería Chilena, Santiago, Chile, Ed. Antártica, No. 175, pp 63–69

Alabert FG (1987a) Stochastic imaging of spatial distributions using hard and soft information. MSc Thesis, Stanford University, p 197

Alabert FG (1987b) The practice of fast conditional simulations through the LU decomposition of the covariance matrix. Math Geol 19(5):369–386

Almeida AS, Journel AG (1994) Joint simulation of multiple variables with a Markov-type coregionalization model. Math Geol 26(5):565–588

Babak O, Deutsch CV (2009) Collocated cokriging based on merged secondary attributes. Math Geosci 41(8):921–926 [7]

Badenhorst C, Rossi M (2012) Measuring the impact of the change of support and information effect at olympic dam. In: Proceedings of the IX international geostatistics congress, Oslo, June 11–15, 2012, pp 345–357, Springer

Boucher A, Dimitrakopoulos R, Vargas-Guzman JA (2005) Joint simulations, optimal drillhole spacing and the role of stockpile. In: Leuangthong O, Deutsch C (eds) Geostatistics Banff 2004. Springer, Netherlands, pp 35–44

Caers J (2000) Adding local accuracy to direct sequential simulation. Math Geol 32(7):815–850

Caers J (2011) Modeling uncertainty in the earth sciences. Wiley-Blackwell, Hoboken, p 229

Deutsch CV (2002) Geostatistical reservoir modeling. Oxford University Press, New York, p 376

Dimitrakopoulos R (1997) Conditional simulations: tools for modelling uncertainty in open pit optimisation. Optimizing with Whittle. Whittle Programming Pty Ltd, Perth, pp 31–42

Dimitrakopoulos R, Ramazan S (2008) Stochastic integer programming for optimising long term production schedules of open pit mines: methods, application and value of stochastic solutions. Min Tech: IMM Transactions Section A 117:155–160

Dowd PA (1994) Risk assessment in reserve estimation and open-pit planning. Trans Instn Min Metall Sect A-Min Industry 103:A148–A154

Froidevaux R (1992) Probability field simulation. In: Soares A (ed) Geostatistics Toria '92, pp 73–83

Geman S, Geman D (1984) (November) Stochastic relaxation, Gibbs distributions, and the Bayesian restoration of images. IEEE Trans Pattern Anal Mach Intell PAMI-6(6):721–741

Glacken IM (1996) Change of support by direct conditional block simulation. Unpublished MSc Thesis, Stanford University

Godoy M (2002) (August) The effective management of geological risk in long-term production scheduling of open pit mines. Unpublished PhD Thesis, W.H. Bryan mining geology research centre, The University of Queensland

Gómez-Hernández JJ (1992) Regularization of hydraulic conductivities: a numerical approach. In: Soares A (ed) Geostatistics Troia '92, pp 767–778

Goovaerts P (1997) Geostatistics for natural resources evaluation. Oxford University Press, New York, p 483

Guardiano FB, Parker HM, Isaaks EH (1995) Prediction of recoverable reserves using conditional simulation: a case study for the Fort Knox Gold Project, Alaska. Unpublished Technical Report, Mineral Resource Development, Inc

Hardtke W, Allen L, Douglas I (2011) Localised indicator kriging. 35th APCOM symposium, Wollongong pp 141–147

Isaaks EH (1990). The application of Monte Carlo methods to the analysis of spatially correlated data. PhD Thesis, Stanford University, p 213

Journel AG (1974) Geostatistics for conditional simulation of ore bodies. Econ Geol 69(5):673–687

Journel AG, Huijbregts ChJ (1978) Mining geostatistics. Academic Press, New York

Journel AG, Kyriakidis P (2004) Evaluation of mineral reserves, a simulation approach. Oxford University Press, p. 216

Journel AG, Xu W (1994) Posterior identification of histograms conditional to local data. Math Geol 26(6):323–359

Leuangthong O, Deutsch CV (2003) Stepwise conditional transformation for simulation of multiple variables. Math Geol 35(2):155–173

Leuangthong O, Hodson T, Rolley P, Deutsch CV (2006) Multivariate geostatistical simulation at Red Dog Mine, Alaska, USA, Canadian Institution of Mining, Metallurgy, and Petroleum, v. 99, No. 1094

Luster GR (1985) Raw materials for Portland cement: applications of conditional simulation of coregionalization. PhD Thesis. Stanford University, Stanford, p 531

Marcotte D (1993) Direct conditional simulation of block grades. In: Dimitrakopoulos R (ed) Geostatistics for the next century. Montreal, Canada, June 2–5, 1993 Kluwer, pp 245–252

Matheron G (1973) The intrinsic random functions and their applications. Adv Appl Prob 5:439–468

Metropolis N, Rosenbluth AW, Rosenbluth MN, Teller AH, Teller E (1953) Equations of state calculations by fast computing machines J. Chem. Phys., 21(6):1087–1092

Oz B, Deutsch C, Tran T, Xie Y (2003) DSSIM-HR: A FORTRAN 90 program for direct sequential simulation with histogram reproduction. Comput Geosci 29:39–51

Pawlowsky-Glahn V, Olea RA (2004) Geostatistical analysis of compositional data. Oxford University Press, New York p 304

Ren W (2005) Short note on conditioning turning bands realizations. Centre for Computational Geostatistics, Report Five, University of Alberta, Canada

Ren W, Cunha L, Deutsch CV (2004) Preservation of multiple point structure when conditioning by Kriging, CCG Report Six, University of Alberta, Canada

Rosenblatt M (1952) Remarks on a multivariate transformation. Ann Math Statist 23(3):470–472

Rossi ME (1999) Optimizing grade control: a detailed case study. In: Proceedings of the 101st annual meeting of the Canadian Institute of Mining, Metallurgy, and Petroleum (CIM), Calgary (May 2–5)

Rossi ME, Camacho VJ (2001) Applications of geostatistical conditional simulations to assess resource classification schemes. Proceedings of the 102nd annual meeting of the Canadian Institute of Mining, Metallurgy, and Petroleum (CIM), Quebec City (April 29–May 2)

Soares A (2001) Direct sequential simulation and cosimulation. Math Geol 33:911–926

Strebelle S (2002) Conditional simulation of complex geological structures using multiplepoint statistics. Math Geol 34:1–22

Van Brunt BH, Rossi ME (1999) Mine planning under uncertainty constraints. Proc of the optimizing with Whittle 1999 conference, (22–25 March), Perth

Verly G (1984) Estimation of spatial point and block distributions: the multigaussian model. PhD Dissertation, Department of Applied Earth Sciences, Stanford University

Verly G (1993) Sequential Gaussian Cosimulation: a simulation method integrating several types of information. In: Soares A (ed) Geostatistics Troia '92, vol 1. Kluwer Academic Publishers, Dordrecht, pp 543–554

Wackernagel H (2003) Multivariate geostatistics: an introduction with applications. Springer, New York, p 387

Resource Model Validations and Reconciliations

11

Abstract

Mineral resource estimates are based on many interdependent and subjective decisions. There is a need to check the fidelity of the models with the available data, ensure that the models are internally consistent, and validate the models with production data if such data is available. This Chapter discusses some commonly used techniques for model validation.

11.1 The Need for Checking and Validating the Resource Model

There are many important reasons that mandate the need for checking and validating resource models. The validation of the resource model has two basic objectives: (a) to ensure the internal consistency of the models, and (b) to provide, if at all possible, an estimate of the accuracy of the model with respect to the predicted variables.

Internal consistency means that all processes that result in a resource model have been executed as intended, and that there are no inconsistencies, explicit errors, omissions or other factors that cause the model to deviate from what was intended. The model should be a fair representation of the available data. To ensure this, checks must be completed at every step of the modeling process, including the assay and composite database, topography, drill hole locations, down-the-hole surveys, geologic coding, geologic interpretation, block model development, grade estimation and resource categorization.

An accurate model is one that reproduces well the actual tonnages and grades mined. This check can only be performed if the mine is operating. Production data may be used to calibrate and improve future updates of the resource model. However, the process is far from simple, as there are multiple complications and potential pitfalls that are discussed later in this chapter.

The motivations for validating the resource model stem from different sources. In all cases, a minimum number of checks should be performed to ensure that the model is adequate and performing as expected, regardless of the purpose of the model. In addition, there may be internal or external factors that contribute and determine the need for additional checking and validation of a resource model, including the participation in this process of independent auditors. Although the motivations may be different, the validation and/or due diligence process is similar in all cases. Increasingly, the mining industry prudently requires and emphasizes validation and due diligence work on resource models (François-Bongarçon 1998; Vaughan 1997).

An internal or external due diligence is sometimes required by the project owner(s) because there is a need to ensure that the model is reliable and provides enough details to make key investment decisions, such as acquisitions or development decisions. An external due diligence or audit is triggered when external financing is sought to develop or acquire new mining assets.

The level of detail required to adequately validate the model is also related to the level of detail of the resource modeling itself and its objective. Determinant factors may include the development stage of the project and the possible due diligence or audit requirements.

11.2 Resource Model Integrity

When discussing the integrity of the resource model, it should be kept in mind that there are a large number of steps and processes that contribute to a resource model; every one of them should be checked to ensure that the final product is reasonable.

M. E. Rossi, C. V. Deutsch, *Mineral Resource Estimation,* DOI 10.1007/978-1-4020-5717-5_11,

11.2.1 Field Procedures

The checks and validation should begin in the field, and should include checks related to sampling, collar locations, topographic and down-the-hole surveys, drilling methods, sample collection and preparation, assaying, and sample quality control and quality assurance programs.

The topographic surface should be checked against the drill hole collars to ensure that there is a reasonable match. As a general guideline for open pit mines, while a difference greater than half the proposed or actual bench height is considered a serious error, many practitioners would only accept vertical differences of 2 m or less. It is best to relate the acceptable error to the level of project development and the detail required by the design engineering. At a feasibility-level stage, accurate estimates of ground or soil movement and infrastructure location would require smaller than 2 m differences; projects at an earlier stage of development can tolerate larger errors. In underground mines, topographic precision is much more critical.

Sometimes topography is derived from widely-spaced survey points or a smoothed version of an aerial photograph with few ground control points. If there is sufficient confidence on the elevation of the drill hole collar as provided by an independent survey, it may be advisable to re-interpolate the topographic surface including the drill hole collar elevations to better reflect local variations in topography.

Another important aspect is the definition of the coordinate system (projection system) used, and the tie point to the local grids. Sometimes the "official" coordinates of the critical tie-in points as provided by government agencies change, as re-surveys using modern techniques are used. The history of the base points used in the survey should be well understood. If there are changes in the coordinates of the base tie-in points, a general project coordinate transformation (translation and/or rotation) may be sufficient to fix the problem.

Drill hole collar locations should be checked for accuracy. Down-the-hole surveys, which measure hole deviation, can also be a source of significant errors. The expected deviation of a hole will depend on the drilling method and the driller's experience, the type of rocks traversed, and whether they are inclined or vertical holes. All drill holes should be surveyed for deviations, and all measurements should be checked for consistency and inaccuracies. Issues to be considered include the measuring device used; whether it is affected by magnetic minerals present in the rock; whether the measurements have been corrected for declination, if it is significant for the project's latitude and time period; whether the azimuth and dip measurements have been taken at sufficiently close spacing; and whether the information has been properly interpreted, analyzed, and incorporated into the database. These issues were discussed in Chap. 5.

The drilling methods used should be documented, recording core or hole diameters, the presence or not of groundwater, and the drilling rate in meters per hour. Comparisons and statistical evaluations of data obtained from different methods and sometimes hole diameters should be part of the data validation work. If significant discrepancies are observed, a detailed study attempting to resolve the issue should be completed. Core or chip recoveries (as percentage or recovered weights, respectively) from the drill holes should be incorporated into the database and statistically evaluated. Relationships between grades and recoveries or sample weights, if any, should be understood and described.

Sample collection, preparation and splitting procedures should be well documented, whether completed in the field next to the drill rig or in the preparation laboratory. The sample chain of custody should also be well documented, and checks performed while the drilling is on-going, if at all possible.

It is important to observe and document if loss of fines occurs, if there is excessive water being used at the time of drilling, and all other issues that may impact sample quality. It is often necessary to recover some of the fines produced at the time of drilling and analyze them to verify whether it has significant mineralization or not.

11.2.2 Data Handling and Processing

Any transformation of the coordinate system used should be checked, also considering the different levels and types of surveying that may have occurred along the life of the project. Additionally, coordinate transformations performed to facilitate or improve the modeling, such as unfolding, should also be checked.

The computerized sampling database is sometimes taken for granted and not thoroughly checked. At best, both the geologic codes and the assayed grade values are electronically recorded at the time of capturing the data. Digital entry reduces the risk of introducing mistakes. In practice, manual entry is still often encountered, and if so it should always include a double-entry procedure to minimize data entry errors.

Minimum checks to be automatically performed within the database should include from-to checks (the "to" value is always greater or equal than the "from" value); out-of-boundary checks for collar coordinates (ensures that no digits are lost or added to coordinate values); stoichiometric checks if applicable (whereby the addition of grades are no greater than a pre-specified value); values-within-range checks (such as all Cu assays between 0 and 100 %); presence of duplicate sample coordinates (to avoid batches of data being entered twice); and others as deemed appropriate. It is important to consider, however, that these checks are a first line of defense against potential errors, but are not sufficient by themselves to ensure database integrity, and thus do not preclude the need for periodic reviews and additional checks.

There should be established procedures for safe keeping and updating the database such that no errors are introduced. All these procedures should be documented in a manual always available as part of the audit trail for future reference. In addition, the protocols and procedures that have been formally established should be periodically reviewed and audited to ensure that they are applied as intended. Record-keeping of these internal reviews provide a very useful log of the evolution of the stored information. Much like an owner's maintenance log of a car, it will increase the value of the project and the reliability of the resource model.

Backups of the electronic database should be preserved in different locations. The database itself should be relational, and preferably not require specialists for custody and maintenance.

Checks against original information should be done by comparing the original assay certificates, as issued by the laboratory and properly signed by a laboratory representative, against the values stored in the database. This check should also be done against the original geologic logs, the original down-the-hole information (certificates or photos, depending on the method used), and the original signed surveyors report for drill hole collar locations. The expected error rates commonly accepted is 1 % or less of all records when comparing the original information and the computerized database. Although practitioners take different approaches, it is common to differentiate consequential and non-consequential errors. More tolerance is applicable when dealing with errors that have little impact.

Bulk density data is often forgotten in the validation process. Chapter 5 discusses in more detail the importance of density data, but in all cases, there should be sufficient number of measurements for each rock type or geologic domain; their location should be well-documented; and the measurement should be of in-situ density. Measurements on crushed material (such as the ones performed by metallurgical laboratories) are not adequate for resource estimation. Voids that may be present in the rock are one of the most common sources of error, and thus the measurement should be taken using a wax-coated method. For some types of deposits, such as massive sulfides or deposits in lateritic or tropical environments with high humidity, bulk density is a key variable that may be a significant source of error.

Details of suggested sample quality assurance and quality control programs were discussed in Chap. 5. The available information should be analyzed well in advance of the completion of the resource model, and while drilling is ongoing. This allows corrective measures, such as re-assaying, to be completed before the modeling process begins. This information should be stored as part of the overall project database.

11.3 Resampling

Cross-validation and jackknife techniques are sometimes used in an attempt to determine the "best" variogram model to use in the grade estimation process. Also, kriging plans are sometimes optimized based on cross validation exercises.

There are several flavors of these methods, the most commonly used requiring that a sample be extracted from the database and its value re-estimated using the remaining samples and the variogram models being tested. If multiple variogram models and estimation strategies are tested, then the one that produces the smallest error statistics can be chosen. As tempting as it sounds, this cross validation method should not be abused, as discussed below.

A more acceptable alternative, but used little in practice, is to discard from the dataset a sub-group of data, and re-estimate or simulate it using the remaining information and the variogram models being tested. This method requires using a well established stationary domain with a good number of samples, such that about 50 % of them can be taken out and still the variogram model and other statistical properties are maintained.

11.3.1 Cross-Validation

This technique, sometimes also called jackknifing, has been used to validate alternative variogram models. The idea is to re-estimate each drill hole sample interval $z(x_\alpha)$ ($\alpha = 1, \ldots, n$) ignoring the sample at that location, and using the other $(n-1)$ samples in the re-estimation. After repeating this process for each sample throughout the domain of interest, a set of n errors $[z^*(x_\alpha) - z(x_\alpha)]$ is available, where $z^*(x_\alpha)$ are the re-estimated values at each location, for which the known assayed value $z(x_\alpha)$ is available. Statistics performed on these errors give an indication of the goodness of the variogram model and kriging plan used in the re-estimation. Typically this method is used to compare two or more alternative variogram models, or alternative types of kriging (ordinary kriging, universal kriging, etc.), or different kriging plans.

The validity and usefulness of this type of cross-validation techniques have been rightly questioned, mainly because the method is not sensitive enough to detect minor advantages of, say, one variogram model over another (Clark 1986; Davis 1987). In addition, the analysis is performed using the set of samples, which does not allow for any definite conclusion about the final block estimates. A ranking of alternative variogram models according to their performance in re-estimating samples will not necessarily correspond to the ranking when performing the final estimation run.

Another potential issue when using this technique is whether the closest samples to the one being re-estimated

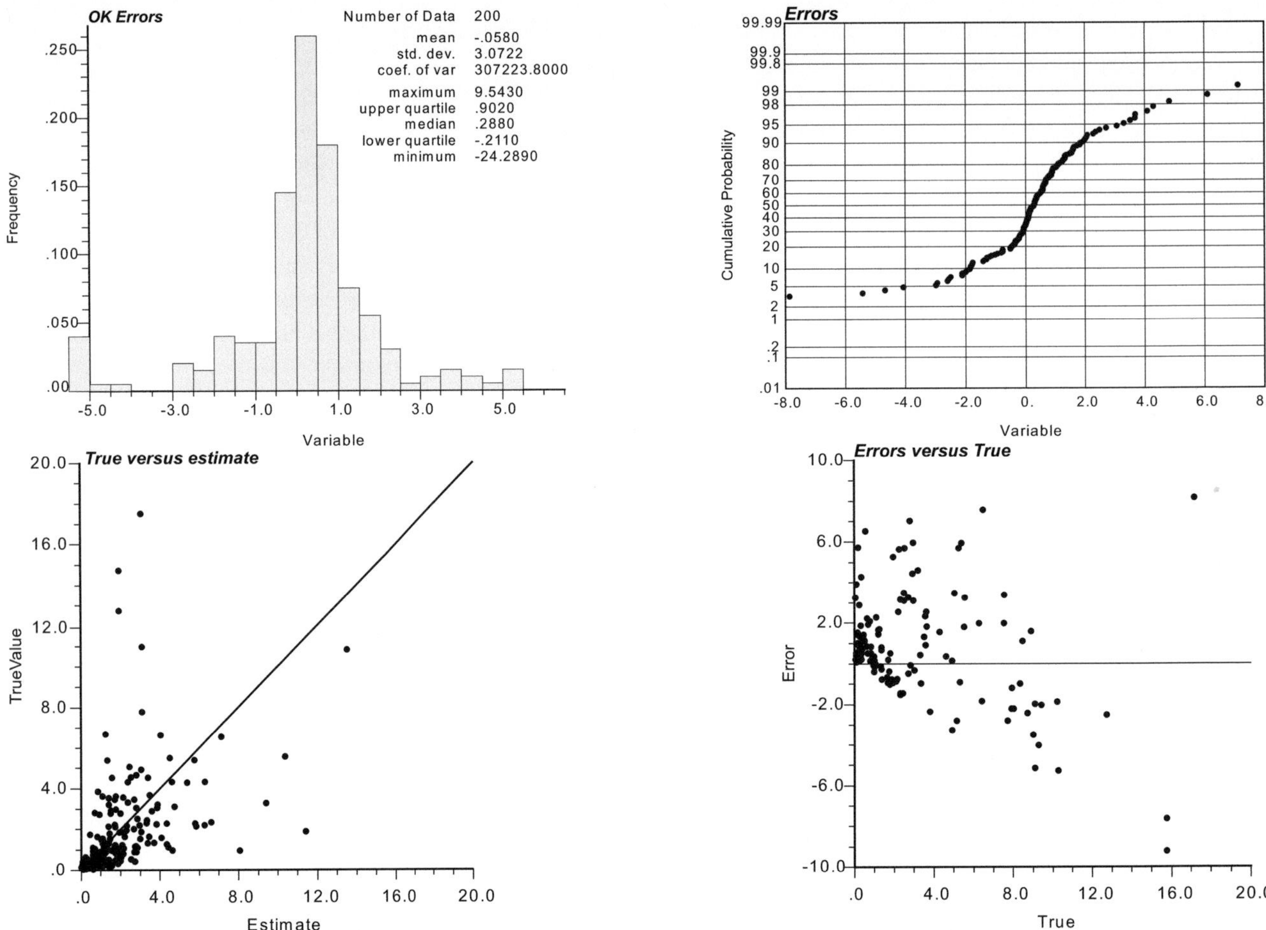

Fig. 11.1 Error checks. Distribution of errors, cumulative distribution function of, True vs. Estimate, Error vs. True

should be ignored in the jackknifing process. Since the variogram models are not very sensitive to small changes (see comments regarding the robustness of kriging with respect to variogram models, Chaps. 6 and 8), the error statistics may be misleading or too similar to provide effective guidance.

Still, the technique may be useful as an indication of how appropriate the variogram models are, in particular when comparing two very different variogram models. Also, the statistics generated are an aid in understanding the results of a geostatistical analysis, and hence it should be seen as another exploratory data analysis tool. It may also detect serious modeling errors and flag numerical or computational problems encountered when solving the kriging equations.

The cross validation method based on splitting the database in two subsets of the drill hole data is a more interesting alternative as long as the domain is stationarity and there is sufficient data to obtain a statistically meaningful set of errors. This cross validation method has the same purpose explained above. Journel and Rossi (1989) describe a case where this cross validation was used to evaluate the differences between ordinary and universal kriging. If a "true" dataset is available, then the results can be compared in terms of distribution of errors, as shown in Fig. 11.1. These comparisons can be show as distribution of the errors (PDF and CDFs) and a cross plot of true vs. estimated values. Also it is useful to look at the relationship between errors and true values.

The magnitude of the errors may be correlated to location, and thus is it useful to obtain a location map of errors, as shown in Fig. 11.2. Cross-validation can thus be considered a rehearsal before the production run.

11.4 Resource Model Validation

The mineral resource model should be validated using statistical and graphical tools (Leuangthong et al. 2004). The suggested checks are useful to ensure the internal consistency of the model, which implies that the model has the expected characteristics; also, that there are no gross or spurious errors, and that all processes were implemented correctly. The resource model validation in fact should cover the data and the geological model used, the estimation domains defined, and the geostatistical model applied.

It is good practice to allow for model iterations in the model building schedule and budget. By changing estimation

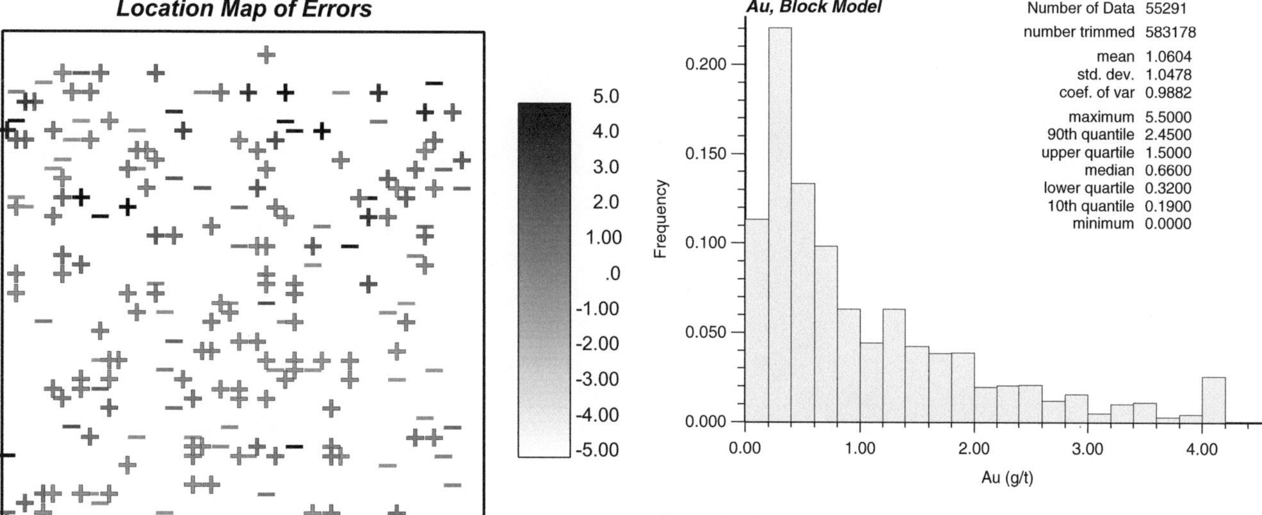

Fig. 11.2 Location map of errors

Fig. 11.3 Histogram of estimated Au grades in g/t

parameters it is possible to better control the estimated grades, including smoothing, contact grade profile reproduction, and global unbiasedness.

11.4.1 Geological Model Validation

The validation of the geologic model should be both statistical and graphical. A discussion on graphical validation is presented later in this chapter.

The geologic model is key in determining the tonnages above cutoff of the resource model. This is because the estimation domains generally define high, medium, low grade, and barren zones. Those volumes with grades mostly higher than the economic cutoff will be sent to the processing plant. If the volumes of these zones have been over or under-estimated, that error will directly translate into errors in the corresponding tonnages above cutoff. In those circumstances, there is very little that geostatistics can do to correct the volumetric bias.

The most common numerical check is by looking at the proportions of each geologic unit in the database compared to the proportions (volumes) of the modeled three-dimensional solids. This is normally done by back-tagging the composites used for resource estimation with the codes from the modeled geologic units. Then, the statistics of each modeled units can be compared to the original, logged intervals that were used to create the model. Some discrepancies are acceptable, as there may be some units too small or too complicated to model, and some intercepts in the drill hole database that are too narrow to be considered. A possible target is to have a better than 90 % coincidence for each geologic unit between the logged intervals and the back-tagged composites, but this percentage will vary depending on how complex the geology is.

Another check that can be performed, similar in concept, is to assign geologic codes to blocks using the Nearest Neighbor (NN) technique. The assumption is that the declustered data (through NN) correctly represent the proportion of each mineralization type and lithology within the deposit. Then it is expected that the corresponding modeled volumes represent similar proportions for each minzone and lithologic unit.

There are a few caveats to this check. First, the assumption is that the mapped and logged meters in the drill hole database are representative. Spatial clustering or spatially non-representative data is one possible source of error. Additional possible discrepancies occurs if changes to the originally mapped drill hole information are made as interpretation progresses, and they are not incorporated into an updated database. Sometimes the decisions are made at the time of interpreting the units, and do not prompt a full-scale re-logging or any changes in the database.

Also, the interpreted volumes may not be representative because some of the units are "border" units in the model. Thus, they may be extended beyond what would be reasonable for other units, as is the case with wall or host rock in Lithology models. The check is applicable for the units that are well delimited within the model and enclosed by the outlying units.

11.4.2 Statistical Validation

The basic statistical analysis compares means and variances of the data and the model, including the spatial correlation models in the case of simulations. In all cases, the drill hole data used in these checks should be the same as the data used to estimate the model, generally composites. This should be done per estimation domain used to condition the estimation or the simulation and using representative (declustered) drill hole statistics for each domain. Figure 11.3 shows an example of histogram of estimated Au values.

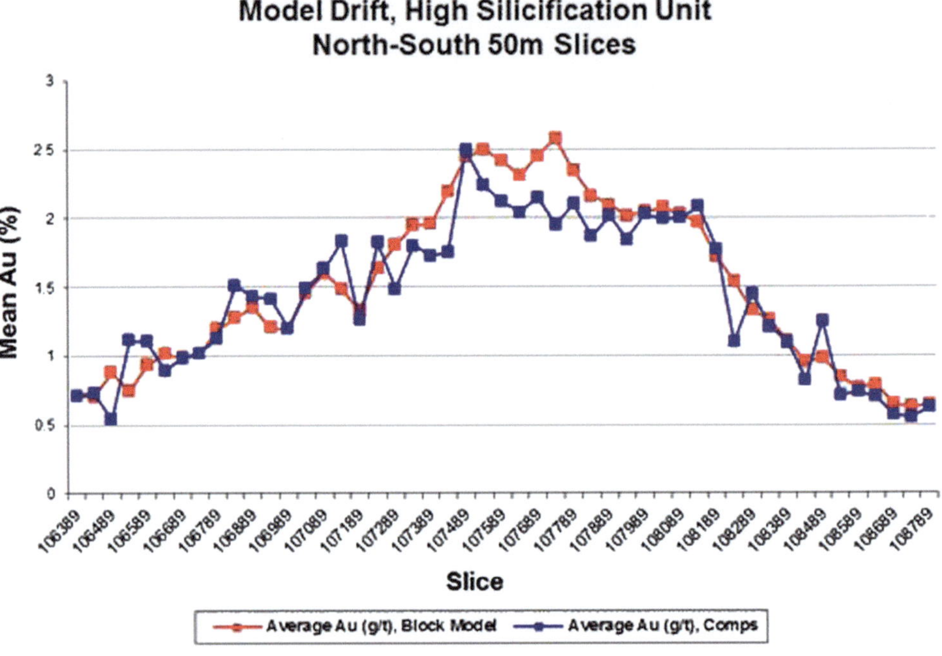

Fig. 11.4 North-South trend of Au grades (%), 50 m slices

The histograms and basic statistics may be compared to the original, declustered drill hole data used to estimate the grades for each domain. This is to check, among other aspects, that the overall means (without applying a cutoff) are very similar, since the estimated grades should be unbiased. Also, the shape of the histogram can provide clues as to the quality of the estimation, in particular if the domain considered is not strictly stationary, with discontinuities in the grade population. Sometimes, a peculiar composite selection method at the time of estimation may cause artifacts on the shape of the histogram of estimated grades, such as creating artificial boundaries. Because of this, it is convenient to always look at the frequency distribution of estimated grades, not just a box plot or a table of main statistics.

It is also important to check the grade trends as observed in the data compared to the model. This can be accomplished by plotting declustered drill hole grades vs. block model averages based on the three main Cartesian coordinates, and considering significantly large volumes at a time. Slices or swaths are usually defined for each main direction. The swath width should be large enough to provide reasonable estimates of the average declustered grade for the slice, which is approximated usually with a NN estimate. Figure 11.4 shows an example.

Another important aspect that the resource model should adequately reproduce is the behavior of grades near contact zones. The model should be checked to ensure that the grade profiles near contacts are reproduced, based on the conditions imposed at the time of estimating. This involves producing contact profiles from the resource model. One such comparison is shown in Fig. 11.5. Note that the block model grades are somewhat smoothed near the contact, and tend to marginally over-predict the grades of Unit 5 while under-predicting the grades of Unit 6. This type of comparison should not be analyzed in isolation, rather as one more piece of the puzzle before deciding to either iterate the grade estimation process changing some parameters, or accepting the model as is.

One of the most important issues to be assessed is an evaluation of the degree of smoothing and conditional bias in the block model, and how it compares to the expected or theoretical smoothing of the actual or planned operation. Excessive smoothing of the model grades amounts to too much internal dilution being added to the model. By using geostatistcal models to predict the expected internal dilution (volume-variance effect), the amount of internal dilution expected for a given Selective Mining Unit (SMU) can be predicted. Therefore, a target or reference grade-tonnage curve can be developed to validate the resource model.

Table 11.1 shows the comparison of the predicted SMU distribution vs. the grade model in terms of means and coefficient of variations, while Fig. 11.6 shows a grade-tonnage curve where both the predicted SMU and the estimated grade model are compared. Notice that, for most cutoffs, the grade model appears to predict slightly higher tonnage and lower grade than predicted for the SMU distribution. In most cases, small differences such as the ones shown are acceptable since the resource model should incorporate other types of dilution, not just within-block, or internal, dilution.

11.4.3 Graphical Validation

It is always good practice to visualize graphically the model obtained using an appropriate scale to observe both the data

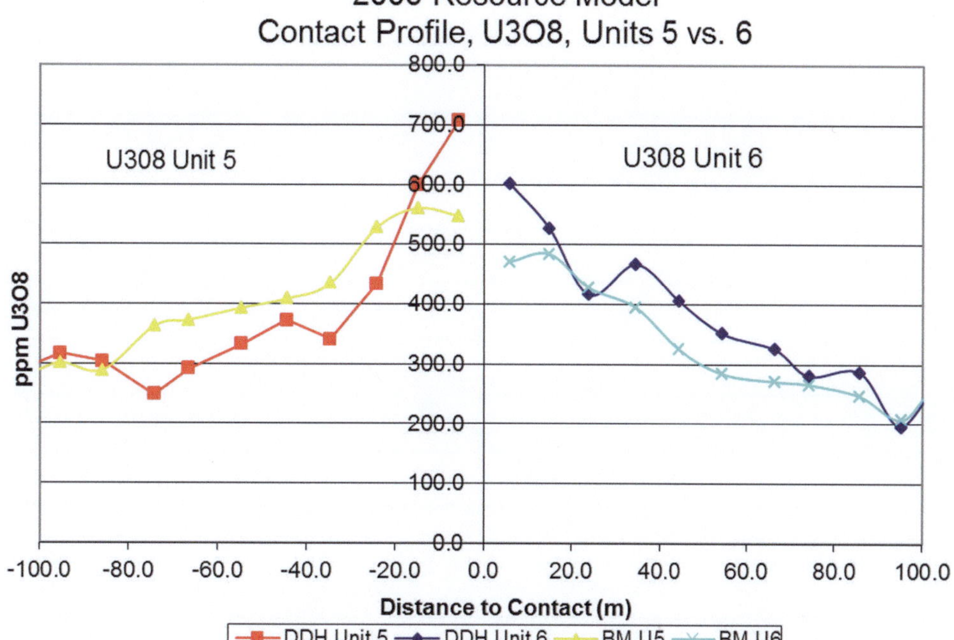

Fig. 11.5 North-South trend of TCu grades (%), 50 m slices

Table 11.1 Example of target vs. achieved basic statistics by estimation domains

Estimation Domain	Kriged Mean Grade	Predicted Mean Grade of the SMU	Coef. of Variation of Block Estimates	Predicted Coef. of Variation of the SMU
1	0.805%	0.882%	0.745	0.698
2	1.553 %	1.673%	0.504	0.470
3	0.231%	0.333%	0.735	0.805
4	1.294%	1.648%	0.579	0.599
5	0.691%	0.823%	0.639	0.649
6	0.509%	0.611%	0.563	0.596
7	0.927%	1.058%	0.607	0.627

and the geologic codes applied. This should be done both on the computer screen, using the visualization capabilities of the mining software used, and on as large a sheet of paper as practical, paper size E being the most common.

Typically, cross sections and plan views of the geologic model and of the block model grade and the composited drill hole data are plotted out. Figure 11.7 shows a detail of a cross section comparing two models of mineralization units for BHP Billiton's Cerro Colorado mine. The two models are compared with different shades for the outlines of the units. The old model is represented in white outline while solid colored outlines represent the updated model: green is oxides, red are enriched supergene sulfides, and yellow is a transitional supergene-hypogene unit. Brown is the leach cap unit, and purple is the hypogene unit, for which the interpreted outlines are not shown here.

This comparison is appropriate when a new, updated model has been interpreted based on additional drill hole data, and it is necessary assess the impact of the new drilling on the old interpretation. Notice that a new drill hole (second from right in the Fig. 11.7) has truncated the upper oxide body (in green) and has created a small isolated supergene sulfide body (in red). The outlines of the previous and updated model are visible.

Figure 11.8 shows an example of a cross sectional view with color-coded block grades, as well as drill hole composite grades. Several surfaces are also shown, including current topography (in green) planned final pit outline (in magenta), and the mineralized envelope that defines the volume filled with blocks. Good practice is to use warmer colors for higher grades, and the same color scheme for both blocks and composite grades.

Detailed review and examination of these plots provides assurance that the estimation or simulation procedure did not produce unexpected erroneous results. It also provides a set of ready plots for future reviews and audits, internal or exter-

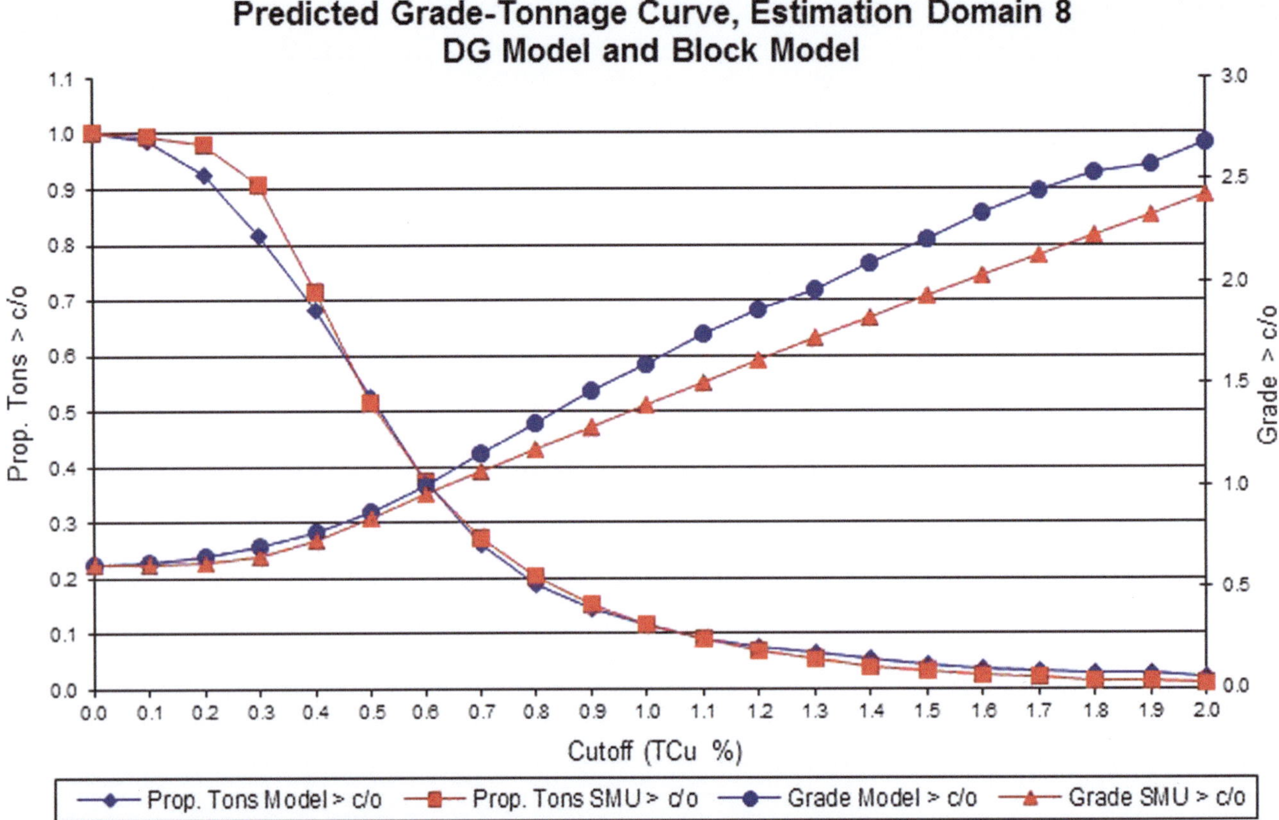

Fig. 11.6 Example of grade-tonnage curves comparison between the discrete Gaussian-predicted model for the SMU, and actual resource model grades

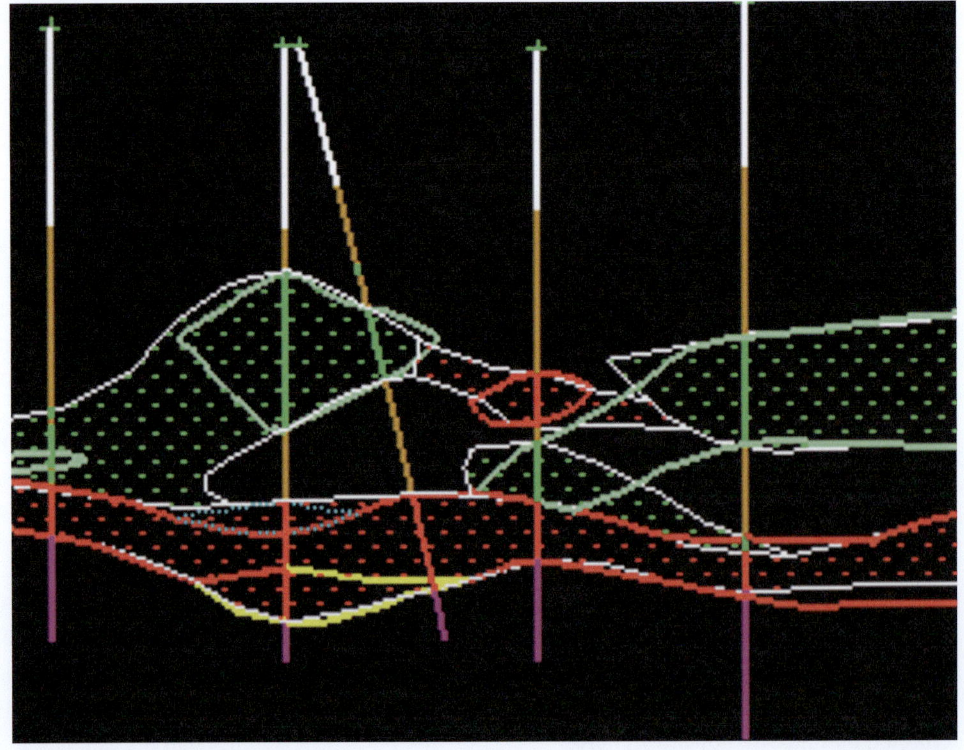

Fig. 11.7 Detail of cross section 82230N of the Cerro Colorado Cu deposit in Northern Chile, showing color coded geologic units for two models being compared and drill hole with the mapped units. Courtesy of BHP Billiton

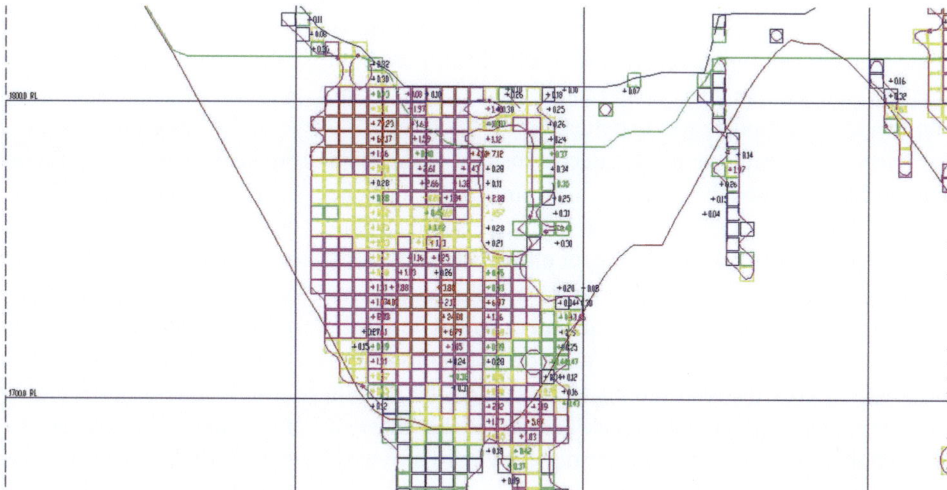

Fig. 11.8 Detail of a cross sectional view showing color coded model on 5 × 5 × 5 m, 5 m composites grades, and mineralized zones. Also the surfaces corresponding to topography at the time of modeling, and predicted final pit are shown

nal, and should be considered part of the final documentation regarding the resource model.

11.5 Comparisons with Prior and Alternate Models

It is always necessary to compare updates with prior models, a visual example of which was shown in Fig. 11.7. The comparison is useful for both operating mines and development projects, since the resources and reserves of any deposit evolve over time because new drilling changes the information available and the quality of the models, and also because the operation is mining out portions of the deposit.

The comparison between models must be done ensuring that the portions of models compared are relevant to each other. For example, care must be taken if the definition of the estimation domains has changed from one model to the next; if the selectivity of the mine or dilution conditions have changed; or if the economic cutoff of the operation has changed with time. These aspects will impact not only the methodology used to estimate the updated resource model, but also its reporting and documentation.

Figure 11.9 shows the histogram of the differences in gold grades (in g/t) between two blocks models, prior and updated, on a block basis. A negative difference means that the updated model has lower grades. In this case, there is little difference between the two. Notice how the histogram quickly highlights the blocks where the differences in grade are most extreme (the maximum and minimum, and the 25th and the 75th percentiles). These blocks can be identified in space and the reasons for the significant differences fully understood.

Another numerical example is shown in Fig. 11.10, where the Q-Q plot of two models for the Escondida mine (characterized by their development date) is shown. Note that this and most other numerical comparisons are most useful when comparing corresponding estimation domains. In the case of

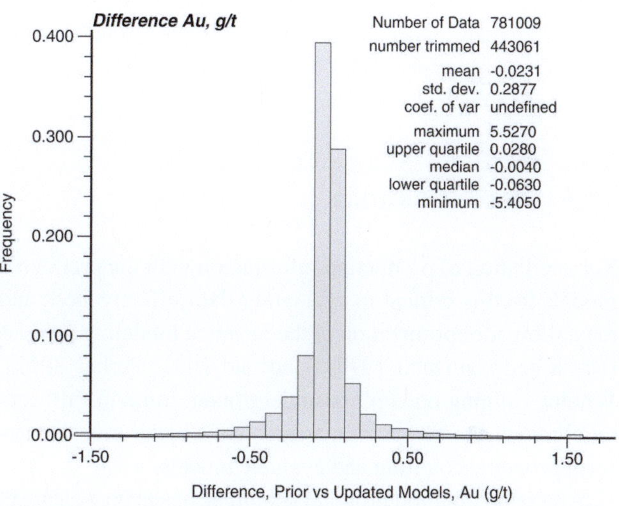

Fig. 11.9 Histogram and basic statistics of the differences between two block models, Au grade in g/t

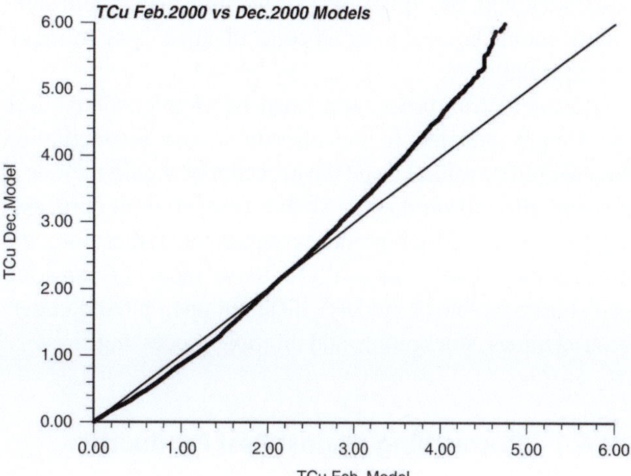

Fig. 11.10 Q-Q plot of the February and December resource models, Escondida, Supergene Sulfide estimation domain, structural blocks 1 and 2

Fig. 11.10, the supergene enriched unit is being compared. Note how the December model is less smoothed (more high and low grades) than the February model. Indeed, one of the objectives of the update was to better control smoothing.

Whenever possible, a detailed decomposition of the main factors that are thought to affect the comparison must be made. All differences must be explained in terms of data density and values, geologic modeling, or other relevant factors.

Another option is to obtain alternative block models for comparison, using a different estimation or simulation technique. Simple models such as a Nearest Neighbor model can be used to check the resource model. Although the models are not comparable block by block because of the different methodology used to obtain the grades, they may provide a general indication that the resource model is reasonable.

Care must be taken to understand and state, where appropriate, the characteristics of each model obtained, and the differences expected to be encountered due to their respective properties. Additionally, a clear definition of an acceptable match is required.

11.6 Reconciliations

Reconciliation of production information with the predictive models used is critical to evaluating their effectiveness and may allow for optimization of the resource modeling process (Rossi and Camacho 1999; Schofield 2001; Parker 2012). Whether mining open pit or underground, mine-to-mill reconciliations can be one of management's better tools to perform proper accounting and evaluate models.

Any reconciliation program should be based on a clear set of criteria and objectives. It should also be executed through a stepwise, logical approach. There are a number of assumptions and key requisites for reconciliations to be effective, and they are not without pitfalls. There are benefits and costs associated with the up-keeping of the information and safeguards should be used to avoid collecting and using misleading information.

Reconciliation procedures must be simple, robust, and specifically adapted to the operation. The reconciliation data should be reliable, and the procedures should include if possible the full production stream (model, mine, processing facilities, and final product comparisons); therefore, the process may involve several predictive models (long-term and short-term block models), different open pit and underground mines, stockpiling, and multiple processing streams.

11.6.1 Reconciling against Past Production

Production reconciliation can be considered an optimization tool (Rossi and Camacho 1999; Schofield 2001). This concept goes beyond common industry practice because reconciliation is typically seen as a material movement and material balances accounting tool. If used as an optimization tool, the basic data used to analyze the performance of the predictive has to be sufficiently accurate and precise.

The main purpose of any reconciliation program in a producing mine is to properly account for all material mined, both ore and waste. But it can also be used to assess the accuracy of the resource and reserves models, therefore allowing for a more accurate valuation of the mining property at all times.

These objectives are interrelated, and they all share some basic requirements if the results are to be meaningful. The most important requirement is, naturally, reliable data. This is not trivial, since many operations do not sample mill head grades and tonnages adequately. Automatic sampling devices placed on the input stream to the processing facility may be expensive, but are invariably well worth the expense. Unfortunately, this is often not realized until well into the operation's life, if at all.

Sometimes there are other issues associated with sampling some of the processing streams. For example, run-of-mine (ROM) material cannot be sampled because the rocks, as blasted, are loaded directly into leach pads. Without further size reduction, sampling of ROM material is impractical. In general, leach operations that stack coarse material do not lend themselves to reliable sampling of head grades, and they must rely on blast hole information (pre-blast) to provide an estimate of grade loaded to heaps. Reliable head tonnages may not be available also, unless a careful program of truck weighing is implemented or each truck is equipped with a weightometer. The weightometer should be calibrated on a regular basis. Some operations, for simplicity, use truck factors, derived from long-term averages of material delivered to the mill. The use of truck factors is unreliable and non-specific to the area or period being mined, and thus should be avoided.

At some operations blast hole data may not be reliable; the operation may not sample blast holes; or may not sample *all* the blast holes available. In underground mines, sampling grades from production stopes is difficult at best, often relying on grab or muck samples to inform stope grades. Bulk tonnage mining methods, such as sub-level or block caving, may sample sufficient tonnage of ore at the stopes' draw points, but this is not always done. These problems should be assessed to evaluate whether implementing a detailed reconciliation program would result in reliable information, suitable for long- and short-term model calibration. If certain changes or additions are required, such as a new sampler for head grades, it is generally feasible to perform a cost-benefit analysis that would allow management to make informed decisions.

In addition to having reliable raw data, it is necessary that the operation's top management be committed and involved to facilitate the necessary coordination among the geology,

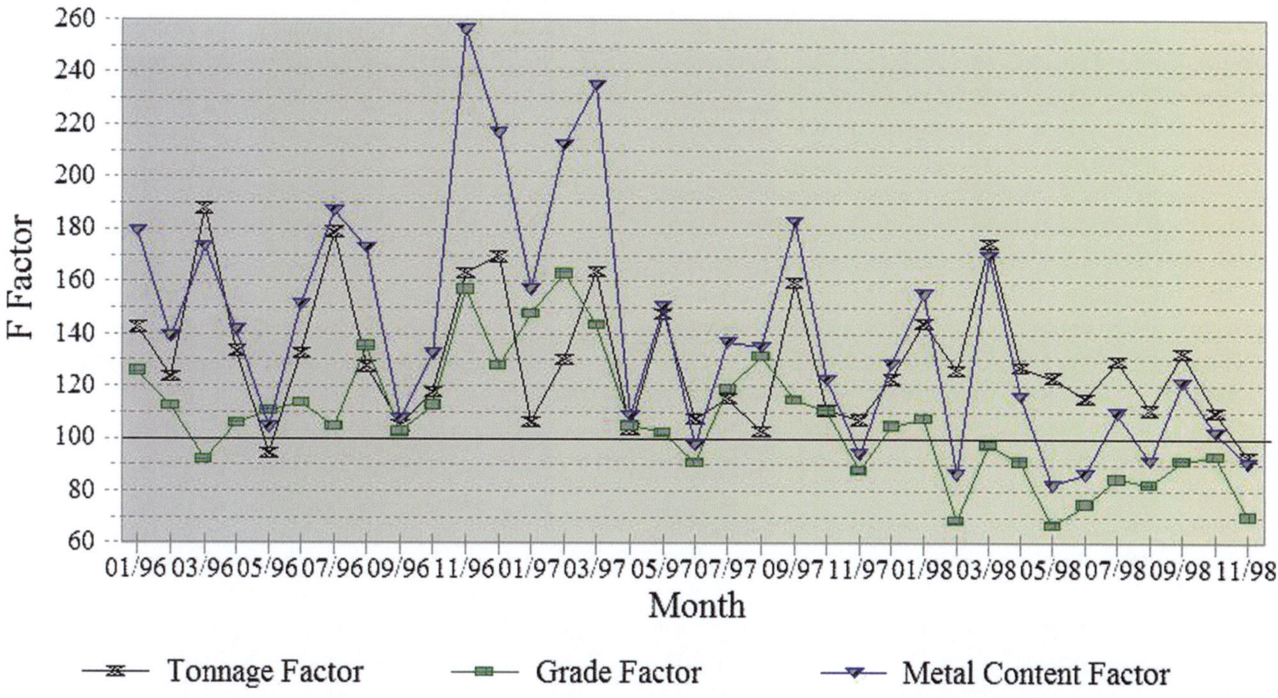

Fig. 11.11 Example of long-term versus mine reported reconciliations. Observe that after 05/1997 the reconciliation changes. The factors have not been adjusted to overall metal production

mining, and metallurgy departments, which includes a clear delineation of responsibilities, thus ensuring proper information gathering and processing.

The first requirement is essential because production reconciliation can quickly become a bone of contention among the groups involved. Reconciliation can be a minefield because of inevitable organizational politics. As readers that are or have been involved in operations would all too easily recognize, metallurgists will tend to blame production reconciliation discrepancies on the mine for lower head grades, while the mining department in turn will tend to blame the geology department for poor predictive models; and the geology department will eventually propose that more in-fill drilling is required to solve the mine to mill reconciliation discrepancies. Therefore, the best way to ensure a good faith effort from all involved is for an Operations Manager to recognize the importance of the issue, and to adequately distribute responsibilities and prioritize the tasks involved in a reconciliation program.

Finally, the procedures and data sources should be maintained constant through time, as much as possible, to allow for relevant comparisons regarding model, mine, and plant performances. There are multiple options for establishing a reconciliation program, but in all cases it should be developed from basic principles as discussed. Figure 11.11 shows an example of monthly reconciliations based on raw, non-adjusted data, and thus more likely better reflecting operations performance. In this case, note how reconciliations improve at a certain point in time after introducing corrective measures in the modeling process.

11.6.2 Suggested Reconciliation Procedures

A simple but systematic reconciliation approach is proposed to compare Long-term block models to Short-term models, grade control models, mine reported, and mill feed information, if available.

The reconciliation procedure is common in industry. Most operations that reconcile production against predictive models do so utilizing some variant of comparison factors, sometimes known as Mine Call factors. The presentation here is based on an expanded scheme from that proposed by Parker (2012).

The performance factors proposed here are intended to separately evaluate the performance of long-term models (resource and reserve block models), short-term models (quarterly or monthly, see Chap. 13), daily grade control models, and dilution and ore losses resulting from mining (operational dilution). These comparisons ideally are anchored in reliable *head grades and tonnages* to the processing facilities. The information should be compared based on a reasonable production period, most commonly on a monthly basis, although there can be exceptions to this.

Comparisons based on longer periods (quarterly, semi-annually, or even yearly) may be necessary and the only available option if the quality of the information is relatively poor, or if the operation operates on a very small scale. Estimating average values of larger volumes of data (periods) is easier than for smaller periods. Also, longer periods may be necessary if the operation handles a significant number of stockpiles, since accurately measuring tonnage and grades drawn from and put into stockpiles is difficult. The suggested method does not include internal reconciliation of the processing facilities (mill balances), which should be an integral part of the reconciliation process, but outside the scope of this book.

The basic information to be compiled includes:

- Tonnage, grades, and metal content of the long-term resource model for the period. This implies obtaining the mine advance positions for the period, and to superimpose them on the block model.
- Similarly, tonnages, grades, and metal content should be obtained from short-term models for the same periods, if they exist.
- Tonnage, grades, and metal content should be obtained from the daily production model (grade control model). This information should be gathered daily, but compiled into the proper reconciliation period (monthly). In open pit mines, the grade control model for the same periods, with all its potential problems and pitfalls, including sample quality and sometimes inadequate modeling techniques, represents the best possible "in-situ" information available, given the typical density of information. Sometimes this is also true for underground mines, but more commonly, the only reliable information is the short-term model from which the final stope designs are made.
- Tonnage, grades, and metal content *as reported by the mine* should also be compiled. The grade usually corresponds to the grade assigned to the extracted panel or stope by the grade control model. This may include some downgrading to consider operational dilution and ore loss, as well as blast movement. Reported tonnage may be from truck weights (preferably avoiding truck factors), or by direct topographic measurement of the extracted volumes. Sometimes the only tonnage available is the one reported by the grade control model.
- Tonnage, grades, and metal content informed as head grades and tons. This should be based on *direct sampling*, as opposed to back-calculated from tailings grades and adjusted recoveries. Back-calculated head tonnage and grades should not be used for model optimization.

There may be stockpiles to be considered in between the mine and the mill. Also, material within the metallurgical stream itself should be accounted for. However, some or all of these stockpiles may not be relevant to the reconciliation program if they are completely "turned over" or replaced within the reporting period. In that case, they can simply be ignored for production reconciliation.

With the information described above, several dimensionless factors can be calculated:

1. F_1 factor, defined for tonnage, grade, and metal content (F_{1t}, F_{1g}, F_{1m}). It is calculated from the corresponding tonnage, grade, and metal of the long-term and short-term models as:

$$F_1 = \frac{\text{Short-term}}{\text{Long-term}}$$

2. F_2 tonnage, grade, and metal content factors (F_{2t}, F_{2g}, F_{2g}). These compare the grade control (production) model versus the short-term model (if it exists), and are calculated as:

$$F_2 = \frac{\text{Grade-control}}{\text{Short-term}}$$

3. F_3 factors (F_{3t}, F_{3g}, F_{3m}) may be defined based on the tonnages, grades, and contained metal of the monthly mine report versus the grade control model. Sometimes mine reports for tonnage and grades are simply taken from the grade control model, and are informed as material sent to the mill. In other instances the mine reports the grade provided by the grade control model (in open pits generally there is no other option, although for underground operations may be based on additional sampling), but the reported tonnage is based on truck weights, counts, or volumetric measurements of the advances. If applicable, the F_3 factors are calculated as:

$$F_3 = \frac{\text{Mine-reported}}{\text{Grade-control}}$$

4. F_4 factors (F_{4t}, F_{4l}, F_{4f}) based on tonnages, grades, and metal content of the "received-at-mill" material versus the mine reported material. The F_4 factors may be calculated as:

$$F_4 = \frac{\text{Received-at-mill}}{\text{Mine-reported}}$$

Not all these factors need to be defined, as for example when no short-term model exists. Note that, as defined, a factor greater than 1.0 implies underestimation, while a factor smaller than 1.0 implies overestimation. From these factors, several performance measures can be readily obtained. For example, to quantify, the performance of the long-term model in terms of tonnage and grade of ore delivered to the mill, the F_{LTM} factor is obtained as:

$$F_{LTM} = \frac{\text{Received-at-mill}}{\text{Long-term}} = F_1 \times F_2 \times F_3 \times F_4$$

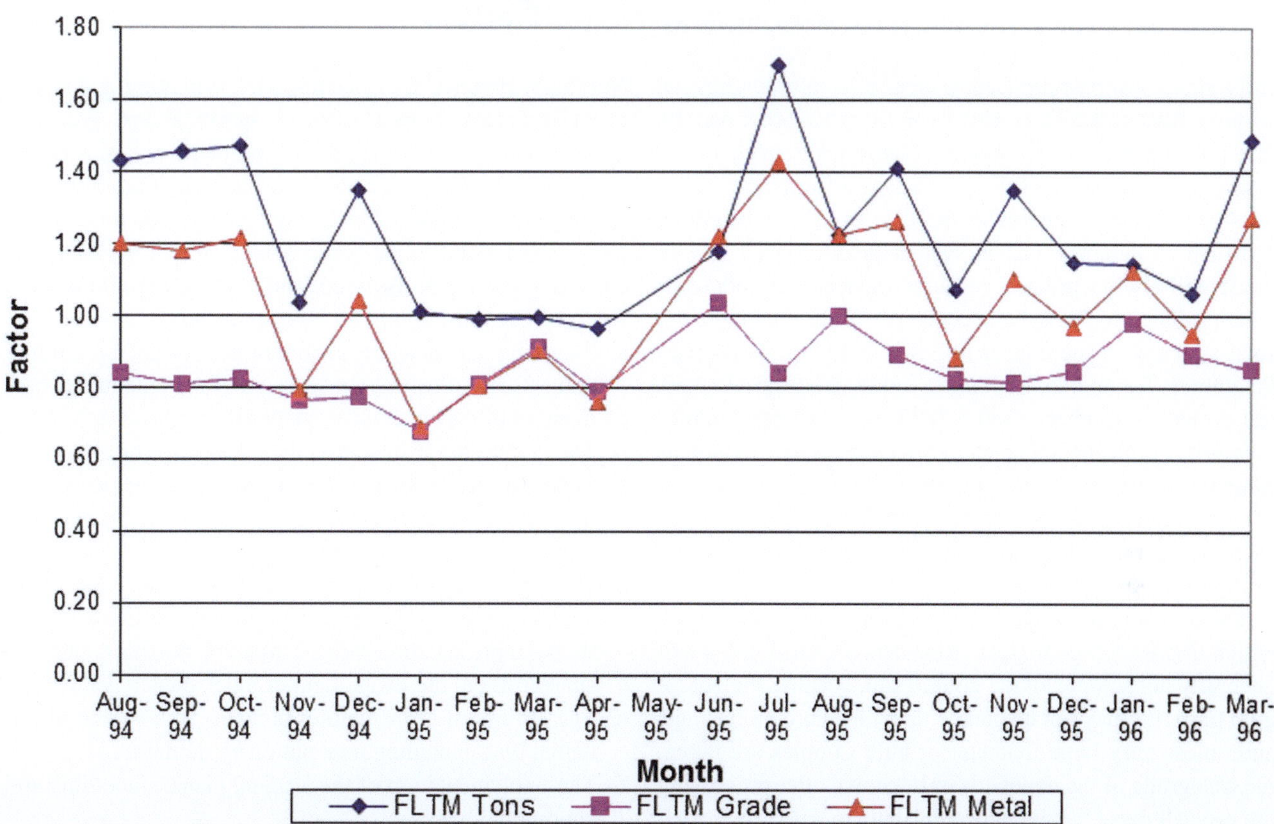

Fig. 11.12 Long-term resource model vs. received at mill reconciliation

F_{LTM} measures how well the resource block model predicts material delivered to the mill, which is the basis for the prediction of future cash flows from the operation. Similarly, an F_{STM} can be defined to quantify the benefits achieved by in-fill drilling, assuming an intermediate short-term model is developed:

$$F_{STM} = \frac{\text{Received-at-mill}}{\text{Short-term}} = F_2 \times F_3 \times F_4$$

To compare the performance of the grade control model versus the material received at the mill, that is, to evaluate mine operating performance and unplanned dilution and ore loss, F_{GCM} can be calculated as:

$$F_{GCM} = \frac{\text{Received-at-mill}}{\text{Grade-control}} = F_3 \times F_4$$

Finally, note that the F_4 factor measures directly the operational ore loss and dilution. This is, naturally, assuming that the sampling points at the mill are deemed reliable for both feed tonnage and grades.

It is important to consider an appropriate time scale for these comparisons. For example, it is not likely that comparing the resource model and the material delivered at the mill on a weekly or bi-weekly basis is relevant. The purpose of the resource (long-term) model is to support long-term mine planning and scheduling, which is generally based on time-units of monthly, bi-annually, yearly, or greater. Therefore it is not appropriate to compare them at a smaller time unit, since the long-term model should not generally be used for small-scale estimation. Similarly, and depending whether stockpiles exist and how large they are, the F_4 factor can be compared on a daily basis, since it measures the mine reported material versus the received at mill material.

Figure 11.12 shows an example of the results of monthly reconciliation for a heap-leach Au operation between the Long-term resource model and the "received at mill" values. Note that monthly factors can vary significantly from one month to the next, which may be mostly a function of production tonnage (smaller mines tend to have larger relative deviations) and geologic complexity. If deemed reliable, graphs like the one shown in Fig. 11.12 should be used to assess the predictive capabilities of the models and to improve future models.

The common expectation is that the resource models generally match past production. This is based on the assumption that if the model does not estimate well the grades and tonnages of past production areas it is valid to question its predictive capabilities. However, there are other aspects that may have to be considered. These are related to the quality and reliability of the production information and to possible systematic biases in the data used. For example, a drill hole database that is systematically biased low in grade will always produce resource models that underestimate production. There have been several cases in industry, such as many of the copper mines in Northern Chile (Los Pelambres, Escondida, Radomiro Tomic, the Gabriela Mistral project, and others), where the Cu grade from samples obtained within the supergene enrichment horizons underestimate the in situ grade. Mineralization is sometimes lost as very fine dust, the most obvious evidence being significant dust plumes light blue or green in color observed when drilling the supergene blanket!

Other examples include the presence of coarse gold, which implies a high nugget effect deposit. Drill holes will generally not sample all the gold in situ, because the coarse gold particles may not be picked up by the core or cuttings, and unless very large diameter or bulk samples are taken, the true grade of the deposit is only known after processing the ore. All these factors may contribute to a poor comparison between the resource model and production information.

Another important issue is to what extent past production can be used to "calibrate" current resource model. The answer depends on the geology of the deposit and the characteristics of the mine operation. It is possible to implement estimation strategies that are intended to reproduce recent production (for example, for the past two years), based on monthly volumes and for each estimation domains defined. The assumption is that the methodology used in successfully re-estimating recent past production should work equally as well in the near future.

When reconciliation data is considered reliable, past production can be used to produce error estimates, i.e., an uncertainty model, for specific volumes and estimation domains. These Mine Call Factors, essentially F_{LTM} above, would be applied to individual estimation domains.

The authors caution against applying MCFs to resource models because they tend to be global, even if developed for each estimation domain. It is preferable to use the Factors to calibrate the resource model, as opposed to using them as correction factors. The main issue is again whether matching past production is relevant to predicting future resources. At least, past production should be used to provide a "reality check" on the models developed, both for resource estimation and simulation and risk analysis.

11.7 Summary of Minimum, Good and Best Practices

Minimum practice for checking and validating a resource model includes thorough use of statistical and graphical tools available to assess the degree of internal consistency of the model. Checking for potential biases is important, both in the geologic model (which may translate into mineralized tonnage bias) and in the grade model. Biases regarding the data used should be assessed, and considering all potential sources. The most important issues include:

a. Sampling and analysis, including the QA/QC program in place, the results, and a discussion of the potential consequences of the observed deviations.
b. The quality of the database, including error rates.
c. Topographical and hole locations errors, including drill hole deviations and its impact on both the interpreted geology and the grade model.
d. Density determinations and issues such as the number and quality of the available samples, the validity of their spatial representation for all estimation domains, and possible errors in the assignment of block density values.
e. The definition of estimation domains as a source of potential bias, including non-stationary domains.
f. The implementation of the Kriging plans, smoothing and conditional biases.
g. Resource accounting, the use of appropriate economic cutoffs, etc.

Also, consideration should be given to the resource classification scheme, and its implication in resource declarations. At a minimum, comparisons should be made between past production (if available) and the resource model, and possible discrepancies explained in detail. There should always be a full documentation and disclosure document, including a full set of sections and plan views at an appropriate scale, showing the geologic model, the data used for interpretation and modeling, the composite dataset used for estimation or simulation, and the block model itself, showing all important variables.

Good practice requires, in addition to the above, a more thorough validation using additional tools. Checks and assessments of accuracy of the predictions are also required. The resource model should be checked using alternate models, comparing to prior models, and detailed production reconciliations, if available. The resource model should be calibrated using a known "truth", for example production data. And a detailed description of the process should be provided as an integral part of model validation.

Best practice consists of, in addition to the above, full use of alternative models available to characterize the quality of the resource model presented. Conditional simulations should be used to provide global and local uncertainty measures, as described before, interpreted here as further validation of the resource model.

In all cases, a detailed audit trail should be documented, corresponding to the level of detail of the resource model and the development stage of the mining project or mine.

11.8 Exercises

The objective of this exercise is to review cross validation and to experiment with ways of checking uncertainty. Some specific (geo)statistical software may be required. The functionality may be available in different public domain or commercial software. Please acquire the required software before beginning the exercise. The data files are available for download from the author's website—a search engine will reveal the location.

11.8.1 Part One: Cross Validation

Provided with enough data, cross validation is a useful exercise. The drillholes are left out one at a time and re-estimated from surrounding drillholes. Blunders in the data or estimation may be detected, as well as obtaining an initial appreciation for the expected degree of uncertainty.

Question 1: Setup to perform cross validation with the `largedata.dat` used in previous exercises. Plot a scatterplot of the estimate versus the true value and a histogram of the errors. Look for data/drillholes where unusually large over- or under-estimates have occurred. Comment on the results.

Question 2: Perform between 2 and 4 reasonable sensitivity runs and document. You could vary the search, the number of data or the variogram model. Comment on the results.

11.8.2 Part Two: Checking Simulation

The goal of simulation is to reproduce the input histogram and variogram. Use the same set of realizations constructed in previous exercises. You may have to recreate some SGS realizations.

Question 1: Plot the histogram over 10 realizations and compare the mean and variance to from the realizations to the original declustered data used in the simulation. Comment on the results.

Question 2: Plot the variogram from 10 realizations (in one principal direction) and compare to input model. Comment on the results.

Uncertainty has a very precise meaning—80 % probability means that a proportion of 0.8 of the values should fall within the 80 % probability interval. For the data in `red.dat`:

Question 3: Construct distributions of uncertainty of the normal scores transform of grade in cross validation mode.

Question 4: Calculate the proportions of true values within fixed intervals of the Gaussian distributions of uncertainty and plot an accuracy plot.

References

Clark I (1986) The art of cross-validation in geostatistical applications. Proceedings 19th APCOM, pp 211–220

Davis BM (1987) Uses and abuses of cross-validation in geostatistics. Math Geol 17:563–586

François-Bongarçon DM (1998c) Due-diligence studies and modern trends in mining. Unpublished internal paper, Mineral resources development, Inc

Journel AG, Rossi ME (1989) When do we need a trend model? Math Geol 22(8):715–738

Leuangthong O, McLennan JA, Deutsch CV (2004) Minimum acceptance criteria for geostatistical realizations. Nat Resour Res 13(3):131–141

Parker HM (2012) Reconciliation principles for the mining industry. The Australasian Institute of mining and metallurgy. Mining Tech 121(3):160–176

Rossi ME, Camacho VJ (1999) Using meaningful reconciliation information to evaluate predictive models, Preprint, SME Annual meeting, March 1–3, Denver

Schofield NA (2001) The myth of mine reconciliation. In: Edwards AC (ed) Mineral resource and ore reserve estimation—the AusIMM guide to good practice. Vic., AusIMM, Melbourne, pp 601–610

Vaughan WS (1997) (July) Due diligence issues for mining investors post Bre-X. Randol conference on sampling and assaying of gold and silver, Vancouver

12 Uncertainty and Risk

Abstract

This chapter shows how multiple realizations can be used to support the assessment of uncertainty and risk.

12.1 Models of Uncertainty

All estimates have some error or uncertainty. Predictions are always inaccurate, with errors stemming from widely spaced data, geological variability, lack of knowledge to determine the best parameters for estimation, approximations made in the estimation procedure, and limitations of the models used.

Although the error will never be known except at locations where data are collected in the future, traditional statistics and geostatistics provide models of uncertainty. Chapters 8–10 discussed estimation, estimation variances, and methods to obtain a conditional distribution of uncertainty for a random variable:

$$F(z;u \mid (n)) = \Pr ob\{Z(u) \leq z \mid (n)\} \quad (12.1)$$

Equation 12.1 is a complete description of uncertainty in the variable z based on our random function model. Obtaining reliable models for the conditional distributions denoted in Eq. 12.1 has proven difficult, particularly for small volumes (one block at a time), as opposed to large deposit-scale volumes.

Early attempts in geostatistics to characterize uncertainty relied on the kriging variance, typically in the form of confidence intervals attached to each estimated block grade:

$$(\mu_{x|(n)} - d) \leq \mu_{x|(n)} \leq (\mu_{x|(n)} + d) \quad (12.2)$$

where d is the difference from the average value that defines the confidence level. For example, $d=2*\sigma$ (twice the standard deviation of the random variable) represents the 95 % confidence level if the distribution has a Gaussian shape (Chap. 2).

The minimized estimation variance or kriging variance can only be equated to a local estimation error if the error distribution is Gaussian and the estimation error does not depend on the actual sample values, a property called homoscedasticity, discussed in Chap. 8. In this case, the estimation variance could be associated to the variance of the error distribution. This is seldom found in practice because most grade distributions are positively skewed and the local uncertainty will depend on the local grades; more uncertainty will be expected in high grade areas. The estimation variance does not provide a reliable uncertainty model for small blocks.

The kriging variance may be used in instances where the distribution is likely to be Gaussian. This may apply if very large volumes of material are considered, since most spatial distributions will tend to become more symmetric, and therefore become more Gaussian-like as more small scale values are averaged together. The reasonable limits of application are not known ahead of time, see Davis (1997) among others.

Other, more recently developed techniques, have attempted to introduce local measures of uncertainty by making the kriging estimation variance data dependent. Most of these techniques have been applied in the context of resource classification (for example, Arik 1999).

Non-linear geoestatistical techniques rely on data transformation to obtain a probabilistic estimate that carries uncertainty (Chap. 9). Except for the case of the indicator transform, the uncertainty model is developed in the transformed space, most commonly Gaussian.

Conditional simulation provides a model of uncertainty at each location by a set of simulated realizations. The uncertainty is better described when a large number of realizations are available, but a relatively small number (say 100) is sufficient to provide a reasonable approximation.

M. E. Rossi, C. V. Deutsch, *Mineral Resource Estimation,* DOI 10.1007/978-1-4020-5717-5_12,

Simulation techniques and the resulting models of uncertainty rely heavily on stationarity; trends and departures from stationarity significantly affect the model of uncertainty and its quality and usefulness. A model of uncertainty based on simulation depends on the Random Function model used. In certain deposits, a Gaussian-based model may be appropriate, while for others a non-parametric technique such as indicator simulations may be preferable. The model of uncertainty will also depend on the number and statistical characteristics of the conditioning data. Therefore, the model of uncertainty is not unique, nor is there an objective or true model of uncertainty: uncertainty is model-dependent. This has been discussed in Journel and Kyriakidis (2004) and Goovaerts (1997) among others.

Typically, simulations models cannot capture all possible sources of uncertainty that exist in a resource model. In this sense, they are incomplete descriptions of the space of uncertainty, and thus it is relevant to discuss how appropriate the conditional simulation model is with respect to the problem at hand.

A practical consequence of the dependence on a model is that the simulation method should be simulated from the same RF model used to obtain the resource model. It is important that they both share the same basic assumptions and implementation parameters; otherwise, the base case resources could be different and the models incompatible.

Sources of Uncertainty Resource models will include uncertainty from many sources. There are several factors that contribute to the overall uncertainty, and they do not necessarily cancel each other out. The sample values themselves have a degree of uncertainty, partly coming from the intrinsic heterogeneity of the material being sampled; however, most sampling errors are due to the sampling process itself. Sampling theory deals with the development of procedures for minimizing sampling variances, although there will always be an error that cannot be fully eliminated. Sample collection, sample preparation, the chemical analysis itself, and the overall data handling are all sources of uncertainty.

The amount of drill hole information available depends on the geology and the project's development stage. Typically, when additional data is included in the model, the uncertainty will tend to decrease. Geologic models are also a major source of uncertainty. Based on sparse drilling, they are representations of mineralization controls but still carrying a degree of uncertainty stemming from mapping and logging; data handling; the interpretative process itself; and the development of the computerized model. Often, the geologic model's uncertainty has the most important impact on the resource model since it heavily conditions the estimated tonnages above cutoff (Fig. 12.1).

There is uncertainty related to the process of grade interpolation including data spacing, kriging method chosen, variogram model and kriging plan. In addition, a correct amount of dilution must be included in order to predict tonnages and grades available at the time of mining. The prediction of recoverable resources and reserves is another significant source of uncertainty for resource models.

The model of uncertainty can also change when different implementation parameters of the geostatistical models are used, as discussed in Chap. 11 and also Rossi (2003), among others. Seemingly minor decisions, such as whether a random path or a multiple grid search for simulating values is used, can impact the resulting uncertainty model. Other parameters typically considered are search radii, number of original data used, number of previously simulated data used, the number of simulations to be run, and the kriging method to be used, among others. One alternative is to assess the uncertainty related to implementation criteria by choosing bounds or "best" and "worst" cases, although the process is subjective and difficult to justify.

There is limited information with large, unsampled areas between data points. There is uncertainty in the statistical parameters such as the overall mean of the deposit. A model of parameter uncertainty is also subjective, but may lead to a more realistic assessment. Some possible approaches to quantify parameter uncertainty include using an analytical model, the conventional bootstrap method or the spatial bootstrap method.

Bootstrapping is a name generically applied to statistical resampling schemes that allow uncertainty in data statistical parameter to be assessed from the data used to calculate the same parameter in the first place. The basic procedure is to draw n values from the original data with replacement, calculate the statistic from the bootstrapped sample, and repeat a number of times to build up a distribution of uncertainty. It is assumed that the input distribution is representative of the overall distribution. If the drawing is done using Monte-Carlo simulation (MCS), then there is an additional assumption that the data are independent.

Assuming that the sample data are independent is not realistic when they are known to be correlated. The spatial bootstrap simulates at the data locations. The uncertainty generally decreases as the number of drawn values (n) increases. The spatial bootstrap requires a variogram for the data set, simulation, and then computation of the mean for each simulated set of data.

12.2 Assessment of Risk

An uncertainty model can be used to characterize risk. It is important to distinguish uncertainty and risk, since large uncertainties, in some cases, may not lead to significant risks. In other situations, small uncertainties may correspond to unacceptable risk.

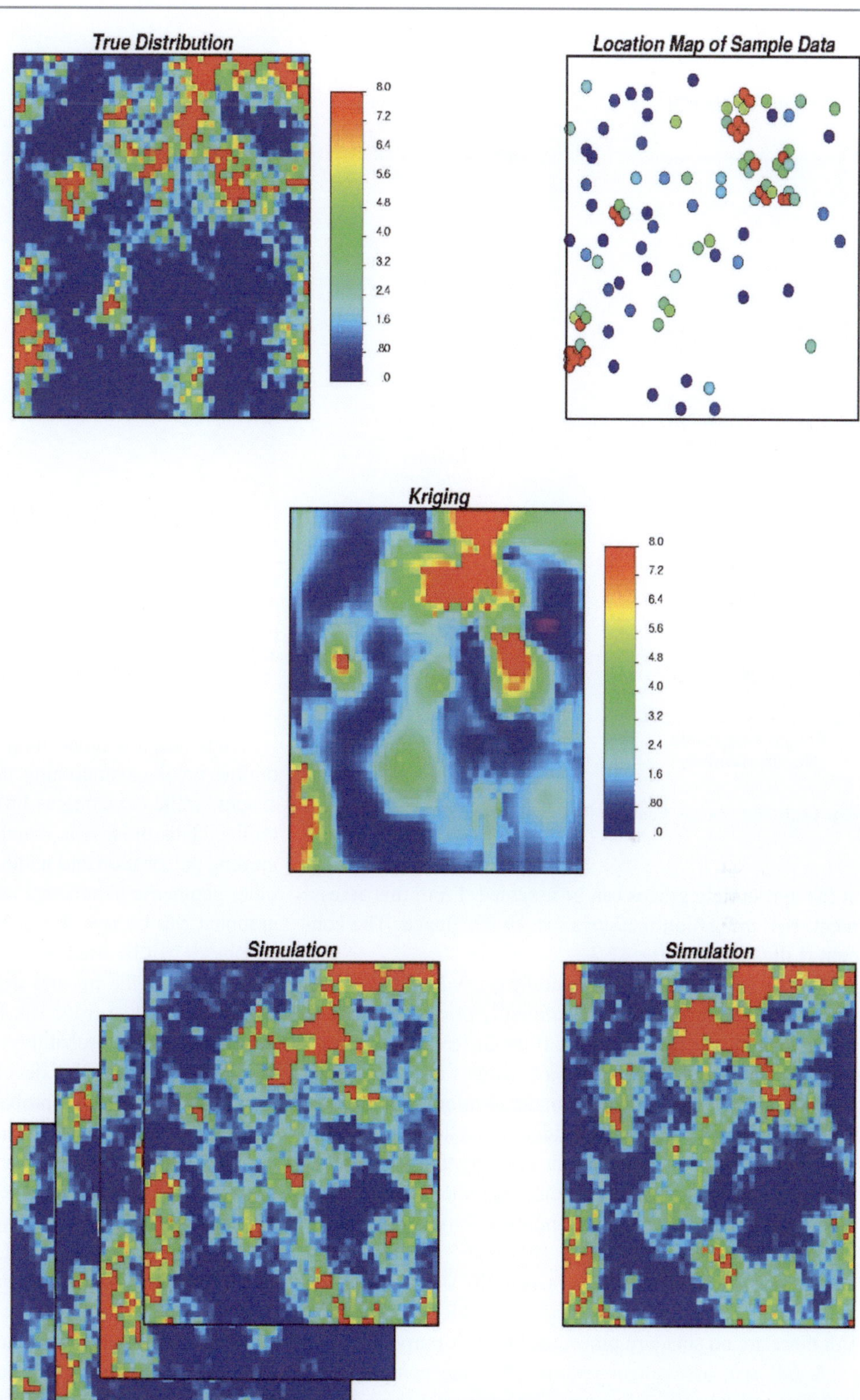

Fig. 12.1 Multiple realizations represent a model of uncertainty of the original variable, while an estimated map does not have an attached uncertainty model

Risk considers the impact of uncertainty on the application being assessed. The concept is summarized as a "Transfer Function" (TF, Matheron 1976), which conceptualizes all processes required to obtain the final product. For example, the TF can represent a pit or a stope optimizer and a production or mine scheduler, used to define mineable reserves. If the uncertainty model is carried through the TF, then the risk of not delivering to the mill the expected number of tons

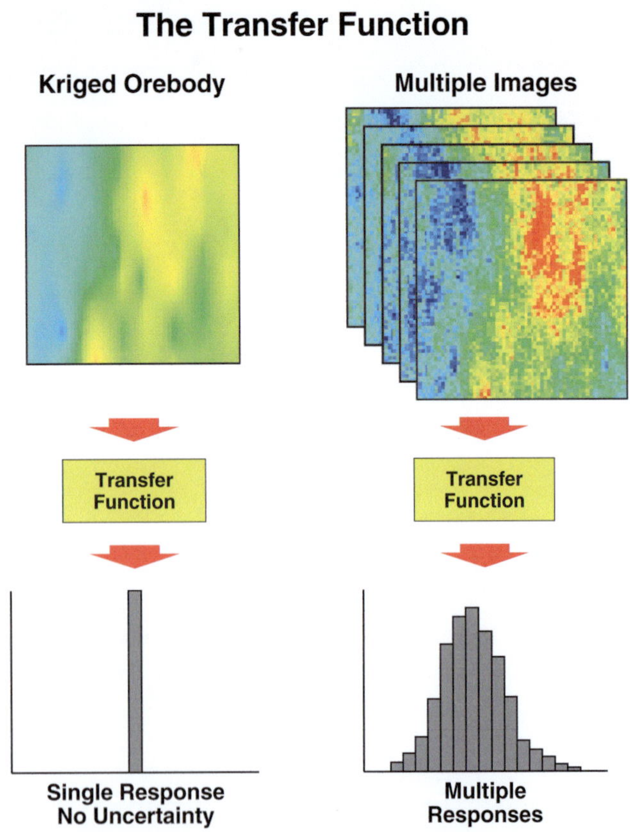

Fig. 12.2 The Transfer Function for estimated and simulated models

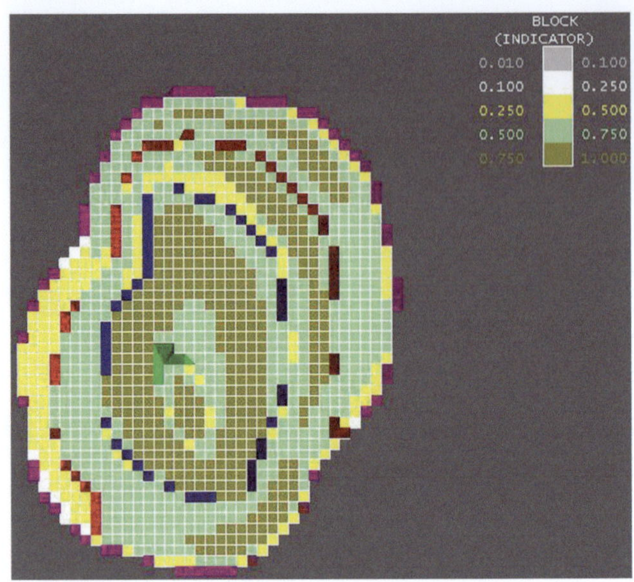

Fig. 12.3 Bench mining probability map. Blocks are coded by probability of being mined. *Magenta*, *blue*, and *maroon* colors indicate the position of the intermediate and final mining walls. (From Van Brunt and Rossi 1999)

at the appropriate grades can be assessed. From this assessment, risk mitigation measures can be developed. The concept is illustrated in Fig. 12.2.

Sensitivity analyses are commonly carried out by mining engineers. The impact of the commodity price or change in the estimated grades is assessed. If the different commodity price or grades result in significant changes in the designed pit walls, for example, then the material in question may be marginal. It is also important to identify areas of the pit that are extracted to contain few erratic or highly uncertain zones of mineralization. Engineers developing mine plans will usually consider simple sensitivity analyses, for example by adjusting the block model grades +10% and −10%. A similar approach is used to analyze the sensitivity of the project to metal prices, operating costs, and other relevant variables. But there are no standard procedures for this purpose.

A full risk assessment requires that the complete uncertainty model (all realizations) be processed through the transfer function; this may involve a full mine planning exercise, including scheduling of ore through the mill for certain periods of the mine life (Jewbali and Dimitrakopoulos 2009). In practice, certain shortcuts are possible, such as processing only the best, worst, and most likely scenarios. These shortcuts have their own pitfalls, including the criteria to rank the realizations.

Producing a detailed mine design from an optimized pit outline involves smoothing the outlines to provide minable shapes, while deviating as little as possible from the optimal outline. This process is manual, and the decisions made regarding the location and width of accesses, ramps, berms, and other geometric parameters required to make the mining operational can be significant. Probability maps by bench and by phases can be used as guides during the final smoothing and design of the pit and definition of the ramp positions. Figure 12.3 (taken from Van Brunt and Rossi 1999) shows a bench map of the probability of each block being mined according to the mine plan developed from the resource model. Developing conditional probability maps such as the one in Fig. 12.3 gives the mine planning engineer an advantage over conventional planning. Risks resulting from highly variable mineralization can be mitigated through the addition of intermediate phases and modifying the position of the pit walls. Also, these maps can be used to target additional infill drilling.

Grade control is an application where risk analysis is used directly to make an economic decision. In this case, the consequences of grade uncertainty are directly evaluated and the optimal choice is made based on the maximum profit or minimum loss choice.

The decision to recover and send to the mill or not a certain panel in the open pit is typically based directly or indirectly on grade estimates, $z^*(\underline{x})$. The loss function $L(e)$ (Journel 1988; Isaaks 1990; Rossi 1999) is a mathematical expression that attaches an economical value (impact or loss) to each possible error, measured in, for example, dollars. By applying a loss function to the conditional probability distribution (Eq. 12.1) derived from the realizations, the expected loss can be found by:

$$E\left\{L(z^* - Z)|(n)\right\} = \int_{-\infty}^{+\infty} L(z^* - z)\cdot dF(z;x\,|\,(n))$$
$$\cong \frac{1}{N_{real}} \sum_{l=1}^{N_{real}} L(z^* - z_l) \qquad (12.2)$$

where N_{real} is the number of realizations and z^* is the retained estimate.

The minimum expected loss can then be found by simply calculating the conditional expected loss for all possible values of the estimates z^*, and retaining the estimate that minimizes the expected loss. As explained in Isaaks (1990), the expected conditional loss is a step function whose value depends on the assumed costs of each bad decision, and the relative costs of misclassification. This implies that the expected conditional loss depends only on the classification of the estimate $z^*(\underline{u})$, not on the estimated value itself, as long as all benefits and costs are constant with respect to grade.

The Loss Function thus quantifies the consequences of false positives and false negatives, weighs the probability and relative impact of each, and then provides the minimum cost solution under the loss model used. For example, the loss incurred when an ore grade panel is sent to the waste dump is a type of lost opportunity cost, measured by the profit that should have been realized. If the same panel is waste, but is sent to the mill, the loss is a combination of the loss incurred in processing material that does not produce the metal to pay for itself, plus the loss derived from the opportunity lost in processing payable material, if any.

Loss functions are in general asymmetrical, since the consequences of under- or overestimation have different costs. In metal mining, where small volumes of ore may have high value, it is typically costlier to send ore to the waste dump than to process waste. Precious and most base metals mines have this characteristic, which is more notable if high economic cutoffs are used. There are other cases where the opposite is true, such as high volume, direct-shipping iron ore mines, who prefer to avoid dilution in the shipment.

Optimal estimates can be derived for a Loss Function if the conditional distribution of the random variable is available. The uncertainty model as described by the realizations provides all the information required to optimize decision-making under uncertainty.

When assessing uncertainty and risk it is also important to consider the scale of interest, i.e., the volume of material being assessed. There are differences between a global, deposit-wide geologic confidence assessment and a more local, mine production-oriented risk assessment. A global confidence measure cannot be used for local, block-by-block risk assessments. A typical example is the resource classification scheme, often used by mining engineers as a measure of confidence on mine schedules, for example on a monthly basis. Resource classification, as discussed below, is generally meant to be a global guideline of confidence, meant mostly for the benefit of shareholders and investors, and should not be used as an uncertainty model to provide a detailed risk assessment of the mine schedule.

Figures 12.4 and 12.5 illustrate how risk may change as a function of the volumes considered. Figure 12.4 shows the monthly probability intervals of Cu grades for an operating copper mine. The graph shows the two values that correspond to the P_{90} (90th percentile) and P_{10} (10th percentile) of the conditional distribution derived from the conditional simulations. It also shows the resource model grade for the same period, as well as the Mine Plan grade, which is generally a lower value than the resource model grade. This is because the mine planner sometimes adds dilution and a safety factor to the grade predicted by the resource model, typically on a monthly basis, not block by block. Mine planners may consider the monthly average grade provided by the resource model as risky, thus penalizing in some fashion the estimate. But the practice is variable and no standard methodology exists. It is dependent on the experience and prejudices of the engineer that defines the budgeted grade.

Figure 12.5 shows a similar graph for yearly periods of a 5-Year Mine Plan. Note that Year 1 in Fig. 12.5 is obtained by simply averaging the grades of the 12 months shown in Fig. 12.4.

Note how Fig. 12.4 shows much more variability than Fig. 12.5. As expected, the smaller volumes represented by the 12 months in Fig. 12.4 are more variable than grades averaged over a yearly volume (Year 1, Fig. 12.5). Also, it is interesting to note that the grades predicted by the resource model and the mine plan do no necessarily fall within the interval defined by the P_{90} and P_{10} limits. This occurs both for monthly and yearly volumes, and more so when considering periods further away in time. This is to be expected, since periods further away in time are likely to have less drilling and thus be more uncertain.

The risk of not achieving the predicted production for each period can be mitigated through further infill drilling. The infill drilling can be directed to those areas with higher uncertainty. A global confidence measure as used on most resource classification schemes would not allow optimization of the infill drilling to that level of detail.

12.3 Resource Classification and Reporting Standards

Public disclosure of estimated resources requires that resource estimates be classified according to degrees of confidence and allocated as measured, indicated and inferred. Reserves must be classified as either proven or probable reserves, derived under certain rules from resource categories. Different resource classification standards are used in differ-

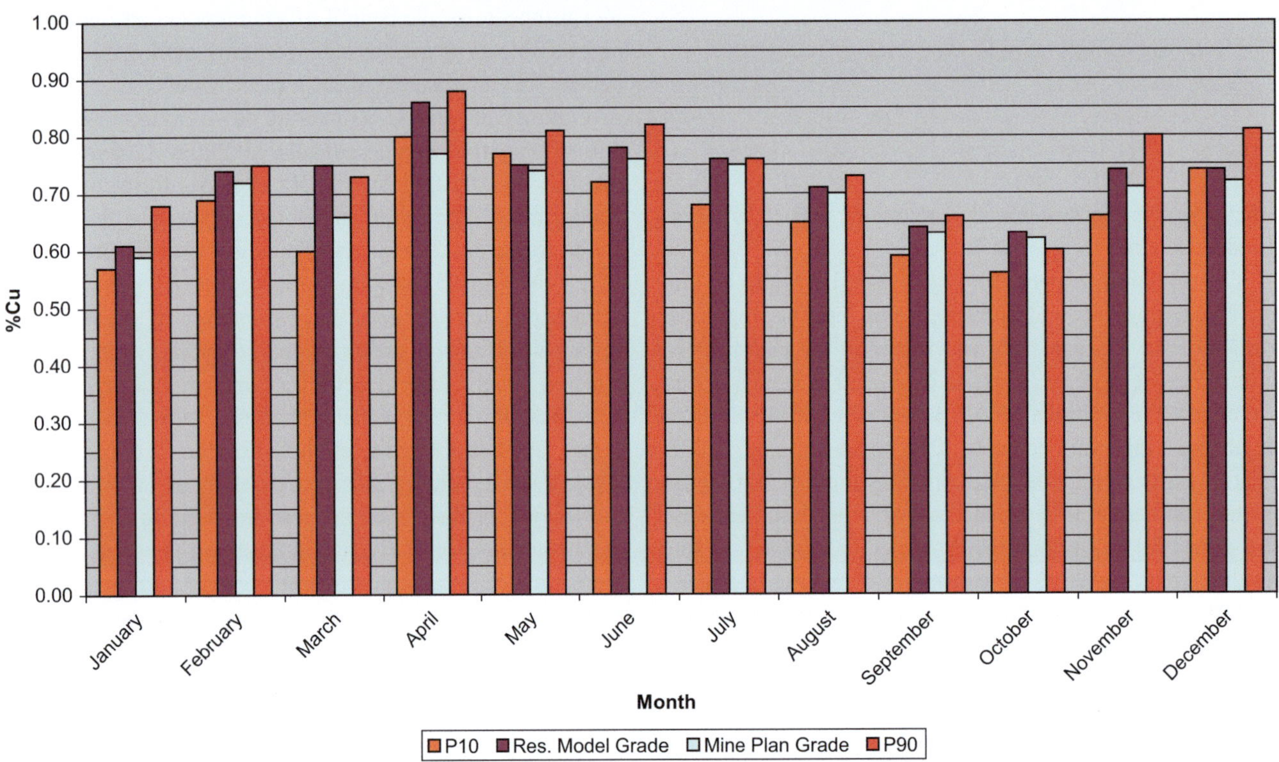

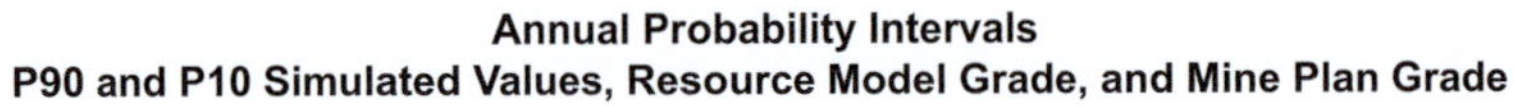

Fig. 12.4 Monthly Probability Intervals for year 1: P_{90} and P_{10} from Simulations, Resource Model Grade, and Mine Plan Grade

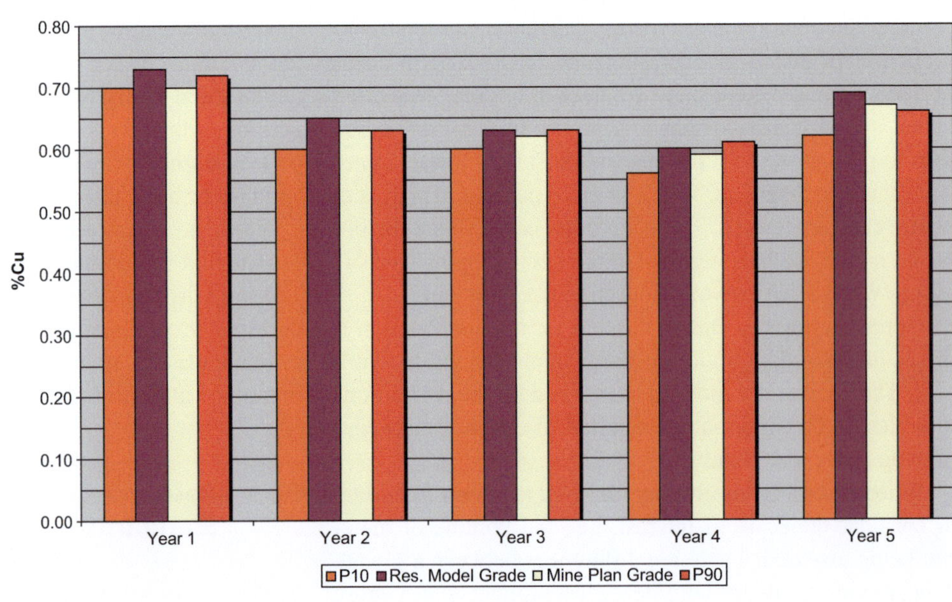

Fig. 12.5 Annual Probability Intervals for 5-year Mine Plan: P_{90} and P_{10} from Simulations, Resource Model Grade, and Mine Plan Grade

ent countries; while fairly similar in intent and form, each has its own particularities. The resource classification schemes are mostly intended to provide protection to the investor, and so are typically enforced by Securities Commissions or other appropriate government agency in each country.

Resource classification guidelines have been developed mostly as a response to the need for transparency in the disclosure of mineral resources. As such, resource classification is not necessarily a technical issue, but rather a self-regulated response of the mining industry for conveying investment risk, and also as a response to some notorious fraud cases. The codes have been developed according to specific needs for each jurisdiction, although all have a general common thread that makes them similar in spirit and in the application of its main concepts. Given the global nature of the mining industry, this commonality has led for a long-standing effort towards internationalization of the codes, unifying some of the details of application, to define a set of worldwide accepted set of definitions, namely the International Standards (Miskelly 2003).

Although the most commonly used codes have attached guidelines to them, they are non-prescriptive in all that relates to technical issues. Thus, the responsibility for the appropriateness of the disclosure is left to the technical competency of the individual(s) signing off on the resource calculations and classification, defined as the Competent or Qualified Person (CP or QP). In this context, the published Guidelines that accompany the different Codes are used to set minimum standards for practice, and are not intended to be used as enforcement tools.

The most widely used codes are the Joint Ore Reserves Committee (JORC, www.jorc.org); the CIM guidelines used in *National Instrument 43-101: Standards of Disclosure for Mineral Projects* (NI43-101) in Canada (www.cim.org); the Securities and Exchange Commision's Industry Guide 7 in the United States (www.sec.gov/about/forms/industryguides.pdf); the SAMREC code in South Africa (www.saimm.co.za/samrec.asp); and the Pan-European Union and United Kingdom's Reporting Code (www.crirsco.com/PERC_REPORTING_CODE_jan2009.pdf).

The JORC code has received broad international acceptance. In Canada, most Provincial Securities Commissions and the Toronto Stock Exchange (TSE) have adopted NI 43-101, which applies to all oral statements and written disclosure of scientific or technical information, including disclosure of a mineral resource or mineral reserve. NI 43-101 defers to the Canadian Institute of Mining, Metallurgy and Petroleum (CIM) for definitions and guidelines. The Council of Mining and Metallurgical Institutes (CMMI), of which CIM is a member, have developed a Resource/Reserve classification, definition and reporting system that is also widely accepted.

In recent years there has been an increased emphasis on the concept of a qualified (QP) or competent (CP) person. The professionals preparing resource models and statements are required to be experts in the field and also in the type of deposit being modeled. Typical requirements are that the individual(s) be members in good standing of recognized professional associations, which includes having approved a State or Provincial-sponsored professional exam, and have no less than 5 years experience modeling the same type of mineral deposits.

As an example, the 2010 CIM guidelines adopted in the National Instrument 43-101 of Canada allows classifying mineralization or other natural material of economic interest as a Measured Mineral Resource by the Qualified Person when the nature, quality, quantity and distribution of data are such that the tonnage and grade of the mineralization can be estimated to within close limits and that variation from the estimate would not significantly affect potential economic viability. This category requires a high level of confidence in, and understanding of, the geology and controls of the mineral deposit.

Mineralization may be classified as an Indicated Mineral Resource by the Qualified Person when the nature, quality, quantity and distribution of data are such as to allow confident interpretation of the geological framework and to reasonably assume the continuity of mineralization. The Qualified Person must recognize the importance of the Indicated Mineral Resource category to the advancement of the feasibility of the project. An Indicated Mineral Resource estimate is of sufficient quality to support a Preliminary Feasibility Study which can serve as the basis for major development decisions.

Mineralization is classified as Inferred Mineral Resource if the quantity and grade or quality can be reasonably assumed, but not necessarily verified. Due to the uncertainty that may be attached to Inferred Mineral Resources, it cannot be assumed that all or any part of an Inferred Mineral Resource will be upgraded to an Indicated or Measured Mineral Resource as a result of continued exploration. Confidence in the estimate is insufficient to allow the meaningful application of technical and economic parameters or to enable an evaluation of economic viability worthy of public disclosure. Inferred Mineral Resources must be excluded from estimates forming the basis of feasibility or other economic studies.

A Mineral Reserve is the economically mineable part of a Measured or Indicated Mineral Resource demonstrated by at least a Preliminary Feasibility Study. This Study must include adequate information on mining, processing, metallurgical, economic and other relevant factors that demonstrate, at the time of reporting, that economic extraction can be justified. A Mineral Reserve includes diluting materials and allowances for losses that may occur when the material is mined.

A Proven Mineral Reserve is the economically mineable part of a Measured Mineral Resource demonstrated by at least a Preliminary Feasibility Study. This Study must include adequate information on mining, processing, metallurgical, economic, and other relevant factors that demonstrate, at the time of reporting, that economic extraction is justified.

A Probable Mineral Reserve is the economically mineable part of an Indicated, and in some circumstances a Measured Mineral Resource demonstrated by at least a Preliminary Feasibility Study. This Study must include adequate information on mining, processing, metallurgical, economic, and other relevant factors that demonstrate, at the time of reporting, that economic extraction can be justified.

Reporting Codes and corresponding Guidelines use vague language in its definitions, as it is difficult to provide a general Guideline applicable to all different types of mineral deposits and resource estimation practices. There is a general tendency to suggest the use of some form of statistical description of uncertainty, if only as an accompanying tool that would clarify the degree of uncertainty.

All guidelines discuss geologic and grade continuity as key components of the classification criteria, sometimes adding modifying factors to adjust to local conditions. It is the QP's decision as to what an acceptable evidence of that continuity is, which may be partly dependent on the QP's prior experience with that type of deposits. In practice, resource classification is often reduced to deciding the criteria to be applied, including continuity, and then finding a method to classify the resources that best captures that basic criteria. A common misconception is that resource classification methods provide an objective assessment of confidence; in fact, the classification is an expression of a QP's opinion.

A common practice is to use some form of distance of drill holes to the estimated blocks. The choice of geometric criteria should be based on common practice for the deposit type, site-specific considerations and an expert judgment of other factors. The benefits of using simple distance measures are that the criteria can be simply stated, it is a transparent and easy-to-understand process, and leaves little room for mischief. Also, it does not depend on the estimation method chosen. Some of the most common concerns stated against these types of methods are that they are overly simplistic, as they fail to fully capture geologic confidence.

Geometric methods for classification generally do not give an actual measure of uncertainty, and if so, only for very large volumes, as with the kriging variance. There is an increasing interest in quantifying uncertainty at different volumes (block by block, if possible), which leads to relevant risk assessments.

Other alternatives encountered in practice include kriging variances, commonly applied early on in geostatistical resource estimation (Blackwell 1998; Diehl and David 1982; Froidevaux 1982; Royle 1977); a combination of distances to drill holes (in a certain pattern); the number of drill holes used to estimate each block; multiple-pass kriging estimation plans to account for density of information and other geologic factors; and possible combinations of these, as well as hand-contouring and smoothing, usually as a post-processing step to any of the above.

There has been a move toward systematic and standard methods to evaluate and present uncertainty (Dohm 2005). Common aspects of uncertainty reporting include specification of the population or sample being considered, measure

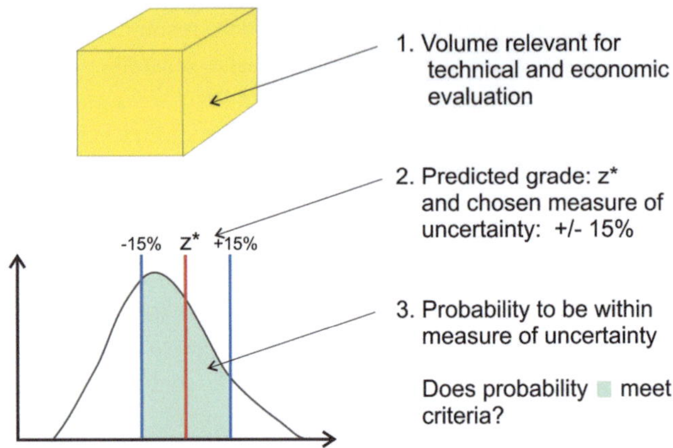

Fig. 12.6 Schematic illustration of the three parameters often used in probabilistic classification schemes: (1) volume related to a production period, (2) precision, and (3) probability to be within the specified precision

of the "+/-" uncertainty, probability to be withing the "+/-" measure of uncertainty, and a list of assumptions and components of uncertainty. There are three aspects to consider in resource classification. They are volume, measure of "+/-" uncertainty, and probability to be within the "+/-" measure of uncertainty. The format for uncertainty reporting is clear and understandable. For example, H.M. Parker (personal comminucation) proposes to classify as measured resources those monthly production volumes for which the true grade is predicted to be within 15 % of the estimated grade 90 % of the time. Quarterly production volumes where the true grade will be within 15 % of the predicted grade 90 % of the time are defined as indicated. There are no established rules or guidelines to decide on these three parameters; this remains in the hands of the qualified person.

Figure 12.6 highlights the three parameters often used in probabilistic classification schemes: (1) volume related to a production period, typically a month or a quarter, (2) the required precision, and (3) the probability to be within the specified precision. The volume need not be a contiguous block, but for simplicity it is often chosen as a simple volume. This can be a significant limitation, because production for any given period will generally come from different areas of the mine, areas that will likely present different geological characteristics, and have been estimated with uneven uncertainty. The second two parameters summarize uncertainty, which can be understood as proportions over a defined population. The probabilistic statement that *there is a 90 % probability that the grade of a monthly production volume be within 15 % of the estimated grade* means that 90 out of 100 true grades of similarly classified monthly production volumes will be within their estimate plus or minus 15 %.

Another alternative is to fix the volume of interest, for example a quarter's production, and then decrease the number of times the true value is expected to fall within the intervals, as shown in the schematic of Fig. 12.7. In this figure measured

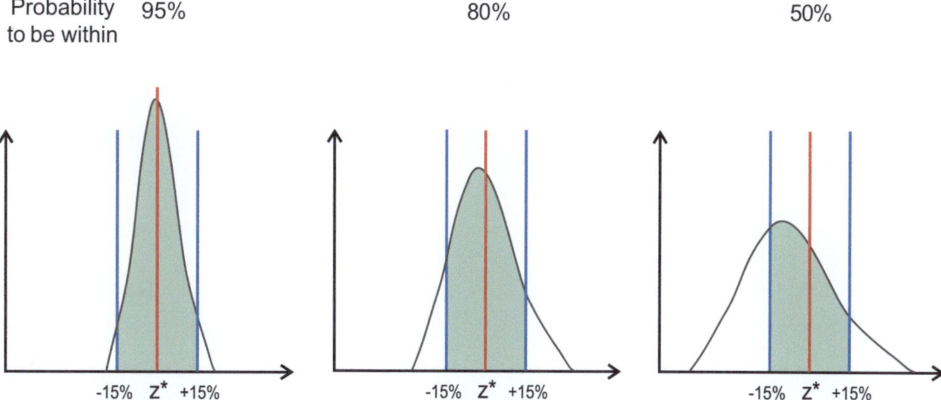

Fig. 12.7 Probability intervals for classification

resources are those for which the expected monthly production is within ±15% of the true value 95% of the time. Indicated resources are those for which the condition is relaxed to 80% of the time, while Inferred only requires that 50% of the time (or production months) the true value be within ±15%.

Uncertainty predictions can be from geostatistical or more traditional methods. If geostatistical procedures are used to construct probability distributions of uncertainty the parameters vary locally and within domains. There are a number of techniques that can be used, but conditional simulation is the best option, since the uncertainty of any parameter of interest can be predicted at different scales by simply averaging up the simulated values.

The uncertainty model can be checked by predicting the uncertainty at locations where there is information from drillholes or past production data. The probability intervals are constructed, counting the number of times that the true values fall within those intervals, thus determining if the predicted percentage is verified.

In any resource estimation work, the purpose of classifying the estimated resources should be clearly stated, and also a clear distinction between geologic confidence (i.e., resource classification) and mining risk assessment should be made. It is tempting to use resource categories as a means to obtain a mine production risk assessment, although they are intended for geologic confidence assessment in a very global sense.

There is no consistent scheme for resource classification for all deposits, although certain common practices can be identified.

12.3.1 Resource Classification based on Drill Hole Distances

Multiple variants of this concept have been used, but in its most simple form the resource is classified based on the distance from the centroid of the estimated block to be to the nearest sample used in the interpolation. Estimated blocks that have close samples nearby will have a higher confidence assigned to them. This is considered a very simplistic method.

Another alternative is to otain the average weighted distance of all samples used to estimate the block. This distance could be anisotropic, following the variogram model ellipsoid and/or the shape of the search neighborhood. It may appear as a reasonable option since *all* samples used in the estimation are considered. This could potentially avoid artifacts related to assigning high confidence to a block estimated with one very close sample and many others much further away. But there are drawbacks with this system, again related to the lack of uncertainty measures and the simple criteria used.

The actual classification of the resources should depend on the distances chosen to characterize confidence, which in turn should be based on geology, drilling density and variogram ranges. Commonly, different estimation domains will have different classification parameters applied to them. Also, a minimum number of samples and drilling density measures are sometimes used, as well as differences in the geologic characteristics in different areas of the deposit.

12.3.2 Resource Classification Based on Kriging Variances

The kriging variance is an index of data configuration. As such, it can be used to rank the resource model blocks based on how much information is used to estimate each block. It can be standardized, for example, to a local mean, such that the resulting relative kriging variance can be used across different grade mineralization zones.

The values for kriging variances that define resource categories are usually related to a pre-specified drill hole configuration, as exemplified in Fig. 12.8. This is an example taken from a porphyry copper deposit in northern Chile. After obtaining a variogram model for each of the three main copper mineralization types present in the deposit, two standard drill hole configurations were used as references to determine resource categories. The kriging variance values for the 5-composite configuration (Case B) defines the limit between measured and indicated for each mineralization type, while

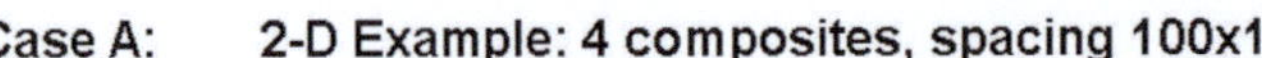

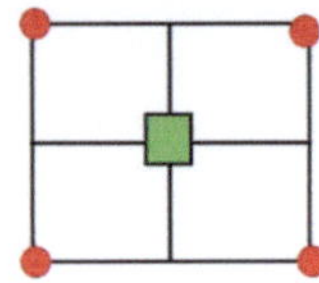

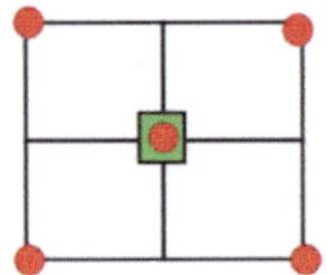

Fig. 12.8 Schematic example of resource classification through kriging variances for reference drill hole configurations

the corresponding kriging variances for the 4-composite configuration (Case A) define the limit between the indicated and the inferred categories. Note that the kriging variances are always used as relative thresholds, since the values themselves do not have any physical or geological meaning.

Other alternatives for defining resource categories can include visual inspection of the kriging variances, although rarely there will be a clear break or indication of kriging variances that can be related to resource classes. Therefore, it is highly dependent on the subjective criteria to define the thresholds for each category. Because of this, the method can be considered equivalent to the distance to the drill hole-based methods, just developed with in a more formal geostatistical framework.

12.3.3 Resource Classification Based on Multiple-Pass Kriging Plans

Another option is to derive the resource classification from multiple kriging passes. Several kriging iterations are done to estimate the model grades using different levels of restrictions, that is, from a more to a less constrained kriging.

The constraints are defined in terms of requisites for an estimate to occur; in the more constrained case, a higher minimum number of samples combined with a larger minimum number of drill holes, and shorter search radii may be used. A smaller number of blocks will be estimated in the more constrained pass, but they will be better informed than blocks estimated in later estimation passes. If the estimation passes are set based on geologic and geostatistical criteria, a flag for each block indicating in which pass it was estimated could be used as an initial indicator for resource classification.

12.3.4 Resource Classification Based on Uncertainty Models

Conditional simulation provides realizations that provide models of uncertainty in a global as well as local sense. These realizations are applicable to both resource classification and mine production risk analysis; however, the use of realizations from which probability intervals can be obtained and used for resource classification is not yet widespread. The resource classification codes, beginning with the JORC code, encourage quantification of uncertainty whenever possible, but they do not mandate it, nor do the corresponding Guidelines suggest specific methodology for such quantification.

Deutsch et al. (2006) argue that the uncertainty models derived from conditional simulations should only be used as a backup to other more simple, geometric methods, such as drill hole distance. Several reasons are given in the paper for this recommendation mostly because the probability intervals are shown to be sensitive to the definition of some of the parameters used to obtain them, as well as the overall model dependency. The uncertainty model is dependent on the specifics of the implementation parameters used in the simulations (Rossi 2003).

Probabilities can be checked using actual proportions, and, whenever possible, this check should be made. Operating mines will generally maintain sufficiently good production records to be able to check actual production tonnages and grades. If the modeled uncertainty can be verified by actual production, then there are several good reasons to rely on the uncertainty model for resource classification: (1) the magnitude of the grades and the local configuration of data are accounted for, (2) the mining volume is explicitly accounted for, and (3) uncertainty is perceived as more objective and transportable to different deposits.

The probability used to define measured, indicated, and inferred resources depends on the mining company's practice. Many will simplistically translate the kind of precision required of other engineering studies and cost estimates during pre-feasibility or feasibility studies into resource classification. Typically, a measured resource would be a quarter known within ±15 %, 90 % of the time; an indicated resource, within ±30 %, 90 % of the time; and inferred, within ±30 % and ±100 %, 90 % of the time. Material known within more than ±100 % will not qualify as resource, and may be flagged (but not publicly reported) as blue sky or potential mineralization.

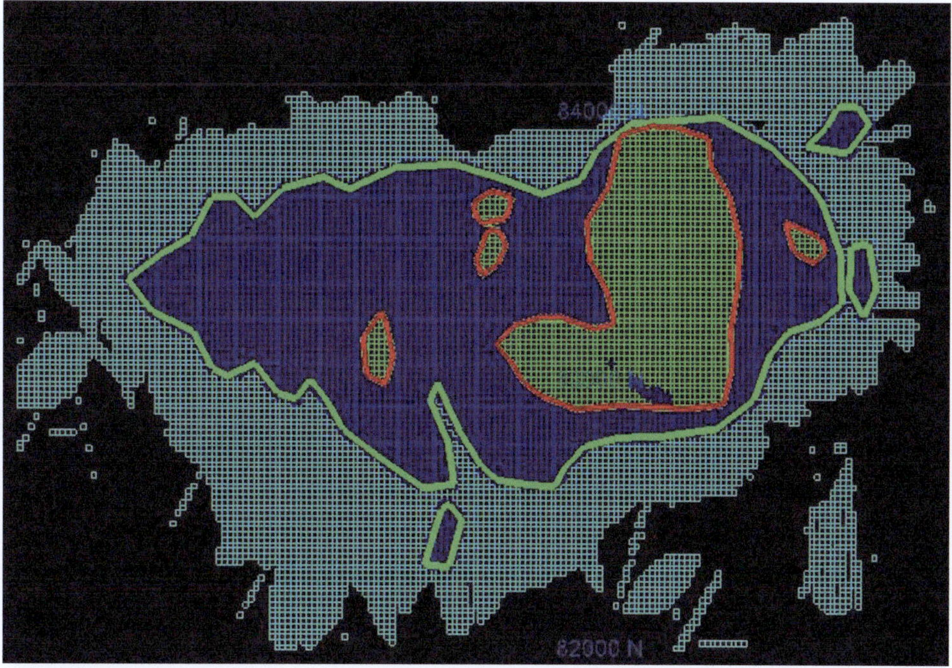

Fig. 12.9 Resource classification contours, Bench 2440m, Cerro Colorado 2003 Resource Model, Northern Chile. *Red* encloses measured material, *green* outline encloses indicated material. Courtesy of BHP Billiton

12.3.5 Smoothing and Manual Interpretation of Resource Classes

Since resource classification is usually performed on a block by block basis, most of the non-probabilisitic methods mentioned above will generally require a posterior smoothing of the resulting volumes, mostly because of the common accepted idea that the classified material should be fairly homogeneous, without intermixing of resource classes over short distances.

This is mostly an aesthetic issue, since classification schemes are meant to provide global indicators of confidence, and not necessarily smooth block-to-block images. Any of the methods described above will likely produce volumes for each resource class that are consistent with the criteria used to specify them. It is common to see in areas with heterogeneous drill hole spacings, variable geologic characteristics and abrupt transitions between the resource classes.

If smooth and contiguous volumes are desired, then manually interpreting the zones, based on the initial definition, is probably one of the most practical means to achieving this. Alternatives could include running a smoothing algorithm that would transform, based on windows of certain sizes, the resource classification of the blocks within to produce more homogeneous volumes. In any case, this should be done with care, not to bias or significantly alter the global volumes defined by the criteria established. There should only be minor corrections for consistency and what may be deemed inconsistent classification classes based on geologic or geostatistical knowledge. It is good practice to check the overall grade-tonnage curves by resource class before and after the smoothing process, to understand the degree of changes introduced.

Figure 12.9 shows an example of smoothing through hand-contouring done at Cerro Colorado, BHP Billiton's porphyry copper operation in Northern Chile. The smoothing was done by interpretating on benches and smoothing out the edges and, in some cases, the intermixing of resource classes. The red outline defines the measured volume, the bright green outline the indicated volume, and the remaining material is classified as inferred. Note how some of the material originally classified as indicated is inside the red outline (central-East portion of the bench), and thus finally classified as measured. Also, there is a small area in this bench to the Northeast of the picture where measured runs directly into inferred, due to a change in the geologic environment.

12.4 Summary of Minimum, Good and Best Practices

Minimum practice for the development of uncertainty models requires the application of simple and more traditional statistical techniques. The scope of application of these models is relatively small, and can only be attached to large volumes. The two most common examples include Resource Classification (for all the methods described, with the exception of conditional simulations), and global confidence intervals derived from the variance of averages for large volumes. Risk assessments are thus limited, and normally qualitative.

Good practice requires, in addition to the above, the development of conditional simulation to obtain realizations of an uncertainty model. This model should be reasonably comprehensive, in the sense of including as many sources of uncertainty as possible, but principally geologic and grade estimation uncertainties. Within these, issues related to dilution should be emphasized, as well as an assessment of the information effect. The resulting model of uncertainty should be checked against actual production, if available, or against some resource model taken as reference or base case. Risk assessments should be fully developed, validated, and documented, with clearly stated objectives.

Best practice consists of, in addition to the above, full modeling of all recognized and quantifiable uncertainties, including those attached to the data, to the sampling and assaying procedures, to the geologic model and simulation domain definition (as above), and the modeling of grade. Conditional simulations should thus be used to provide both global and local uncertainty measures, and a full description of the resource model. However, the exclusive use of probabilities for resource classification is not recommended. An arbitrary choice of probabilistic criteria will often lead to unreasonably large or small volumes in each category. It is however advisable to apply geometric criteria for resource classification, with or without smoothing out the zones with mixing of resource classes, and provide further support through a probabilistic analysis. The probabilistic analysis may cause the competent person to reconsider their geometric criteria, but the geometric criteria are used for disclosure.

If, however, the possibility exists of reliably validating the uncertainty model obtained from the conditional simulations through mine production, then it is reasonable to use the probabilistic intervals as basic definition for resource classification.

12.5 Exercises

The objective of this exercise is to review aspects of uncertainty and risk assessment together with loss functions and decision making. Some specific (geo)statistical software may be required. The functionality may be available in different public domain or commercial software. Please acquire the required software before beginning the exercise. The data files are available for download from the author's website—a search engine will reveal the location.

12.5.1 Part One: Sampling Uncertainty

The objective of this exercise is to experiment with different uncertainty sampling and sensitivity assessment approaches. Available methods for these two purposes can vary greatly depending on whether one is interested in sampling efficiency and/or realistic uncertainty assessment accounting for dependency structures. The set of tools we will explore in this exercise applies different methods that satisfy these two features in varying degrees. Consider a simple calculation of oil in place (OIP) that depends only on a few input parameters:

$$OIP = 6.2898 * GRV * \varphi * (1 - Sw) / FVF$$

where GRV is the gross rock volume, φ is the porosity, Sw is the water saturation, and FVF is the formation volume factor. The constant 6.2898 is a metric conversion factor to relate cubic metres to stock tank barrels. Suppose that each of the input variables can be described as a random variable: All variables are normally distributed with the following mean and variance values:

Variable	Mean	Variance
GRV	79 million cubic meters	5 million cubic meters
ϕ	17%	$5\%^2$
S_W	11%	$9\%^2$
FVF	1.3	0.2

Question 1: Using Monte Carlo simulation, draw 100 realizations for each input parameter and then calculate the corresponding OIP for each realization. Plot the distribution of uncertainty about OIP.

Question 2: Consider now partitioning each of the input distributions into ten different partitions (you can set the thresholds at the deciles). Apply latin hypercube sampling (LHS) and calculate OIP (you should only need to draw 10 realizations for each input and ensure that you only draw from each partition once). Plot and comment on this distribution of OIP.

Question 3: Suppose now that there is a relationship between φ and Sw, which can be described as bivariate Gaussian with correlation of 0.5. Given that there is no longer independence between all the input variables, describe how you would implement a Monte Carlo approach (similar to Question 1) to account for the impact this relationship has on uncertainty in OIP. If you have time, you may wish to implement this and compare against the distribution in Question 1.

Question 4: Perhaps the most common approach to sensitivity analysis is the vary one at a time approach. This requires keeping all the input variables at the base case value (usually the mean), and then for one input variable, choose say the p10 and p90 of that input variable and evaluate the impact on OIP. Plot this impact as a tornado chart by ordering the input variables in descending order of impact.

Question 5: Consider now varying each input variable (keep all other variables at the base case) by changing its value by ±5 % increments from the base case value until say ±20 %. For each case evaluate the change in OIP, and plot this as a spidergram.

Question 6: Rather than changing each input variable by a percentage difference from the base case, change each input by a set of percentages. For this, consider evaluating OIP as you change an input variable based on its deciles. Now plot this result in a similar format to a spidergram, and comment on any differences you notice from the spidergram in the previous question.

12.5.2 Part Two: Loss Functions

The consequences of over and under estimation are often not the same. The two common loss functions, however, are symmetric.

Question 1: Prove that the mean of a distribution always minimizes the mean squared error loss function, that is, a loss function where the loss increases as a square of the error for both over and under estimation.

Question 2: Prove that the median of a distribution always minimizes the mean absolute error loss function, that is, a loss function where the loss increases as the absolute value of the error for both over and under estimation.

Question 3: The L-optimal value is a specific quantile of the distribution of the penalty for over and under estimation is both linear with different slopes. The 0.5 quantile or median is optimal if the slopes are the same. What is the quantile for arbitrary (different) slopes for over and under estimation?

References

Arik A (1999) An alternative approach to resource classification. In: APCOM proceedings of the 1999 Computer Applications in the Mineral Industries (APCOM) symposium, Colorado School of Mines, Colorado, pp 45–53

Blackwell GH (1998) Relative kriging errors—a basis for mineral resource classification. Explor Min Geol 7(1,2):99–106

Davis BM (1997) Some methods for producing interval estimates for global and local resources. SME Annual Meeting, SME 97(5), Denver

Deutsch CV, Leuangthong O, Ortiz J (2006). A case for geometric criteria in resources and reserves classification. Centre for Computational Geostatistics, Report 7, University of Alberta, Edmonton

Diehl P, David M (1982) Classification of ore reserves/resources based on geostatistical methods. CIM Bull 75(838):127–136

Dohm C (2005) Quantifiable mineral resource classification—a logical approach. In: Leuangthong O, Deutsch CV (eds) Geostatistics Banff 2004, 1:333–342. Kluwer Academic, Dordrecht, p

Froidevaux R (1982) Geostatistics and ore reserve classification. CIM Bull 75(843):77–83

Goovaerts P (1997) Geostatistics for natural resources evaluation. Oxford University Press, New York, p 483

Isaaks EH (1990) The application of Monte Carlo methods to the analysis of spatially correlated data. PhD Thesis, Stanford University, p 213

Jewbali A, Dimitrakopoulos R (2009) Stochastic mine planning: example and value from integrating long- and short-term mine planning through simulated grade control. In: Orebody modelling and strategic mine planning 2009, Perth, pp 327–334

Journel AG (1988) Fundamentals of geostatistics in five lessons. Stanford Center for Reservoir Forecasting, Stanford University, Stanford

Journel AG, Kyriakidis P (2004) Evaluation of mineral reserves: a simulation approach, Oxford University Press, New York

Matheron G (1976) Forecasting block grade distributions: the transfer function. In: Guarascio M, David M, Huijbregts C (eds) Advanced geostatistics in the mining industry. Reidel, Dordrecht, pp. 237–251

Miskelly N (2003) Progress on international standards for reporting of mineral resources and reserves. In: Conference on resource reporting standards, Reston, 3 October

Rossi ME (1999) Optimizing grade control: a detailed case study. In: Proceedings of the 101st annual meeting of the Canadian Institute of Mining, Metallurgy, and Petroleum (CIM), Calgary, 2–5 May

Rossi ME (2003) Practical aspects of large-ccale conditional simulations. In: Proceedings of the 31st international symposium on applications of Computers and Operations Research in the Mineral Industries (APCOM), Cape Town, 14–16 May

Royle AG (1977) How to use geostatistics for ore reserve classification. World Min 30:52–56

Van Brunt BH, Rossi ME, (1999) Mine planning under uncertainty constraints. In: Proceedings of the optimizing with Whittle 1999 conference, Perth, 22–25 March

13 Short-term Models

Abstract

Most mineral resource estimates are not final. They are interim estimates modified by more information as it becomes available. At the time of actual mining, or just before mining, the nature and requirements of estimation is different. Results that are accurate over a longer time scale are no longer sufficient. This Chapter explains considerations for short and medium term mine plan models.

13.1 Limitations of Long-term Models for Short-term Planning

Resource models are said to be long-term when they are used for long term mine planning, such as Life of Mine (LOM) plans. When a feasibility study is prepared for a new mining project, a mining schedule needs to be prepared to estimate future cash flows from the operation. The LOM plan is based on a reserve model, in turn converted from the resource model. It provides estimates of tonnage and grade for each period involved through to the end of the life of the mine. Often, the LOM plan is scheduled according to variable units of time. For example, it may be that for the first two years of the operation, the schedule is monthly; the following two years it may be based on semi-annual volumes; and from the fifth year until the end of the mine life it may be yearly.

Long-term models are based on widely-spaced drilling, which is gradually filled in as the project advances. The long-term models are usually updated on a yearly basis with information gathered from new drill holes. More accurate forecasts in the short term are often needed as well. It is tempting to use the existing long-term resource model for shorter term predictions. However, because of the dynamics of the operation, the long-term model quickly becomes outdated.

Long-term models are by construction designed to provide global estimates with acceptable accuracy. Global estimates are understood to correspond to volumes equivalent to a year or longer. Therefore, it cannot be expected to perform as well on a block by block basis, or even for a small volume. Sometimes reasonable accuracy is obtained from long-term models for volumes smaller than a year, particularly for disseminated-type deposits, deposits with very simple geology, and grade variables that do not exhibit high spatial variability. In the case of a new operation, the long-term model will generally be based on relatively tight drill hole spacing (infill) covering the initial years of operation, designed to accurately estimate the payback period.

Updating the long-term model is required virtually in all mine operations for several reasons. The most important reason is the need to improve accuracy for Medium- and Short-term mine plans. These plans would correspond, for example, to yearly budgets and quarterly forecasts of mine production and corresponding cash flows.

For month to month mine planning, the model's reliability is increased through infill drilling. The additional drilling will result in improved accuracy of the resource model for the near-term mine operation. Updating the long-term model with the new data and subsequently updating the corresponding mine plans results in less uncertainty about the operation's short-term cash flow.

The definition of "medium" and "short" term models varies widely from one mining company to the next, and also from one geographic area to the next. In many cases, a "short-term" model is in fact a grade control model, the daily ore/waste selection process. In this book, a medium-term model will be any model that is meant to provide estimates on much smaller volumes than the long-term resource model, and is also short-lived. It generally means a volume equivalent to one to six months production, although it depends on the type of mining performed. The models developed for daily ore/waste selection and weekly mine plans are always called here short-term or grade control models.

M. E. Rossi, C. V. Deutsch, *Mineral Resource Estimation,* DOI 10.1007/978-1-4020-5717-5_13,

13.2 Medium- and Short-term Modeling

Updating the long-term resource model using in-fill data implies repeating many of the steps described in previous Chapters. This is regardless of whether the task involves estimation of values, estimation of distributions, or simulations. However, some special considerations are required, particularly if production information is used.

One of the most difficult aspects of updating short-term models is updating the geologic model and estimation domains using production data. In practice, face, bench, or stope mapping from underground workings and a description of blast hole cuttings or production drill holes may be available, but seldom used. This is partly due to data quality, and also to the tight timeframe involved.

The grade model can be updated using both infill drill hole data and production data. The use of blast holes can be controversial for several reasons, including perceived sampling quality, and discrepancies of its grade distribution compared to the exploration drill holes grade distribution. Despite the difference in quality of the individual samples (drill holes vs. blast holes), often the much larger number of blast holes available compensate for the poorer precision of the individual sample. The key to using blast hole samples is that there should be no significant bias.

A different issue is the estimation strategy. The implementation of any estimation method should consider the possibility of blast holes overwhelming the infill drill holes in certain areas; thus, an adequate estimation strategy should carefully consider how blast holes are used.

In all cases, the medium- or short-term block model should be updated only for the relevant portion of the deposit, for example, corresponding to the next three months of production. An example is given here of a medium-term model prepared for the Escondida copper mine in Northern Chile and courtesy of BHP Billiton. It illustrates a practical application of the process.

13.2.1 Example: Quarterly Reserve Model, Escondida Mine

At Minera Escondida in early 2002, medium-term planning was required on 13-week intervals, since this was the forecast period used, and updated on a monthly basis. Therefore, the quarterly planning cycle was in fact a monthly moving-window that represented the planned mined volumes three months at a time. In order to develop a practical methodology and demonstrate the usefulness of updating the long-term resource model, an initial study was developed that consisted of the following:

1. Develop a Sequential Gaussian conditional simulation model and comprising the volume corresponding to the previous year of production, FY01, (July 1, 2000–June 30, 2001) was prepared. The simulation grid was $1 \times 1 \times 15$ m, and was used as a reference to compare the alternative models and methodology developed. The simulation model not only honored the histogram and variogram models of the conditioning data, but reflected actual production figures. The simulated variables were Total Copper (TCu), Sulfuric acid-soluble Copper (SCu), Arsenic (As) and Total Iron (Fe). The conditional simulation model is not described here in detail, as it was only used as a reference.
2. The volume to be mined in the following quarter was defined, and a reserve block model is created within it. The blocks can be the same size as the long-term resource model blocks, or smaller if the additional infill and/or blast hole data justifies it. In the case of the initial study, for each month of the FY01 period, a quarterly model was defined based on actual mined out volumes.
3. The geologic model is updated monthly using information from bench and face mapping, as well as blast hole cuttings. For example, when completing the quarterly model for the month of January, the planned mining volumes corresponding to the months of February through April are considered, and the geologic information available up to December 31 is used.
4. The grade models (TCu, SCu, As, and Fe) were updated using infill drill holes and blast holes through the previous month. The same methodology as used for the long-term resource model is applied, except that smaller blocks sizes were used as warranted by the additional drill holes available. The long-term block model is $25 \times 25 \times 15$ m, while the quarterly model is based on $12.5 \times 12.5 \times 15$ m blocks; therefore, within each block of the long-term model there are 4 blocks of the quarterly model. It is always convenient to define the quarterly block model in a manner consistent with the geometry of the long-term model, such that comparisons can be easily made.
5. The quarterly models are compared with the long-term resource model and with the reference simulation model to quantify the improvements obtained. In the case of the routine, operational procedure, the comparison is done against the monthly reconciliation figures for the prior months, such that a closer control of the long- and the medium-term models is maintained.

The long-term resource model historically underestimated mine production, particularly in-situ TCu grade. The resource estimation methodology was partly to blame, but even after improving the estimation methodology, the resource model still had a small TCu deficiency. This deficiency was traced to a lower-than-expected TCu grade in the exploration drill holes, mostly those drilled using conventional rotary techniques, but also present in reverse circulation holes and in, to a lesser degree, existing diamond drill holes.

The under-representation of TCu grades in the drill holes was explained by the loss of high grade chalcocite (copper sulfide), sometimes present in non-crystalline form, and easy to wash away during the drilling process. Shorter infill drill holes were less likely to loose such material, and so were the blast holes, because of their larger diameter, large numbers, and also awareness of the problem. To improve the short-term grade and tonnage estimates, it was important to incorporate the most recent production information and local geologic mapping.

Another important requisite is that the quarterly model be obtained in a short time, hopefully in two or three days of work, and without requiring significant additional resources other than those already available. An additional requirement is the company's goal: to obtain a model with ±5% accuracy on a monthly basis for both copper grades and tonnages above economic cutoff.

The database used for the study and quarterly model updates is the same 15 m composites database used to estimate the long-term resource model. This included the more recent infill holes, and also the addition of the current blast hole database. The blast holes represent the grade of a full 15 m bench.

13.2.2 Updating the Geologic Model

Since the production geology (bench, face, and blast hole cuttings mapping) was done by a different group of geologists than those that map the exploration and infill drill holes, a prior step of consolidating and homogenizing nomenclatures and coding was required.

The lithology, alteration, and mineralization type models were updated from the existing geologic model (used to estimate the long-term resource model) only within the volume corresponding to the next three months of production. An additional area surrounding this volume was also re-modeled to allow the "tie-in" of the long-term geologic model with the more detailed Short-term model. The updating of the geologic model was done by modifying the existing interpretation from the long-term resource model on plan view. The polygons were adjusted bench by bench, from which three-dimensional solids were created. It is not necessary to apply the same level of detail as for the long-term model (see Chap. 3), since the update is an adjustment of a prior interpretation. If unexpected geologic features are encountered, then it would be necessary to review the original geologic interpretation.

Figure 13.1 shows an example of the resulting Total copper (TCu) estimation domains for Bench 2845. The larger blocks are the long-term resource model blocks, the smaller ones correspond to the same definition of estimation domains, but after updating for the quarterly model. The area shown is the complete volume planned to be mined in this bench in the period considered. Note that there is generally good agreement between the two models of estimation domains, although there are differences near contacts.

The TCu, SCu, Fe and As grades were estimated using the same methodology as used in the long-term resource model, i.e., ordinary kriging, and using the same kriging plans. The data used was all data available, including blast holes. The estimation was done on three different estimation passes, which helped control the influence of each data type. Blast holes were used only in the first pass, using the smallest search neighborhood and more data restrictions before a block could be estimated. This restricted the influence of the more abundant blast hole data.

Figure 13.2 shows for the same area in bench 2845 the estimated TCu grades for both the long-term and quarterly block models. Grades are color-coded according to the legend shown. Note that there are some differences which are significant for short-term planning, and mostly near contacts. The differences are both gains and losses. The quarterly model better delineates areas of high and low grades. For example, observe at the northern tip of the area shown (North of coordinate 108,000N) where the quarterly model predicts a NW-trending higher-grade narrow structure higher than 3% TCu, and shown in orange. This high grade corridor was not predicted by the long-term resource model.

Overall, results from the medium-term model are as expected. The use of infill drilling and blast hole increases the grade and metal content of the reserve model, and also increases its variability. The local definition of geology and grade increases also the confidence level in the estimated values. The Quarterly model is less smoothed than the long-term model.

Figure 13.3 shows the comparison of the grade-tonnage curves of the long-term (LT) and quarterly (QT) models. Note how both models have very similar tonnages above cutoff, but the QT model presents slightly higher grades for most cutoffs. The cutoffs of interest are 0.7% TCu (direct mill feed) and 0.3% TCu (marginal stockpile).

Figure 13.4 shows the grades for the two models by bench averages, for the Quarterly period beginning February 2002. Note that most benches have very similar estimated grades, although there are some where the overall average is somewhat different. This is particularly the case for Bench 2845, the grades shown also in Fig. 13.2.

Figure 13.5 shows the comparison of the relative differences of monthly TCu grade averages of the LT and QT models for the three-month period beginning in February 2002. They are compared to the conditional simulation reference model, which was calibrated to production data. Negative errors imply underestimation of TCu grades for the month.

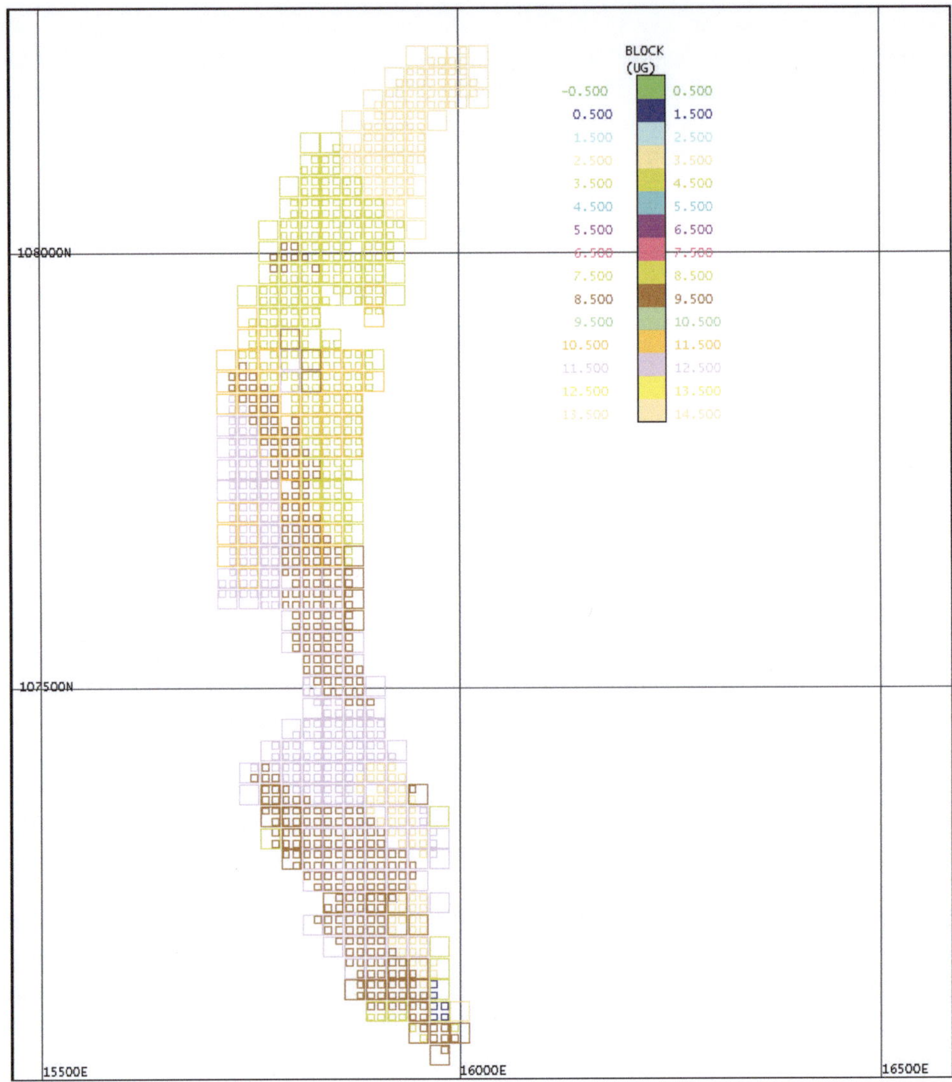

Fig. 13.1 TCu Estimation Domains for the long-term and quarterly models, Bench 2845, Escondida Mine

Note how the QT model monthly averages approximate much better the corresponding grades predicted by the reference model for most months. Although the reference model is only another model (based on a single conditional simulation), by construction represents well the production grades from previous periods. The QT model, based partly on blast holes, is also expected to be a better predictor of production grades.

13.3 Selection of Ore and Waste

The process of ore/waste selection at a mine, or grade control, whether underground or open pit, is the most geological important decision at the mine. The final, irreversible decision as to what is ore and what is waste is made. In open pit mines, the decision is generally made on a daily basis, and commonly based on sampled blast hole information. In the case of underground mines, the process may be based on infill drilling and completed at the time of defining the stopes to be mined (short-term mine planning) as usually the complete stope is classified as either ore or waste. Any mistakes that may occur at this decision point are not only irreversible, but also cannot be compensated by other types of errors, as is sometimes the case with resource estimation.

Grade control is key to the mine's profitability because the resource is finite, and the time of selection is the last opportunity that the mining company has to realize its expected revenue. It is also used to maximize resource recovery, or more frequently in the Western world to optimize recovered dollar value. Also, the processing plant usually works better when a constant grade is fed to it. Sometimes stockpiling is necessary to avoid fluctuating grades. There are four areas of interest in grade control: classification, cutoff grade, loss functions for grade control, and the consideration of non-free selection.

Classification is the process of deciding where to send the mined out material. A block is selected as ore if the revenue from processing it as ore exceeds the cost of mining it as waste. As discussed in Chap. 7, the calculation of cut-

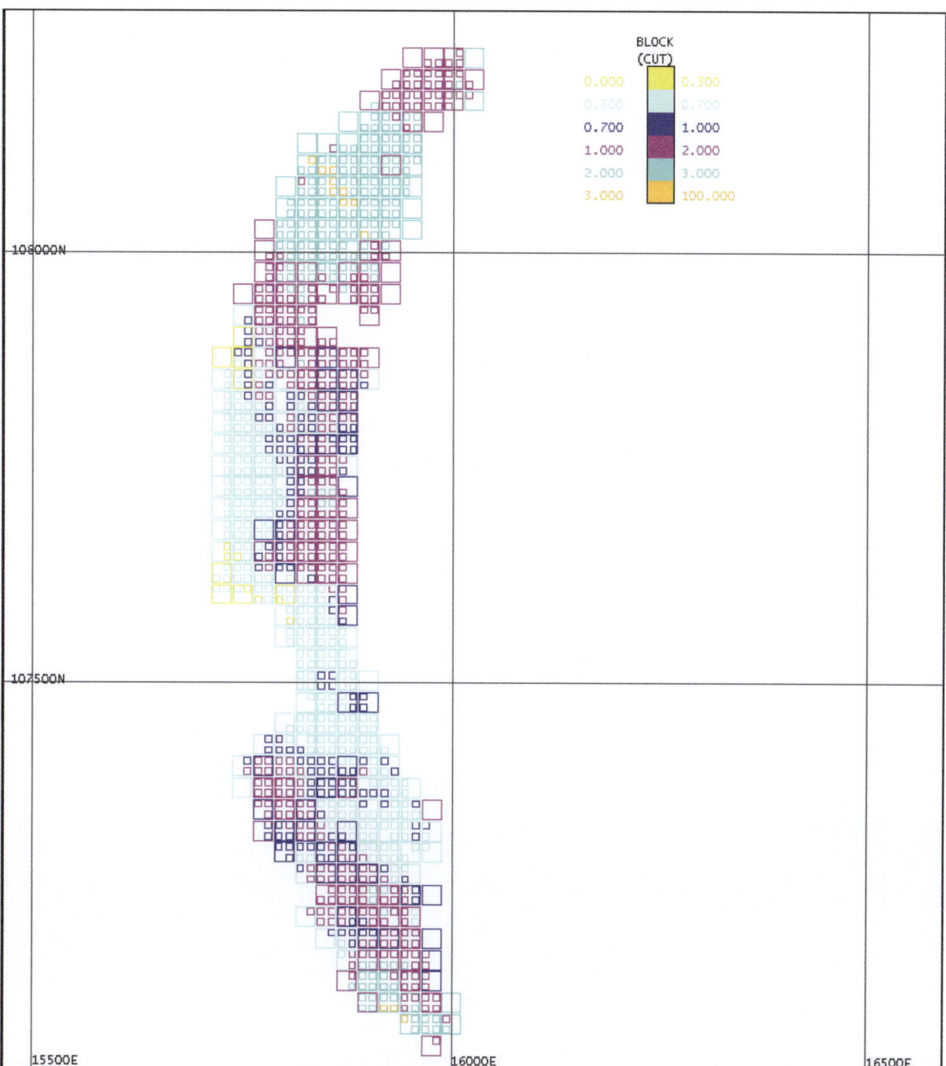

Fig. 13.2 TCu grades for the long-term and quarterly models, Bench 2845, Escondida Mine

off grades may be complex and site specific. Many different costs and variables may come into play. One possible definition of a processing (also called marginal or in-pit) cutoff grade is:

$$z_c = \frac{c_t + (c_o - c_w)}{pr} \quad \left[z_c = \frac{c_t}{pr} \right]$$

where c_t is the unit treatment (milling) cost; c_o is the unit ore mining cost; c_w is the unit waste mining cost; r is the metal recovery factor; p is the unit metal price; and z_c is the grade that makes revenue nil. In this marginal cutoff equation, costs such as General and Administration (G&A) and mining costs are not considered, only the additional costs that may exist when mining ore as opposed to waste. This cutoff grade is applicable when the operation has already committed to moving the material. The only remaining decision is whether it is sent to the waste dumps, stockpiled, or processed.

Grade control attempts to minimize miss-classification. The basic issue is shown in Figure 13.6, where a scatterplot of unknown true values for each block are plotted against the corresponding estimated grades. The most important task in grade control is to avoid as much as possible sending material to the wrong destination.

Chapter 7.4 discussed the issue from the point of views of the Information Effect, including perfect and imperfect selection. In traditional geostatistical literature the term imperfect selection is used to signify that the decision is based on estimates of grade, and without the knowledge of the true values. Perfect selection is thus impossible, because we can never know the true in-situ grades.

Another consideration is that free selection is impossible. Ore and waste blocks cannot be selected independently of each other during mining. This causes dilution and ore loss. There are also other practical (operational) factors affecting the decision, including how exactly the ore/waste markers have been laid out in the extraction area; a certain amount of unavoidable dilution (unplanned operational dilution); and mistakes made at the time of extraction, including some as simple as sending the loaded truck to the wrong destination.

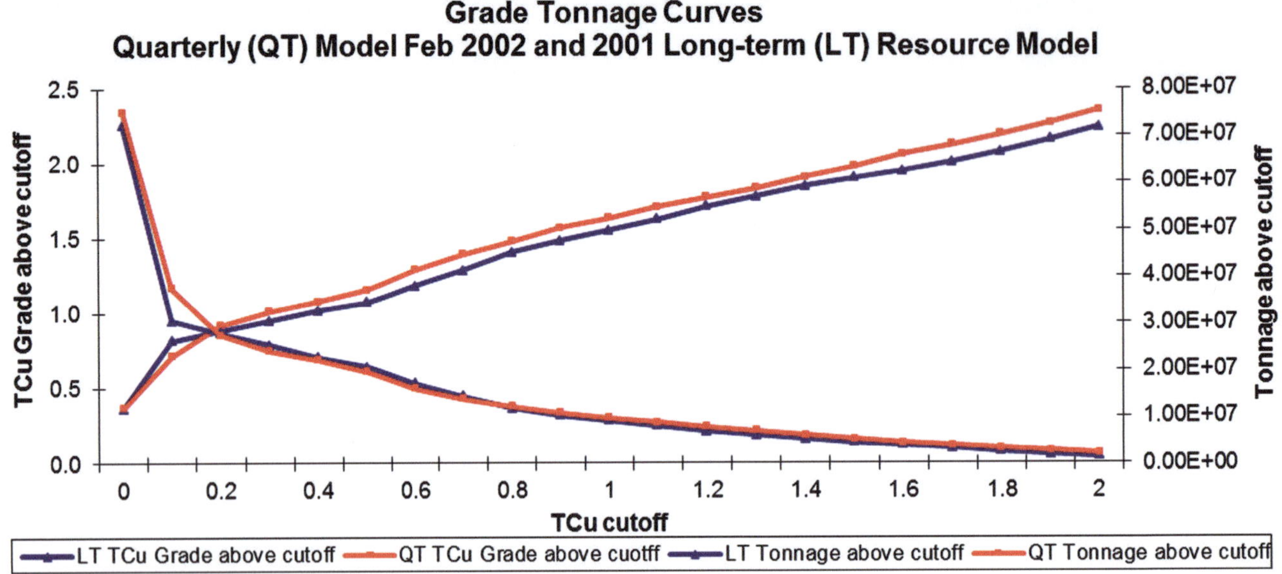

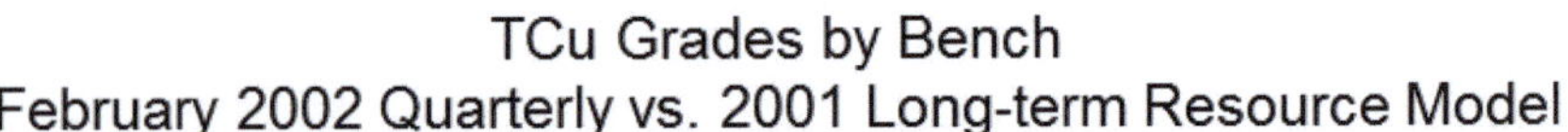

Fig. 13.3 Grade-Tonnage curves, 2001 Long-term (LT) and February 2002 Quarterly (QT) resource models. Note how the QT model has higher grade and less tonnage than the LT model for most cutoffs

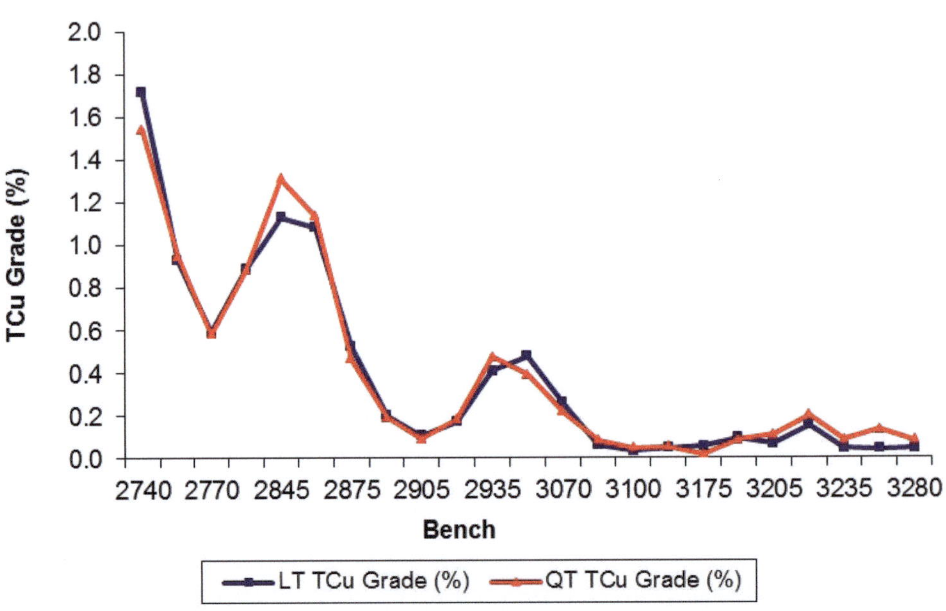

Fig. 13.4 Total copper grades by Bench, 2001 Long-term (LT) and February 2002 Quarterly (QT) resource models

In general, sampling errors, estimation errors, limited information, and operational constraints result always in ore loss and waste dilution, which in turn leads to economic losses. These losses can be serious enough to make the operation unprofitable.

One example was the Hartley platinum mine in Zimbabwe, which produced its first concentrate in 1997 and closed in 1999 after what were deemed to be insoluble geologic problems and low mine productivity (Matthey 2001). Hartley is located within the Great Dyke, a geological feature running roughly north-south through the heart of Zimbabwe for about 550 km. The platinum group minerals occur in a layer known as the Main Sulphide Zone, which is typically about 3 m thick. However, the economic mining width may be as little as 1 m, depending on grade, metal prices and the chosen mining method. The reef is difficult to mine because it is not visible to the naked eye. This can lead to significant unplanned dilution and ore loss, which reduces head grades.

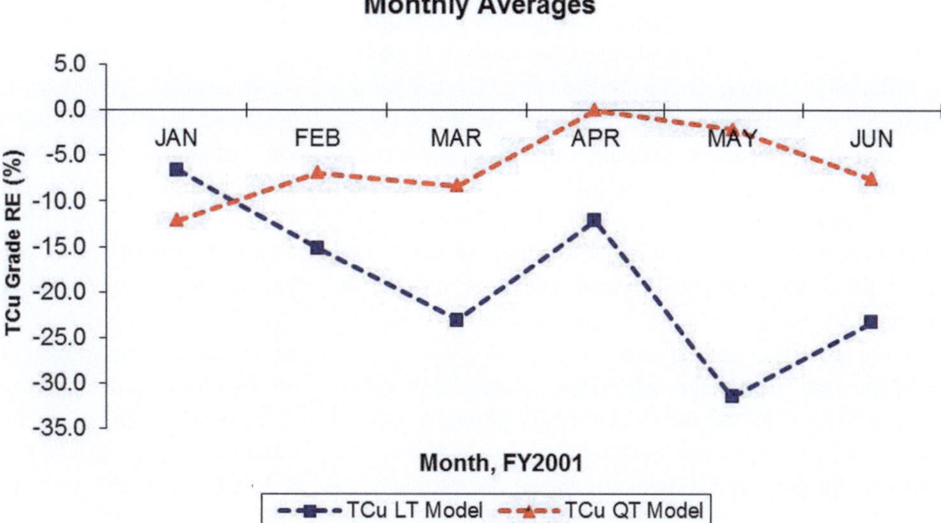

Fig. 13.5 Relative errors, Long-term (LT) and Quarterly (QT) models vs. Reference model calibrated with production data

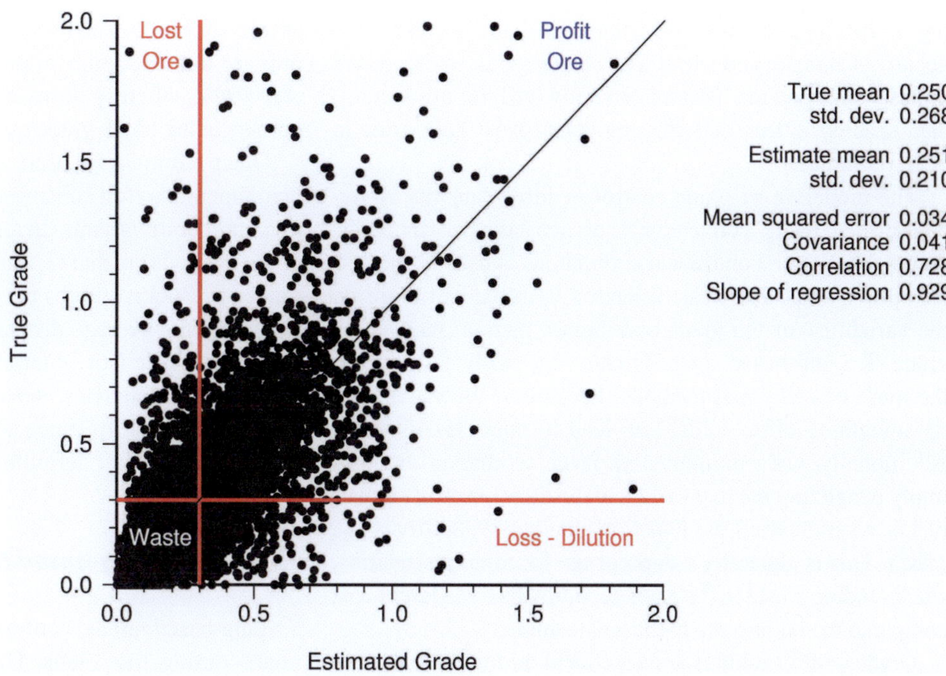

Fig. 13.6 Miss-classification in grade control

Grade control methods should attempt to minimize all possible sources of error, and not just the error prediction of the in-situ grade. Grade control should always be viewed as a complex process in which at least three basic aspects must be considered: data collection and quality; grade control model to determine ore and waste boundaries; and operational procedures and constraints, including mining methods, mining practices, and operational culture.

Firstly, data collection and data quality are always important, but it becomes even more critical when operational constraints limit the time and availability of sampling crews. Thus, the quality of the samples used to make the decision is impacted. Secondly, the samples are modeled to provide a prediction of grades, block dollar values, and other important attributes. The actual selection of ore and waste is based in those estimates. And third, all related operational procedures should be considered and controlled. The grade control method should consider the type and limitations of available sample data, the geotechnical and blasting conditions, and also the operational constraints that may render certain grade control practices non feasible.

Data collection and quality is highly dependent on the mining method, and to some extent the geometry of the orebody being mined. In open pit mines, blast holes are the most common source of data for grade control. Ocassionally, reverse circulation (RC) grade control drilling is done.

The additional cost of the dedicated RC drilling should be paid for by the increased economic benefit of the improved grade control, since almost always blast holes still need to be drilled for blasting. Grade control using RC drilling is a fairly common practice in gold mines in Western Australia and parts of Africa. It is generally applicable if the ore is of high intrinsic value (such as high grade Au) and if the higher-grade distribution is sub-vertical. Unfortunately, not all operations perform a detailed cost-benefit analysis of the use of RC drilling for grade control. The costs of using RC drilling may be higher than the economic benefits derived from the improved grade control.

In the case of underground mines, mining methods are much less flexible and therefore there is generally little or no opportunity for ore and waste selection at the time of extraction. When a stope is defined as being ore, typically the complete stope is considered ore (with the planned and unplanned dilution as encountered). This implies that the grade control data is actually the data used to design the stopes during short-term planning. In such case, infill drilling is used to decide what is ore and waste. The challenge for underground mines is thus greater, because generally infill (or production) data spacing is less than the equivalent blast hole grids in open pit mines.

The modeling of grade control or infill data can be accomplished using conventional or geostatistical methods. Among the latter, conditional simulations is usually the better option, since ore/waste selection is dependent more on the variability of the grade distribution than on its average grade. Kriging-based methods can very easily fail (as can the more conventional methods) because of its characteristic smoothing effect which can lead to miss-classification. Additionally, using minimum-variance estimation methods imply penalizing the over- and underestimation errors equally, i.e., a symmetric Loss Function (Journel 1988; Srivastava 1987). This is generally inappropriate for mining scenarios, since sending waste to the plant generally has a different cost compared to sending ore to the waste dump.

Grade control models are dependent on mining practices and methods. It is possible that more detailed and sophisticated grade control methods can provide a better ore/waste selection, but the mining method has to able to capitalize on that opportunity. It may be an overkill to develop and implement a sophisticated grade control method if the mining method and operational practices are not good enough to take advantage of the additional level of detail.

13.3.1 Conventional Grade Control Methods

Conventional methods used for grade control include blast hole averaging, inverse distance methods, and nearest-neighbor-based methods. For the mathematical description of the methods the reader is referred to Chap. 8. Here the more common industry practices are discussed.

Unfortunately, even after major technological advances in many aspects of grade control including geostatistical modeling, most operations still do not fully appreciate the importance of grade control, and devote insufficient resources and thought to this task. The flexibility that open pit mines generally enjoy is not always fully utilized. Many operations work with very simple methods that are not optimal. This is also true for underground mines. Indeed, it is more difficult to perform effective grade control in underground mines because of operational constraints, but still, too few operations have profited from modeling advances over the last 20 or 30 years.

In open pit mines, probably the most commonly used method to predict in-situ grades is a simple arithmetic average of the available blast holes. A block model is defined, generally with the block size similar to the blast hole spacing, and the predicted block grade is the arithmetic average of the blast holes that fall within the block. Multiple variants exist, as for example the "four-corner" average method, popular in some gold mines in Northern Nevada (Douglas et al.1994), whereby the average of the four blast holes at the corners is the block grade estimate.

Other commonly used methods include the nearest-neighbor method and inverse-distance methods, implemented in a number of variants. In all cases, the main characteristics of the methods are that (a) a simple estimator is used to assign grades to blocks, and (b) the blocks are relatively large with respect to the average distance between sample points. The second characteristic is unjustifiably common, and a major source of inaccuracies, since the data density is generally sufficient to justify much smaller blocks. Smaller blocks would lead to better definitions of ore and waste boundaries.

13.3.2 Kriging-based Methods

Kriging-based grade control became popular in open pit mines during the 1980s. Different types of kriging algorithms were used, but most commonly ordinary and indicator kriging were applied, for example in gold mines in Northern Nevada.

In the case of ordinary kriging, the application of the method is similar to those described as conventional methods above. Ordinary kriging is used to provide an estimate of grades, based on which the selection panels are drawn. The advantages of kriging over other estimation methods were discussed in Chap. 8 and include the minimization of the estimation variance. In practice, kriging has been only marginally more successful at grade control compared to conventional methods because of the inherent smoothing and the use of inadequate kriging plans. Also, the minimization of the estimation variance is not optimal for grade control (Srivastava 1987).

Multiple variants of the indicator kriging approach have been used. A common application considers a single indicator estimated at the ore/waste boundary of interest, thus providing the probability of any block or point within the blast being ore or waste. Generally point kriging is performed, usually at a larger-than necessary grid spacing. Occasionally, block kriging may be done, ignoring the fact that the average of estimated probabilities within a block is not the same as the point probability derived from the ore/waste indicator (Chap. 9). Nonetheless, the practice is to analyze equal-probability contour lines for several values and decide based on visual observations which one adjusts better to prior production. Commonly, in gold operations that use this method, probabilities of being ore of about 30–40 % are used to define ore/waste boundaries.

A method that has proven successful in several operations is the "Breakeven Indicator Method" (BEI), as described in Douglas et al. (1994). It was implemented first at Independence Mining Company's Jerritt Canyon, north of Elko, Nevada, in the early 1990s.

The BEI grade control method uses a combination of both indicator and grade kriging. An ore/waste indicator variable is used to predict the probability of ore occurrence at a given location $P_o(x)$, which is obtained by kriging the ore/waste indicator variable. The ore-grade blast holes are then used to krige an ore grade $Z_o(x)$ for the location x. Similarly, the waste-grade blast holes are used to krige a waste grade, $Z_w(x)$, for the same location. Then, the expected revenue is estimated from the kriged probability $P_o(x)$ and ore and waste grades:

$$E(R) = P_o \cdot R(Z_o) + (1 - P_o) \cdot R(Z_w) \qquad (13.1)$$

The revenue function is traditionally calculated as

$$R = (gold\ price) * (metallurgical\ recovery) * (grade) - (costs)$$

where "costs" generally imply metallurgical processing costs only. The method offers the flexibility of adding additional costs if desired, to work on what would amount to a higher ore/waste cutoff grade.

If the expected revenue from Eq. 13.1 is negative, the material at the location is waste. If the expected revenue is positive, the material at the location is ore. If the grade of ore is high, the corresponding revenue will be high, allowing for a block with a low probability of being ore to be sent to the mill. In this case, the ore pays for large amounts of waste, which ensures all high grade ore is recovered. Alternatively, if the ore grade is low, the revenue will tend to zero and the estimated probability of ore will have to be close to 1: the lower grade ore will not pay for much overbreak. Thus, the method requires that the low grade be most surely higher than the economic cutoff. This can be seen by calculating the probability that corresponds to the economic breakeven cutoff, $E(R) = 0$:

$$P_o(BE) = \frac{-R(Z_w)}{R(Z_o) - R(Z_w)} \qquad (13.2)$$

The method should be applied on small blocks, one third to one half of the blast hole spacing, allowing the grade control engineer to define dig lines based on revenues. The BEI method is designed to improve grade control performance most along contacts of ore/waste zones. If the panels to be mined are very large (wide), the ratio of contact surface area per ton of ore is small. The opposite is true for panels that are narrow for which this method would provide the most improvements.

If compared to the single indicator kriging method outlined before, the BEI is equivalent to working on a variable probability of being ore, which is dependent on the revenue function defined.

13.3.3 Example: Grade Control Study

A comparison of several grade control methods was performed for the copper-molybdenum Ujina open pit mine in Northern Chile. It is summarized here, courtesy of Compañía Minera Doña Inés de Collahuasi (CMDIC). The company mines a Cu-Mo porphyry deposit with a significant Cu enrichment blanket, which was the main target of mining at the time. As a massive, disseminated-type deposit, it could have been assumed that grade control is a simple process; however, there are factors that made grade control at Ujina a complex process.

The differences observed among the methods tested will be larger if the grade distributions being modeled are more variable. Also, if there are many different possible destinations for ore and waste, the grade control process is more complicated: the grade ranges that are used to separate the material become narrower. Table 13.1 shows the possible destinations for ore coming out of the Ujina pit at the end of 1999.

A quick inspection of Table 13.1 suggests that a large degree of accuracy and precision is required of the grade control method, since the mining method and metallurgical processing requirements are very specific.

The methods tested included the inverse distance cubed (ID^3) as used at the time by the mine; ordinary kriging (OK); the breakeven indicator method described above (BEI); and the maximum revenue method, based on conditional simulations and loss functions as described further below. Only a short summary of a long and detailed study is presented here

Table 13.1 Material type classifications as of December 1999, Ujina

Material Type	*Dispatch Code*	*Destination*	*Description*
High-grade sulfide	SAL	Stock 1	TCu >= 2.0%
Medium-grade sulfide	SME	Stock 2	1.0% =< TCu < 2.0%
Low-grade sulfide	SBA	Stock 5	0.8% =< TCu < 1.0%
Marginal-grade sulfide	SMR	Stock 4	0.4% =< TCu < 0.8%
Sub-marginal grade sulfide	SSM	Stock 6	0.2% =< TCu < 0.4%
High As sulfide	SAS	Stock 3	As > 100 ppm y TCu >= 1.0%
High-grade Oxides	OXA	Stock 10	TCu >= 1.0%
Medium-grade Oxides	OXM	Stock 12	0.6% =< TCu < 1.0%
Low-grade Oxides	OXB	Stock 11	0.3% =< TCu < 0.6%
Low-Oxi	OXL	Stock 30	TCu >= 0.2%, with clays and Fe oxides
Mixed	MIX	Stock 13	Mixed, TCu > 0.7%
Waste Rock Types	IGS, IGC, RIO, SUE, PLR, OTR	Waste dumps	Waste, TCu < 0.2%

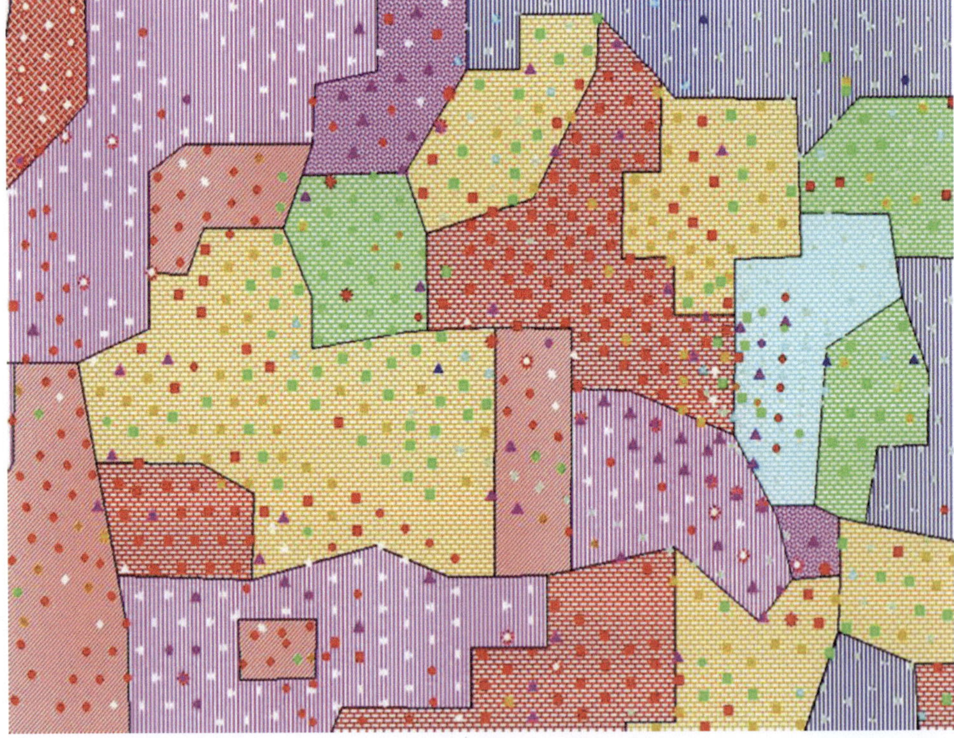

Fig. 13.7 Blast holes, color- and shape coded by destination, and grade control panels based on ID[3] interpolation. Blast hole spacing is approximately 8 × 8 m, and the area is 250 m per side. Blast holes and panel hatching represents Stocks 1 through 6 in Table 13.1

to illustrate the performance of different grade control methods, even in deposits with relatively low variability.

Figure 13.7 shows a small area of Bench 4270 with the Total Copper (TCu) blast hole grades and selection panels as defined by ID^3, which was the method used by the operation. Figure 13.8 shows the same area with panels as defined by the BEI method. And finally, Fig. 13.9 shows the comparison of the panels defined based on these two methods. In this area only sulfide material was present, corresponding to destinations (Stocks) 1 through 6 in Table 13.1. These figures demonstrate that, locally, the differences among the different grade control methods can be significant.

The comparison among the four methods tested was made against a reference model corresponding approximately to

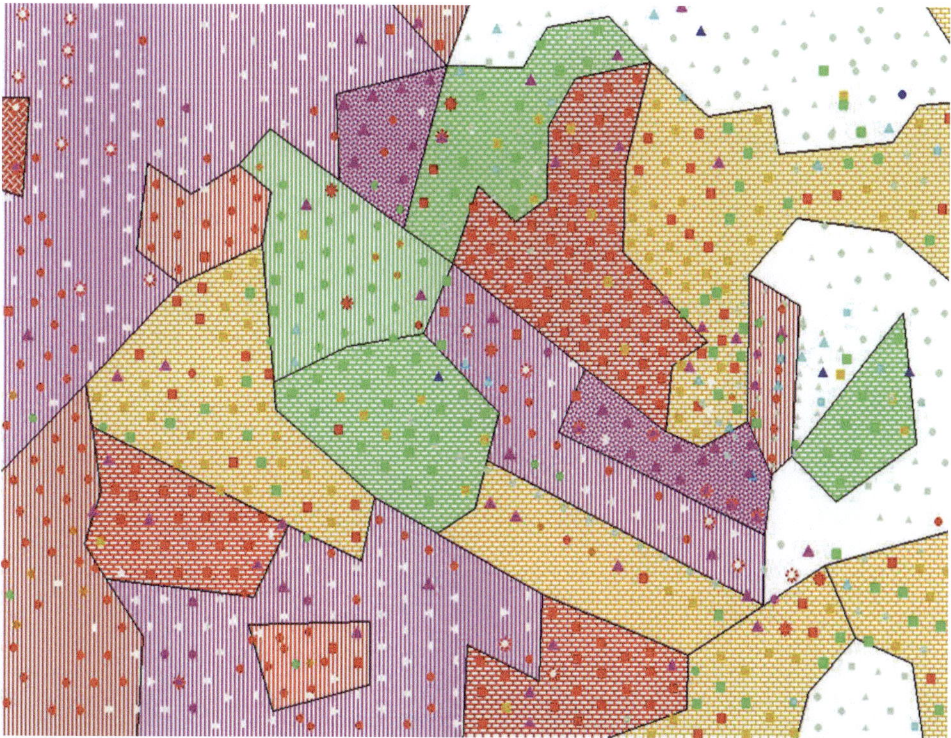

Fig. 13.8 Blast holes, color- and shape coded by destination, and grade control panels based on the BEI method. Same area as Fig. 13.7

Fig. 13.9 Comparison of grade control panels according to the ID[3] used by the mine and the BEI methods, same area as Figs. 13.7 and 13.8. Note the sometimes very different selection panels

two years production from the open pit. The reference model is a single realization of a Sequential Gaussian simulation for all variables involved, and adjusted to production data. The same areas were re-modeled based on the available blast hole database, and selection panels for each destination re-drawn according to the results of each method tested. The study involved development of an appropriate revenue function, consideration of mining practices and constraints, and compared alternative methods to the actual grade control panels developed by the mine using ID[3].

Table 13.2 TCu performance factors of the ID^3 method by destination relative to the SGS reference model

Destination	Tonnage (Dest./Reference)	TCu Grade (Dest./Reference)	Cu Metal Content Cu (Dest./Reference)
SAL	1.10	0.91	0.99
SME	1.16	1.06	1.22
SBA	0.18	1.15	0.21
SMR	0.50	1.36	0.68
SAS	0.55	1.02	0.56
OXA	1.29	0.85	1.10
OXB	1.16	1.08	1.25
OXL	0.44	1.54	0.68
MIX	0.52	0.90	0.47
TOTAL	1.16	0.84	0.98

Table 13.3 TCu performance factors of the BEI method by destination relative to the SGS reference model

Destination	Tonnage (Dest./Reference)	TCu Grade (Dest./Reference)	Cu Metal Content Cu (Dest./Reference)
SAL	1.10	0.92	1.00
SME	1.09	1.00	1.09
SBA	0.45	1.01	0.45
SMR	0.43	1.01	0.44
SAS	0.87	0.95	0.82
OXA	1.13	0.93	1.05
OXB	1.98	0.98	1.94
OXL	1.49	1.41	2.10
MIX	0.71	0.78	0.55
TOTAL	1.11	0.89	0.99

Only the results for the ID^3 and BEI methods are presented here. The simulation-based method produced similar and slightly better results compared to the BEI method, but it is more complicated and slower to implement. The OK method produced marginally worse results.

Tables 13.2 and 13.3 show the relative performance of the ID^3 and BEI methods with respect to the reference model for tonnages, TCu grade, Cu metal content, and revenues. The closer the value to 1.0, the better the method reproduces the to reference model, and, by extension and within the approximations of the reference model calibrations, actual production. A factor greater than 1 implies overestimation with respect the reference model. The destinations corresponding to waste, SSM, and OXM are not shown due to the low tonnages produced within the evaluation period. The overall ore and marginal ore production for the period was about 59.5 million tons, so the statistical mass available for comparison is significant.

Note how for most destinations and variables considered, the BEI method is superior. Recall that a 1 % difference between the two methods represents close to 600,000 metric tons of ore, or about 10,000 metric tons of contained Cu. Considering the depressed Cu prices at the time, a 1 % difference in contained Cu represented about US$ 16 million. At 2013 copper prices, the dollar value of the difference would be between US$ 70 and 80 million. In most cases, even though the differences in percentage points may be small, they represent significant economic improvements given the size of the operation.

The added economic benefit of the BEI method results from virtually no additional expenditure, since all operational practices remain the same. Also, the panel drawing process is facilitated by the use of smaller blocks and less sharp corners (Figs. 13.6 and 13.7). This in turn results in less unplanned operational dilution, because the shovels will extract the material following more faithfully the delineated zones. Although real, this effect is more difficult to quantify.

13.4 Selection of Ore and Waste: Simulation-based Methods

The objective of the simulation-based methodologies is to optimally select ore from waste according to different optimality criteria. Also, it provides more flexibility to handle several destinations for recoverable material, including ore blending with different metallurgical responses. Minimum-variance algorithms such as kriging have traditionally been the optimization criteria in most geostatistical applications, but are not always appropriate (Srivastava 1987).

In open pit and underground grade control, optimization should always be based on maximizing the economic value of the recovered material. The material selected for metallurgical processing should provide the maximum possible economic benefit given all operational constraints. Other possible optimization criteria, such as maximizing resource utilization, is not applicable in the case of grade control, since the decision is short-term in nature, and aims at making the most out of the current operation on a daily basis.

Loss Functions can be used to optimize based on pre-determined functions that assign value to estimates, or equivalently, costs to mistakes. They were described in Chap. 12, and further reading can be found in Journel (1988), Isaaks (1990), and Goovaerts (1997). Conditional simulation is used to provide a model of uncertainty that can be used to optimize grade control. One alternative is the Minimum Loss/Maximum Profit method as presented below, which has been implemented with success in several open pit operations. The expected profit calculation is

$$P_{ore} = \sum_{\text{all realizations}} \left[- \underbrace{c_o}_{\text{ore mining cost}} - \underbrace{c_t}_{\text{milling cost}} + \underbrace{prz^{(l)}(\mathbf{u})}_{\text{revenue}} \right]$$

$$P_{waste} = \sum_{\text{all realizations}} \left[- \underbrace{c_w}_{\text{waste mining cost}} - \underbrace{c_{lo}}_{\text{lost opportunity}} \right]$$

establish c_{lo} by calculating revenue if it were milled

$rev = prz^{(l)}(\mathbf{u})$

$$c_{lo} = \begin{cases} 0 & \text{if } rev < 0 \\ rev & \text{if } rev > 0 \end{cases}$$

13.4.1 Maximum Revenue Grade Control Method

The Maximum Revenue grade control method is a two-step procedure, first outlined by Isaaks (1990), and applied with success at some mine operations, for example Aguilar and Rossi (1996). Initially, a set of conditional simulations is obtained from the blast hole data available. These conditional simulations provide an uncertainty model for grades at any specific point within the blast. Second, an economic optimization process is implemented using loss functions to obtain the optimal ore/waste selection. The Loss Function quantifies the economic consequences of each possible decision.

The simulations are used to build models that reproduce the histogram and spatial continuity of the conditioning data. By honoring the histogram, the model correctly represents the proportion of high and low values, the mean, the variance, and other statistical characteristics of the data. By honoring the variogram, it correctly portrays the spatial complexity of the orebody, and the two-point connectivity of low and high grade zones. These are critical variables for the optimization of ore/waste selection because it depends on accurately predicting the variability of high to medium to waste grade transitions.

Typical grade control simulation grids can be 1 m by 1 m by bench height (corresponding to the sampled blast hole column). These are used directly in obtaining the uncertainty model for ore/waste selection panels. Larger grid sizes may be used and sometimes required because of time or general computer hardware limitations, still providing reasonable estimates when enough simulated points are included within the selection panels.

Given that conditional simulation models are sensitive to departures from its stationarity assumption, it is critical that they be controlled by geologic models. The use of geologic boundaries may introduce issues of ergodicity, which should be carefully handled. A constantly updated geologic model, in addition to constant geologic control at the pit is required to ensure that the uncertainty models derived from the conditional simulations are realistic and also representative of local geology.

Other important aspects include the behavior of the high-grade population, which is required to control the simulated high grades, see Parker (1991) and Rossi and Parker (1993). Issues such as limiting the maximum simulated grade should be carefully considered, since it may significantly impact the selection panels. The issue should be resolved through calibration with existing production data.

A small number of realizations, perhaps 20 or 30, are typically used. This reflects practical limitations, since grade control is a process that has to be completed in a short period of time; but it may also be a sufficient number of simulations to adequately describe the model of uncertainty, given the data density available.

Recall that the model of uncertainty provides the probability of that node in the grid of being above (or below) any grade z:

$$F(z;x \mid (n)) = \Pr ob\{Z(x) \le z \mid (n), \alpha = 1,\ldots,n\} \qquad (13.3)$$

where $F(z;\underline{x}|(n))$ is the cumulative frequency distribution curve for each point $\underline{x}$ of the simulated grid and obtained using the (n), $\forall = 1, \ldots, n$ conditioning blast holes.

In grade control, the selection decision (which material is ore and which is waste) has to be based on grade estimates, $z^*(x)$, while still attempting to minimize miss-classification. Since the true grade value at each location is not known, an error can and will likely occur. The loss function attaches an economical value (impact or loss) to each possible error, as described in Chap. 12.

The minimum expected loss can be found by calculating the conditional expected loss for all possible values for the grade estimates, and retaining the estimate that minimizes the expected loss. In grade control, the expected conditional loss is a step function whose value depends on the operating costs (Isaaks 1990). This implies that the expected conditional loss depends only on the *classification* of the estimate $z^*(x)$, not on the estimated value itself. For example, the loss incurred when a block of leach ore is sent to the mill is a function of the difference in processing costs related to both leach and mill; it will, of course, also depend on the *true block grade,* but not on the *estimated block grade* value itself.

13.4.2 Multivariate Cases

Grade control in the presence of multiple variables introduces additional challenges that can be easily handled. The Ujina open pit example briefly discussed above is in fact a multivariate grade control issue. There are multiple variables that add to the value of each parcel of material (copper and molybdenum), and also multiple variables that detract from its worth, such as Arsenic or the presence of clays. The multiple variables can all be mine products, or a combination of mine products, metallurgical performance variables, and contaminants in general.

In cases where there are spatial relationships between the variables of interest, then either co-estimation or co-simulation (Chaps. 8–10) can be performed. This is most important when simulating for grade control, since modeling relationships among different variables is consequential. In Chap. 14 two multivariate simulation case studies are presented.

13.5 Practical and Operational Aspects of Grade Control

There are many operational aspects that need to be considered for an effective grade control. The most important are (a) the relationships between the grade control activity and mine planning; (b) the practicality of obtaining representative samples; (c) time constraints, always present in any operation; the daily production target is the operation's main driver which does not allow for detailed modeling and planning work; (d) the gathering and use of geologic data; (e) the appropriate staking of the ore/waste zones; (f) the control of the mining process; (g) the destination of each truck or load of material; and (h) the accounting of material movement and overall reconciliations.

Each one of the aspects mentioned deserves detailed discussions and are outside the scope of this book. However, they are highlighted here to remind the reader that adequate grade control involves multiple areas of an operation, and cannot be developed in isolation from other aspects of the mine. Issues related to material accounting, particularly volumes or tonnages extracted and mine-to-mill reconciliations are among the most important. As argued in Chap. 11, they can also be the basis for model performance evaluations.

Operational details, sometimes seemingly trivial, can have a significant impact on the bottom line. Without pretending to be exhaustive, some illustrative examples mostly applicable to open pit mines are:

- Sufficient laboratory capacity to provide the assays' results in the required amount of time, usually 24 h or less for 200 to 300 samples or more;
- Traffic and destination control in the pit, particularly if truck dispatch systems are not available; in areas where manual labor is relative cheap, it is common practice to place an individual at the pit exit to verify that trucks go to the correct destination;
- Truck weighing, as a control to truck factors and volumetric measurements;
- If visual indicators of ore are available (such as green or blue oxide Cu minerals), mine geologists should visit daily the waste dumps, to ensure that the operation is not misplacing the ore loads; also, a 24-h operation should have adequate artificial lighting in the pit, more so when visual aids are used in grade control.
- The amount of broken ore in the pit should be sufficient to feed the mill for a few days; an operation where loading is always pressuring for more blasting goes counter to good grade control practices.
- Confirm the in-situ bulk density of material loaded; the operation should monitor in situ density variations, sometimes taking bulk samples from the pit. Also, consider the estimate of humidity in the rock, which is generally a simple global estimate. These estimated values affect the conversion of volumes into tonnages, with a direct impact in the accounting of metal moved.

Semi-Automatic Dig Lines Definition A computational algorithm can be used to develop semi-automatically dig lines (Neufeld et al. 2005). While it is unlikely that all issues will be solved, always presenting the optimal solution, the process of defining dig lines can be sped up. It is expected, though, that a degree of manual intervention and validation will always be required.

The process of automatically defining dig limits is based on pre-defined operational and selection criteria. Figure 13.10 shows two cases for dig limits. The model used to define the ore/waste selection panels is the same in both cases; the difference is how much one dig limit considers the ability of the mining equipment to mine to the exact limits defined.

The optimal dig limits can be posed as an optimization problem. Sequential annealing (see Chap. 10) can be applied by defining the objective function as:

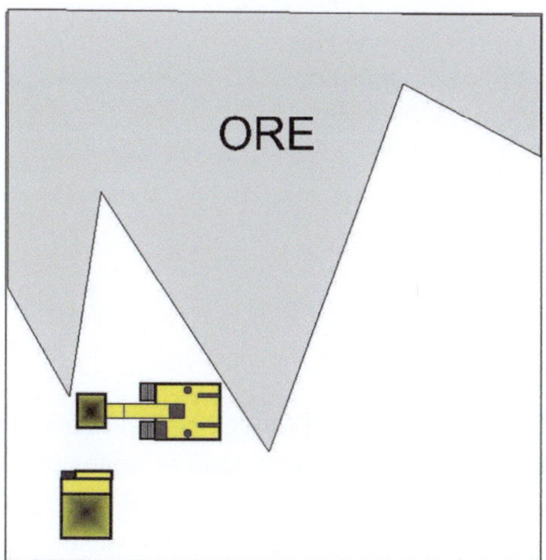

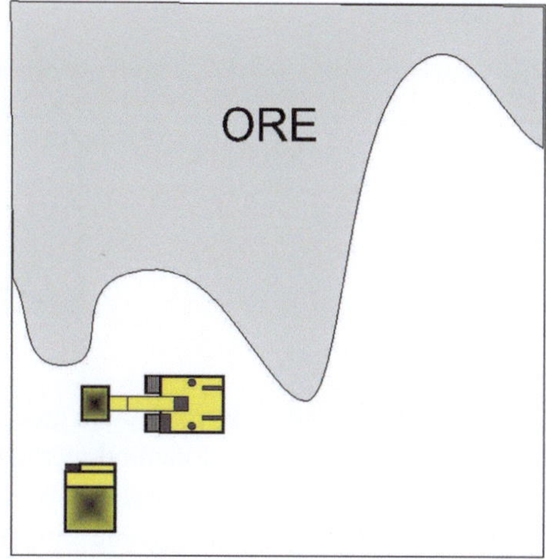

Fig. 13.10 Comparison of two ore/waste dig limits. The left option is more precise, but less realistic, and impossible for the shovel to dig to. Therefore, a large amount of unplanned dilution would be expected. The right option is a smoother dig limit, easier to dig for the shovel, but that it may be sub-optimal, depending on the characteristics of the mining equipment

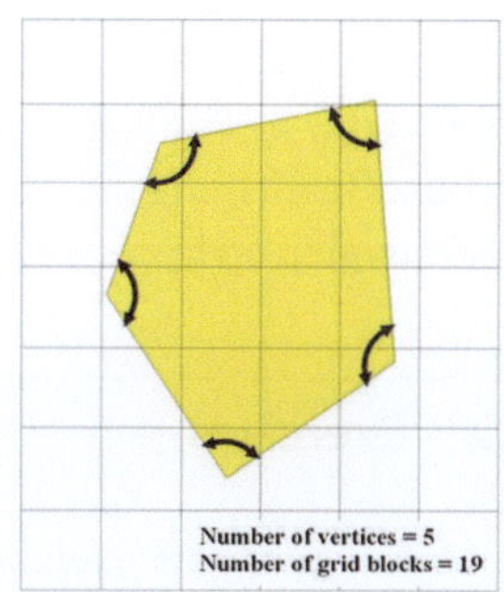

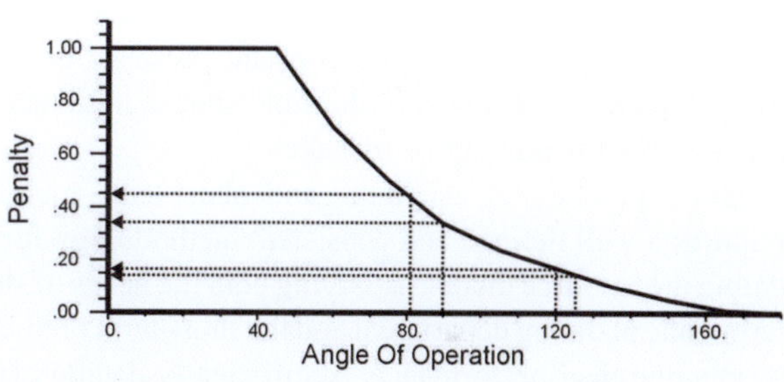

Fig. 13.11 Example of an ore polygon, with 5 vertices and affecting 19 blocks. The penalty assigned is a function of the angle of operation of the shovel

$$O_{global} = O_{profit} - O_{digability}$$

The initial profit is calculated as the sum of all fractional blocks that are considered ore (profitable):

$$O_{profit} = \sum_{ix=1}^{nx} \sum_{iy=1}^{ny} frac_{(ix,iy)} \cdot \overline{P}_{(ix,iy)}$$

where P represents the profit assigned to each block in the model, and "*frac*" represents the volume within each profitable block.

The initial digability is calculated based on the characteristics of the mining equipment, taken for example from an equipment curve, and interpreted as the sum of the penalties for each angle in the ore/waste polygon, see Fig. 13.11:

$$O_{digability} = \sum_{iv=1}^{nv} pen_{iv}$$

Using simulated annealing, the vertices and angles can be moved within a small circle (tolerance) to change the angle that it defines, and thus changing the penalty and overall profitability. A vertex is randomly selected and moved within a small distance (see Fig. 13.12). New profit and penalties are calculated, and the new objective function obtained. The results are sorted into accepted or rejected perturbations based on its impact on the objective function, and the process is iterated until convergence is achieved.

The dig limit selection algorithm can be made semi-automatic if the option of an additional constraint is added manually, allowing for the technician to account for the limitations of mining equipment and the value of the material. The dig limit algorithm works by systematically giving up ore or taking in additional waste to pay for the increased digability, i.e., less sharp angles defining the corners of the ore/waste selection panels.

13.6 Summary of Minimum, Good and Best Practices

At a minimum, all short-term models should be updated to include new data that becomes available. Proper procedures for validation and checking should be in place, and the complete sequence of updating the model should take less than a

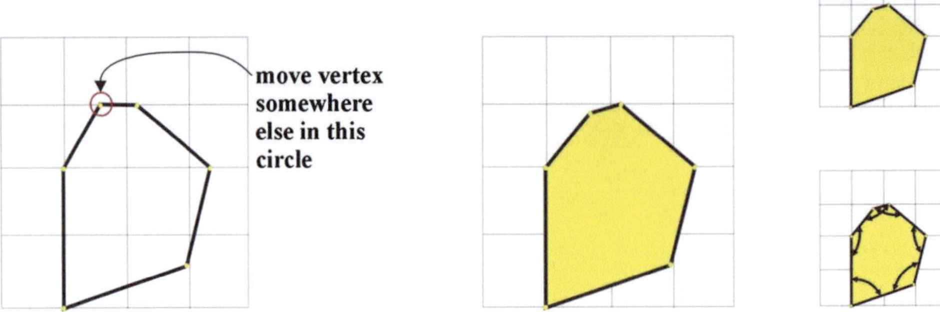

Fig. 13.12 A vertex is randomly selected and moved, a new shape obtained and new profit and penalties obtained

week of work. The ore/waste selection process will normally be based on a conventional method, perhaps some form of Kriging duly restricted with geology. Blast hole sampling should routinely provide acceptable samples for ore/waste selection. Information from relevant prior blasts should be used in defining current dig lines. Geologic mapping should aid in the daily task of defining the dig lines, which is generally a manual operation. Proper material accounting, reconciliation procedures, and constant presence and control by the mine geologist in the field should minimize the probability of making gross mistakes.

Good practice of medium- and short-term modeling requires a well defined and consistent methodology for updating the resource model, satisfying both the needs of short-term mine planning department and the short-term prediction of metallurgical performance. A sufficiently detailed study would have determined all the important implementation parameters and methodological details, including the procedures required to update the geologic model. The short-term models should be produced at regular time intervals, be always reconciled with recent past production, and compared against the original long-term resource model for the same areas. The model updating process should be semi-automatic, although always fully validated. Good practice in ore/waste selection requires the recognition of the limitation of selecting on grade, and therefore the use of an optimal selection method, with consideration of the basic economic parameters. Dig lines are usually hand drawn, and control and accounting procedures are strict. Reconciliation is usually kept on a blast-by-blast basis, and reported monthly.

Best practice in medium- and short-term modeling, in addition to the above, involves using conditional simulation models to provide for the uncertainty model and the risk assessment that short-term mine planners need. Other aspects of the model updating should be similar to what is defined as good practice, but the models are more likely to be simulation models. Similarly, the ore/waste selection should have been fully optimized, including the possibility of automatically drawing dig lines on a daily basis. In all cases, reconciliation procedures should be in place, and should be used to feed back and maintain an optimum implementation of the method as mine conditions change.

In addition, best practice in long- and medium-term modeling involves the development of dynamic models, which are constantly updated, not only in terms of grade estimation, but most importantly in terms of the geologic model. Production data and infill drilling are used with production mapping (drift or bench) to update on a regular basis portions of the long-term model that is therefore constantly up to date. It amounts to merging the medium and long-term model into a single model, updated, for example, on a monthly basis.

13.7 Exercises

The objective of this exercise is to review some concepts related to grade control. Some specific (geo)statistical software may be required. The functionality may be available in different public domain or commercial software. Please acquire the required software before beginning the exercise. The data files are available for download from the author's website—a search engine will reveal the location.

Consider the molybdenum data in `bh-data.dat`. You will be asked to conduct a full geostatistical study from histograms through simulation. The exercise will go quickly because the data are closely spaced and reasonably well behaved.

Question 1: Plot a location map and histogram of the Mo data. Comment on the spacing of the data. Your final estimation/simulation model should be at a spacing of about 1/3 to 1/2 of the blasthole spacing. We will not consider any volume averaging in the simulation. Decluster the data if you consider it necessary.

Question 2: Calculate and fit the variograms of the molybdenum grade and estimate a model with ordinary kriging. Perform cross validation if time permits and ensure that no conditional bias exists in the estimates.

Question 3: Calculate and fit the variograms of the normal scores transforms of molybdenum.

Question 4: Simulate 100 realizations of the grade. Plot the average grade and four realizations to verify that the simulated realizations are reasonable. The average grade model should look very close to the kriged model created previously.

Question 5: Calculate the expected profit assuming a cost/price/recovery structure that will give about 50 % ore in the model area.

Question 6: Establish initial polygon limits for an ore/waste interface. Optimize the dig limits for different digability settings.

References

Aguilar CA, Rossi ME (1996) Método para Maximizar Ganancias en San Cristóbal, Minería Chilena, Santiago, Chile, Ed. Antártica, No. 175, pp 63–69, January

Douglas IH, Rossi ME, Parker HM (1994) Introducing economics in grade control: the breakeven indicator method. Preprint No 94–223, Albuquerque, February 14–17

Goovaerts P (1997) Geostatistics for natural resources evaluation. Oxford University Press, New York, p 483

Isaaks EH (1990) The application of Monte Carlo methods to the analysis of spatially correlated data. Ph.D. Thesis, Stanford University, p 213

Journel AG (1988) Fundamentals of geostatistics in five lessons. Stanford Center for Reservoir Forecasting, Stanford

Matthey J (2001) Special report: Platinum 2001

Neufeld CT, Norrena KP, Deutsch CV (2005) Guide to geostatistical grade control and dig limit determination. Guidebook series, vol 1. Centre for Computational Geostatistics, Edmonton

Parker HM (1991) Statistical treatment of outlier data in epithermal gold deposit reserve estimation. Math Geol 23:125–199

Rossi ME, Parker HM (1993) Estimating recoverable reserves: is it hopeless? In: Forum 'Geostatistics for the next century', Montreal, 3–5 June

Srivastava RM (1987) Minimum variance or maximum profitability. CIMM 80(901):63–68

14 Case Studies

Abstract

This chapter presents a few real-life examples of different types of Mineral Resource Estimation and Geostatistical studies. They are intended to illustrate some of the techniques described in the book, but are not exhaustive in scope or content.

14.1 The 2003 Cerro Colorado Resource Model

The Cerro Colorado Mine is located at an elevation of 2,500 m, on the western slopes of the Andes in Northern Chile. The mine is located about 120 km east of the city of Iquique (Fig. 14.1), on the western edge of the Domeyko Cordillera.

The Cerro Colorado mine is a bio-heap leach, solvent extraction-electro winning (SX-EW) copper operation, has been in production since 1994. It started up as a small open pit operation, producing in the order of 20,000 t of cathode copper. As of 2003, and after several expansions, the mine was producing about 120,000 t of cathode copper per annum. Originally operated by the Rio Algom mining company, later became part of the assets transferred to the Billiton mining company as part of its merger with Rio Algom, and in 2001 became part of the BHP-Billiton base metals group. Figure 14.2 shows an aerial view as of 1999 of the mine operation. The overall drill hole spacing at the time of preparing this resource model was a nominal 100 × 100 m, with some areas infilled to 50 × 50 m.

The work described in this section was completed in conjunction with Cerro Colorado Geologists., Chief Geologist, and Mine Superintendent. The BHP-Billiton Base Metals organization is gratefully acknowledged for support and permission to publish this work.

14.1.1 Geologic Setting

The Cerro Colorado deposit is located within a north-trending belt of Eocene to Oligocene-age porphyry Cu deposits (Cepeda et al. 1982; Campbell 1994). A thick sequence of Cretaceous andesite tuffs, flows and agglomerates of the Cerro Empexa Formation underlies most of the Cerro Colorado region. Multiple phases of tonalite, granite and quartz monzonite porphyry were emplaced into the volcanic rocks in late Cretaceous and early Tertiary time. Copper mineralization is related to late-stage intrusives of tonalite and quartz monzonite microbreccia. Andesites and intrusives are covered by a locally thick sequence of Pliocene ignimbrites and gravels of the Altos de Pica Formation.

Copper occurs in a series of sub-horizontal layers of supergene oxide minerals and supergene sulfides. Mineralization occurs in andesite and porphyry along an east-west to northeast trend. Copper mineralization extends at least 2,000 m East-West and from 1,000 to 2,000 m North-South. The orebody is covered by post-mineral gravels and ignimbrites of the Altos de Pica Formation, except for some oxide ore exposed in a nearby gully (Quebrada de Parca).

Mineralization is thickest in two distinct areas, forming the West and East deposits. Mineralization in the West deposit is generally centered along the southern margin of an east-west trending body of porphyry where elongate fingers of microbreccia are developed in porphyry and andesite. The East deposit is located along the southern flank of a northeast-trending body of quartz monzonite microbreccia. The deposit is centered in an area where abundant, small apophyses of porphyry intrude andesitic tuffs and porphyritic flows.

14.1.2 Lithology

The present mapping and logging system uses five lithological codes: Colluvium, Ignimbrite, Porphyry, Breccia, and Andesite. No distinction is made of the different porphyries that have been identified in the deposit. Colluvial boulder and conglomerate deposits of recent age overlay

M. E. Rossi, C. V. Deutsch, *Mineral Resource Estimation,* DOI 10.1007/978-1-4020-5717-5_14,

Fig. 14.1 Location map of the Cerro Colorado Mine in northern Chile. Courtesy of Compañía Minera Cerro Colorado S.A.

volcanic ignimibrite (ash-flow tuff). The total thickness of un-mineralized cover, including conglomerate, ignimbrite and colluvium deposits, varies from 30 m over the east half of the East deposit to more than 100 m over the West deposit, with an average thickness of post-mineral cover of about 60 m.

The andesites consist of lithic, lapilli tuff, porphyritic flows and coarse, flow breccias and agglomerates. Volcanic rocks dip to the south at a low angle. Quartz monzonite, quartz-monzonite microbreccia and tonalite porphyry form complex bodies intruding andesites in the West and East deposits. Above an elevation of about 2,600 m, quartz monzonite and quartz-monzonite microbreccia are present as large east-west trending bodies in the West deposit and as northeast-trending shapes in the East deposit. Breccia bodies coalesce below an elevation of 2,550 m, forming a northeast-trending body extending from the western edge of the West deposit to the northeast end of the East Deposit, surrounded by numerous apophyses of tonalite porphyry. The number and extent of tonalite porphyry intrusives increases with depth, eventually merging below an elevation of 2,450 m to form a large stock surrounded by smaller apophyses of porphyry and intermixed with bodies of quartz-monzonite microbreccia.

Fig. 14.2 Aerial view towards the N-NE of the Cerro Colorado operation, January 1999. Courtesy of Compañía Minera Cerro Colorado S.A.

Mineralization principally occurs in andesite, tonalite porphyry and within small breccia bodies. The margins of larger breccia bodies may be preferentially mineralized. The siliceous cores of breccia units are normally weakly mineralized. It is believed that intense hydrothermal alteration and brecciation of the quartz monzonite culminated with deposition of Cu mineralization.

14.1.3 Alteration

Several hypogene alteration phases have been identified: potassic, phyllic and propyllitic alteration zones. There is also silicification of quartz-monzonite microbreccia and the margins of intrusive bodies, which may have preceded Cu mineralization, as well as supergene alteration. Potassic hypogene minerals are preserved generally below the 2,450 m elevation where supergene leaching is less advanced, and is more evident in quartz monzonite, quartz-monzonite microbreccia and tonalite porphyry in the East deposit compared to the West deposit. Andesite, quartz monzonite, microbreccia and tonalite are altered to sericite/muscovite, quartz and pyrite in the phyllic zone. Phyllic alteration overprints earlier potassic alteration and is overprinted itself by the formation of kaolinite and alunite during later supergene leaching. Propyllitic alteration is evident in outcrops within the Quebrada de Parca and in some the outermost drill holes. Andesites are the most common rock affected. The andesite is strongly chloritized and contains disseminated and veinlet epidote. The groundmass of quartz-monzonite microbreccia is strongly silicified locally. Silicification of andesite is also evident along contacts with tonalite porphyry and quartz monzonite. Silicified units could be weakly or strongly mineralized.

14.1.4 Mineralization Types

Supergene alteration formed during leaching of sulfide-rich rock. Acids generated by the interaction of ground water and sulfides leached Cu from sulfides and re-deposited it as oxide Cu minerals above the water table and as supergene Cu sulfides below the water table. The resulting rock is relatively soft and commonly has vuggy veinlets of quartz, kaolinite, alunite, limonite and jarosite.

Hypogene mineralization (HYP) is represented by pyrite accompanied by chalcopyrite and bornite. Hypogene sulfides are deposited in all rock types as disseminations or in veins and veinlets with quartz-feldspar-biotite, but are best preserved under the base of the supergene layers or in veinlets in the transition zone between supergene sulfides and hypogene. The average Cu grade of HYP is from 0.20 to 0.30 % (Campbell 1994).

Supergene alteration and mineralization occurs in four main zones, consisting generally from top to bottom of: (a) leached zone (LIX) in which acid leaching has removed most or all Cu mineralization; (b) oxide mineralization consisting almost entirely of oxide Cu minerals (OX); (c) supergene Cu sulfide mineralization, dominated by the presence of chalcocite and lesser covellite (S); and (d) a transitional zone consisting of a mixture of supergene Cu sulfides and hypogene Cu sulfides (MSH).

Supergene oxide and sulfide mineralization occur in multiple, sub-horizontal layers and vertical pods that extend over a total distance of about 2,700 m East-West and 2,000 m North-South. The top of the largest layers of oxide mineralization is at an average elevation of 2,500 m, or from 50 to 200 m below the surface. Smaller, less continuous pods of oxide mineralization are present above this elevation. The leach cap above Cu oxides is 25–75 m thick.

Oxide (OX) is defined as material in which the ratio of sulfuric acid-soluble Cu (SCu) to total Cu (TCu) is 0.5 or greater. Chrysocolla is pervasive in the oxide Cu zone and forms the dominant ore mineral. Brochantite, libethenite, malachite, pseudomalachite, paratacamite, cuprite and tenorite can also be present (Cepeda et al. 1982).

Supergene sulfide (S) is defined as material for which the ratio of sequential Cu (sulfuric acid-soluble plus cyanide acid-soluble) to total Cu is 0.8 or greater; and the ratio of sulfuric acid-soluble to total Cu (SCu/TCu) is less than 0.3. Chalcocite and lesser covellite replace or occur as coatings of hypogene pyrite, chalcopyrite and bornite.

The transition zone (MSH) is defined as mineralization with a ratio of sequential Cu to total Cu of from 0.5 to 0.8. Cu is present as chalcocite, covellite, chalcopyrite and bornite. Chalcocite and covellite are preferentially present in veins and fractures, but can occur as disseminations.

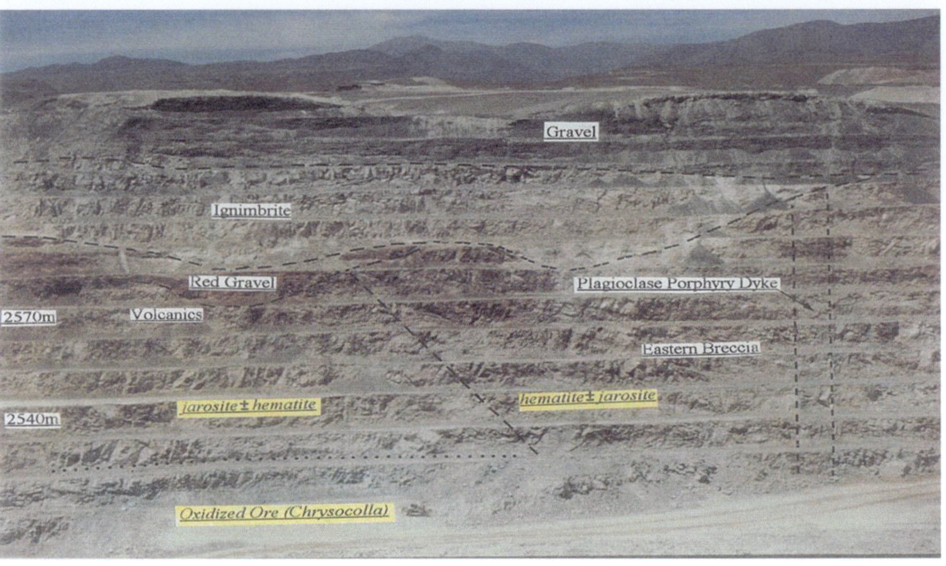

Fig. 14.3 Photograph showing some of the rock types and oxide mineralization, along with contacts and some local faulting. Benches are 10 m high

Figure 14.3 shows a picture of the upper section of the open pit wall, with elevations, main lithology zones, mineralization types (LIX and OX), and some structural features annotated.

14.1.5 Structural Geology

Cepeda et al. (1982) identified numerous, northwest-trending, post-mineral faults in surface mapping and exploration adits and cross TCus. These generally strike N60 to 70W and dip vertical to 70° east. No orebody displacement is apparent. The northeast elongation of the East deposit suggests that northeast-trending pre-mineral faults may have been present. Abrupt contacts between leach, oxide and sulfide bodies also suggest that northwest-trending faults or fracture zones may have influenced the development of these units.

14.1.6 Database

The database includes a mixture of older holes used for the original feasibility study and later in-fill and short-term drill holes, intended to support detailed Mine Planning. Blast hole data were used in a limited fashion for calibration and reconciliation of the resource model to past production. Through statistical analyses, the different data types and sources were shown to be consistent.

The database contains 1,575 drill holes, including Reverse Circulation (RC) and Diamond Drill Hole (DDH) types. Sampled intervals are generally 2 m, except in overburden, where no sample is taken. The sample's lithology, alteration, and mineralization type is logged from the samples. Total Cu (TCu) and Soluble Cu (SCu) are assayed. Some older holes do not have geologic information.

There are a total of 56,018 sample intervals with TCu grade. Figure 14.4 shows the histogram of the TCu sample values, while Figs. 14.5 and 14.6 show the distribution of TCu grades in oxide and supergene sulfide mineralization, respectively. Note that the overall average of the TCu values in the database is 0.62 % TCu; the overall average of the oxide grades is 0.88 % TCu and represent about 30 % of the total dataset; for sulfide samples, the average is even higher (1.20 % TCu), representing about 21 % of the total number of samples. Other sulfide mineralization, such as hypogene (or primary) mineralization, has much lower grade (0.33 % TCu, almost 15 % of the database). At Cerro Colorado, primary mineralization has no economic significance, since the operation only processes leachable mineralization, i.e., oxides and supergene sulfides.

Tables 14.1, 14.2 and 14.3 show the basic statistics for TCu samples by lithology, mineralization type, and alteration, respectively. Finally, Table 14.4 shows the basic statistics for SCu according to mineralization type.

The following conclusions can be drawn from the tables above:

- There are 1,991 samples without TCu values and without mineralization type; there are also 311 samples that have no lithology, and 2,571 samples for which there is no alteration available.
- Most of the TCu grade exists in porphyry, andesites, and breccias. There is very little mineralization in gravels, ignimbrites, and dikes.
- The Leach mineralization type is essentially barren, with an average TCu grade of 0.1 %, although there are isolated high values, up to 3.74 % TCu. These high grades are attributed to oxide mineralization that cannot be recovered in the mine's SX-EW plant, generally called "black" oxides, and include minerals such as cuprite, native Cu, and Cu Wad.
- The overwhelming majority of samples present phyllic alteration. There will be limited use for alteration as a mineralization control.

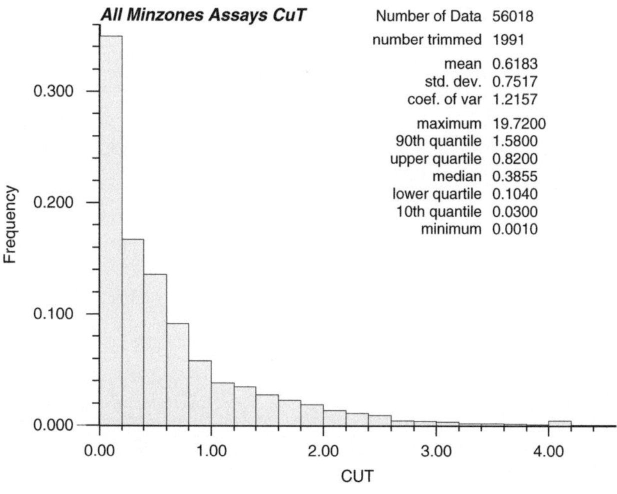

Fig. 14.4 Histogram and basic statistics of all samples, total Cu grade

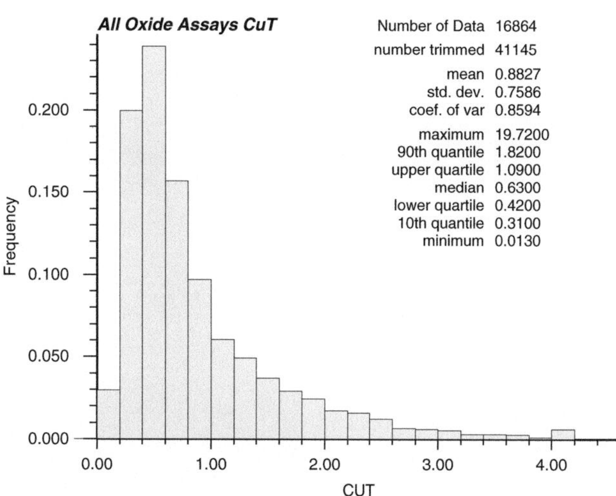

Fig. 14.5 Histogram and basic statistics of all samples, total Cu oxide grade

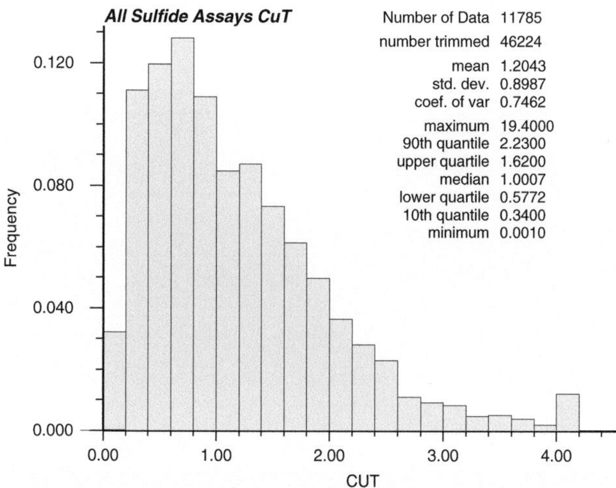

Fig. 14.6 Histogram and basic statistics of all samples, total Cu sulfide grade

- From Table 14.4, it can be seen that the oxide and sulfide mineralization are the only two types that have significant soluble Cu (SCu) grades.

14.1.7 Estimation Domain Definition

The definition of the grade estimation domains was done following the general guidelines suggested in Chap. 4. In the case of Cerro Colorado, the estimation domains (or geologic units, "UGs") are based on a combination of mineralization type, lithology, and alteration, and are intended to capture the mineralization controls. It was recognized that mineralization type is the major mineralization control and introduces mineral processing considerations in the resource model. For example, hypogene mineralization with Supergene Sulfide mineralization have different ore genesis, spatial distributions, and they will be treated separately in the mine plan. The final estimation domains used are constrained by the amount of information available, which requires that some data be grouped. The main estimation domains defined are:

1. Domain = 0 (leach) includes all samples mapped as mineralization type "leach", and also all other samples that are mapped as gravels or ignimbrites. This is waste material.
2. There are three oxide domains. The first has better oxide grades with phyllic, argillic, and silicified alterations, and within intrusive rocks (porphyries and breccias).
3. A second oxide grade spatial distribution is found within the andesites. These have lower, but still interesting Cu grades.
4. The third oxide domain is a lower grade domain, and corresponds to all oxide mineralization with potassic and cloritic alterations. This mineralization is located on the fringes of the deposit, and is volumetrically disjointed and overall small in comparison to the two previous domains.
5. The fourth and fifth domains are supergene sulfide mineralization mapped with phyllic, silicified, and argillic alterations. The porphyries have better grades and are spatially distinct from the breccias located in the central and southern areas of the East sector of the deposit.
6. Supergene sulfides with phyllic alteration in andesites tend to have low grades, and are found towards the edge of the deposit, around the rocks that intruded them. This is also true of the supergene sulfide mineralization that has potassic and cloritic alteration, and thus they are grouped in a single domain.
7. The mixed sulfide-hypogene mineralization was estimated separately, despite being a set of spatially disjointed bodies and with small overall volume. This is necessary because this transition zone is distinct from either sulfide or hypogene mineralizations.

Table 14.1 TCu samples statistics by lithology

	TCu – Lithology	*N*	*% of Total Samples*	*Max*	*Mean*	*Median*	*Standard Deviation*	*Coefficient of Variation*
0	**Gravel**	48	0%	1.440	0.280	0.205	0.271	0.969
1	**Ignimbrites**	145	0%	3.020	0.071	0.009	0.284	4.010
2	**Porphyry**	27,251	49%	14.700	0.586	0.370	0.704	1.201
3	**Breccia**	12,050	22%	9.999	0.555	0.330	0.697	1.255
4	**Andesite**	16,137	29%	19.720	0.706	0.430	0.838	1.188
5	**Dikes**	76	0%	2.970	0.424	0.249	0.482	1.138
	TOTAL	55,707						

Table 14.2 TCu samples statistics by mineralization type

	TCu – Mineralization Type	*N*	*% of Total Samples*	*Max*	*Mean*	*Median*	*Standard Deviation*	*Coefficient of Variation*
0	**Leach**	17,794	32%	3.745	0.100	0.060	0.128	1.276
1	**Oxides**	16,864	30%	19.720	0.883	0.630	0.759	0.859
2	**Sulfides**	11,785	21%	19.400	1.204	1.001	0.899	0.746
3	**MSH (Mixed)**	1,292	2%	3.630	0.501	0.450	0.355	0.709
4	**Hypogene**	8,283	15%	4.493	0.335	0.310	0.266	0.793
	TOTAL	56,018						

Table 14.3 TCu samples statistics by alteration

	TCu – Alteration	*N*	*% of Total Samples*	*Max*	*Mean*	*Median*	*Standard Deviation*	*Coefficient of Variation*
1	**Phyllic**	40,444	72%	19.720	0.648	0.400	0.769	1.186
2	**Argillic**	5,550	10%	7.300	0.363	0.104	0.599	1.651
3	**Potassic**	4,115	7%	3.800	0.524	0.430	0.441	0.842
4	**Cloritic**	2,435	4%	3.380	0.338	0.280	0.341	1.010
5	**Silicified**	903	2%	13.800	0.697	0.460	0.820	1.175
6	**Alunite**	-	0%					
	TOTAL	53,447						

Table 14.4 SCu samples statistics by mineralization type

	SCu – Mineralization Type	*N*	*% of Total Samples*	*Max*	*Mean*	*Median*	*Standard Deviation*	*Coefficient of Variation*
0	Leach	17,792	32%	3.745	0.037	0.013	0.073	1.973
1	Oxide	16,864	30%	15.060	0.694	0.490	0.634	0.914
2	Sulfide	11,785	21%	4.480	0.145	0.110	0.144	0.994
3	MSH (Mixed)	1,292	2%	0.810	0.062	0.046	0.080	1.277
4	Hypogene	8,283	15%	2.930	0.036	0.021	0.103	2.882
	TOTAL	56,016						

8. All hypogene mineralization was grouped into the final estimation domain. This was done so because the mine treats this material as waste.

The set of estimation domains are presumed stationary, and form the basis of the TCu and SCu grade estimation.

14.1.8 Database Checking and Validation

The database was thoroughly checked and validated. After an initial spot check, it was determined that the database required full validation. To fully check the Cerro Colorado database, a surface 80 m above the pit's topography as of January 2003 was obtained and used to select the data lying below the surface. Samples 80 m above current topography will not have any significant impact in the resource model, and so were deemed lower priority. The following approach was taken:

1. The complete list of intervals to be checked consisted of 286 drill holes, or 60 % of the database. For each one of these drill holes, the computerized assay information was checked against the original laboratory certificates. If the

laboratory certificate did not exist or was lost, the check was done against the interval grade value annotated on the log sheet. The checks also included comparing the computerized lithology, alteration, and mineralization type against the original log sheet. The drill hole collar coordinates and down-the-hole survey information were also checked. This check was done for over 30,000 intervals, taking about one calendar month for three people.

2. All information related to each drill hole was placed in a binder, and the binder catalogued and stored for future reference. The binder also included a summary of the drill hole's existing information, as well as the missing information. This was done to ensure that future changes and additions to the database could be easily tracked and documented, also leaving the proper audit trail for future third parties' reviews.
3. Besides checking against the original laboratory and log sheets, other consistency checks were completed:
   a. Checking that the Soluble Cu was not greater than the Total Cu (SCu≤TCu, if not, SCu=TCu).
   b. Checking whether duplicate intervals were present (these are sometimes generated at the time of data input).
   c. The "from" and "to" metrage of each down-the-hole interval has to be consistent with adjacent samples.
   d. The names of the drill holes have to be exactly the same in all data tables, so that information can be correctly linked.
   e. Duplicate collar coordinates are only allowed for those drill holes that were started with one drilling method, and were completed with another. Typically, the RC method is used in the upper portion of the drill hole, before entering significant mineralization, and then converted to a diamond drill hole, thus obtaining core in mineralized intervals.
   f. Two drill holes were discarded because their collar coordinates were clearly erroneous, and no information was available to verify their exact location.
   g. The coordinate system used for the drill holes was also checked; the older drill holes were located using local mine coordinates, while later drill holes were located using truncated UTM coordinates.
   h. The values assigned to intervals with missing Cu assays, non-sampled intervals, and with Cu values below laboratory detection limit were also checked.

This checklist serves as an illustration of the type of work involved in ensuring database integrity, which in some cases could entail a significant amount of time. Each project will be different, with the corresponding database will have its own peculiarities; this list does not include all possible aspects that should be validated.

In addition to the validation work described above, the database was also modified because a significant number of older drill holes were remapped from the samples still available in the core shack. This remapping effort was intended to unify the logging criteria used earlier in the project's life, thus making the geologic information more consistent with the current geologic understanding of the deposit. The remapping was specifically focused on the lithology and alteration information, since they are often difficult to map consistently.

14.1.9 Comparison of Drill Hole Types

Another aspect of the database validation relates to the different quality of information obtained from different drill hole types. For any given interval, Reverse Circulation (RC) drilling may result in a different grade than would have been obtained from diamond drilling. This relates to the drilling method, the sampling method, and to different drill hole diameters (support) involved.

Another concern is the different drilling campaigns performed at different times. The differences may result from a combination of different factors:

4. *Changes in drill hole type*. Initial campaigns generally use faster and cheaper drilling methods than those that are intended for resource delineation, that require more accurate samples. This also occasionally applies to the laboratory techniques used in the early sampling. Examples of these differences are abundant across the mining industry, as for example the use of percussion holes in the San Gregorio Au vein-type deposit (Uruguay), or the use of a rotary, open drill rig mounted on a tractor in the case of the Michilla Mine, used to detect satellite mineralization away from the main deposit.
5. *Differences in personnel involved.* Classically, different geologists will see and map differently the same drill hole, sometimes resulting in geologic descriptions that are significantly different. This was the case at Cerro Colorado.
6. If there is a significant time lag between the different campaigns, then the technology for drilling, sampling, sample preparation, and assaying could change. This is typical of projects that have been known for a number of years, but that, for whatever reason, have not been developed in a timely manner. Again, there are many examples in industry, including the Pueblo Viejo Au deposit (Dominican Republic), or CODELCO's Radomiro Tomic Cu deposit in northern Chile, originally drilled in the 1950s, and eventually put into production in the mid-1990s.
7. Geologic knowledge about a deposit naturally evolves with time, as more information becomes available. This is another common source of discrepancies between earlier exploration and development geologic work and the understanding of the deposit geology after the mine enters production. This is also the case at Cerro Colorado, as well as many other mines around the world.

An important issue was the systematic higher grades observed from production information compared to prior re-

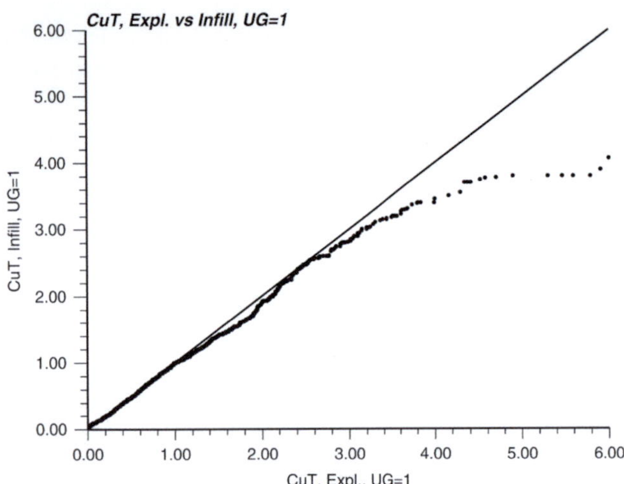

Fig. 14.7 Q-Q TCu plot, showing exploration vs. infill drill holes, Estimation Domain 1 (Oxide)

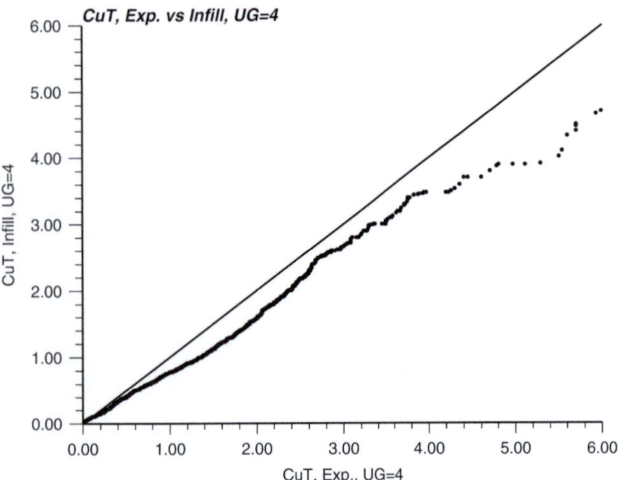

Fig. 14.8 Q-Q TCu plot, showing exploration vs. infill drill holes, Estimation Domain 4 (Sulfide)

source models obtained from exploration drill holes. It was hypothesized that there may be differences in the Cu grades obtained by drill holes form the earlier exploration campaigns (pre-operation), approximately to a 100 × 100 m grid, compared to later in-fill drill holes, used for detailed mine planning and budgeting, and drilled approximately on a 50 × 50 m grid.

Since there were no twinned exploration holes, it is difficult to perform a statistical comparison that would unequivocally indicate whether older drill holes are biased with respect to newer information. Globally within each estimation domain, however, it is evident that the initial drill holes (on a 100 × 100m grid) reported higher Cu grades than the in-fill drill holes. Figures 14.7 and 14.8 show the Quantile-Quantile (Q-Q) plots for the two TCu grade distributions of the most important oxide and sulfide domains. On the *X*-axis the grade distributions from the original exploration drill holes

Table 14.5 Density values used in prior resource models

Description	*Dry Density*
Ignimbrites/Gravel	2.23
Leach	2.30
Oxides	2.40
Sulfides	2.43
MSH and Hypogene	2.42
Other Unclassified Material	2.30

is shown, while the distribution of the TCu grades from the in-fill drill holes is shown in the *Y*-axis.

No corrections of any type were performed, since it is not clear which of the two drill campaigns tend to be closer to the true in situ grade. Several options were available, but in general it would be expected that infill drilling is potentially less accurate, because it is faster and the samples may be of lesser quality. This has not been demonstrated, but if it were the case, then using infill drilling provides a degree of conservatism to the overall estimate.

14.1.10 Laboratory Quality Assurance–Quality Control (QA-QC)

A complete QA-QC follow-up of the laboratory assays was done as the drilling campaign progressed. Details of this QA-QC program are not given here, but it followed the concepts discussed in Chap. 5. In general, TCu analyses are more accurate and precise than SCu assays, due to the nature of the assaying methods involved. In both cases, however, the results were satisfactory and within acceptable standards for this type of deposit.

14.1.11 Topography

The topography used in the model is based on an aerial photogrammetric survey, which was tied to control points in the field. It is considered that the topographic surface is accurate to ±2 m, both in the horizontal and the vertical directions.

Additionally, the pit surface as of January 31, 2003, was taken as the actual pit surface for calculation of remaining resources. This is in addition of the already mentioned auxiliary surface created 80 m above current pit surface, used to select drill holes and blast holes for various data validation exercises, model calibrations, and reconciliations.

14.1.12 Density

The density database available consists of 1,591 samples taken from drill hole data, and tested for density in several campaigns during the late 1990s. Table 14.5 shows

Table 14.6 Relative percentage comparison in drill holes vs. interpreted model, mineralization type

DRILL HOLES			*BLOCK MODEL*	
Mineralization Type	*Total DDH Meters (m)*	*Percentage*	*Volume (m3)*	*Percentage*
MSH	2081.25	3.3	7521500	3.5
Oxide	36090.61	57.6	132463000	61.8
Sulfide	24442.86	39.0	74468500	34.7

the arithmetic average of the density values for each unit described, which approximately coincides with the estimation domains defined above. Using arithmetic averages by geologic unit is a standard procedure in mining, particularly in circumstances where there are relatively few density values available. Also, density should not vary abruptly in space. Spatial coverage of the density values has to be fairly uniform and different grade ranges need to be sampled to capture the variations in density values as a function of grade. Additional density data would allow proper domaining and estimation using Ordinary Kriging, for example.

14.1.13 Geologic Interpretation and Modeling

Some of the considerations that affected the choice of modeling methodology included:

a) The software used in the modeling, which required the use of solids, obtained through either wireframing techniques or extrusions of sectional polygons.
b) The bench height, set at 10 m, which is also the height of the blocks of the resource model.
c) The time and human resources available to execute the work, which led to simplifying the methodology for the least important geologic variables.

Only those geologic variables with significant potential impact on the resource model and volumetric representation in the database were modeled. Considering the lesser importance of lithology and alteration, compared to mineralization type, as mineralization controls, a more detailed modeling process was used for the latter. The following steps were used to build the geologic model:

1. Development of E-W cross sections, spaced 50 m apart, and posting all available drill hole samples that exist within ±25 m of the section plane. This was done for each variable separately.
2. Interpretation of the drill hole intercepts as sectional polygons.
3. Extrusion of the interpreted alteration and lithology polygons to obtain solids, and then cutting these solids with orthogonal constant-elevation planes, corresponding to mid-bench elevations. In the case of mineralization type, the orthogonal planes used corresponded to N-S longitudinal sections every 50 m, again with a ±25 m tolerance. This added interpretation step was done because of the importance of mineralization type as mineralization control.
4. For alteration and lithology, the bench polygons resulting from cutting the E-W solids were re-interpreted using the 10 m composite intercepts available. Additionally, where available, pit and blast hole was used to refine in detail the interpreted polygons. These final polygons were then extruded in the vertical direction to produce the final solids.
5. The last step of the interpretation of mineralization type was similar to Step 4 above, obtaining the final interpreted solids from refined mid-bench polygons, in turn obtained from the polygons cut from N-S sections, and refined using drill hole composites and production mapping for the final interpretation.

The upper portion of the Leach unit was not explicitly modeled. It was left as the by-default unit, and then overprinted with the other modeled units. There were, however, a number of leached bodies internal to other units that had to be modeled explicitly.

14.1.14 Volumetric and Other Checks

Two different comparisons were made to check whether the interpreted volumes were globally unbiased.

The block model assignments of each geologic variable were checked on screen and on paper (sections and plan views plotted at a 1:500 scale) against the interpreted solids. The second check is a comparison between the volume of the solids and the representation (relative percentage) in the database of each unit. There should be no significant volumetric bias after interpretation. Drill hole clustering and subjectiveness in the mapping and logging process may render the geologic data less reliable.

Table 14.6 shows, as an example, the relative percentage in drill holes (logged meters vs. total meters) for each mineralization type (main mineralized units), compared to the total and relative percentages of the corresponding interpretation. Results are quite satisfactory.

The 2003 interpretation was compared to the previous geologic model. Figure 14.9 shows the interpreted mineralization type for Section 82230N, where the different units are color-coded (brown = leached; green = oxide; red = sulfide; yellow = MSH, and purple = hypogene). The 2003 model units are represented with the dotted fill, while the

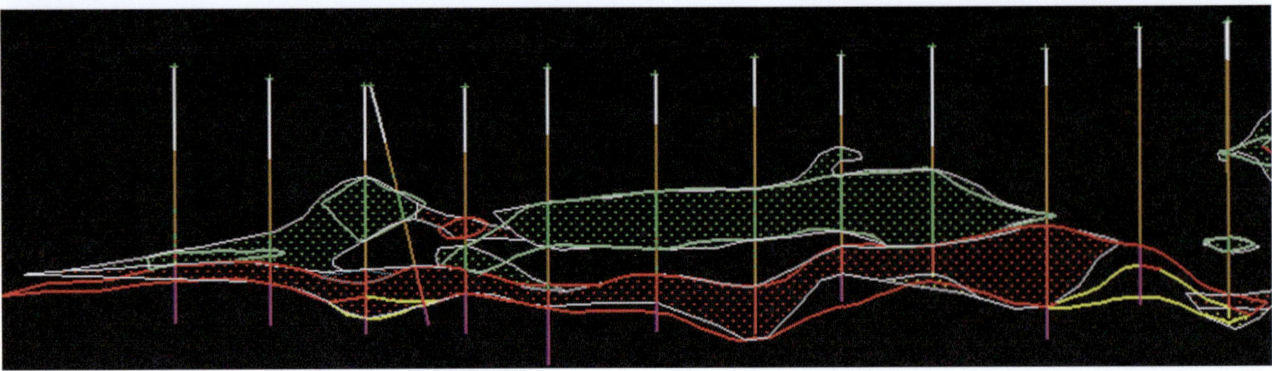

Fig. 14.9 Mineralization type section, 82230N, 2000 (*no fill*) and 2003 (*dotted fill*) geologic models

2000 model units do not have any fill. By observing simultaneously both models, the impact of the new drilling and the quality of the interpretation work can be assessed.

14.1.15 Exploratory Data Analysis

14.1.15.1 Compositing

The original samples were composited down-the-hole at a nominal 10 m interval. Down-the-hole composites were chosen because some of the drill holes are inclined 60° or less and bench compositing is not appropriate.

The original samples were back-tagged with the modeled geology, and the 10 m composites obtained by breaking the composites at the estimation domain contacts (UGs). This resulted in a total 11,809 10 m composites, of which about 12 % are less than 10 m in length. After statistically verifying that the length of the composite is not correlated with grade, only 88 composites less than 2 m were discarded from the database.

Histograms and basic statistics were obtained for TCu and SCu for the different estimation domains. Figures 14.10 and 14.11 show that the grade distributions are fairly typical of porphyry copper deposits.

The corresponding probability curves are shown in Figs. 14.12 and 14.13 respectively. Note that both curves show breaks and departures from a straight-line fit, which may imply multiple populations.

14.1.15.2 Declustering

The drill hole data are spatially clustered as a result of the historic drilling campaigns, and also due to the presence of an underground tunnel, from which several holes were drilled. Declustering techniques were applied (see Chap. 2) to obtain an unbiased prediction of the global mean and also to estimate the expected SMU-support grade–tonnage curves for each domain. The cell declustering method (Deustch 1989) was applied to the 10 m composites for each estimation domain. After analyzing the results for multiple cell sizes, a 100 × 100 × 30 m

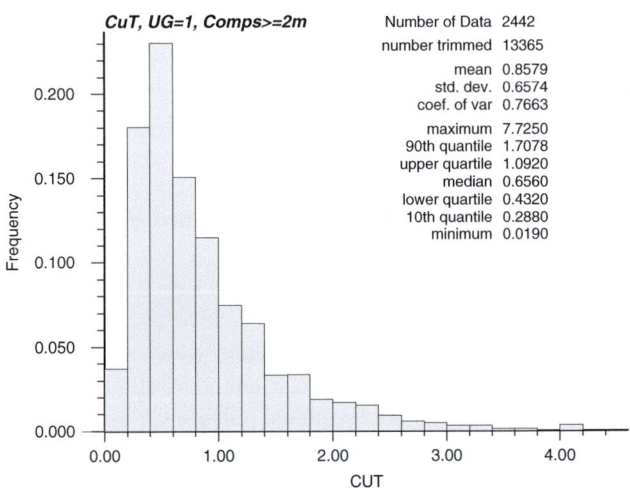

Fig. 14.10 TCu histogram and basic statistics, oxide composites

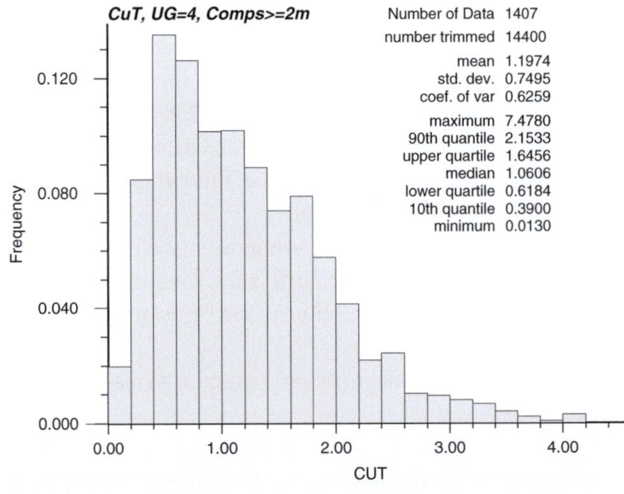

Fig. 14.11 TCu histogram and basic statistics, sulfide composites

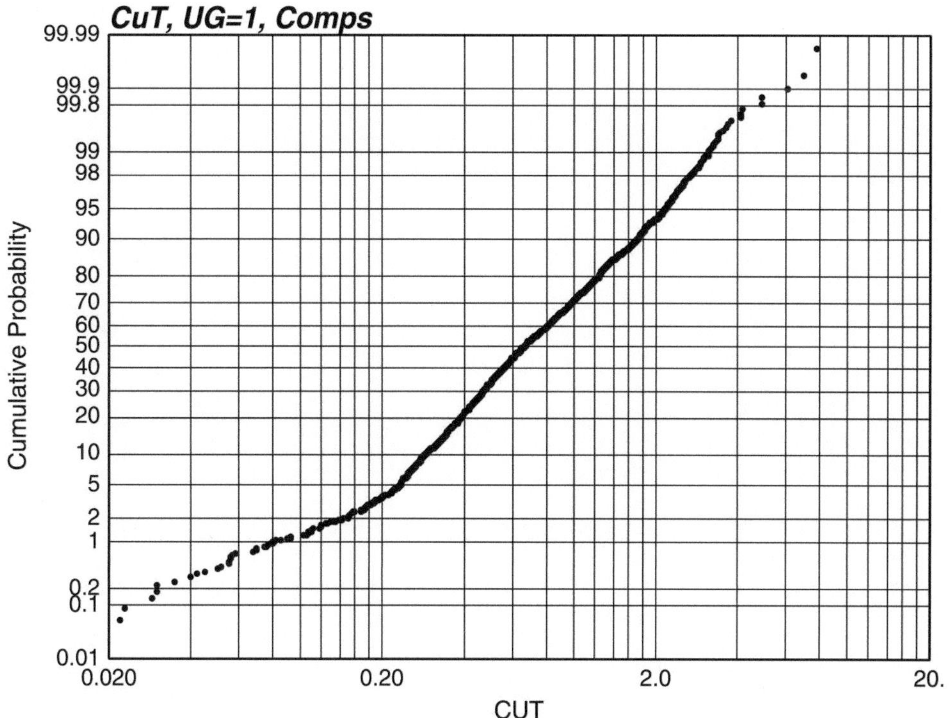

Fig. 14.12 TCu probability plot, oxide composites

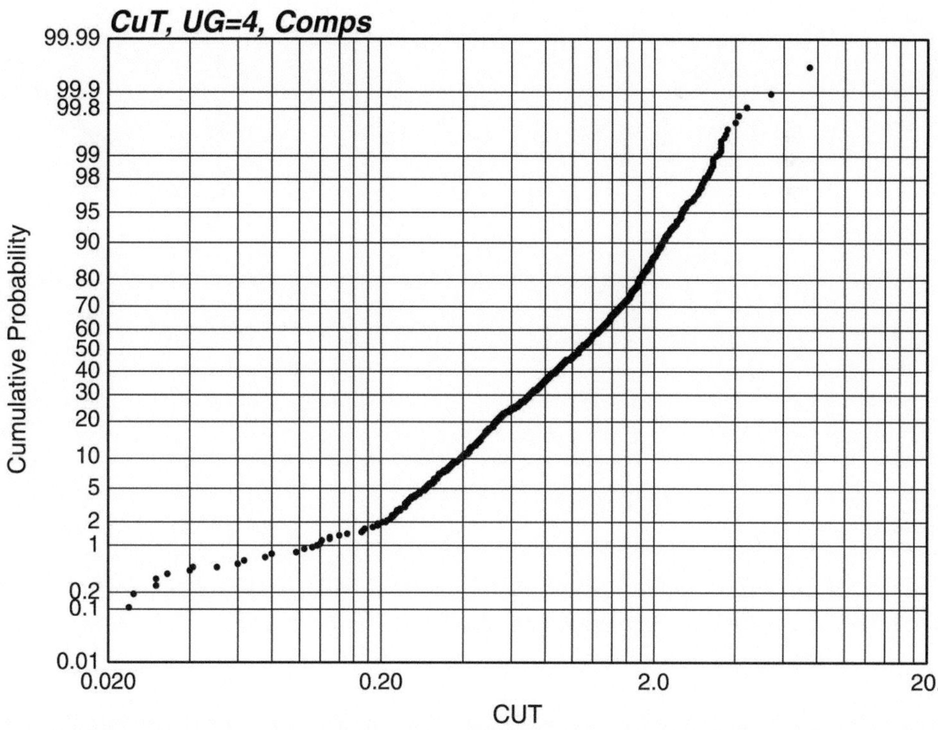

Fig. 14.13 TCu probability plot, sulfide composites

cell was chosen as the optimal one for declustering, since it corresponds to the initial pseudo-regular drilling grid.

Figures 14.14 and 14.15 show declustered histograms of TCu for oxides (UG 1) and sulfides (UG 2) respectively. Note how, in both cases, the statistics have changed significantly. The declustered TCu mean of the first oxide unit (Domain 1) is 0.77% (Fig. 14.14), compared to 0.86% TCu for the non-declustered data. In the case of the first sulfide unit (Domain 4), the declustered TCu mean is 0.96% (Fig. 14.15), while the non-declustered mean is 1.20% TCu. Generally, the higher the grade of the estimation domain (UG), the more severe the impact of clustering is. Logically, the high grade

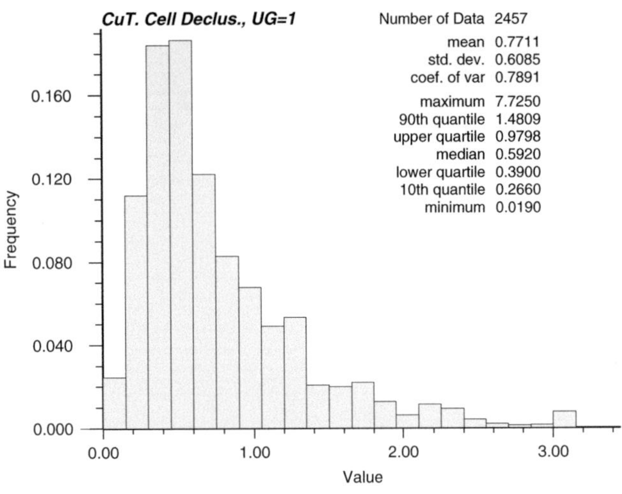

Fig. 14.14 TCu histogram and basic statistics of declustered oxide composites

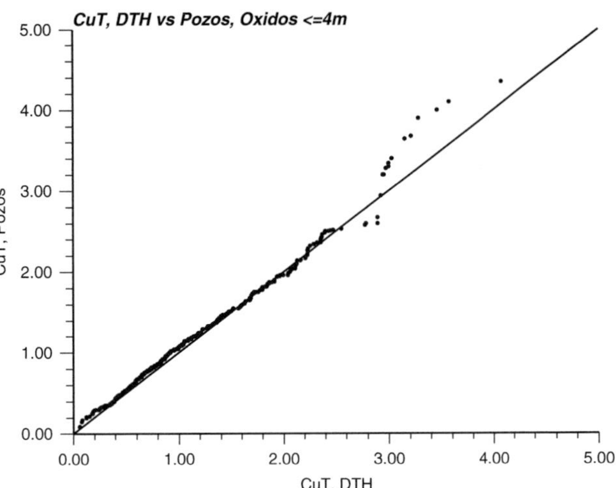

Fig. 14.16 Q-Q plot, TCu, DDH 10 m composites vs. blast holes, oxide

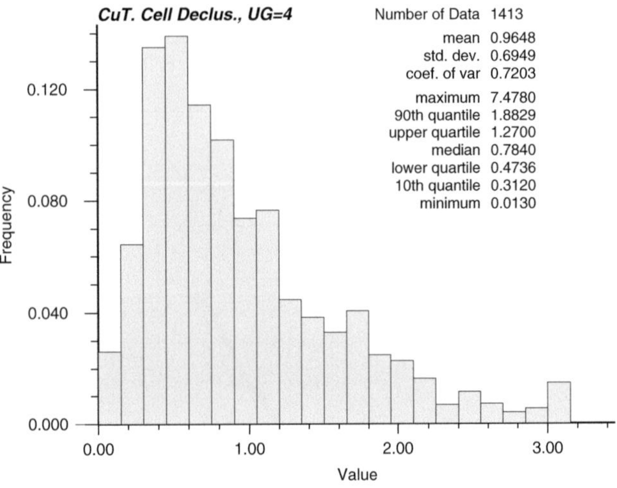

Fig. 14.15 TCu histogram and basic statistics of declustered sulfide composites

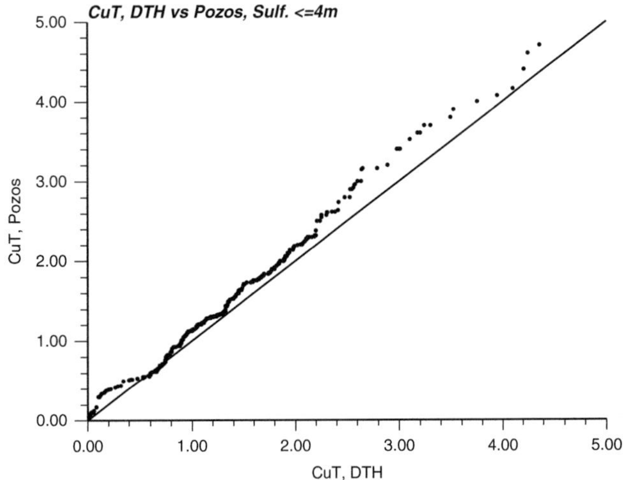

Fig. 14.17 Q-Q plot, TCu, DTH 10 m composites vs. blast holes, sulfide

areas are of interest for infill drilling. The lower grade, marginal mineralization and waste units are more likely to be drilled only in the initial drilling campaigns, which are more regular.

14.1.16 Comparison Between Composites and Blast Hole Data

Blast hole data are used to select ore and waste and are sampled on 10 m benches. There are several statistical tools that can be used to compare two distributions (Chap. 2); however, the first major decision is which data should be compared. A set of paired ("twins") data was obtained by searching in three dimensions for composites that are located a maximum distance of 4 m from a blast hole. The resulting pairs can be compared with a Q-Q plot.

Different types of drill hole data were compared separately to blast holes, since the database contains diamond drill holes (DDH), reverse circulation holes (RC), and open percussion holes (DTH). Additionally, the analysis was done separating oxide and sulfide mineralization. Three examples of the comparisons are shown in Figs. 14.16–14.18. Blast hole data tend to have higher grades, although the differences are small.

14.1.17 Contact Analysis

The analysis of grade profiles across boundaries between estimation domains is often referred to as contact analysis. Its purpose is to understand which contacts are hard, in the sense that grades on either side of the contact are very different, or soft, where the grade transition from one domain to the next is smooth. An example of a hard contact in porphyry-type deposits is typically the interface of the leached

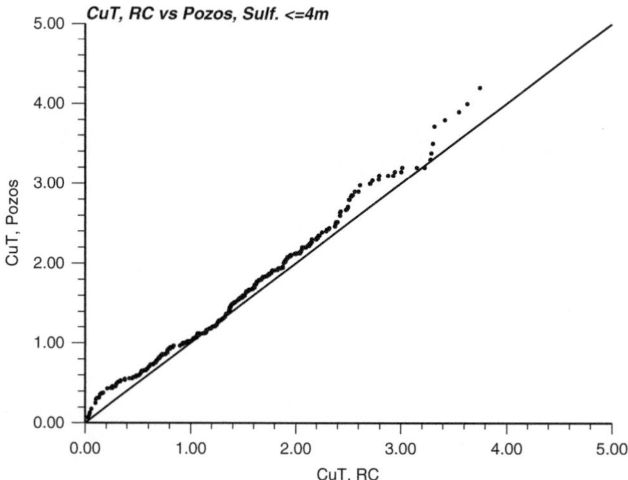

Fig. 14.18 Q-Q plot, TCu, RC 10 m composites vs. blast holes, sulfide

(barren) cap with the oxidized mineralized zones. Within 1 or 2 m in the vertical direction, grades can change from barren (<0.1 % TCu) to 1 %. Soft contacts may exist between high and low secondary enrichment units, where the grade transition is much smoother, to the point that, based on grade alone, it is difficult to define the contact itself.

Figure 14.19 shows an example of a soft contact. The grade profile corresponds to two sulfide estimation domains, and shows the average grade of all composites in each side of the contact by class distance, in this case 20 m. Figure 14.19 shows that, despite the fact that the overall averages are different, locally and near the contact the grades are similar and also somewhat depleted with respect to the global average. Therefore, for these estimation units, it is reasonable to use composites across the contact to estimate the grades of either unit within a limited distance.

14.1.18 Correlogram Models

The correlogram function (Srivastava and Parker 1988) was chosen to characterize spatial continuity at Cerro Colorado because experience has shown it to be more robust than the traditional variogram with respect to outliers and trends, being therefore generally easier to model. Correlograms were calculated for TCu and SCu in each Domain using 10 m composites and blast hole concurrently. Down-the-hole correlograms were used to define the nugget effect and short scale continuity of the correlogram models. Experimental correlograms were obtained in 37 directions using SAGE2001 (Isaaks 1999). Some observations from the correlogram models were:

- Three rotation angles were used to define the directions of anisotropy. In this particular case, there is sufficient information (drill holes and blast holes) to warrant such detailed correlogram models.
- The models were discussed with the local geologists, who reconciled them with their own experience and knowledge of the deposit. The main directions of continuity confirm current geologic knowledge. In addition, for most Domains there is evidence of a short-scale anisotropy that has a different direction than the longer-scale continuity. This is interpreted as a mixture of geologic controls (structures) at different scales, as confirmed by local geologists.

Figure 14.20 shows the experimental correlograms and models for Domain 1, for three directions that are close to the main directions of anisotropy. As shown, two exponential structures have been modeled with an anisotropy ratio of 5:1 in the N-S direction compared to the E-W direction, and even slightly more anisotropic with respect to the vertical direction. For this particular Domain, the first structure showed slightly more continuity on a direction dipping approximately 20°, while the second structure did not show a dip. Shallow-angle, cross cutting structures explained the slightly different anisotropy direction of the first structure.

14.1.19 Change of Support to Estimate Internal Dilution

The Cerro Colorado open pit mine works on a 10 m-bench height and uses large equipment, such that the accepted selective mining unit (SMU) is 20 × 20 × 10 m. This SMU size defines, in theory, the target distribution of estimated grades that should be achieved to accurately estimate recoverable resources and reserves. Although the SMU is a convenient concept, it is important to not lose sight of its limitations, recall the discussion in Chap. 7.

The coefficient of variation ($CV = \sigma/m$) is a useful measure of variability that can be used to characterize the variability of the SMU distribution. This theoretical SMU distribution can be used as a reference distribution and compared to the estimated block model grades. A calibration of the estimated grade distribution can be performed to better match the theoretical SMU distribution.

The dispersion variance is found for the assumed SMU through either (i) the correlogram or variogram models developed for each Domain; or, (ii) as an "experimental" dispersion variance. For the second option, blast holes or other production data should exist at a grid sufficiently smaller than the SMU defined. If so, the available grades are simply averaged into the SMU sizes.

The Discrete Gaussian (DG) method was used at Cerro Colorado to derive a SMU distribution of TCu grades for each Domain. The method was mentioned in Chap. 7, and is described in detail in Journel and Huijbregts (1978). The three elements that are needed for the DG model to predict a grade-tonnage curve for any SMU are (i) the correlogram models for TCu and for each Domain; (ii) the

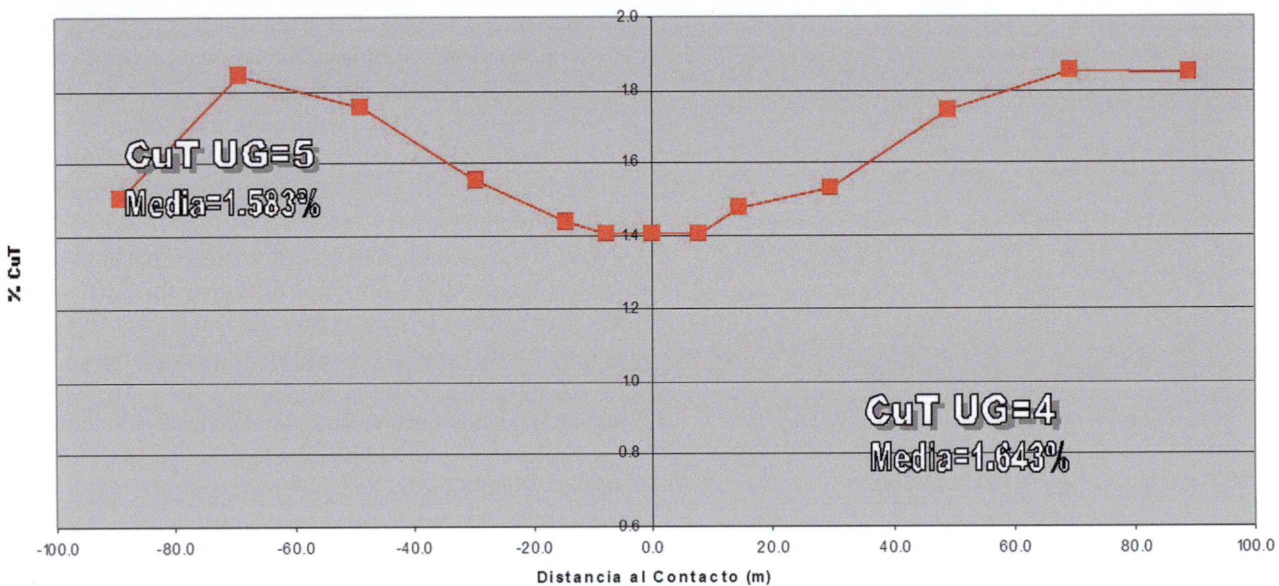

Fig. 14.19 Average grade profile at the contact between Domains 4 and 5

Table 14.7 Parameters of the predicted SMU distribution for the main units, discrete Gaussian model

TCu Domain	*Dispersion Variance, 20x20x10m SMU*	*Predicted TCu mean SMU grade (DG Model)*	*Predicted SMU Standard Deviation (DG Model)*	*Predicted SMU CV (DG Model)*	*CV of the 10m Declustered Composites*
1	0.7101	0.693%	0.4678	0.6756	0.7891
2	0.5133	0.7769%	0.4719	0.6074	0.7564
3	0.7451	0.5183%	0.3054	0.5893	0.6191
4	0.4709	0.7389%	0.3877	0.5247	0.7203
5	0.5707	0.8160%	0.5326	0.6527	0.7672
6	0.7524	0.7289%	0.4866	0.6676	0.7219

definition of the SMU size, in the case of Cerro Colorado as a 20 × 20 × 10 m block, for which, and based on the correlogram models, a dispersion variance and a VCF can be obtained; and (iii) the declustered 10 m composite database for each Domain.

Table 14.7 shows the predicted CV (or target CV) for each Domain and 20 × 20 × 10 m SMU size. Table 14.7 also shows the dispersion variance of the blocks and the CV of the original declustered 10 m composites.

As expected, the predicted CV corresponding to the SMU distribution is lower for every Domain compared to the original composites. For Domain 3, low-grade oxide mineralization, the predicted SMU CV is lower, but fairly similar to the original CV, an indication that there is little mixture of grades within that Domain, evidenced by a highly continuous correlogram model. The opposite can be said about the higher grade units, such as Domain 4.

14.1.20 Predicted Grade-Tonnage Curves for TCu, Cerro Colorado

The predicted grade-tonnage curves from the DG-corrected distribution were obtained for the mineralized units shown in Table 14.7. If these curves are similar to those obtained from the estimated resource model, then it can be said that the estimates incorporate an appropriate amount of volume-variance correction. However, it should always be remembered that the within-block mixing of grades is only one source of dilution and ore loss.

Figure 14.21 shows the grade-tonnage curve for the predicted SMU distribution and the original 10 m declustered composites for Domain 1, higher-grade oxides. Note that the tonnage above cutoff is expressed as a percentage of the total meters of drilling above cutoff, such that a comparison can be made. For a 0.5 % TCu cutoff, the DG model predicts

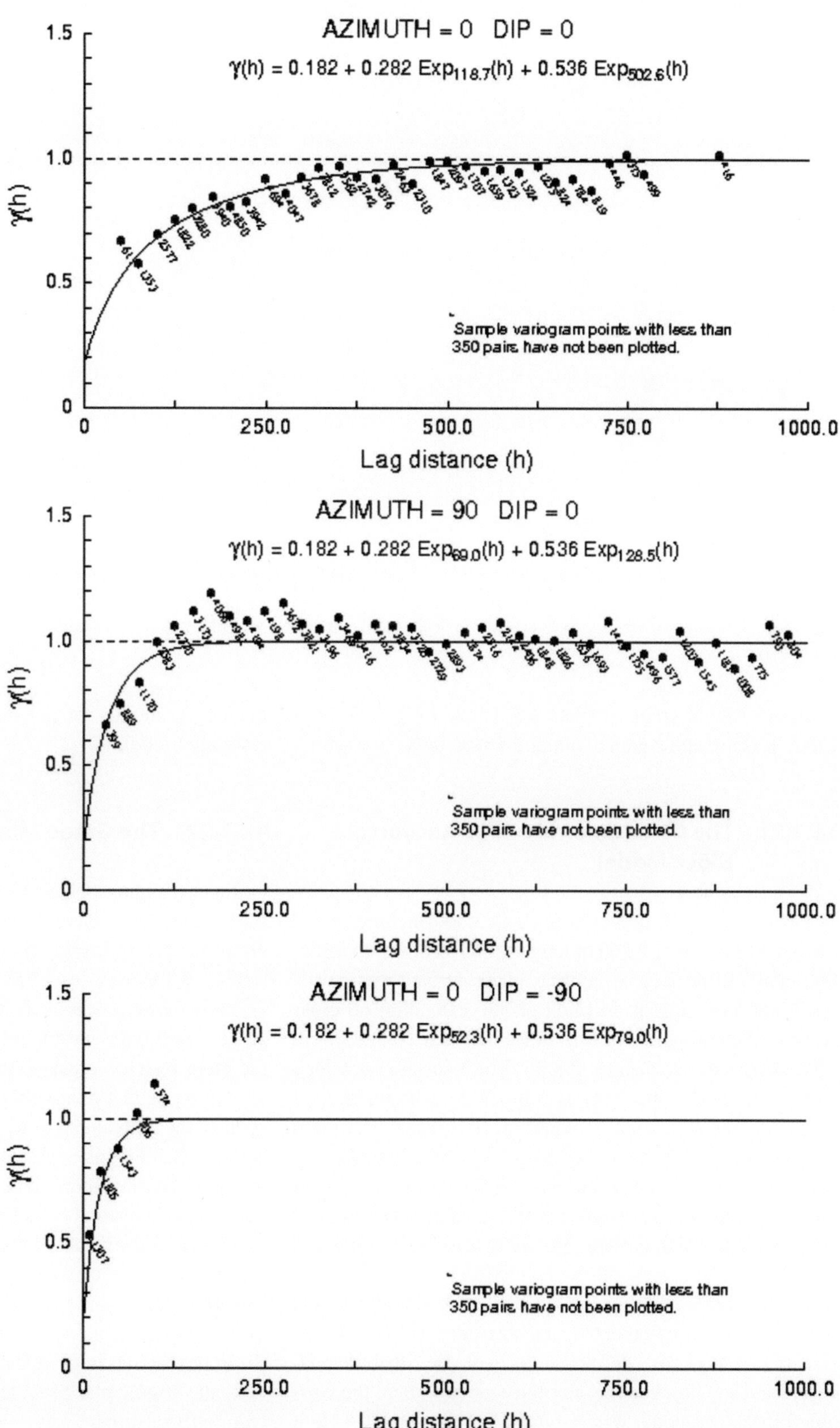

Fig. 14.20 Three experimental correlograms and their models, TCu, Domain 1

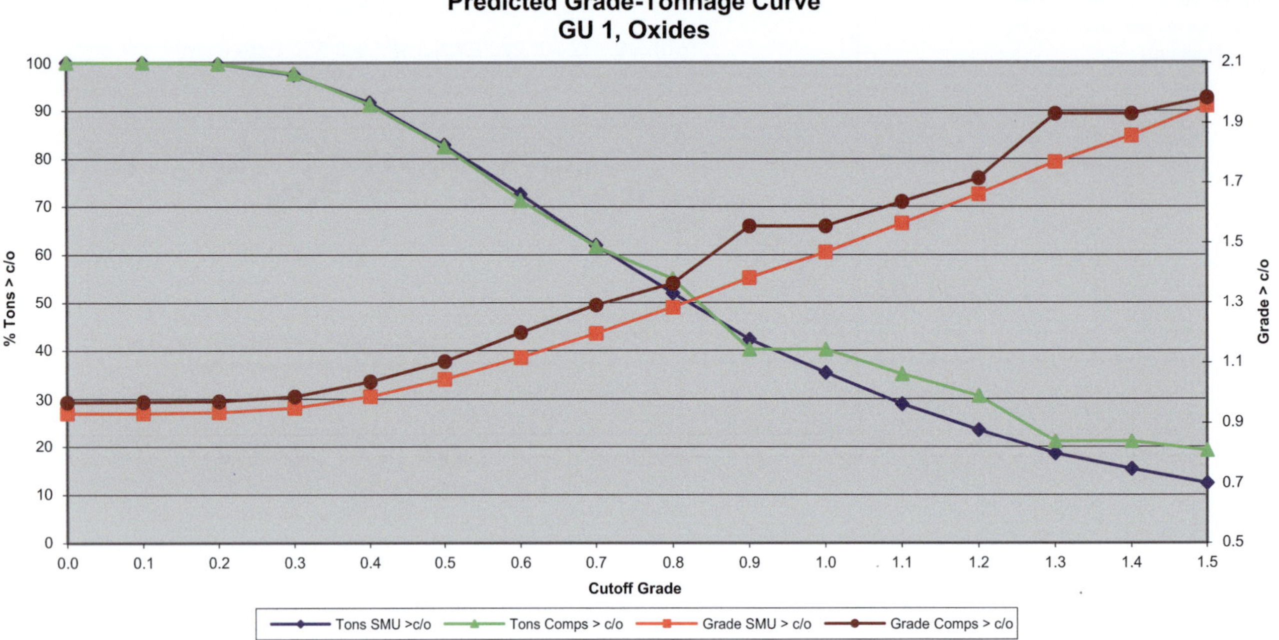

Fig. 14.21 Predicted Grade-Tonnage curve, 10 m declustered composites vs. SMU (from the DG model), TCu, Domain 1 (Oxides)

a relative loss in grade of about 6 % (from 1.104 % TCu to 1.045 % TCu), and a gain of about 0.5 % in tons.

14.1.21 The Cerro Colorado 2003 Resource Block Model

The resource block model defined is 3,600 m long in the East-West direction, 2,200 m long in the North-South direction, and 710 m deep. A limiting lower surface was developed based on current drill hole depth, extending on cross sections the end points of drill holes between 20 and 30 m. This lower surface ensures that all drill holes are contained within the model, and also avoids undue extrapolation of drill hole grades at depth. A block size of 20 × 20 × 10 m was deemed appropriate for the available drill hole spacing. According to Journel and Huijbregts (1978), block size should correspond to approximately 1/3 to 1/2 of the average drill hole spacing, which is about 50 × 50 m after infill drilling is completed.

To incorporate geologic contact dilution, a partial block definition was implemented. An auxiliary variable-block size model was first defined, and geologic attributes assigned to each block in this auxiliary model. Then, the variable-block size model was re-blocked to the 20 × 20 × 10 m block model defined above, and percentages of Domains were calculated for each large block. Thus, a detailed definition of the local geologic contacts is obtained, with the final grade of the block estimated based on a weighted average of the estimated grades for each Domain. In this manner, geologic contact dilution is incorporated into the resource model.

14.1.22 The Grade Model

Ordinary Kriging (Chap. 8) was used to estimate the grade for each Domain, using the corresponding correlogram models and applying kriging plans and criteria derived from the statistical analyses described above.

The kriging process was implemented in three estimation passes, each with varying restrictions in the kriging plan. In the more restrictive pass (Pass 1), shorter search radii were used along with a relatively large minimum number of composites; the second and third passes progressively relax these conditions, although in all cases a minimum of two drill holes is required for any block to be estimated. The purpose of these passes is to (a) estimate the blocks that are well informed with more local information, while the opposite is true for blocks in outer areas of the deposit. This results in a block model distribution with higher variance than otherwise, and thus provides better control over the smoothing effect of kriging; and (b) to provide an initial indicator for resource classification, since the kriging plan provides the summary of all parameters required to estimate each block, and it is domain-specific, which implies that the local geology is taken into consideration. This is accomplished by storing a flag in the block model indicating in which pass the block was estimated.

The kriging plans were the result of a calibration or cross validation exercise using production date. The existing blast

Table 14.8 TCu Kriging plans, by domain, 2003 resource model, final iteration

Domain	Pass	Y'/X'/Z' search radii (m)	Search Rotation Angles.	Min. No Of Comps.	Max. No Of Comps.	Min. No Of Octants.	Min. No of Composites per Octant.	Max. No of Composites per Octant.	Composites used.
0	1	30/40/55	-102/27/22	4	10	4	1	4	0
	2	45/60/120	-102/27/22	3	12	3	1	4	0
	3	100/200/250	-102/27/22	2	12	2	1	4	0
1	1	55/30/45	13/0/74	4	8	3	1	4	1,2
	2	120/50/75	13/0/74	3	10	3	1	4	1,2
	3	400/140/350	13/0/74	2	12	2	1	4	1
2	1	35/55/25	-56/3/-3	4	8	3	1	4	1,2
	2	75/120/40	16/41/-13	3	10	3	1	4	2
	3	380/430/180	16/41/-13	2	12	2	1	4	2
3	1	65/35/50	15/35/-66	4	8	3	1	4	3
	2	120/45/90	15/35/-66	3	8	3	1	4	3
	3	420/170/370	15/35/-66	2	10	2	1	4	3
4	1	55/25/50	47/11/79	4	8	3	1	4	4,6
	2	120/90/30	15/1/6	3	8	3	1	4	4,6
	3	390/390/140	15/1/6	3	10	2	1	4	4
5	1	55/45/25	16/-1/3	4	8	3	1	4	4,5,6
	2	120/75/30	16/-1/3	3	8	3	1	4	4,5,6
	3	380/160/430	-59/22/-16	3	12	3	1	4	5
6	1	35/65/45	-5/8/80	4	8	4	1	4	5,6
	2	40/120/80	-5/8/80	3	10	3	1	4	5,6
	3	420/370/170	-38/-2/14	3	12	3	1	4	6
7	1	55/45/30	0/0/0	4	8	3	1	4	7
	2	120/80/40	0/0/0	3	10	3	1	4	7
	3	420/370/150	0/0/0	2	12	2	1	4	7
8	1	75/30/50	-89/86/-22	4	10	3	1	4	8
	2	140/90/60	42/-9/-8	3	12	3	1	4	8
	3	400/350/150	42/-9/-8	2	16	2	1	4	8

holes within a pre-defined volume representing approximately the last two years of production where averaged into $20 \times 20 \times 10$ m blocks. These same blocks were re-estimated iteratively, using alternative kriging plans, until a reasonable match between kriged estimates and block-averaged blast holes was achieved. It is advisable to validate the kriging plans any time that a reference grade distribution is available, which is different than using geostatistical cross validation methods. Classical cross-validation methods do not directly validate block estimates, and have been questioned as tools for validation of kriging parameters (Davis 1987; Solow 1990).

Block kriging was performed by discretizing the block with $4 \times 4 \times 1$ points. Note that, in the vertical dimension, no discretization is necessary because 10 m composites were used to estimate 10-m high blocks. Octant searches were used, this to aid kriging in the declustering process. The search ellipsoids were oriented according to the correlogram models, although the search anisotropies were less marked. This is done to ensure using composites from directions of lesser continuity, which contribute useful information and could be discarded if a very strong anisotropic search is used. The variable orientations of the search ellipsoids correspond to the orientations of the correlogram model structures. This is because it is assumed that the anisotropy observed in the first correlogram structures is mostly influenced by superimposed short-scale controls, such as cross faulting. Therefore, the kriging passes with smaller search radii are influenced by the first structure of the correlogram model, while the kriging passes that use longer search radii are mostly influenced by the second structure.

The estimation parameters used for each Domain for TCu are shown in Table 14.8. In most cases, all boundaries between Domains were treated as hard boundaries, using only composites from that domain to estimate within the domain (see last column in Table 14.8). However, there were some

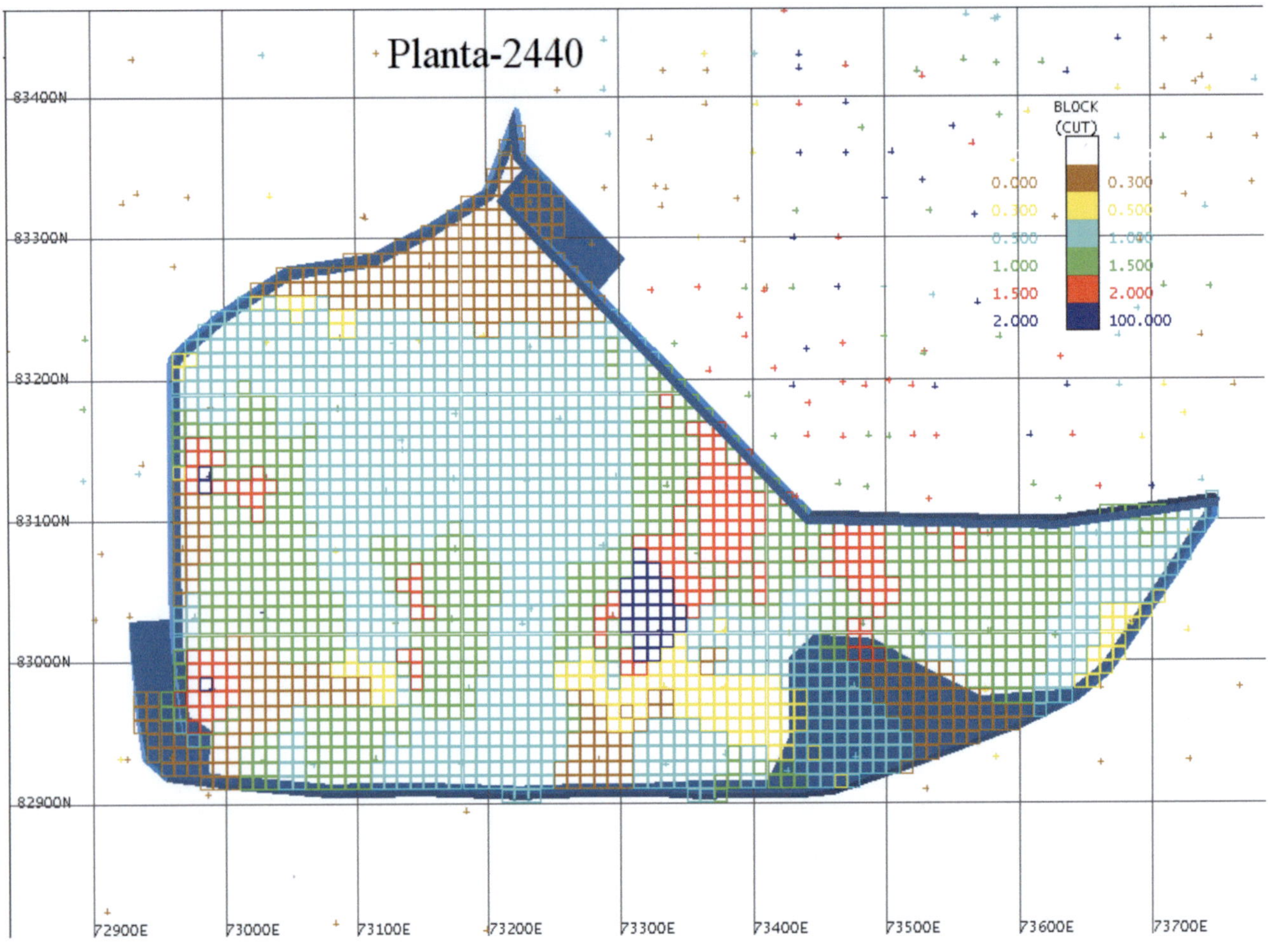

Fig. 14.22 Partial view of estimated TCu grades, Bench 2440

exceptions. The rotation angles used to specify the search ellipsoids are given according to the following convention:

- Θ_1 = left-hand rotation of the *X-Y* axes about the *Z*-axis, clockwise is positive (azimuth).
- Θ_2 = right-hand rotation of the Y'-Z' axes about the X'-axis, positive is up (dip).
- Θ_3 = left-hand rotation of the X'-Z' axes about the Y'-axis, positive is down (plunge).

Figure 14.22 shows a partial view of the estimated grades for Bench 2,440 m, in the southern area of the deposit. Note in this view a NNE-trending zone of higher grades (in red and blue, TCu > 1.50 %), which corresponds to one of the major structural controls in the deposit.

14.1.23 Resource Classification

The method used to classify resources is based initially on the kriging passes described above. These passes provide a measure of the quantity and quality of information used to estimate each block. If a block has been estimated in the more restrictive kriging pass, then it has closer and better information than any block estimated during the second or third passes. If the passes (kriging plans) are related to and influenced by geologic knowledge, then they can be interpreted in the same general sense that is required by current resource classification systems. Additional aspects to consider are:

- If the kriging passes are to be used as the basis for resource categorization, then basic geologic and geostatistical criteria must be met. In the case of Cerro Colorado, this is the case because the search radii used in pass is related to the correlogram ranges and anisotropies. The kriging passes are different for each Domain (just as the correlogram models are), and therefore they reflect the different geologic characteristics of each Domain.
- The use of kriging passes as a basis for resource categorization implies more complexity than most of the other options used in practice. Criteria such as the minimum distance or the kriging estimation variance are simpler, in the sense that they rely on a smaller number of variables to determine the category of each block. Also, it is some-

Table 14.9 Global resources, oxides, supergene sulfides and MSH only

Cutoff (% TCu)	*Metric Tons (x1000)*	*TCu %*	*SCu %*	*Metal TCu (Metric Tons, x1000)*
0.0	509,571	0.768	0.374	3,916
0.3	486,154	0.797	0.387	3,877
0.5	376,097	0.907	0.432	3,411
0.8	186,690	1.187	0.528	2,217
1.0	111,051	1.389	0.562	1,542

times difficult to relate kriging estimation variance values to specific geologic or distance factors. Interestingly enough, the conceptually more complex (and complete) kriging passes scheme is much easier to implement.

- Regardless of whether kriging passes, kriging estimation variance, or any other method is used, it is not recommended that the initial determination of the resource categorization be its final version. A posterior semi-automatic processing of the initial category can be used to smooth out locally patterns of categorization that are not sensible. At the same time, it allows to impose additional constraints.

The post-processing of the initial resource categories was done based on the kriging passes by manual interpretation on bench maps. The purpose is to smooth out areas with high mixtures of categories. This process injected a degree of geological continuity into the categorization pattern, and also allowed for addition of special constraints. For example, the mixed sulfide-hypogene unit (MSH) is very small and shows very limited geological continuity. All MSH blocks initially estimated in the first kriging pass were re-categorized as "indicated". Another specific, somewhat arbitrary, constraint was that all mineralization below the 2,200 m elevation was categorized as inferred, given the few drill holes that reach that deep in the deposit.

Figure 14.23 shows the resource classification of the blocks in Bench 2,440. The red envelope encloses the measured blocks; the green polygon encloses the indicated blocks; and the remaining blocks are inferred.

The resource tabulation for the block model just described is shown in Table 14.9, for total and soluble Cu (TCu and SCu).

14.1.24 Estimation of Geometallurgical Units

The geometallurgical units (GMUs) represent volumes of expected similar response to metallurgical treatment. The GMUs were defined as a combination of mineralization types and alteration types. These are the two most important factors in predicting recovery by bioleaching, the metallurgical process used at Cerro Colorado.

The estimation of the GMUs is based on assigning the majority code of the alteration and mineralization types to each block, in addition to an indicator-based estimation of Point Load values. This hybrid method was applied as follows:

1. Initially, the GMUs corresponding to Waste, Hypogene, and MSH were defined based on the majority code of the mineralization type in each block.
2. Oxides and supergene sulfides were subdivided according to the presence or not of argillic alteration.

14.1.25 Estimation of OXSI/OXSA and of SNSI/SNSA

OXSI and OXSA are the two GMUs defined for oxides, with and without argillic alteration, respectively, while SNSI and SNSA are the codes given to the two corresponding supergene sulfide subpopulations.

The presence or absence of argillic alteration for each oxide and supergene sulfide block was estimated from the original database instead of direct geologic interpretation. Two indicator variables were defined, one for oxides and the other for supergene sulfides, see Eqs. 14.1 and 14.2. The indicator variable takes the value of 1 if there is no argillic alteration (hard rock, OXSI and SNSI), and 0 if the material has argillic alteration (softer rock, OXSA and SNSA).

$$I_{OX}(\underline{x};z) = \begin{cases} 1 \text{ if } z(\underline{x}) \in OXSI \\ 0 \text{ if } z(\underline{x}) \in OXSA \end{cases}$$

$$I_{OX}(\underline{x};z) = \begin{cases} 1 \text{ if } z(\underline{x}) \in OXSI \\ 0 \text{ if } z(\underline{x}) \in OXSA \end{cases}$$

The estimation of the two indicators is done independently using ordinary kriging of the indicator variable, and applying the corresponding indicator variograms. The oxide blocks are kriged using only the oxide data, while the sulfide blocks are kriged using only the sulfide data. The kriged value for each variable and for each block will be between 0 and 1, which can be interpreted equally as a proportion of the block or a probability of the block being one or the other category.

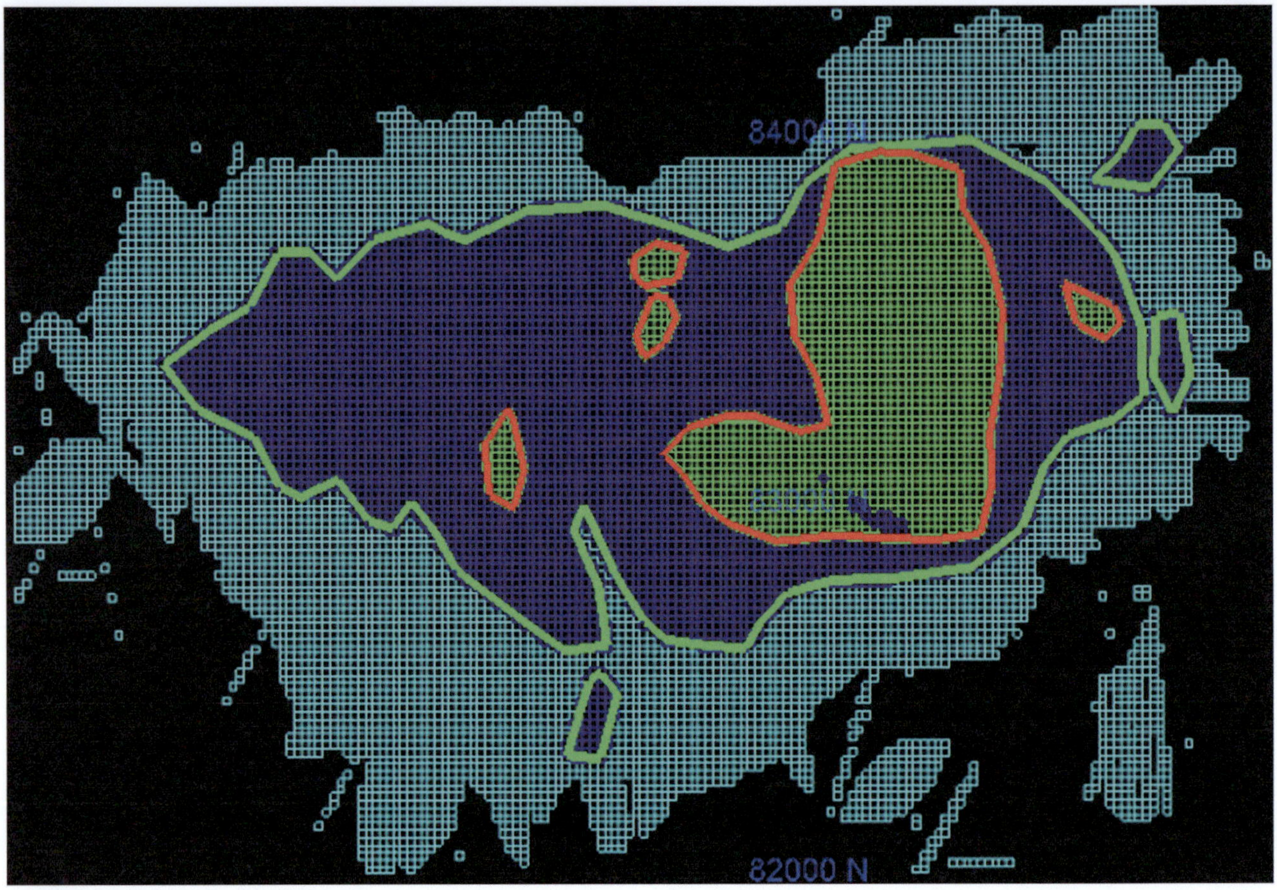

Fig. 14.23 Resource classification, Bench 2440m

Each block is coded as OXSI or OXSA if oxide, and SNSI or SNSA if sulfide, if the estimated indicators is greater or equal to 0.5 (hard rock, no argillic alteration), or less than 0.5 (soft rock, with argillic alteration), respectively.

14.1.26 Estimation of Point Load

The Point Load test measures resistance of the rock (drill hole diamond core) to axial compression. There is known relationship between Point Load and the Geometallurgical Unit, demonstrated by tests from production areas, where samples are taken and its Point Load determined. These values have been used to date mostly to determine optimal parameters and design of the blast patterns, sequences and timing, explosive charges, etc.

There were 1,591 Point Load values available in the drill hole database to build this model, the same intervals for which specific gravity was determined. There is a very good correlation between Point Load and the presence of argillic alteration, which allows for an indicator model to be used to ensure consistency. Figures 14.24 and 14.25 show the values

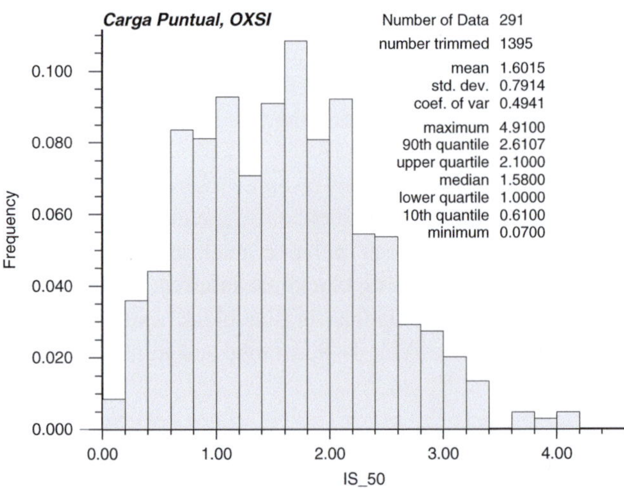

Fig. 14.24 Histogram and basic statistics, Point Load, OXSI

of Point Load for drill hole intervals logged as OXSI and OXSA, respectively.

Limiting values describing the relationship between Point Load and the GMUs were defined based on field experience:

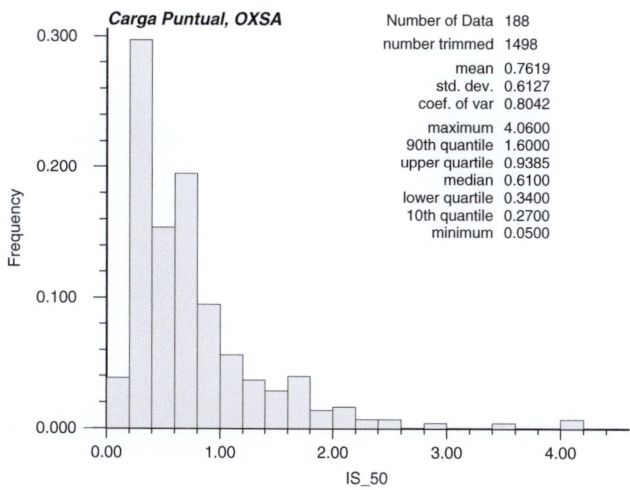

Fig. 14.25 Histogram and basic statistics, Point Load, OXSA

1. OXSI should not have Point Load values less than 0.8.
2. OXSA should not have Point Load values greater than 1.3.
3. SNSI should not have Point Load values less than 0.9.
4. SNSA should not have Point Load values greater than 1.0.

Figures 14.26 and 14.27 show the GMU final model (after correction for the estimated Point Load values) and Point Load model itself for Bench 2400, respectively. Observe how both models share the same general spatial distribution and trends, imposed by the variograms of the OXSI/OXSA and SNSI/SNSA indicators and the corresponding Point Load, respectively.

14.1.27 Resource Model Calibration

Resource Model calibration of the estimation process is intended to reproduce, as closely as possible, past production data. In this sense, it is a form of cross validation, since "known" values are used as reference to improve the resource model. The reference is a block model obtained from blast hole information for a pre-determined production period.

A calibration volume was established based on a surface approximately 80 m above the current topography (as of 01/31/03). Blocks corresponding to extraction periods from 02/01/02 through 01/31/03 were flagged from available triangulations. Thus, all blocks within the calibration model could be identified by month, year, and within the calibration volume.

The calibration volume was estimated using only the 10 m drill hole composites (no blast hole data) and kriging plans defined previously. Estimation "errors" were obtained

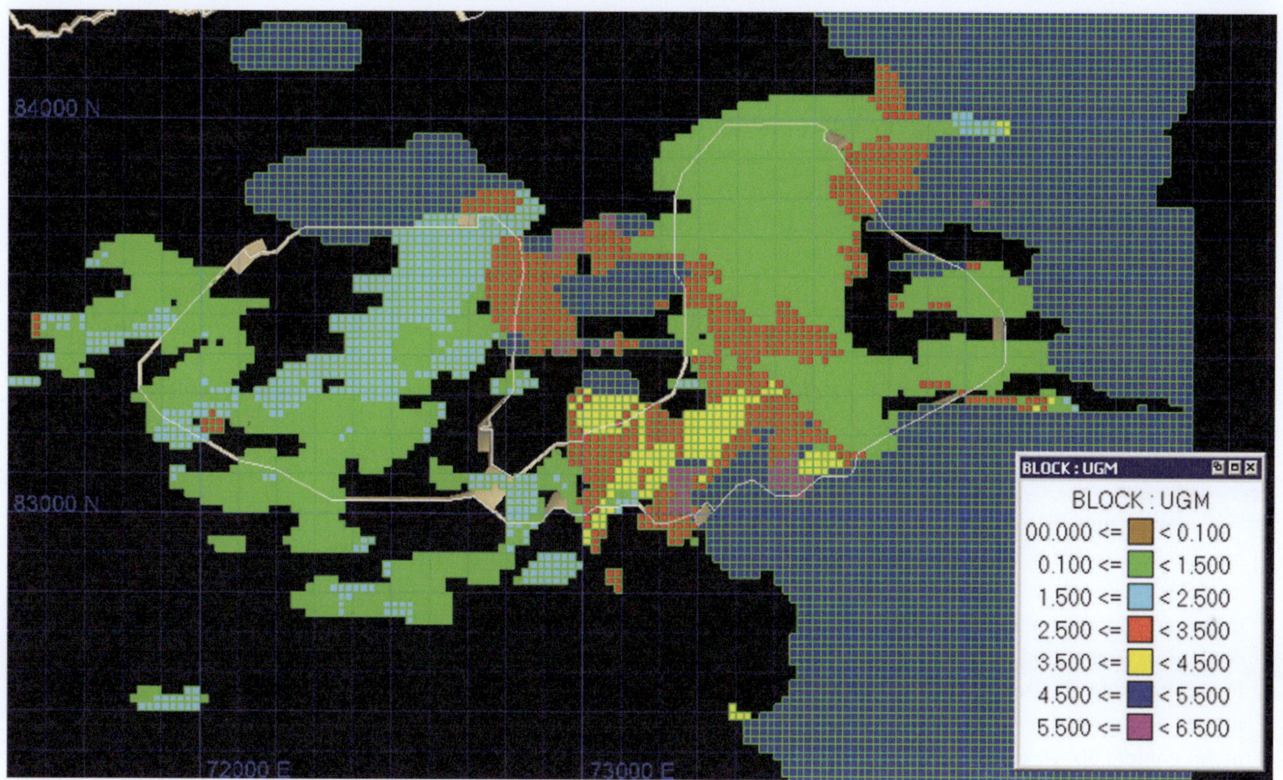

Fig. 14.26 GMU final model, Bench 2400m

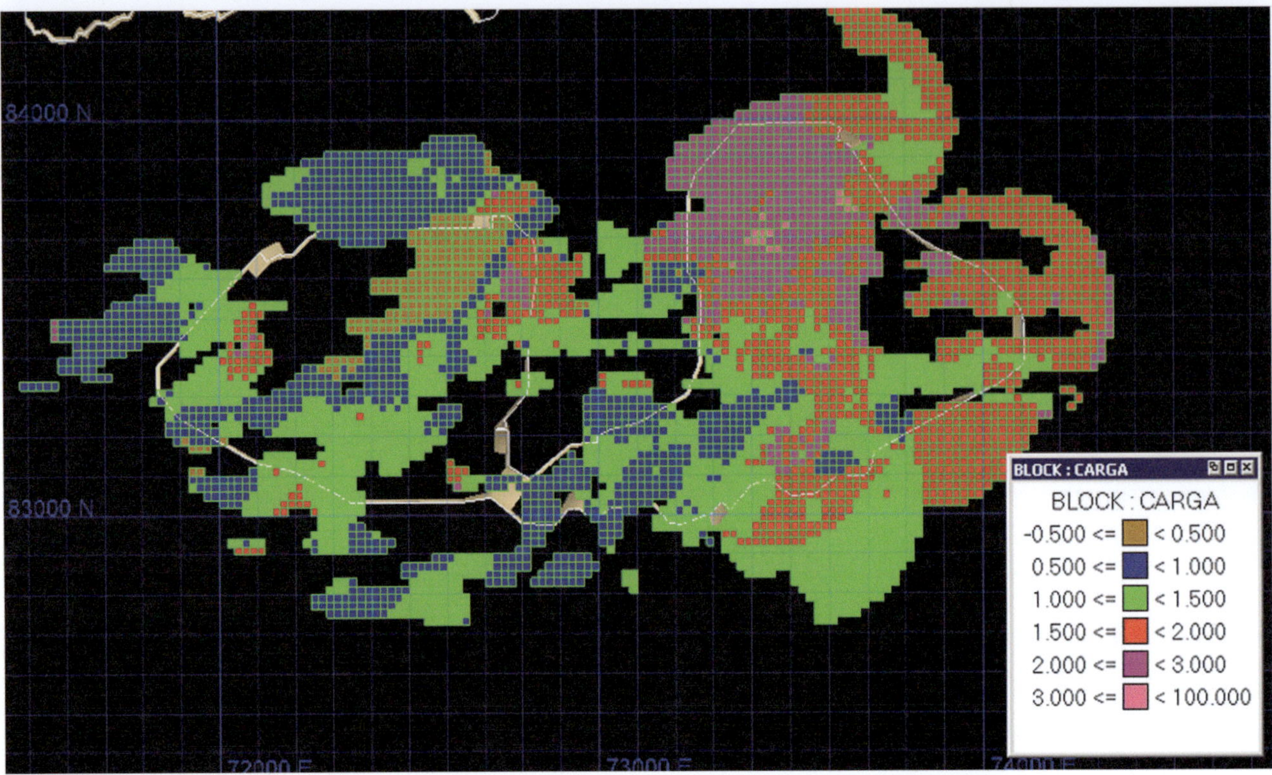

Fig. 14.27 Point load model, Bench 2400m

from comparing the two models by periods (month, year, and global within the 80 m extraction volume). Based on these comparisons, certain kriging parameters were modified, and the resource model re-estimated. A number of iterations were completed until it was decided that no further optimization of the resource model was possible.

An assumption is made that blast holes models provide a reasonable reference model for calibration. It can be argued that individual blast holes are less reliable than individual composite samples, but in the case of a block model this problem diminishes (as long as there are no significant biases), due to the averaging of the individual blast holes.

Another important aspect of the calibration process is to decide what an acceptable match of the reference model is. In the case of Cerro Colorado, the following acceptance criteria were used:

1. For comparisons on a monthly basis, at least 10 of the 12 months had to be within 10% relative to the reference model for tonnage and grade above economic cutoffs. For these same months, variations of 5% or smaller for metal content were required as well.
2. For comparing the annual volume, the accepted deviation was 5% (relative) with respect to tonnage, grade, and metal content above economic cutoffs, and applied on an individual estimation domain (Domains) basis.
3. For the overall production volume, up to 5% deviation for tonnages and grades were defined as acceptable, while metal content was required to be within 3% for all and each of the main mineralized units (Domains).

Table 14.10 shows, as an example, the annual comparisons for each Domain for iteration #5 of the 2003 resource model. The table shows grade, tonnage, and metal content of the Resource Model ("Model"), grade, tonnage, and metal content of the blast hole model ("Reference"), and the relative differences, which are positive if the resource model is larger than the reference model (overestimation). Observe how most of the criteria described are met for the economic cutoffs. Some of the Domains show very good comparisons (1 and 4, in particular). The largest relative differences are found in Domain 6 (medium to low grade sulfides), but largely due to a total of only 200,000 tons produced in the year, equivalent to less than 2 weeks production, or about 20 blocks in the resource model.

An important conclusion from this calibration exercise is that it is possible to re-estimate immediate past production to an acceptable precision with drill hole data only.

14.1.28 Statistical Validation of the Resource Model

Whenever production data is available, validation of the block model generally requires that (a) the resource model is consistent with the assumptions and parameters applied,

Table 14.10 Iteration #5, annual comparison, by domains

Cutoff	Model TCu(%)	Model Tons (x1000)	Model Metal	Reference TCu(%)	Reference Tons	Reference Metal	Grade Diff.(%)	Tons Diff (%)	Metal Diff (%)
Annual Domain 1									
0.0	0.84	3472	29165	0.86	3472	29859	-2.3	0.0	-2.3
0.3	0.84	3420	28728	0.86	3452	29687	-2.3	-0.9	-3.2
0.5	0.90	2988	26892	0.93	2980	27714	-3.2	0.3	-3.0
Annual Domain 2									
0.0	0.80	1300	10400	0.79	1300	10270	1.3	0.0	1.3
0.3	0.80	1300	10400	0.79	1296	10238	1.3	0.3	1.6
0.5	0.86	1128	9701	0.90	992	8928	-4.4	13.7	8.7
Annual Domain 3									
0.0	0.69	1040	7176	0.67	1040	6968	3.0	0.0	3.0
0.3	0.69	1040	7176	0.67	1040	6968	3.0	0.0	3.0
0.5	0.78	748	5834	0.80	656	5248	-2.5	14.0	11.2
Annual Domain 4									
0.0	1.20	1700	20400	1.20	1700	20400	0.0	0.0	0.0
0.3	1.20	1700	20400	1.20	1696	20352	0.0	0.2	0.2
0.5	1.22	1660	20252	1.23	1632	20074	-0.8	1.7	0.9
Annual Domain 5									
0.0	0.89	896	7974	0.86	896	7706	3.5	0.0	3.5
0.3	0.94	828	7783	0.90	836	7524	4.4	-1.0	3.4
0.5	1.03	708	7292	1.06	640	6784	-2.8	10.6	7.5
Annual Domain 6									
0.0	0.79	204	1612	0.67	204	1367	17.9	0.0	17.9
0.3	0.79	204	1612	0.68	200	1360	16.2	2.0	18.5
0.5	0.80	200	1600	0.72	172	1238	11.1	16.3	29.2
Annual Domain 8									
0.0	0.62	140	868	0.63	140	882	-1.6	0.0	-1.6
0.3	0.62	140	868	0.63	140	882	-1.6	0.0	-1.6
0.5	0.66	116	766	0.68	112	762	-2.9	3.6	0.5

i.e., it is internally consistent, and that (b) the model predicts reasonably well past production, which is defined by acceptance criteria such as the ones described above.

The resource model should be internally consistent: The estimated block values should behave as expected, with no anomalous values and consistent with the methodology applied.

The resource model should be unbiased: The global grade averages for each Domain (at a 0.0% TCu cutoff) should be similar to the *declustered* means of the corresponding 10 m drill hole composites. For example, Fig. 14.28 shows the histogram of TCu estimated block values for Domain = 1 (Oxides) of the resource model, which should be compared to Fig. 14.14 (histogram of the declustered 10 m Domain 1 composites). The resource model for Domain 1 is unbiased, with a global mean aver-

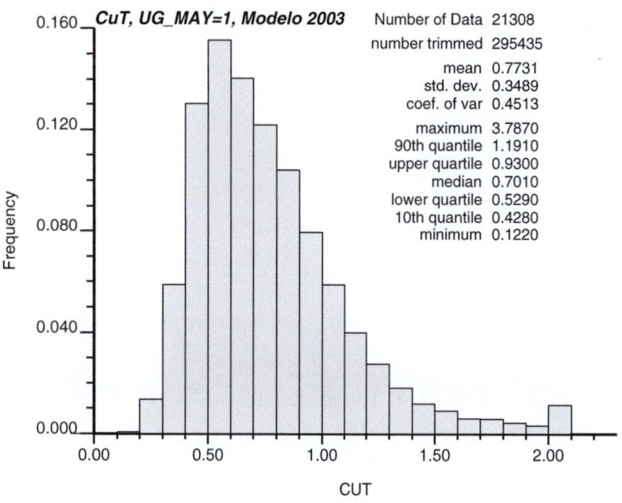

Fig. 14.28 TCu Histogram, Domain = 1, Oxides, Resource Model 2003

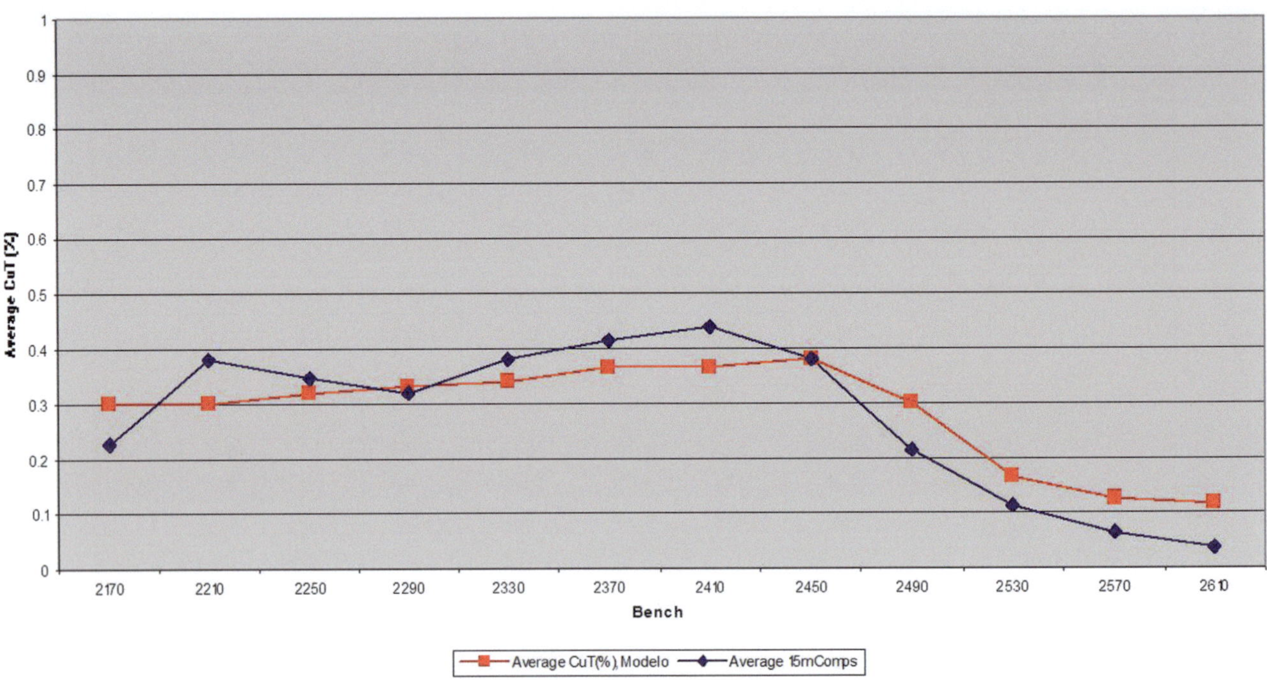

Fig. 14.29 Drift analysis, Resource Model 2003 vs. declustered 10 m composites, vertical direction

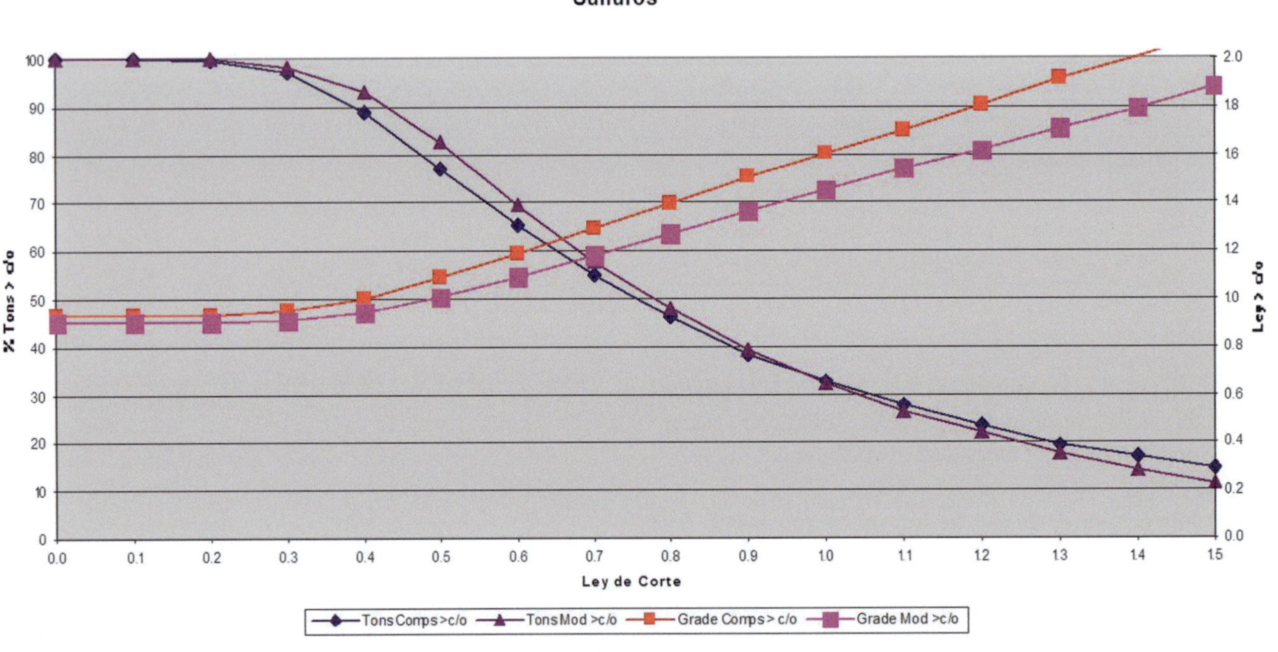

Fig. 14.30 Grade-tonnage curves, 2003 Resource Model vs. DG-predicted SMU Model, all sulfides

age of 0.773 % TCu, compared to 0.771 % TCu for the declustered composites (Fig. 14.29).

Swath or Drift (trend) plots is a one-dimensional graph that shows average grades of the block model and of the declustered composites in certain directions of interest. In practice, for deposits that are massive such as Cerro Colorado (porphyry copper) drift analysis is performed along the three Cartesian coordinates. For a vein- or reef-type deposit, the directions of interest could be along strike and down dip. In the case of Cerro Colorado, the drift in the Easting, Northing, and Elevation directions were analyzed by obtaining the average TCu grade in 100 m slices defined in each horizontal direction, and 20 m (two benches) in the vertical direction. This was done both for the blocks and for the declustered 10 m composites. Figure 14.29 shows the drift of all the estimated block grades in the vertical direction, which compare well with the corresponding declustered composites.

Volume-variance check: The grade-tonnage (GT) curves from the resource model were checked against the expected GT curves, as predicted by the DG model, and corresponding to the SMU-size blocks. Figure 14.30 shows the GT curves for the resource model and for the DG-predicted SMU model, for all the sulfide units combined (Domains 4–6). The correspondence is good, showing that, at a 0.5% TCu cutoff, the resource model has 5% more tons and 9% less grade (is more diluted than predicted by the DG model). A perfect match is not desirable, since the volume-variance correction includes only internal dilution, and the expected grade-tonnage curve is just another model. The resource model should include more dilution than what is predicted by the volume-variance correction, specifically geologic contact dilution and some additional operational (planned and unplanned) dilution.

Software check: The software used for kriging, regardless of whether it is a commercial package or an in-house developed software, should be checked in detail, to ensure that the program actually does what it advertises. For example, an alternative kriging program can be used to estimate several blocks of the most important estimation domains, applying the same correlogram models and kriging plans as used in the resource model. Often this check is performed by third-party reviewers and auditors, but it should be part of the validation protocol of every mining company.

14.1.29 Visual Validation of the Resource Model

The resource model should be checked visually to ensure that there are no inconsistencies in the estimation process or other evident errors or omissions. Each estimated grade in the blocks should be explained by composites surrounding the block, the correlogram models, and the kriging plan used. When performing this check, it is advisable to keep in mind the definition of the estimation domains, and whether soft or hard boundaries where used.

To perform this check, several sets of plan views and cross sections were plotted, showing the estimated TCu and SCu grades for each block, the resource classification code, the composites used in the estimation, and the three-dimensional solids that represent the estimation domains. After completing these checks, the full annotated sets of plan and sectional views should be left as an historic backup, and for third parties to analyze if needed.

This case study represents a very classical application of the principles of mineral resource estimation documented in this book. The following case studies illustrate complementary and different techniques for special applications.

14.2 Multiple Indicator Kriging, São Francisco Gold Deposit

The São Francisco gold deposit is currently owned by Aura Minerals, Inc., and is located in the State of Matto Grosso, approximately 560 km west of Cuiabá, in South-Central Brazil. It is one of several gold deposits on an N-S trending belt.

The São Francisco (SF) deposit occurs in the Fortuna Formation of the Aquapei Group, which is composed of fine- to coarse-grained meta-arenites, with meta-pelites, and occasionally meta-conglomerates. The rocks in the area are folded, faulted, sheared and fractured within a series of broads anticlines and synclines that can be traced over several kilometers, with fold axes striking on a northwest-southeast direction, and gently plunging north 5–10°.

Gold mineralization occurs in epigenetic quartz-filled shear zones generally along the foliation that is oriented parallel or sub-parallel to the axes of the folds. It also occurs in later, flat to shallow dipping quartz veins filling fracture zones and cutting bedding and primary foliation of the host rocks. Gold is frequently coarse, with nuggets several millimeters in diameter in quartz veins. It also occurs within limonite boxworks after pyrite and arsenopyrite. High-grade gold mineralization also occurs where narrow 1–5 cm long quartz veining is intense and it crosscuts bedding, producing a type of stockwork mineralization.

14.2.1 Database and Geology

As of December 2001, the database consisted of over 460 inclined and vertical surface drill holes. The sampling interval is mostly 2 m down the hole, with some of the drill holes inclined about 60° to the Northeast, while others are inclined 60° to the Southwest.

The geologic logging system implemented at São Francisco include descriptions for lithologies, sulfide content, presence of quartz veins and sericitic bands, presence or absence of kaolin, hematite, gold nuggets, and the degree of silicification. The most significant characteristic is hydrothermal alteration, which is the basis for the geologic envelope defined (HAZ envelope). Mineralization occurs when the

Table 14.11 Indicator classes and class mean, Haz-Hi, Haz-Lo, and SAP domains

Class	*Class Mean*	*% Metal*
0.0-0.2	0.086	5.16%
0.2-0.3	0.244	3.90%
0.3-0.5	0.387	6.83%
0.5-0.8	0.631	7.21%
0.8-1.2	0.976	7.02%
1.2-1.8	1.455	6.48%
1.8-3.0	2.294	7.24%
3.0-5.0	3.843	5.95%
5.0-10.	7.108	10.22%
10.0-20	13.969	9.83%
>20.0	44.877	30.17%

Class	*Class Mean*	*% Metal*
0.00-0.25	0.053	19.97%
0.25-0.5	0.346	12.23%
0.50-0.8	0.622	8.55%
0.80-1.2	0.966	6.60%
1.20-1.8	1.463	6.03%
1.80-3.0	2.253	6.23%
3.00-4.5	3.716	5.98%
>4.50	14.330	34.40%

Class	*Class Mean*	*% Metal*
0.0-0.2	0.074	8.36%
0.2-0.3	0.247	6.54%
0.3-0.5	0.387	11.09%
0.5-0.8	0.629	11.13%
0.8-1.2	0.986	12.12%
1.2-1.8	1.455	10.19%
1.8-3.0	2.214	11.70%
3.0-6.0	4.113	15.16%
>6.0	13.927	13.69%

rock has been logged as altered, but it is only a necessary, not sufficient, condition for ore grade mineralization to occur.

Two-meter down-the-hole composites were generated without restriction to lithology or alteration. Short composites may preserve better the high grade tail of the distribution, and thus are preferred in this case to avoid as much as possible grade smearing and dilution when compositing.

14.2.2 Geologic Modeling

The São Francisco geologic model is used to define estimation domains, and consists of three main elements, as delineated by three-dimensional wireframes: the high-grade alteration envelope (HAZ-Hi), a low-grade envelope (HAZ-Lo), and the saprolite (SAP) zone, which was interpreted from drill hole logs.

The definition of the HAZ-Hi envelope was based on certain indicators of mineralization as described in the geologic logs, which are: (1) medium to high (>50%) content of quartz, (2) medium to high (>50%) sulfide (pyrite) content, (3) gold assay greater or equal than 0.40 g/t (this value corresponds to the 60% cumulative probability of the entire Au distribution), (4) the presence of coarse gold (nugget), kaolin, or hematite, and (5) the intercept length.

The following criteria must be met for any interval to be defined as part of the HAZ-Hi zone:

1. Three samples or 6 m length minimum with two or more acceptable indexes (one if it is high sulfide or high gold grade), in a mineralized zone which has at least 2 intercepts at maximum 35 m apart. The structural orientation of the mineralized zones must be shallow dipping to NE in the open pit mineralization (extensional veins zone) or steeply dipping to NE in the Deep South mineralization (extensional + shear veins zone).
2. Six samples or 12 m length minimum with two or more acceptable indexes (one if it is high sulfide or high gold value) to create discontinuous ore zone. Discontinuous ore zones may be extruded to half distance to the next section. Structural orientation must be the same as for the continuous mineralized zone defined above.

The following criteria must be met for any interval to be defined as part of the low-grade (HAZ-Lo) alteration zone:

1. Metasandstone lithology with pellite lenses.
2. In the conglomerate and basement lithologies the alteration boundary must be defined by the presence of sulfide. The low grade mineralization envelope must contain all the high HAZ zones. Low HAZ envelope boundaries must be steeply dipping or vertical.
3. Intercept length must be three samples or 6 m with one or more acceptable index (metasandstone lithology with pellite lenses, or presence of sulfide).

Figure 14.31 shows the histogram and basic statistics for all 2 m Au composites, as well as the defined domains. There are 30,546 2 m Au composites in the database, with a positively skewed Au distribution and few samples representing a significant high-grade population. The average overall Au grade is 0.260 g/t, with a standard deviation of 2.78 g/t, and a coefficient of variation (CV) of 7.88. Approximately 75% of the composite data is below 0.21 g/t (which is below the projected economic cutoff), and only 10% of the data is above 0.573 g/t.

The HAZ-hi envelope has an average grade of 0.81 g/t, although with still a significant proportion of low grade composites, while the SAP zone shows a lower average grade at 0.47 g/t Au, but a larger proportion of medium- and high-grade composites, and therefore a lower CV. Finally, the HAZ low-grade envelope shows an even higher CV, with a 0.23 g/t Au average grade, and only 10% of the composites above 0.37 g/t Au.

Figure 14.32 shows a cross section with interpreted alteration envelopes and corresponding drill holes. The 2 m composites within these wireframes were used in grade estimation.

14.2.3 Class Definition for Multiple Indicator Kriging

The definition of the thresholds (or indicators) and corresponding classes chosen to perform the MIK varied according to the estimation domain considered. Table 14.11 shows the class defined by the Au Indicators, the corresponding de-

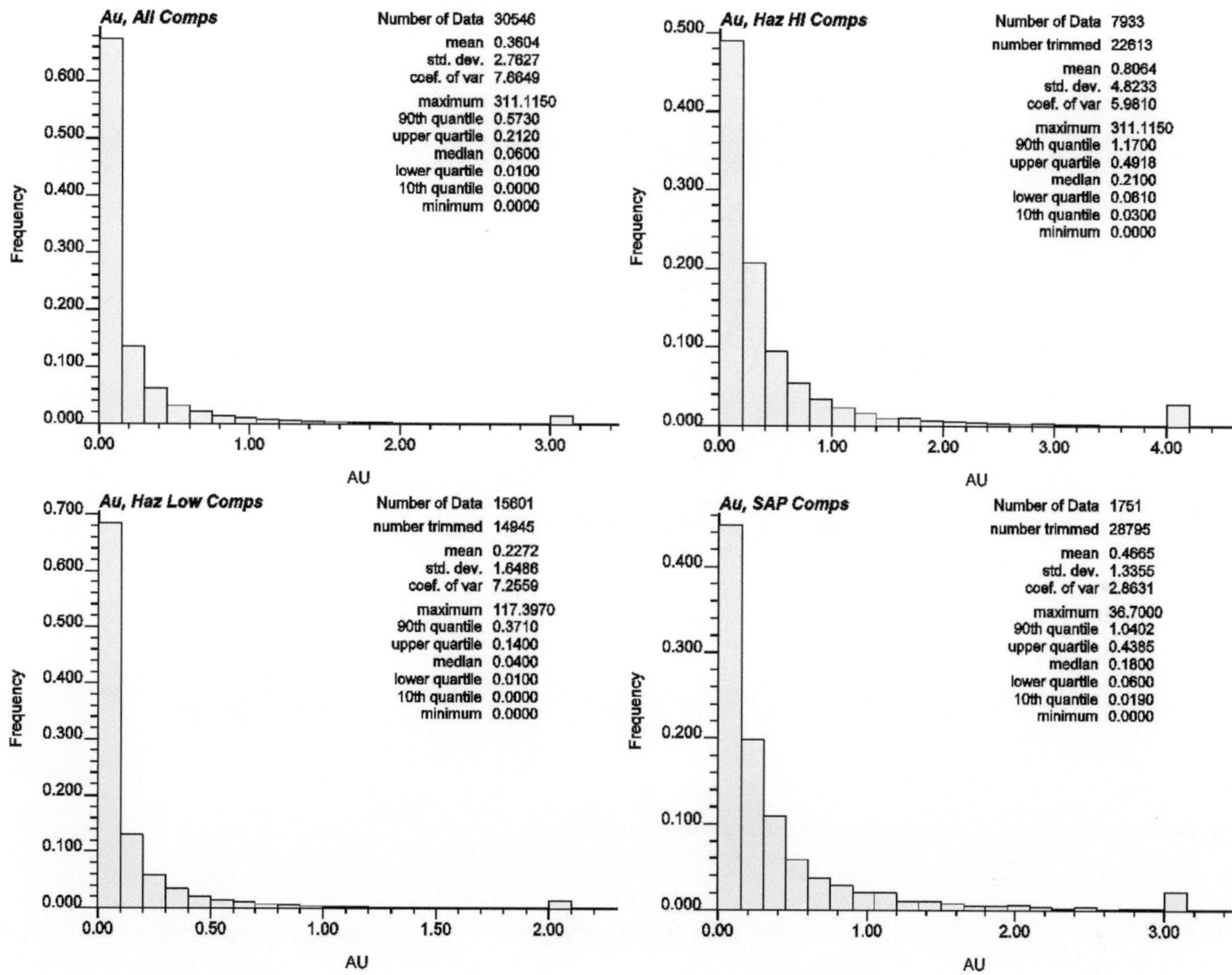

Fig. 14.31 Histogram and basic statistics. All 2 m Au composites (*top left*); High-grade HAZ zone (*top right*); Low-grade HAZ zone (*bottom left*); and SAP zone (*bottom left*)

clustered class means, and the percentage of total contained metal for each class with respect to the total quantity of metal (QM) in the database, expressed as *(g/t)*l,* where *g/t* is the Au grade, and *l* is the composite length. Note how the highest grade classes defined for all three domains contain a significant percentage of the total metal in the zone.

The cell-declustering technique (Deustch 1989) was used to assess the degree of clustering in the deposit. Figure 14.33 shows the histogram and basic statistics of the cell-declustered 2 m composites for the Haz-Hi zone, and should be compared to Fig. 14.31. The clustering effect is significant, particularly for the higher grade units.

14.2.4 Indicator Variograms

The indicator thresholds were used to define the corresponding set of indicator variograms. In most cases, a total of 37 experimental variograms were obtained for each indicator threshold, although for some of the very high thresholds, with few composites in the class, the variogram model was assumed to be a pure nugget effect.

Since there is a large number of indicator variogram models to consider, one for each threshold of each estimation domain, only a few model summaries are shown here as examples. The main conclusions are:

1. As expected, variograms are more continuous for lower indicators, with lower nugget effects.
2. Nugget effects increase and ranges decrease with increasing thresholds, i.e., there is de-structuring of the variogram for higher thresholds.
3. Indicator variograms at the 3.0 g/t or 5.0 g/t thresholds show an almost pure nugget effect, and therefore all variogram models for higher thresholds are assumed to be also pure nugget effects. It should be noted that these indicator thresholds are important, even though their variogram models are pure nugget effect, because they provide critical control of very high-grade values.

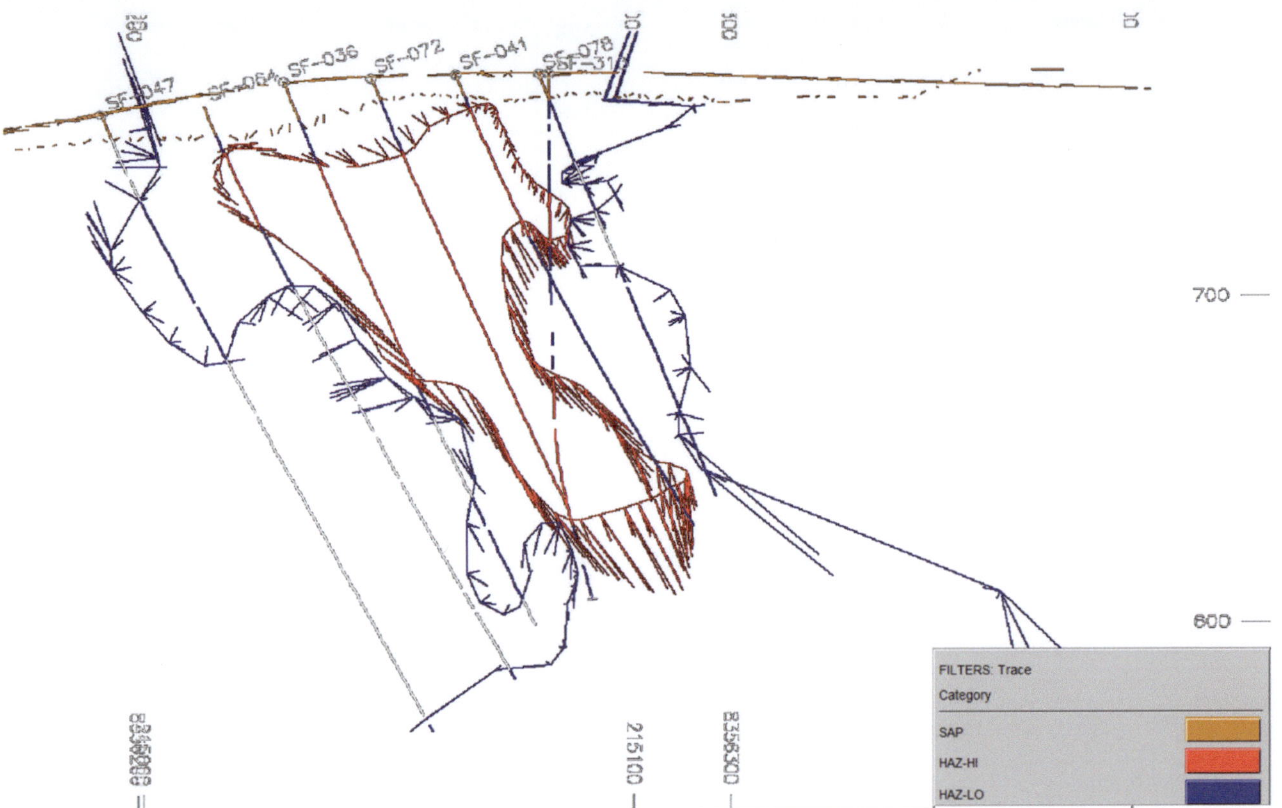

Fig. 14.32 SF cross section showing interpreted alteration envelopes and drill holes

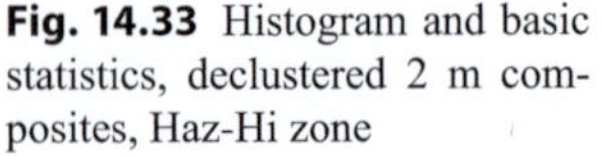

Fig. 14.33 Histogram and basic statistics, declustered 2 m composites, Haz-Hi zone

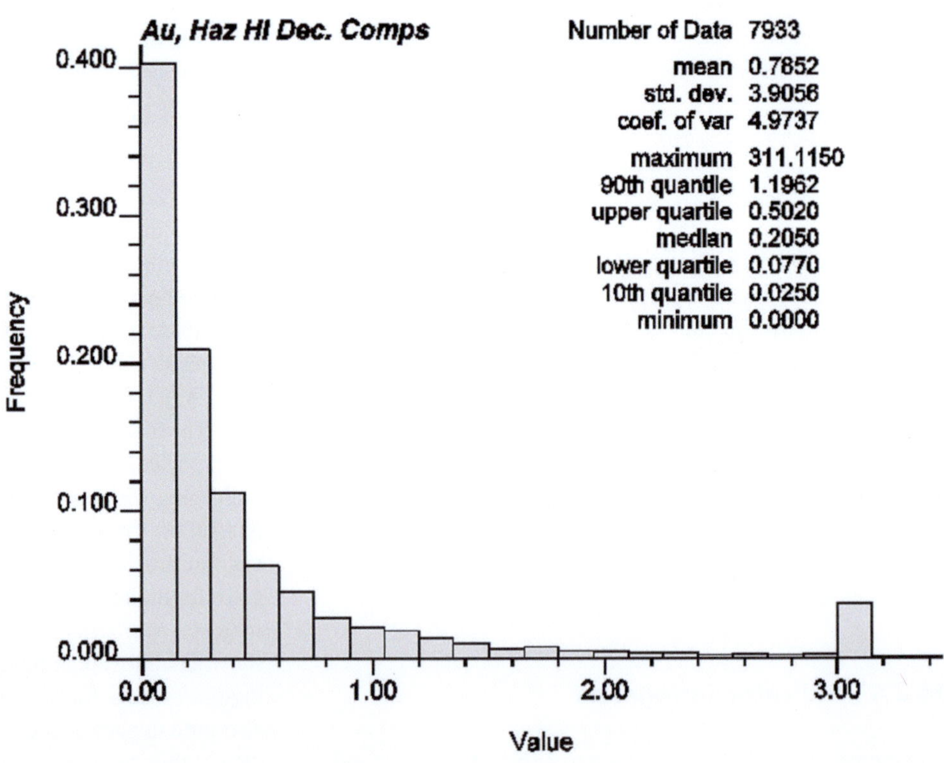

Fig. 14.34 Three main directions of anisotropy, Indicator Variogram Model, 0.2 g/t Au, HAZ-Hi. The directions are N30W (*top*), N60E (*middle*), and vertical (*bottom*). Note that the relative nugget plus the first structure represent over 93 % of the total variance, which means that the range at which kriging will be effective is much smaller than the range of the variogram model

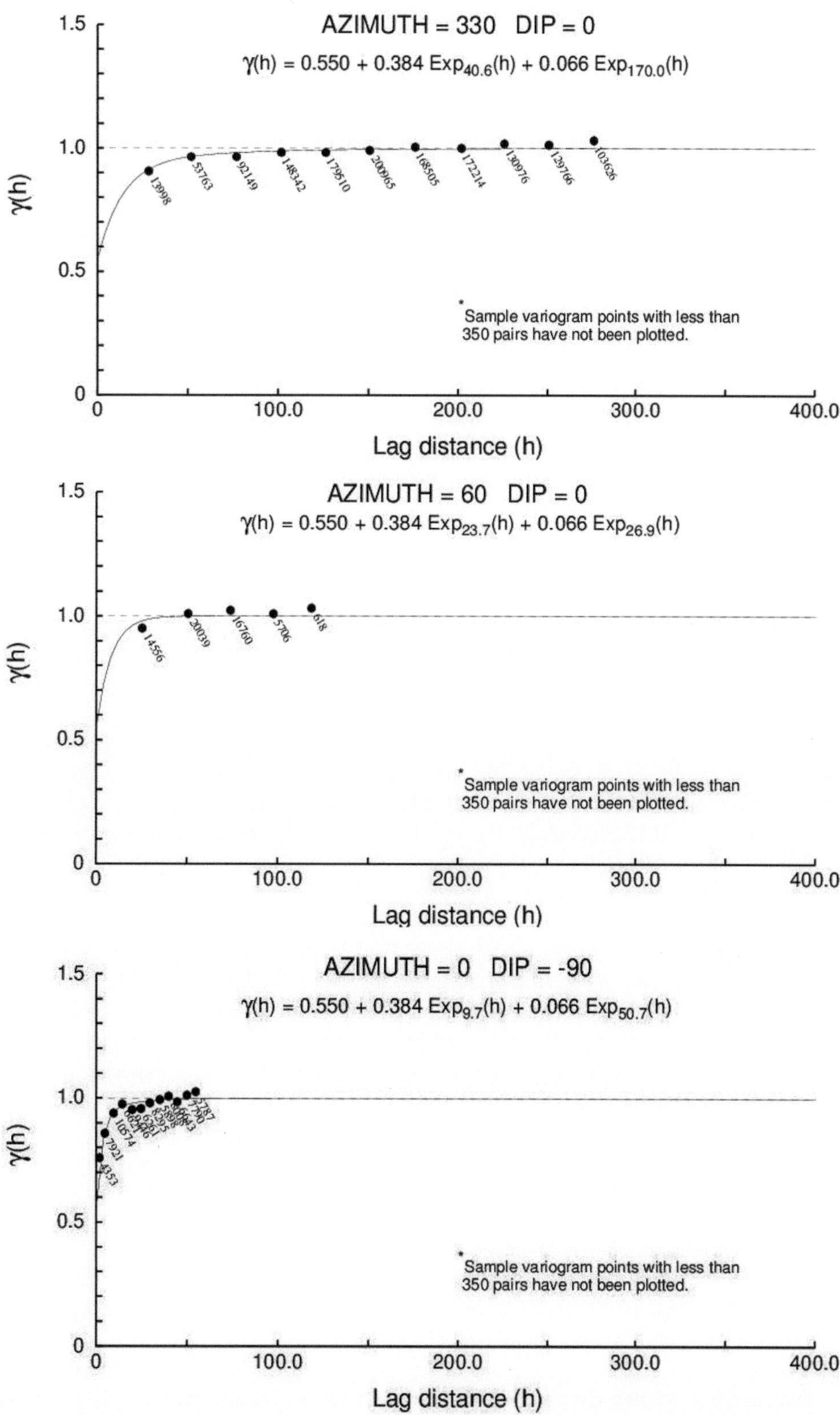

Table 14.12 MIK kriging plans by estimation domain, São Francisco

Estimation domain	Pass	Rotated *Y* search (m)	Rotated *X* search (m)	Rotated *Z* search (m)	Min no. comps	Max no. comps	Rotation angles (after rotation) $\Theta_1/\Theta_2/\Theta_3$
HAZ-Hi	1	20	10	20	5	8	−30/−60/0
	2	40	20	40	4	8	−30/−60/0
	3	130	65	130	3	10	−30/−60/0
SAP	1	20	10	20	5	8	−30/0/0
	2	40	20	40	4	8	−30/0/0
	3	130	65	130	3	10	−30/0/0
HAZ-Lo	1	20	10	20	5	8	−30/−60/0
	2	40	20	40	4	8	−30/−60/0
	3	150	75	150	3	10	−30/−60/0

4. There is a change in continuity and orientation at or about the 0.8–1.2 g/t thresholds, particularly for mineralization within the Haz-Lo envelope. It is likely that it corresponds to the mixture of populations within the Haz-Lo, which would include a more disseminated or stockwork-style zone, with the occasional presence of the narrow veining characteristic of the higher-grade central portion of the orebody (Haz-Hi).
5. For most thresholds, 70–80 % of the total variability is reached at or less than 30–40 m. This implies that for any kind of kriging, the weights assigned to data beyond this distance will be minor, and approximately the same in all directions.

Figure 14.34 shows, as an example, the directional fits for the three main directions of anisotropy for the 0.2 g/t Au indicator variogram, HAZ-Hi domain.

The indicator models (all exponential structures) behave as expected, confirming general directions of anisotropy expected from geologic knowledge and also consistent with those developed in prior resource models. The 0.2 g/t Au indicator shows the general N-W trend of the data, while cross-cutting features drive the 3.0 g/t indicator model (for the first variogram model structure), while the much less important (in terms of contribution to the overall sill) second model structure following the general N-W trend as well.

14.2.5 Volume-Variance Correction

Consideration of a volume-variance correction is necessary because ore will be mined on volumes different than the volume of the composites used in grade estimation, and in general also different than the volume of each block in the resource model.

The Selective Mining Unit (SMU) for the operation is expected to be 10 × 10 × 5 m (500 m^3). This is based on the equipment size and characteristics of the open pit operation. In order to achieve the expected volume-variance correction, two approaches can be considered: (a) implement a more restrictive estimation, whereby the smoothing of the block model grades is controlled through the kriging plan; the e-type estimates are thus used without further corrections; and (b) an affine correction applied to the estimated distribution quantiles. This is done prior to deriving the e-type point estimates.

The first approach assumes that if the variance of the distribution of estimates is similar to the predicted variance of the SMU distribution, then an appropriate amount of internal dilution has been incorporated into the resource model. If so, the grade-tonnage curves obtained from the block model will approximate the expected grade-tonnage curves of the SMUs. The second approach relies on a direct correction of the estimated quantiles. This approach was not attempted here, mostly because of the lack of production data to help in the calibration of the method.

The Discrete Gaussian method (DG) was used to obtain the theoretical (target) grade-tonnage curves from the 2 m composites. Using the appropriate correlogram models, and considering the 10 × 10 × 5 m SMU, block variances were found to be 20.9 % for SAP, 25 % for Haz-Hi, and 17.1 % for Haz-Lo, respectively, of the original composite variances. These values imply that the variance reduction is very significant, which makes the accurate prediction of diluted resources and reserves difficult, particularly for small volumes. For example, in the case of HAZ-Hi, at an Au cutoff of 0.4 g/t the expected internal dilution (from a 2 m composite to a 10 × 10 × 5 m block) is about 35 % in grade and 17 % in tonnage, while for HAZ-Lo the expected grade dilution is about 12 %, with a slight increase in tonnage. The expected internal dilution is more significant at higher cutoffs.

14.2.6 Block Model Definition and Multiple Indicator Kriging

The block size chosen for the São Francisco resource block model was 10 × 10 × 5 m, intended to reflect the drill hole spacing available. It is considered adequate for the available drill hole spacing, and a reasonable compromise between drilling density and model resolution. It happens to be also the assumed size for the SMU block, but in fact there need not be any relationship between the two.

The resource model is defined by the three triangulations that represent the estimation domains (HAZ-Hi, HAZ-Lo,

Table 14.13 Total resource by category, MIK Model, São Francisco

Cutoff (g/t)	Tons (x1000)	Au Grade (g/t)	Tons (x1000)	Au Grade (g/t)	Tons (x1000)	Au Grade (g/t)	Ounces Au (x1000)	Tons (x1000)	Au Grade (g/t)
0.00	48625	0.34	97710	0.41	146335	0.38	1798	207065	0.18
0.10	28894	0.53	62615	0.60	91509	0.58	1695	77111	0.38
0.13	**24398**	**0.60**	**55574**	**0.66**	**79971**	**0.64**	**1651**	**60495**	**0.45**
0.20	17934	0.76	43746	0.79	61680	0.78	1556	35155	0.66
0.40	**9142**	**1.22**	**23360**	**1.24**	**32503**	**1.23**	**1288**	**11891**	**1.43**
0.60	5645	1.68	14238	1.72	19882	1.71	1091	8150	1.87
0.80	3952	2.10	10116	2.14	14068	2.13	962	7350	2.00
1.00	3192	2.39	8197	2.43	11388	2.42	886	6879	2.08

and SAP). The Haz-Lo triangulation constrains extrapolation towards the edges of the deposit. The block model was built using sub-blocks to better adjust the model to the interpreted wireframes. The edge blocks can be as small as $5 \times 5 \times 2.5$ m. After grade estimation, re-blocking incorporates geologic contact dilution in the model, which is significant when a Haz-Hi block is in contact with a Haz-Lo block.

The Multiple Indicator Kriging method, described in Chap. 9, estimates a distribution of possible values (a cumulative conditional distribution function) based on the indicator thresholds defined above. The point conditional distribution is obtained by kriging each indicator independently, and is later post-processed to provide a block estimate. The block estimate is the sum-product of the estimated probability for each class multiplied by the declustered average grade for each class. For the São Francisco block model, the average grade was obtained for each discretization point within each $10 \times 10 \times 5$ m block, and then averaged up to provide the estimated block grade. This is known as the "e-type" estimate, which is obtained as follows:

$$z^*(\mathbf{u}) = p_1(\mathbf{u}) * c_1 + p_2(\mathbf{u}) * c_2 + \cdots + p_n(\mathbf{u}) * c_n$$

Here $z^*(\mathbf{u})$ represents the block estimate, $p_i(\mathbf{u})$ represents the probability for each class defined, and c_i represents the mean grade assigned to that class. Note that the MIK model constructed in this way does not have any explicit allowance for the volume-variance effect. This is approximated using the constrained kriging methodology described above.

14.2.7 MIK Kriging Plans and Resource Categorization

The kriging plan implemented for the MIK model has the following characteristics:

- Three passes were implemented to estimate each of the three domains. Table 14.12 shows the details for each domain. The search radii was defined according to resource classification criteria, see discussion below.
- The minimum and maximum number of samples required for estimation varied according to the pass, as this was one parameter used to control smoothing.
- An octant search was used in all cases, since it aids in declustering the estimated values.
- The anisotropic search ellipsoids used varied according to the estimation domain. The search ellipsoids approximately follow the general orientation of each Domain. The rotation convention used is a left hand (LH) rotation around the Z-axis first, then a right hand (RH) rotation around X', and finally a RH rotation around Y'.

The resources of the São Francisco deposit were classified on a block-by-block basis, using as basis the defined kriging passes (Table 14.12), which indicate the configuration and quantity of information used to estimate each block. A flag is stored in the block model indicating whether the block was estimated on the first, second, or third pass, and this flag thus corresponds to the measured, indicated, and inferred categories. No further correction was required, since the geometry of the deposit and the geologic domains resulted in a smoothly varying resource class.

Blocks estimated within a 20 m-search radius in the strike and down-dip directions and 10 m in the across-strike directions were classified as measured. Blocks estimated with a 40 m search radius in the strike and down-dip directions and 20 m in the across-strike directions were classified as indicated. Finally, all other estimated blocks beyond 40 m are classified as inferred. Figure 14.35 shows a cross-sectional view of the estimated grades in the model.

14.2.8 MIK Resource Model: Grade-Tonnage Curves

Table 14.13 shows the estimated resources by category and for several cutoffs as of December 2001. At a 0.4 g/t Au cutoff, the measured plus indicated diluted resource stands at 32.5 million tons at a 1.23 g/t average grade, for about 1.288 million ounces of contained gold. There are an additional 57.4 million tons in between the 0.13 g/t and the 0.40 g/t Au cutoffs which is considered Run of Mine (ROM) material.

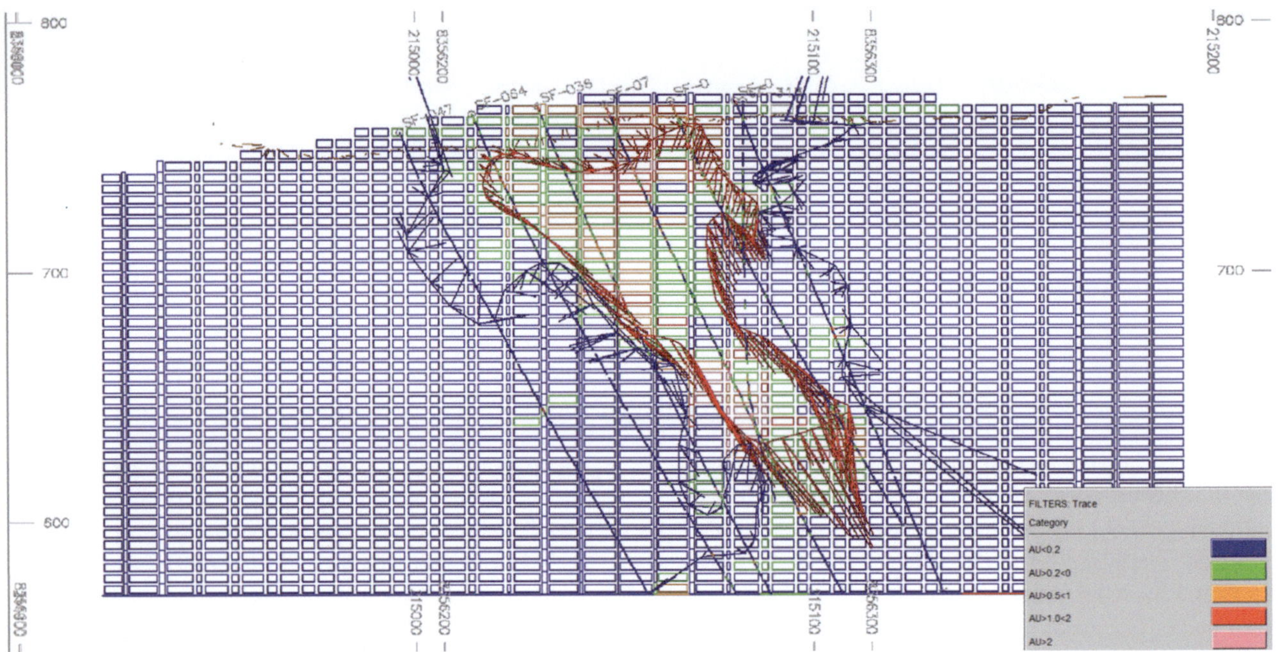

Fig. 14.35 Cross-sectional view of the MIK grade model with 2 m composites used for grade estimation. Warm colors represent higher-grade mineralization

Several statistical and visual checks were performed on the resource model. The block model was reviewed on sections and plans, looking at estimated block grades, the composite data, and the envelopes used to define the volume within which the interpolation took place. The intent of the check is to ensure that every block grade can be explained as a function of the surrounding composites, the variogram models used, and the kriging plan applied.

There is significant mineralization at lower elevations, even below the current pit limit. This is considered a significant risk factor because the optimized pit may reach greater depths based on higher-grade ore with less drill hole information. It is necessary to confirm this mineralization with further drilling.

The smoothing effect of kriging can change significantly the shape of the optimal pits, providing a falsely optimistic or pessimistic picture of grade continuity. Proper consideration of high-grade continuity and the volume-variance effect at São Francisco is extremely important. Figure 14.36 shows that the reblocked MIK model is more diluted compared to the predicted Discrete Gaussian model, tracking the SMU grade-tonnage curves by cutoff region. The differences observed are partly due to the resource model including not only internal dilution, as the predicted SMU distribution does, but also geologic dilution. Nonetheless, there is a clear indication of the difficulties of adequately estimating the internal dilution for this deposit.

Other statistical checks not presented here indicate that the MIK model behaves as expected, is internally consistent with the assumptions and data used to build the model, and has no obvious anomalous values. It is globally unbiased compared to the declustered composites, and internally consistent with the data and correlogram models used to create it.

14.3 Modeling Escondida Norte's Oxide Units with Indicators[1]

The Escondida Norte deposit is owned by BHP-Billiton, and is located 5 miles north of the main Escondida mine and mills. Geologically, the Escondida Norte deposit is the eastern portion of the Zaldívar deposit, currently mined by Compañía Minera Zaldívar (100 % owned by Barrick Gold, Fig. 14.37).

A Porphyry copper deposit such as the Escondida Norte deposit has, characteristically, an oxide zone above the enriched sulfide zones, often separated into high enrichment and low enrichment blankets. This zonation of mineralogy zones stems from the position of the water table which controls supergene mineralization events. A description of this type of deposit and mineralization zones can be found in Guilbert and Park (1985).

Some mineralization units of the Escondida Norte deposit are small relative to the drill hole spacing available, and thus difficult to model using traditional interpretations based on cross sections and plan views. This is the case for the oxides and other units above the Top of Sulfides (TOS) surface. The TOS surface separates the oxidized portion of the deposit, above where acid-soluble copper grades (SCu) are signifi-

[1] BHP-Billiton is gratefully acknowledged for allowing publication of this case study. Geologists R. Preece (BHP-Billiton) and J.L. Céspedes (independent consultant) were responsible for large part of the work described here.

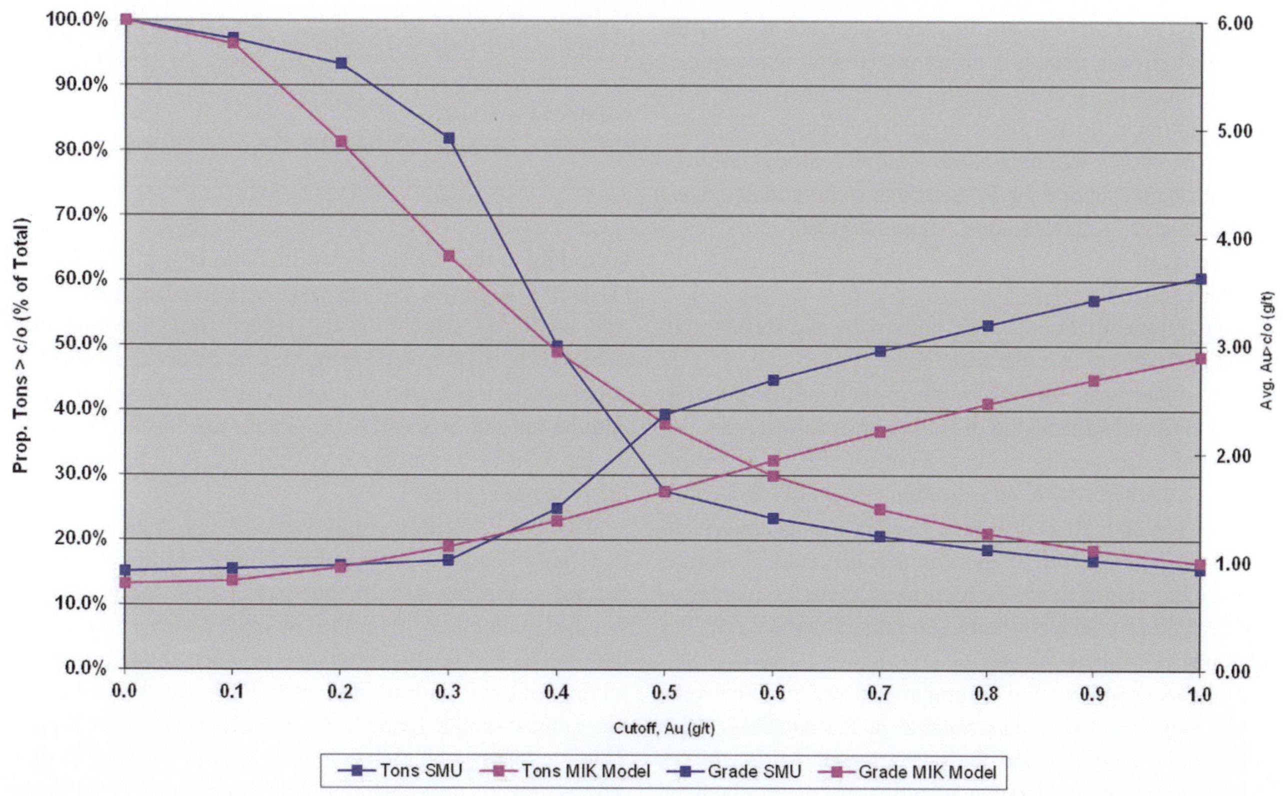

Fig. 14.36 Grade-tonnage comparison, MIK Model vs. Theoretical SMU Model, HAZ-Hi domain

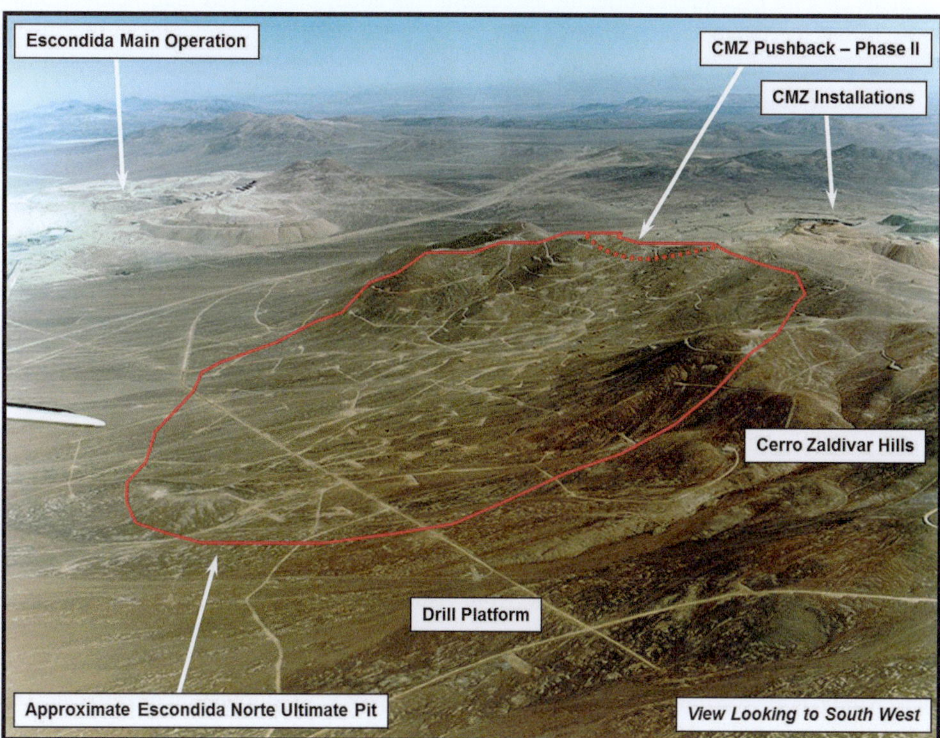

Fig. 14.37 Aerial view of the Escondida Norte Deposit, and the Escondida and Zaldívar (CMZ) mines, looking towards the South-South West

cant, from the sulfide-rich secondary enrichment blanket below. The TOS surface can be defined in absolute terms, i.e., there are no sulfide mineralization above it; or it can be defined in relative terms, i.e., there are very few sulfide minerals above it, generally either mixed with oxide minerals or with minerals characteristic of the leached zone. This last definition is the one that was applied at Escondida Norte.

For the 2002 Escondida Norte Resource Model, all mineralization units above the TOS surface were modeled using an indicator Kriging technique. These included:

- Leach cap;
- Oxide mineralization;
- Mixed mineralization, where both oxide and sulfide minerals are observed in the core; and,
- Partial leach mineralization, where both sulfide minerals and evidence of leaching of Cu minerals is observed.

These units are usually no more than a few tens of meters across in diameter with very erratic, structurally controlled limits. Even on a 50 m drill hole spacing the accurate interpretation and modeling of these units is difficult. Therefore, an indicator approach can be used to estimate the likelihood that each block contains a particular mineralization unit. The main phases of the work are:

1. *Database confirmation*: a comparison was made between the logged information available in the database (from geologic mapping) and the definition of oxide, mixed, and sulfide mineralization resulting from the chemical analyses of Total and Soluble Copper (TCu and SCu). This comparison is intended to ensure the consistency between geologic mapping and sample chemical assays. In the case on Escondida Norte, the agreement was good, such that the logged mineralization units were partially used to define mineralization units.
2. *Definition of estimation units*: the second stage involves the definition of the estimation domains (Chap. 4). This process defines the areas that are considered homogenous for the purposes of estimating oxide and sulfide indicators.
3. *EDA and variogram models*: all necessary statistics and variography was performed for each of the indicators defined (Chaps. 2 and 6).
4. *Block model definition*: a block model covering the necessary volume with an appropriate block size was chosen.
5. *Indicator Kriging and final mineralization unit assignment*: the indicators defined on the available drill hole data were used to krige the probability of each mineralization unit.
6. *Model checks and validations*: significant checking and validation was performed before accepting the proposed probabilistic mineralization model.

The comparison of logged oxidation units from visual observations and chemical SCu/TCu ratios was performed using original samples. For each sampled interval in the drill holes there will be a logged mineralization unit and a corresponding SCu/TCu ratio. The mineralization units were re-defined combining the assayed values and the logged information according to the criteria shown in Table 14.14. In this table, "CHMIN" signifies the chemically defined mineralization unit while "LOGMIN" is the logged mineralization unit. Table 14.14 shows, for example, that oxide mineralization is defined as a soluble Cu grade greater or equal than 0.2 %, and the ratio of SCu to TCu is greater than 50 %. The limits on the SCu/TCu ratio are derived from definitions used by the processing plant.

Table 14.14 Definition of mineralization units

Condition	CHMIN Code	Mineralization unit
LOGMIN = "LEACH" & TCu < 0.2 %	1	Leach
LOGMIN ≠ "LEACH" & TCu < 0.1 %	1	Leach
SCu ≥ 0.2 % & SCu/TCu > 0.5	2	Oxide
CHMIN > 1 & SCu/TCu ≤ 0.15	4	Partial leach
CHMIN ≠ 1,2, or 4, & SCu/TCu > 0.0	5	Mixed oxide-sulfide

The basic statistics of the leach, partial leach, and oxide material are similar, except that the ratio of the chemically defined oxides has a sharp boundary at 50 %, while there are samples geologically described as oxides with a ratio of less than 50 %. The largest difference between the two is in the mixture of oxide and sulfide minerals. There are significantly more mixed samples chemically defined than there are logged. This is explained by the geologist's natural tendency to classify as mixed those samples with approximately the same number of oxide and sulfide particles. If one or the other is clearly prevalent, then the geologist will probably classify the unit as oxides or sulfides, according to the majority observed.

Given that the distributions of both SCu/TCu ratios can be considered similar, the decision was made to use the logged database, relying on the geological definition of the mineralization units.

The next step was the definition of the estimation domains, a process similar to what was described in Chap. 4. The estimation domains result from a combination of lithology and alteration units, as well as structural zones within the deposit. Figure 14.38 shows the structural zones defined in the Escondida Norte deposit. Table 14.15 shows the definitions in terms of oxide and sulfide indicators. The combined indicators define the unit in question: for leach, both indicators have to be 0, i.e., there are no sulfide or oxide minerals; while for mixed, both types of minerals need to be present. There are a total of 5 oxides and 3 sulfide Domains.

After defining the estimation domains, the exploratory data analysis and variography was completed for each indicator and for each domain. This defined the estimation strategy and the indicator kriging estimation plans for each unit.

Indicator kriging was completed using the oxide and sulfide indicators defined in the database, i.e., a weighted linear combination of zeros and ones that result in estimated values between 0 and 1. These interpolated values can be interpreted either as the probability of each estimated block of having

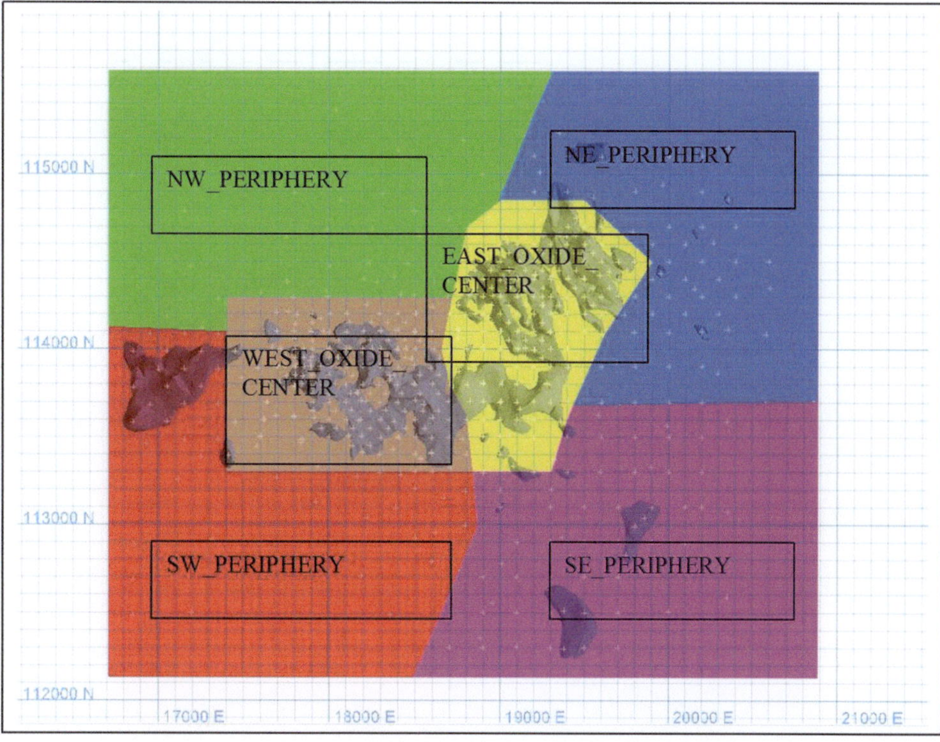

Fig. 14.38 Structural domains for the Escondida Norte deposit. Some estimated oxide bodies are also shown (in *purple*)

Table 14.15 Oxide and sulfide indicators

Indicator	Leach	Oxide	Partial leach	Mixed
Oxide	0	1	0	1
Sulfide	0	0	1	1

Table 14.16 Rules for assigning mineralogy units to estimated blocks

Condition	Numeric code	Mineralization unit
IKOX<0.5 y IKSUL<0.5	1	Leach
IKOX≥0.5 y IKSUL<0.5	2	Oxide
IKOX<0.5 y IKSUL≥0.5	4	Partial leach
IKOX≥0.5 y IKSUL≥0.5	5	Mixed

oxide or sulfide minerals or as the proportion of the blocks with oxide or sulfide mineralization.

For simplicity, a single mineralization unit code was assigned to each estimated block above the top of sulphides (TOS) surface. Therefore, after the kriging runs were completed, a set of rules for assigning the final mineralization unit to the block was enforced. Table 14.16 shows these rules used to define the final mineralization unit in each block above the TOS in the Escondida Norte deposit.

IKOX and IKSUL represent the interpolated value of the oxide and the sulfide Indicators, respectively. If both values are below 50 % (simultaneously), then the block is defined as leached material. If both values are equal or above 50 %, then the probability of the block having both (simultaneously) oxides and sulfides minerals is high enough for that block to be defined as mixed mineralization. If one of the indicators is above 50 % and the other one below, then the block is defined as either oxide or partial leach, depending on which indicator is higher. Several comments are relevant:

1. The rules in Table 14.16 are subjective; although logical and based on the most favorable likelihood for defining the mineralization units, it is still imperfect, since the indicator kriging model will tend to smooth out sharp variations and geological transitions.
2. The indicators are treated independently. This is a limitation of the method, since it is known that the presence of sulfide and oxide minerals above the secondary enrichment blanket are controlled by similar geological factors, most importantly the position of the water table through time. A better alternative may be to use multiple indicator kriging (MIK). In MIK all indicators would be kriged simultaneously and no decision regarding the sequencing of estimation would be needed.[2]
3. The comparison of the average of the estimated indicators and the average resulting from a nearest-neighbor assignment show an acceptable agreement with the exception of one domain of the oxide indicator model.

Figure 14.39 shows a perspective view of the resulting oxide mineralization model, along with the structural domains defined. This example illustrates one possible methodology for developing geologic models in circumstances where (a) the drill hole geologic information is too sparse to be

[2] In subsequent years this modeling methodology was upgraded to a full MIK estimation of the above-TOS units.

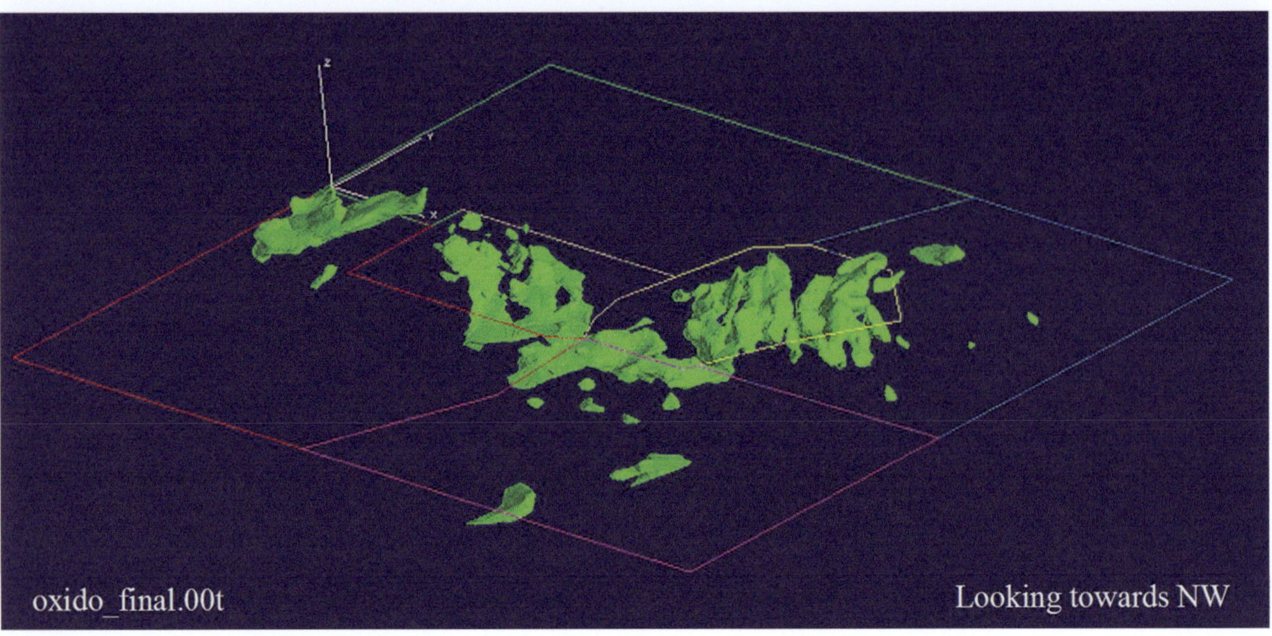

Fig. 14.39 Perspective view of the probabilistic oxide mineralization model, shown with the structural domains

confidently used to build a deterministic modeling through direct interpretation; and, (b) a simple probability-based estimate of presence/absence of a given unit suffices for grade estimation and resource modeling.

14.4 Multivariate Geostatistical Simulation at Red Dog Mine[3]

The Red Dog mine is the world's largest Zn producer. The deposit consists of sulphide ore zones in sedimentary exhalative deposits, with several payable metals and multiple ore types. Zinc recovery is sensitive to deleterious elements such as Ba and Fe. An important goal of the study is to improve the understanding of the complex relationships between these key elements.

Red Dog is located in the DeLong Mountains of the Brooks Range, approximately 90 miles north of Kotzebue, Alaska, USA. The property is owned by the Northwest Alaska Native Association (NANA) Regional Corporation, and the mine is operated by Teck Cominco Limited. There are five known deposits in the Red Dog District. Four (Main, Aqqaluk, Paalaaq and Qanaiyaq) occur in the immediate vicinity of the original discovery; while Aŋarraaq is approximately 7 miles to the north. The mine assays for as many as 10 variables; the four primary ones being Zn, Pb, Fe and Ba.

This study characterizes seven different elements: zinc (Zn), lead (Pb), iron (Fe), barium (Ba), soluble lead (sPb), silver (Ag), and total organic content (TOC), within eight different domains using a joint simulation approach based on stepwise conditional transformation (SCT). The SCT transform is a multivariate data transformation technique that partitions the data into several classes and transforms each class to a standard normal distribution. Geostatistical models were constructed for each variable within the eight domains, and subsequently assembled to give 40 realizations for six 25 ft benches. Gaussian simulation permits reproduction of the input data, original histogram and variogram of the transformed scores; the benefit of using SCT is that the resulting models also respect multivariate relations locally and globally.

Recovery is adversely affected by the presence of high barite and other deleterious minerals and ore textures. The existing long term resource model was constructed by independently kriging the four main variables. The purpose of using a multivariate approach to jointly model the key elements is to improve the prediction of Zn recovery and to understand the spatial variability of the important variables.

In this case study, the stepwise conditional transformation (SCT) (Leuangthong 2003; Leuangthong and Deutsch 2003) was chosen to model those inter-relationships among the different variables, because of its flexibility and its ability to model non-linear features in bivariate relationships. SCT produces Gaussian variables that are uncorrelated.

[3] Teck Cominco Limited is gratefully acknowledged for allowing the publication of this case study. This case study is based on the paper "Multivariate Geostatistical Simulation at Red Dog Mine, Alaska, USA", by Leuangthong, O., Hodson, T., Rolley, P., and Deutsch, C., originally published in the CIM Bulletin, May 2006. The Canadian Institute of Mining, Metallurgy, and Petroleum (CIM) is also gratefully acknowledged for allowing partial reproduction of the paper.

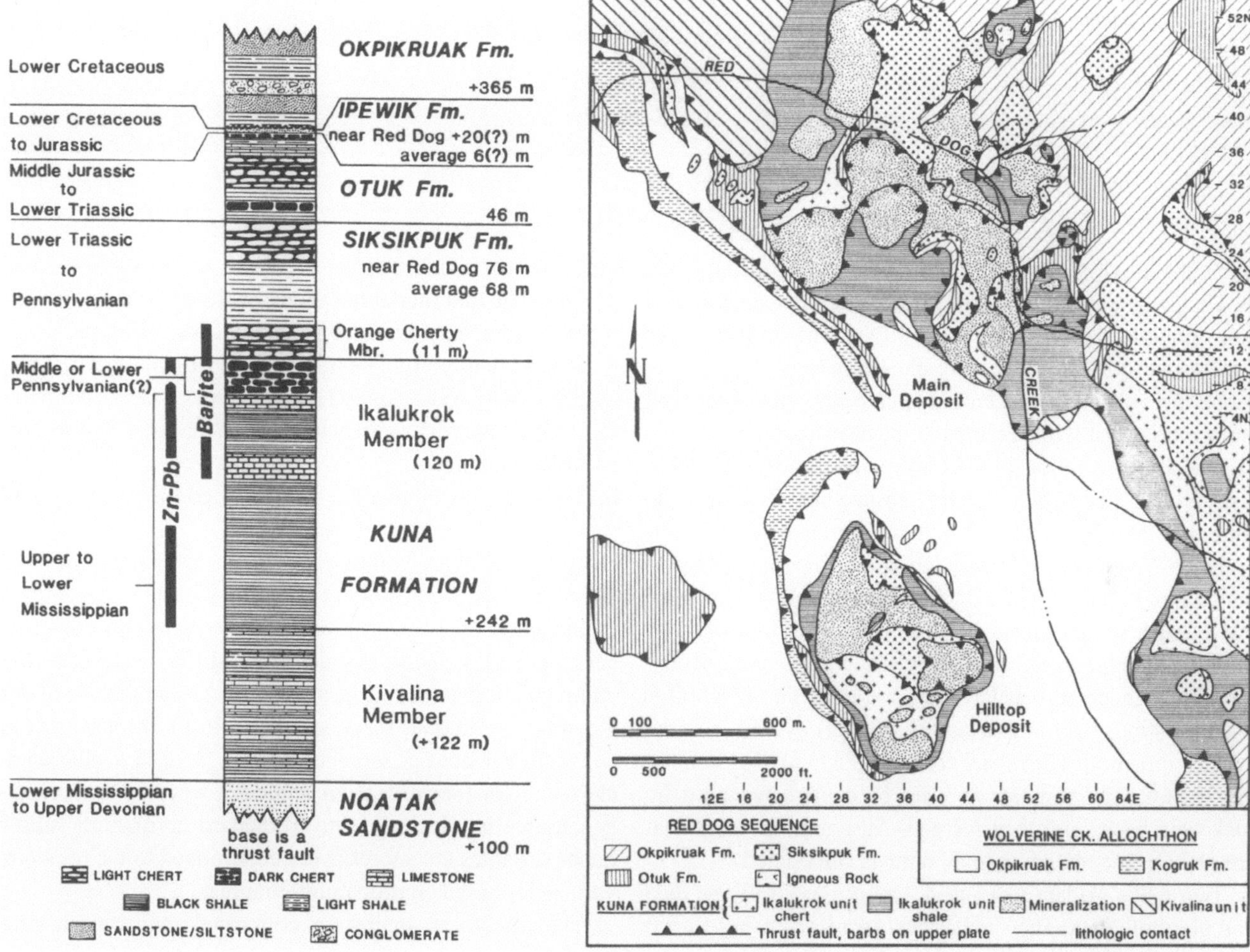

Fig. 14.40 Stratigraphic section of the Red Dog sequence (left) and bedrock geology for the Main and Qanaiyaq (Hilltop) deposits (right). (Moore et al. 1986)

14.4.1 Geology and Database

The Red Dog deposits are SEDEX, zinc-lead-silver deposits hosted in Mississippian- to Pennsylvanian-age black shale. The deposits are found in the De Long Mountains, which are made up of eight stacked and folded allochthons, sedimentary material that has been moved (thrusted) from its original position. The six structurally lowest allochthons are composed of Devonian through to Cretaceous clastic and chemical sedimentary rocks, while the two upper allochthons are of Jurassic and older age and are made of mafic to ultramafic igneous sequences (Moore et al. 1986).

Mineralization is found in the second lowest allochthon hosted by black siliceous shale and chert of the Ikalukrok unit of the Kuna Formation. The stratigraphic footwall is an interbedded, light gray, calcarenite and dark gray calcareous shale, the Kivalina unit. The deposits themselves are a strata-bound accumulation of siliceous rock, barite and sulfides. The hanging wall unit to the mineralization is a silica- and sulfide-poor barite of the lower Siksikpuk Formation of Pennsylvanian to Permian age (Moore et al. 1986). A stratigraphic section and geologic map can be seen in Fig. 14.40.

The main deposit is a nearly flat, elongated stack of three significantly mineralized lenses. It extends 1,600 m in a northwest direction, varies in width from 150 m to 975 m and is up to 135 m thick. The main deposit consists of two major and one minor mineralized plates and their associated overlying waste rocks. The upper plate is a flat-lying sheet of Kivalina unit limestone and shale, Ikalukrok unit siliceous shale and sulfide-bearing barite rock. The median plate contains most of the resources in this zone and consists of a sequence of massive to semi-massive sulfide rock, sulfide-bearing silica rock and sulfide-bearing barite rock. The mineralized portion of the median plate is capped with a sequence of shale and chert of Siksikpuk, Otuk and Okpikruak units. The lower plate mineralization in the Main deposit consists of sulfide-veined, silicified, Ikalukrok shale, semi-massive to massive sulfides, and sulfide-bearing barite rock.

Table 14.17 Stepwise conditional transformations for Red Dog

Transform No.	*Variable*	*Conditioning Variables*
1	Zn	
2	Pb	Zn
3	Fe	Zn, Pb
4	Ba	Zn, Fe
5	sPb	Zn, Pb
6	Ag	Zn, Pb
7	TOC	Zn, Fe

This case study is limited to eight geological domains corresponding to four different ore type units in the Upper and Median plates. These were chosen because they correspond to a volume that includes both recently mined material and material that will be mined in the near future.

The existing grade models were independently kriged onto $25 \times 25 \times 25$ ft blocks. The geostatistical models were simulated at $12.5 \times 12.5 \times 12.5$ ft resolution, and later upscaled to 25 ft blocks. The simulation is on a point-scale support. Thus, simulating at a finer resolution and then averaging to larger blocks shows the variability of the block grades more accurately; change of support is addressed numerically.

The six modeled benches span a volume that is 4,500 ft wide (Easting) by 4,500 ft long (Northing) by 150 ft in the vertical direction. The model consists of a total of 1,555,200 grid points. The simulations were be constructed on a by rock type basis, and then merged. All comparisons shown are global, corresponding to all domains combined.

Two types of data were used: composited drillhole data and blasthole data. The simulations were done using 12.5 ft composites. The geology model, originally coded on the 25 ft blocks, was recoded into the 12.5 ft model.

There were a total of 9,847 12.5 ft composites available for the eight domains of interest. The term drillhole (DH) refers to the 12.5 ft composites. DH data are at a nominal 100×100 ft spacing. For these same domains, there were 58,566 blasthole (BH) data available for model validation. BH data are spaced data 10×12 ft spacing, and represent the entire 25 ft bench.

14.4.2 Multivariate Simulation Approach

Conditional simulations were performed for seven variables: Zn, Pb, Fe, Ba, sPb, Ag, and TOC. These seven variables were modeled for each rock type using Gaussian simulation with stepwise conditionally transformed variables. The main steps of the simulation are:

1. Data declustering to obtain representative distributions for each variable.
2. Transform data with SCT (Leuangthong 2003; Leuangthong and Deutsch 2003; Luster 1985; Rosenblatt 1952) to obtain independent Gaussian variables.
3. Calculate and model the directional variograms for each of the transformed variables within each rock type.
4. Independently simulate transformed variables via sequential Gaussian simulation (Isaaks 1990).
5. Back transform the simulated values (back-SCT) to return to original units.
6. Validation of simulation results to confirm data, histogram and variogram reproduction.

Once all variables within all domains were modeled, the block models were merged to form multiple realizations of the study area for uncertainty assessment and post-processing. The methodology described differs only from the common geostatistical Gaussian simulation in the use of the stepwise conditional transformation (SCT), in place of the conventional normal score transform. SCT is a multivariate Gaussian transformation approach whereby the primary variable is transformed to a standard normal; all subsequent variables are successively conditioned to the previous variable(s) based on probability binning.

The transform applies to collocated multivariate data making them independent prior to simulation. Cross variograms between transformed variables should be checked to verify that spatial correlations are approximately zero. After such verification, independent Gaussian simulation can proceed. Back transformation restores the complex relationships between the multivariate data.

The need to consider seven variables simultaneously for any one rock type poses a practical problem. The multivariate stepwise conditional transform would require 10^7 composites in order to have a minimum of 10 data per probability class. To overcome this problem, a *nested* application of the stepwise conditional transformation was implemented. Inference of a trivariate distribution would require approximately 10^3 or 1,000 data to define the conditional distributions with a minimum of 10 data. This is more reasonable given the number of composites available.

The transformation ordering for the stepwise conditional transform will affect the reproduction of the variogram of the simulated values. Thus, the most important variable or the most continuous variable should be chosen as the primary variable (Leuangthong and Deutsch 2003), which, for Red Dog, is Zn. To account for the other six variables, sets of transformations were proposed (see Table 14.17).

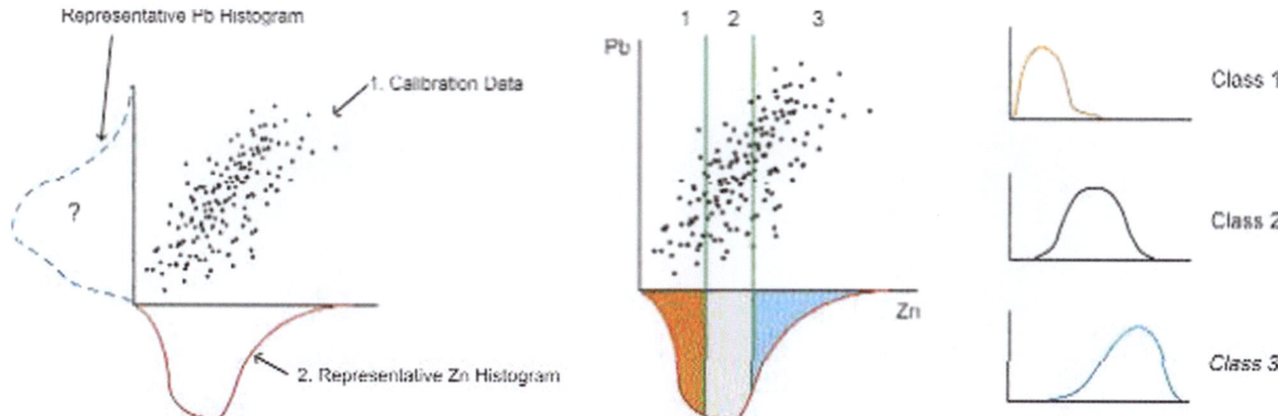

Fig. 14.41 Schematic illustration showing multivariate calibration data, declustered Zn and Pb histograms to be determined (left). The division of multivariate calibration data into multiple classes, with distributions on the right representing the conditional distribution of Pb for each class (right). Weights applied to conditional Pb distribution all shown in light blue, gray, and orange shaded regions of the Zn histograms (Leuangthong et al. 2006)

The transformation order reflects the significance of each variable. All variables other than Zn (main variable) and Pb (main secondary variable) were transformed conditional to either Zn and Pb or Zn and Fe. The choice of the secondary variable in each transform order reflects the relationship between the secondary and tertiary variable. This is not measured by the correlation coefficient alone; non-linearities and constraint features (if present) need to be examined as well in crossplots between the different elements. The determination of the secondary and tertiary variable was based on careful assessment of the relevant bivariate and trivariate distributions.

Declustering was performed to compensate for preferential drilling. Given the multivariate nature of this dataset and the intended application of a multivariate transformation technique, declustering must be consistent between all variables. Declustering was performed within domains, and the declustered Zn distribution obtained using the accumulated weights from kriging within a rock type. This approach considers drill hole data redundancy through the spatial variability model and by domain.

Secondary variables were declustered through a bivariate calibration of the Pb distribution using both the representative distribution of Zn and the crossplot of Zn and Pb. Specifically, the declustered Zn distribution is divided into a series of classes and the corresponding conditional distributions of Pb are determined. The declustered Pb distribution is then constructed by accumulating all the conditional distributions weighted by the declustered Zn probability for the corresponding class (see Fig. 14.41). For all tertiary variables, the same rationale was applied, and the declustered distributions for Fe through TOC were determined using the declustered distributions of the two dependent variables plus the trivariate calibration data.

The Stepwise Conditional Transformation was then performed on the declustered distributions. Figure 14.42 shows the scatterplots of the variables resulting from the first transform sequence of Zn, Pb and Fe (see Table 14.17). The transformed variables are independent and multiGaussian, which translates to a circular shape in the crossplot. From Fig. 14.42, crossplots with the third variable (Fe, in this case) show some banding; however this is simply a numerical artefact of having many classes and consequently fewer data within each class (Leuangthong 2003). This banding does not impact data reproduction. Independence of the transformed variables means that each variable can be simulated independently.

Variograms were then calculated and modeled for each of the transformed variables. Figure 14.43 shows an example of the horizontal and vertical variogram models for the stepwise conditionally transformed Zn, Pb, Fe and Ba for one rock type. Note that secondary and tertiary variables exhibit relatively high nugget effect; this is explained by the independence imposed by transforming each class separately (Leuangthong and Deutsch 2003).

Sequential Gaussian simulation was then independently performed for the seven transformed variables. A total of 40 realizations were generated for each variable within each domain. Only those blocks belonging to the specific rock type were simulated. Each realization was then *back transformed* to the original units of the data. The back transformation for each simulated realization is also conditional. For example, the back transform of Fe is conditional to the simulated values for Zn and Pb.

The simulations were thoroughly checked to ensure reproduction of (1) the composite values at their respective locations, (2) the histogram and associated summary statistics, and (3) the variograms in Gaussian space of the stepwise transform scores. For this multivariate simulation, the multivariate relations were also checked. The simulated models were then upscaled to $25 \times 25 \times 25$ ft blocks to compare to the existing long term model.

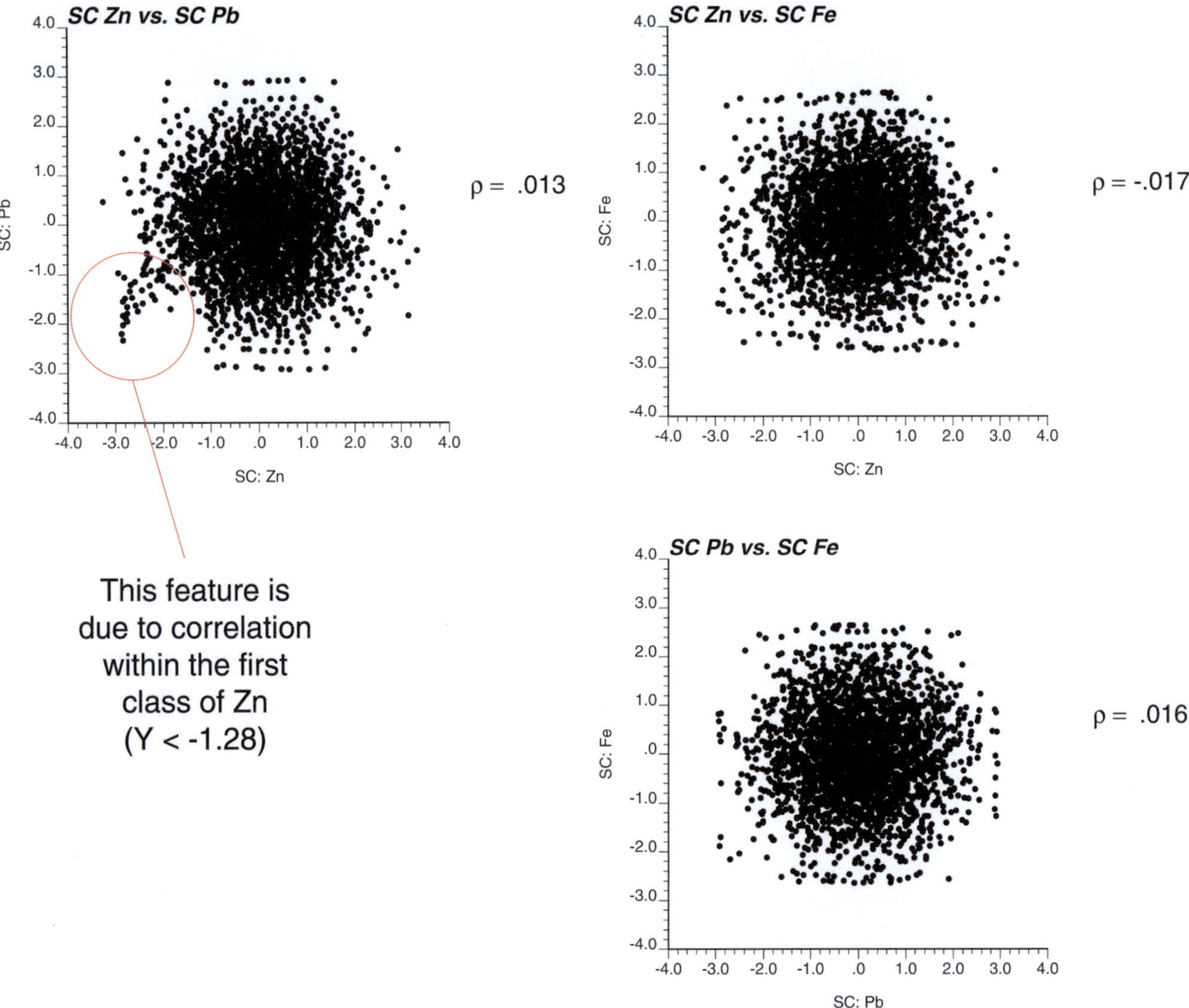

Fig. 14.42 Crossplot between stepwise conditionally transformed variables for Zn, Pb, and Fe. Zn was transformed first, then Pb was transfor med conditional to Zn, and finally Fe was transformed conditional to both Zn and Pb (Leuangthong et al. 2006)

Figure 14.44 shows a comparison of the crossplot reproduction from simulation to those crossplots from 25 ft composites and the existing long term resource model. In general, the simulated realizations reproduce the trivariate relations with comparable variability to the 25 ft composites; the corresponding plots from the existing long term model shows similar bivariate relations but with noticeably reduced variability. Recall that it is this variability between the multiple elements that was impacting the Zn recovery, and provided the motivation to undertake such a case study.

Once all simulated models were generated and validated by rock type, a single realization for each variable was obtained by merging the simulated properties from each rock type. With these multiple realizations (see Fig. 14.45), the uncertainty at any location and/or region can be assessed.

14.4.3 Profit Comparison

In practice, multiple variables are estimated independently with ordinary kriging. It is interesting to address the impact of the multivariate simulation approach using the stepwise conditional transform relative to the conventional practice of kriging. Note that this exercise is for illustrative purposes only, prices and recovery functions have been greatly simplified for this specific comparison.

The idea is to apply a profit function to obtain a true profit dataset for Red Dog. A subset of the reference data will be extracted and used to model grades using both kriging and simulation. The profit function will be applied to these grade models. Based on the expected *profit* from each approach, each location within the model will be classified as either ore or waste. The true profit at each location is known, so the profit differential from each model can be calculated.

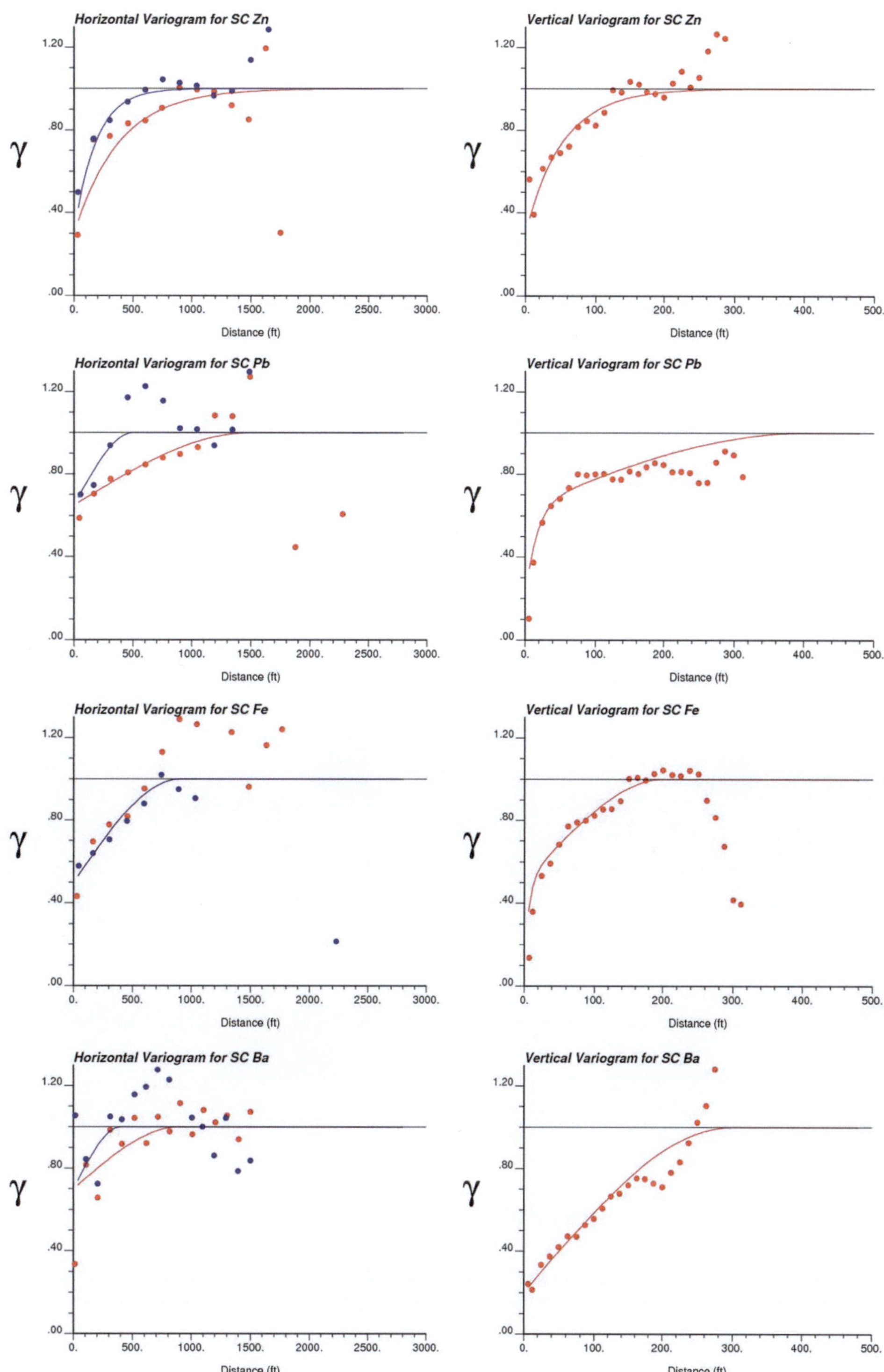

Fig. 14.43 Horizontal (left) and vertical (right) variograms for stepwise conditionally transformed Zn, Pb, Fe, and Ba for one domain

14.4.4 Profit Function

A simple profit function was developed to account for Zn and Pb grades, recovery functions and metal prices. Penalty functions to account for the impact of Ba and Fe on Zn recovery were incorporated. Pb recovery was considered constant.

The kriging and simulation models provide the metal grades. The metal recoveries for both Zn and Pb were calculated as Red Dog's five year average recovery (1998–2002) based on Teck Cominco's financial report (Teck Cominco 2003). These were 83.6 % Zn recovery and 58.7 % Pb recovery. The penalty functions, constructed to mimic the impact of Fe and Ba on Zn recovery (decreasing functions on a scale

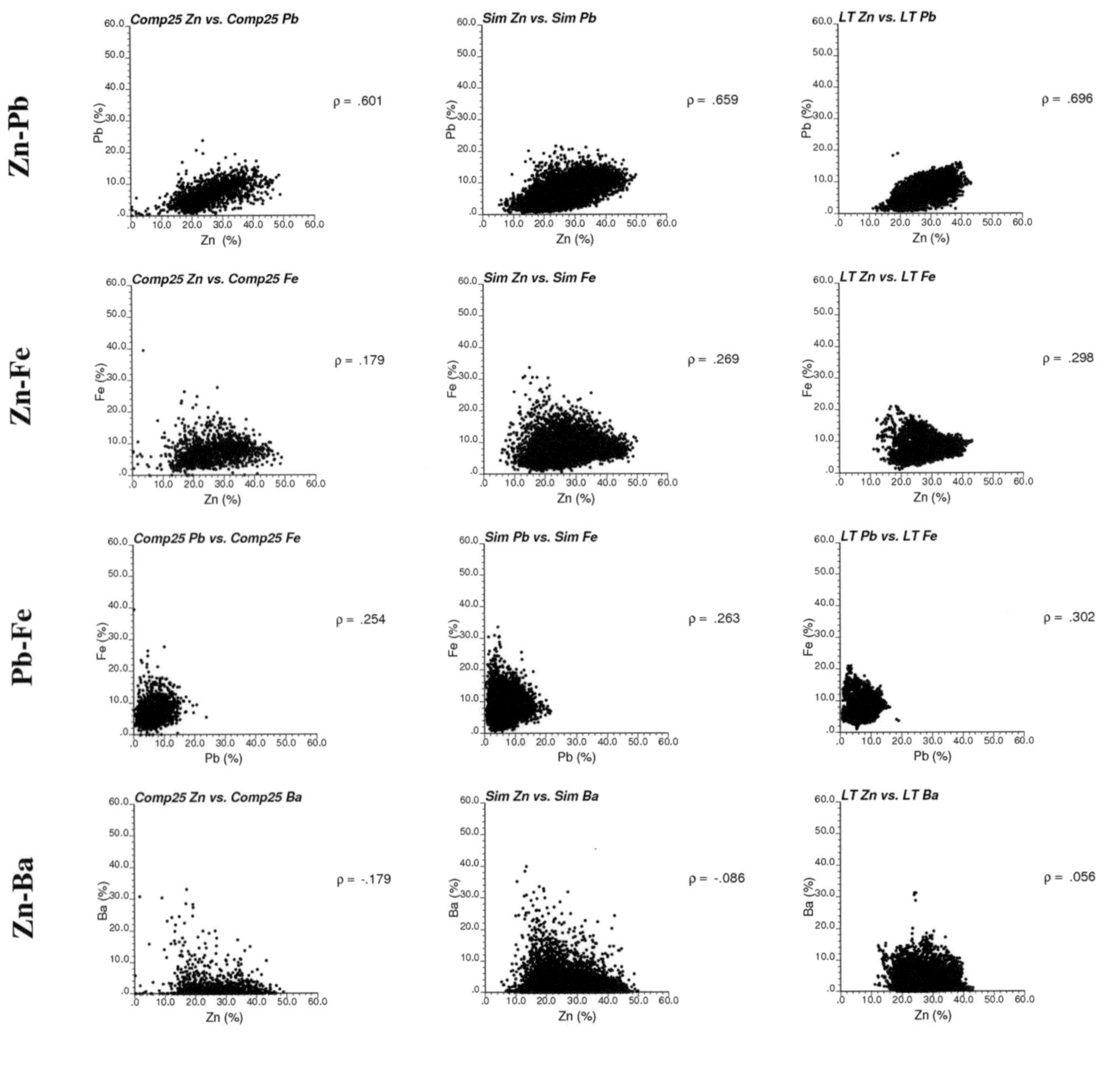

Fig. 14.44 Comparison of multivariate features reproduction for Zn-Pb (top row), Zn-Fe (second row), Pb-Fe (third row), and Zn-Ba (bottom row). Cross-plots using 25ft composites are shown on the left column; from the upscaled simulations are shown in the middle; and from the available long term resource model are shown in the right column

of 0 to 1.0), were used to determine a multiplicative factor for the maximum Zn recovery of 83.6%; in this way, high Fe or Ba content would result in reduced Zn recovery. The price for Zn was chosen to be $ 680/ton of Zn, and the price for Pb was chosen as $ 380/ton of Pb; both prices were approximated based on the metal prices from the London Metal Exchange in 2003. In order to yield approximately 50% ore and 50% waste classification, the cost per ton mined was chosen arbitrarily.

14.4.5 Reference Data

The density and number of BH data is sufficient to be considered as a reference data set. Only a small area was modeled, chosen to be in a marginal zone, where ore/waste classification based on the models would have the largest impact.

Figure 14.46 shows the available BH data in the chosen region of 400 × 400 ft in the Bench 850, and the subset of data extracted from this region. The available data consists

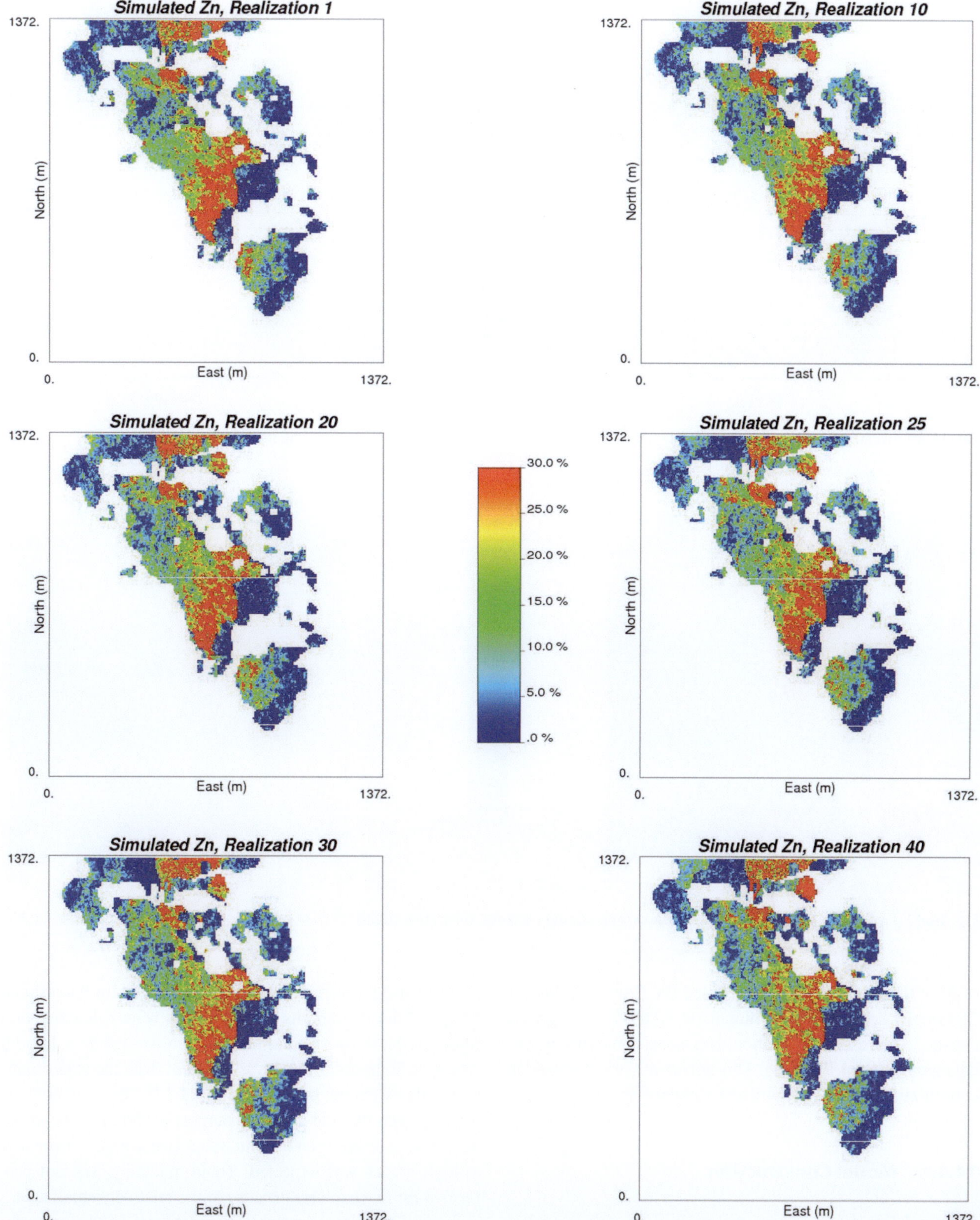

Fig. 14.45 Simulated realizations of Zn at 12.5 ft grid resolution

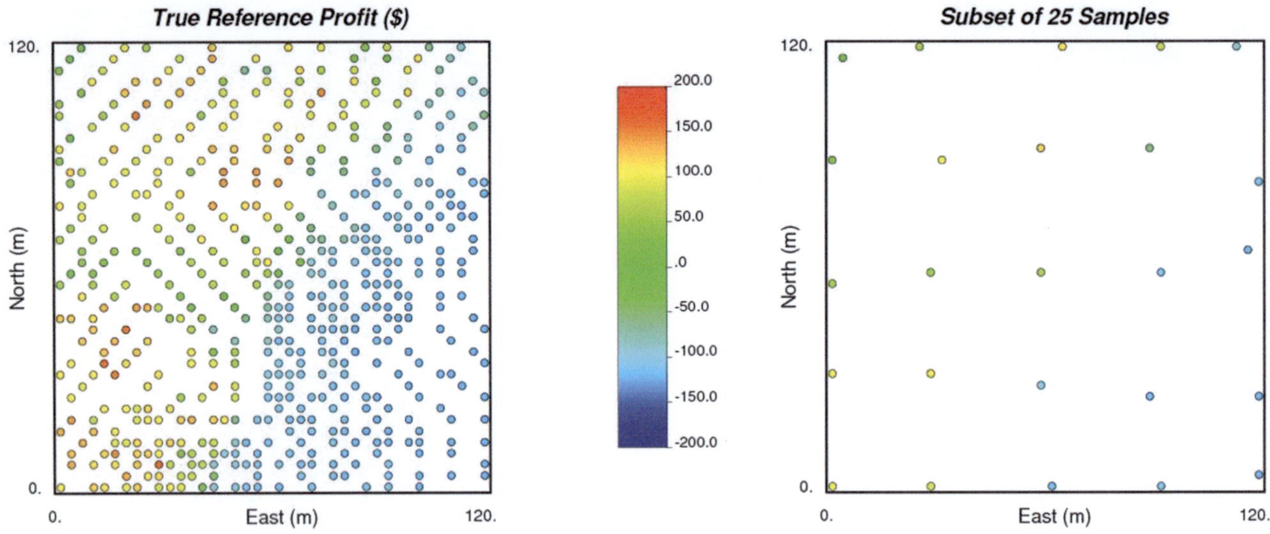

Fig. 14.46 Location map of reference BH data (left) and sampled BH data (right) for use in comparing model approaches

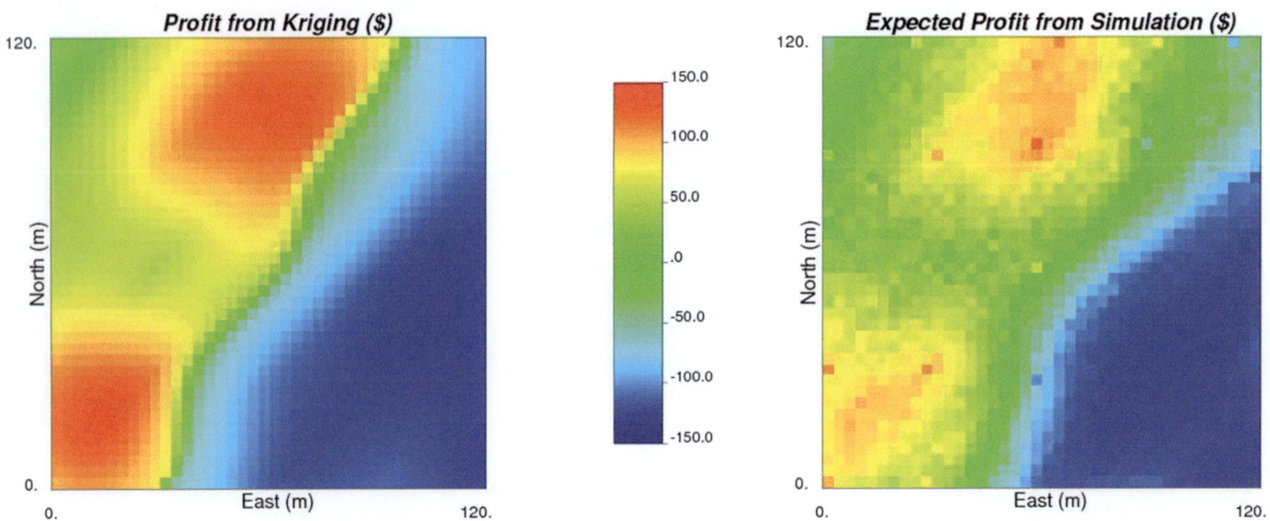

Fig. 14.47 Profit maps for ore/waste classification from kriging (left) and simulation (right)

of 532 BH samples of Zn, Pb, Fe and Ba. From this dataset, 25 samples separated at a nominal 100 × 100 ft spacing were chosen to represent exploration data, consistent with the DH data available for Red Dog. This subset of data was used as conditioning data for kriging and simulation.

14.4.6 Model Construction

The model grid was chosen to be 10 × 10 × 25 ft, which is similar to the 10 × 12 × 25 ft spacing of the BH data. A total of 1,600 grid points were modeled. Further, variograms for both approaches were calculated and fitted using the reference 532 BH data. This filtered out the influence of poor variogram inference due to scarce data.

The variograms for kriging were calculated for the original data, and the variograms for simulation were calculated and fitted for the stepwise conditionally transformed data. In both sets of variograms, a trend was apparent from the experimental points extending beyond the sill of 1.0. This was not surprising given that the area was purposely chosen to be in the transition zone between ore and waste, hence a trend from low to high grades was expected. Trend modeling was not performed for this exercise because of the relatively small area.

For kriging, each variable was estimated independently using ordinary kriging. For simulation, the stepwise conditionally transformed variables were independently simulated using sequential Gaussian simulation to generate 100 realizations of the grades and subsequently back transformed to the original units of the data.

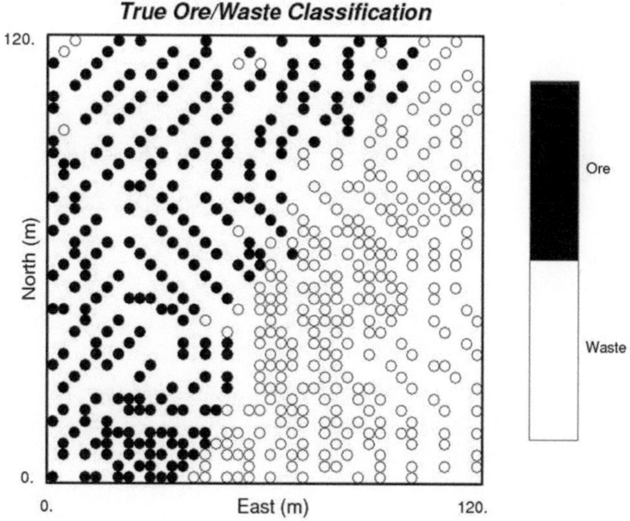

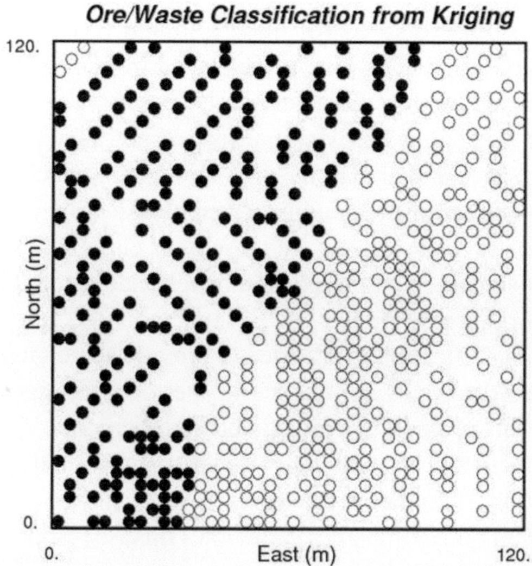

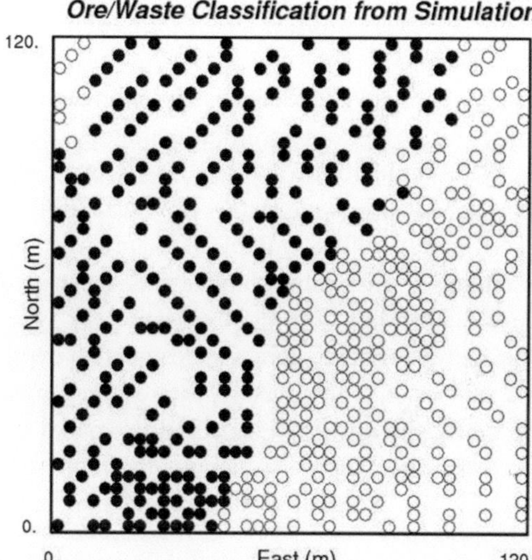

Fig. 14.48 Comparison of true ore/waste classification (top) and the classification from kriging (bottom left) and simulation (bottom right) at data locations

14.4.7 Results

These grade models were then processed by applying the profit function at each location within the model. Although 100 realizations of profit were available from simulation, the ore/waste classification was based on the expected profit map obtained by calculating the expected value of profit at each location. Figure 14.47 shows the profit map obtained from simulation and kriging.

Although 1,600 locations were modeled, only the 532 points corresponding to locations where true data were available can be compared. At these locations, the true profit was known. The profit model from kriging and expected profit model from simulation were used to classify the 532 locations as either ore or waste. Figure 14.48 shows the comparison of the ore/waste classification of the 532 locations from the true reference data to the kriging and the simulation approaches. Overall, both approaches clearly show the waste and ore regions; relatively few blocks were misclassified.

Table 14.18 shows the summary of the ore/waste classification from both kriging and simulation relative to the true classification. Table 14.18 shows that the kriging approach resulted in a total 7 % of blocks that were misclassified, compared to the 6 % misclassified by simulation. From the true profit, 251 blocks (47 % of the true data) were classified as ore; simulation correctly classified ore for 98 % of those blocks while kriging correctly classified 90 % of those blocks.

For those blocks classified as ore, the profit of ore mined as a result of the classification from each method was com-

Table 14.18 Ore/waste classification summary of kriging (left) and simulation (right) relative to true ore/waste classification

Kriging	TRUE Ore	TRUE Waste
Ore	225	11
Waste	26	270

Simulation	TRUE Ore	TRUE Waste
Ore	246	27
Waste	5	254

pared with the true profit of $ 7.89 M. The results from such a comparison showed that the simulation approach yielded $ 7.28 M while kriging yielded $ 7.06 M in profit. Although these profit values appear high for the relatively small area of a single bench, the relative percentage increase in profit is the key result. Multivariate simulation resulted in 92 % of the true profit relative to the 89 % yielded by kriging. In practice, this 3 % difference may translate to several millions of dollars in increased profit if a larger area and multiple benches are considered.

Conventional Gaussian cosimulation approaches are sufficient for straightforward multivariate problems; however, for the complexity of the Red Dog deposit, these common approaches are inadequate. The availability of multiple metal grades within multiple domains warrants some consideration of the relationship between these grades, and how these relationships change from one rock type to the next. The approach shown here was designed to explicitly address this key issue. Consequently, the resulting models not only reproduce the univariate data and its spatial variability, but taken together, they also honour the multivariate relations between the different metals/minerals within the different domains.

14.5 Uncertainty Models and Resource Classification: The Michilla Mine Case Study

Geostatistical simulation provides a model of uncertainty at different stages of a mining project and for different types of risk assessment. Simulation has been used for grade control in daily operations, to assess the uncertainty of minable reserves at the project's feasibility stage, and to assess mineralization potential in certain settings. Other applications include assessment of recoverable reserves, resource and reserve classification, and drill hole spacing optimization studies. All large-scale applications of conditional simulations intend to benefit from a model of uncertainty that describes the variability observed in the data and its impact on the process assessed.

In this case study, a resource classification derived through more conventional methods is compared to probability intervals that result from the simulation model. Resource classification is an exercise intended for public disclosure and thus considers large volumes. It is global in nature. They should not be used to provide a technical answer on a local scale, such as risk assessment of mine schedules. This case study considers two different mining methods (open pit and underground), which implies that recoverable reserves are assessed based on different Selective Mining Units (SMUs). Probability intervals derived for these different SMUs are contrasted to the uncertainty model developed from the classification scheme used by the operation to report resources and reserves. The study assesses the risk of not achieving predicted tonnages and grades within a mine plan, which is based on selecting measured and indicated blocks only.

There are several aspects of resource classification schemes that should be emphasized to better understand the motivation for this case study:

1. Resource classification is intended to provide some measure of the degree of confidence in the resource statements. In this sense, it is a global uncertainty model. The same can be obtained deriving probability intervals from the conditional simulation models.
2. Internally, mining companies sometimes misuse resource classifications as risk assessment tools, although the manner in which this is done varies widely among geologists, mine planners, and mine management. This stems from the temptation to use resource classification codes on a block by block basis, or at a more local scale than warranted.
3. Despite the existence of codes and guidelines that may give the appearance of objectivity to the process of resource classification, Competent Persons would use different resource classification schemes for any given deposit. Management's perception of risk is likely to be different. Technical personnel will generally disagree about the application of the Resource Classification scheme used, and how to inject the different levels of confidence into the mine plan and projected cash flow.

This general lack of understanding of the purpose and meaning of resource classification schemes can be mitigated by using a geostatistical model of uncertainty. Although they are not objective, a more detailed description of the predicted uncertainty for each particular block, phase, zone, or

Fig. 14.49 Location map of the Michilla district and mines

geologic region of the deposit allows for a better understanding and use of the classification scheme implemented.

In addition, basic questions such as "how different is measured from indicated?", or "does measured mean 0% error?", or "how different is measured in Zone A compared to measured in Zone B?" can have a quantitive answer, as provided by the simulation model.

14.5.1 The Lince-Estefanía Mine

The Lince-Estefanía mine is located within the district of the same name, some 120 km to the north of Antofagasta, Chile. The mine is operated by Minera Michilla S.A., producing 50,000 t of cathode Cu per year from both an open pit (Lince, using a 10 m bench height) and an underground mine (Estefanía, mostly cut-and-fill with 5 m lifts) as of 1999. The mine is located approximately at 900 m above sea level. Figure 14.49 shows the location of the district in northern Chile.

The district geology shows a very significant strato-volcanic sequence, called La Negra Formation, regionally dipping 30° to the NW, and composed of a series of andesites and volcanic breccias of different characteristics. The andesites vary from afanitic to porphyritic, intermixed with the volcanic breccias. Mineralization is hosted in this volcanic sequence, where the porous breccias are the most favorable mineralization hosts (Ferraris and Di Biase 1978).

The resulting mineralized bodies ("mantos") are ellipsoidal in shape, of variable dimension and grades, and generally concordant with stratification. It is difficult to predict the existence and the size of each manto, as well as its grade distribution. Generally, the mantos are small, 4–5 m thick, with length and widths of up to 40 or 50 m, and many being less than 25 m. Mineralization is mostly Atacamite, a Cu hydroxide, with some Chrysocola, as well as sulfide Cu mineralization at depth: chalcocite, covellite, bornite, and chalcopyrite. Grades within the mantos are typically 1 to 5% Total Cu (TCu), with up to 10% TCu. The cathode plant receives a head grade of 1.6% Cu. Declared resources in late 1999 were, at a 0.5% Cu cutoff, approximately 63 million tons with a 1.44% TCu and 0.86% Soluble Cu (SCu) grades. This is distributed in various zones and at different depths, and includes all drill-indicated resources to that date.

The main subzones within the deposit are separated into amenable to open pit (Lince), or underground (Estefanía). In addition, within the open pit mine there are several mining areas, such as Lince, D4 Zone, and Hilary; within the underground mine, zones are delimited by mining extraction sequence, and are named using letters and numbers, such as A1, B2, D3, and so forth. There are at least 17 areas of interest within the open pit and underground mines. Resources were classified following the guidelines of the 1999 JORC Code, and result in about 21% of the total resources being classified as measured, 64% classified as indicated, and the remaining 15% classified as inferred.

In early 2000 an infill drilling campaign was completed, and the existing drillhole database updated. The geologic model was updated, and a new resource block model was obtained. The grade model was obtained using multiple indicator kriging, and providing an e-type estimate for each block. Following the completion of the Resource Model and its classification into Measured, Indicated, and Inferred, a conditional simulation model was implemented to assess uncertainty and risk. This conditional simulation model was a Sequential Indicator Simulation (SIS, Alabert 1987) model.

The simulation model focused on areas and phases to be mined in the upcoming 5 years (both from the open pit mine and for the underground mine). The simulated realizations

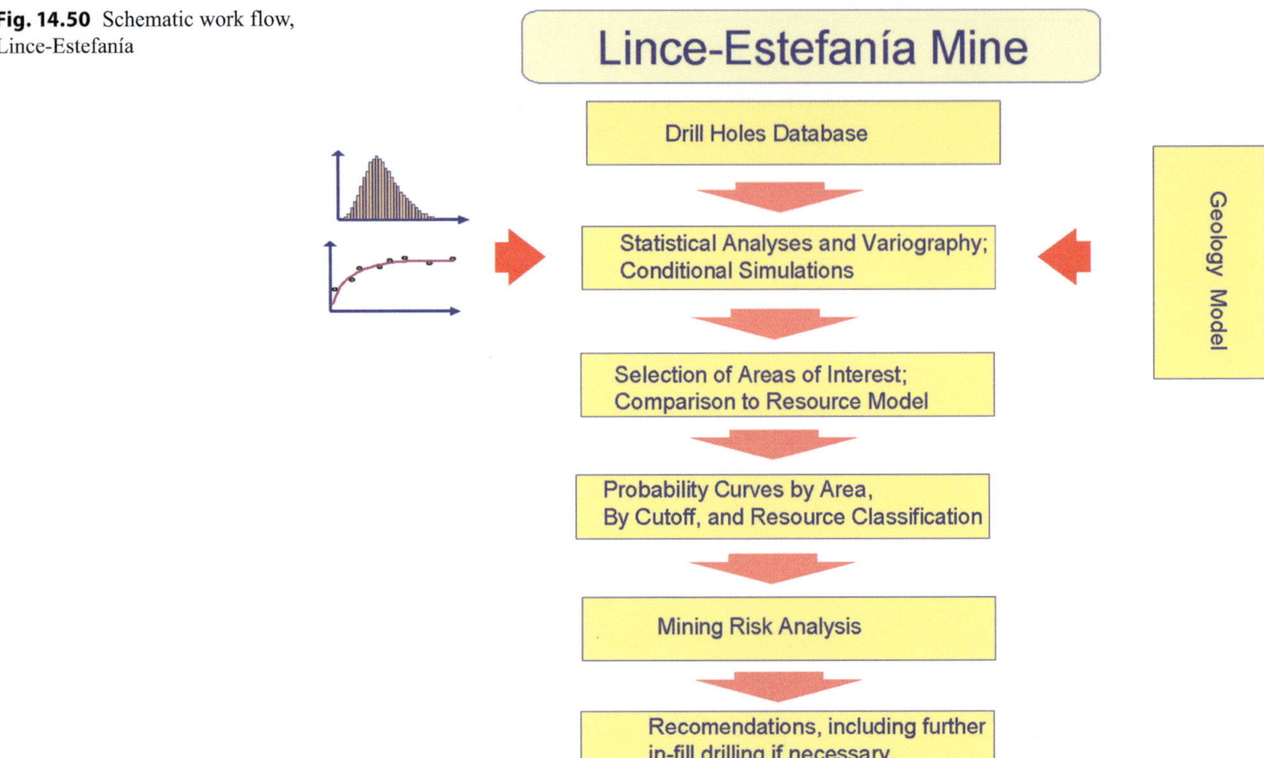

Fig. 14.50 Schematic work flow, Lince-Estefanía

were reblocked and assigned to the same blocks used in the Resource Model. This allows for a direct comparison between the two models, and to assign an uncertainty model to the predicted grade. Figure 14.50 is a schematic representation of the general work flow at Minera Michilla.

14.5.2 Developing the Model of Uncertainty

The geologic model includes the definition of a mineralized envelope, defined at 0.1 % TCu. The purpose of the mineralized envelope is to define the volume within which mineralization can exist; outside this volume, there is no mineralization. In this sense, this is a geochemical-type boundary, whose purpose is to avoid overestimation of the kriging process into barren areas. This is necessary because mineralized bodies can abruptly end due to a post-mineral fault, or without any obvious reasons. This same envelope was used as an external hard limit to define the overall volume of the simulation model.

In addition, there are three major geologic units (GUs) defined (in fact, lithological groupings), which are used to define estimation domains at the time of grade estimation. The units are volcanic breccias (generally mineralized), andesites (could be mineralized, but generally barren or poorly mineralized mantos), and intrusives (including tectonic breccias), which are mostly barren, but could occasionally present significant mineralization.

Five-meter down-the-hole composites were used for both estimating and simulating the resources. The composites are tagged according to the interpreted geologic units, and the length is chosen due to the selectivity required in both the underground and open pit mines. In the case of the open pit, even though the nominal bench height is 10 m, the operation works by cleaning out areas of waste surrounding the mantos, and sometimes mining partial, 5 m benches where deemed necessary. Blast holes are sampled at half-benches, that is, there are two 5 m blast hole sample for every hole drilled to the 10 m bench. These 5 m blast hole data were used in the open pit area (Lince) to improve the definition of the variogram models and to better condition the simulation in mined-out areas and close to the current topographic surface at the time.

The SIS technique requires the definition of indicator thresholds in order to discretize the original grade distribution. The definition of these indicators was identical for the estimation and the simulation models, and included the following indicator classes: from 0.0 to 0.19 % TCu; from 0.2 to 0.49 % TCu; from 0.5 to 0.79 % TCu; from 0.8 to 0.99 % TCu; from 1.0 to 1.19 % TCu; from 1.2 to 1.49 % TCu; from 1.5 to 1.99 % TCu; from 2.0 to 2.99 % TCu; from 3.0 to 4.99 % TCu; from 5.0 to 6.99 % TCu; from 7.0 to 9.99 % TCu; and greater than 10.0 % TCu. The corresponding declustered class means were obtained for each class, using the median of the last class to avoid over-estimation of the high grade portion (outlier control).

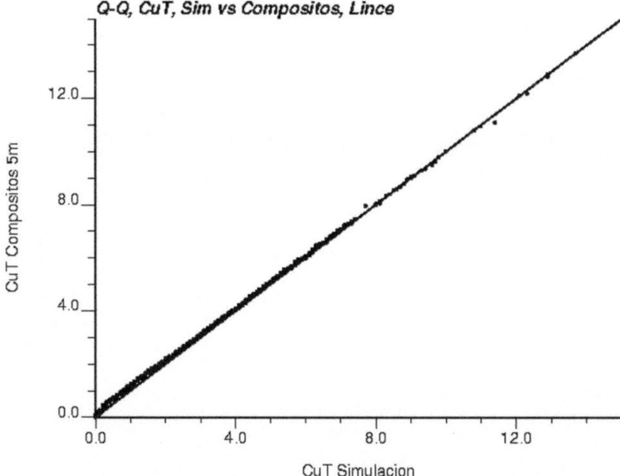

Fig. 14.51 Comparison of simulated values vs. 5 m composites and blast holes, TCu, Lince

14.5.3 Indicator Variograms for TCu and by Geologic Unit

The indicator method requires that an indicator variogram model be obtained for each of the 11 thresholds defined above. In addition, there should be one set of indicator models per GU considered, and per subzone or area in the deposit. So in total 33 indicator variogram models are required per subzone of the deposit. In addition, three major zones were defined for the purpose of variogram modeling (Lince, D4 and Hilary combined, and Estefanía), resulting in a total of 99 variogram models for the resource estimation and conditional simulation model. Some observations regarding these variogram models:

- The variogram models used for the resource estimation are the same as those required for the conditional simulation model. That is, this part of the work (as all other required statistical work) is only done once, and used for both the resource estimation and the conditional simulation models.
- As expected, the variogram models for lower indicator thresholds (waste or low grade) are more continuous than the higher grade thresholds.
- As the indicator thresholds are increased, the overall spatial correlation decreases, evidenced, for example, by the increase of the nugget effect. This corresponds to the intuitive notion that higher grade mineralization has less spatial correlation than the more pervasive, lower grade mineralization.
- There can be differences in anisotropy angles and ranges for different grade ranges, as is the case for Lince-Estefanía. This is explained by different geologic controls affecting parts of the grade distribution differently, for example with a specific set of structures controlling the higher grade mineralization. Therefore, modeling different anisotropies for different grade ranges is entirely appropriate, since they have been validated with geologic knowledge, ascribing them to different mineralization controls.

14.5.4 Conditional Simulation Model

The conditional simulation model was obtained on a $5 \times 5 \times 5$ m grid, and the volume modeled is such that it results in a model with more than 40 million nodes. Although not every simulated node is retained in the end, the model was run for the complete grid, although including the restriction that at least one real 5 m composite is in the neighborhood of the node before the simulated value can be obtained. After the simulated model is obtained, it is restricted to the areas where the 0.1 % TCu mineralized envelope exists, as mentioned above. In this exercise, and largely due to time constraints, 10 grade realizations were run.

The model is based on a two-stage simulation. First, the geologic model was simulated using SIS on the categorical variables that define the lithology package (Volcanic Breccias, Andesites, or Intrusives). The output from this stage is a model representing the probability that each unit exists at a given node. Due to logistics and time constraints, the simulated lithology was used as prior probability distribution to condition grade estimation (as opposed to use as a direct conditioning of the grade simulation).

The purpose of this first stage is to inject into the model the relationship between grade and lithology. For example, a node with a high probability of being a Volcanic Breccia is more likely to have better grades than one with a high probability of being an Andesite. This relationship between lithology and grade is input into the grade simulation as prior distributions of possible Cu grades for each node simulated. The local declustered statistics of the 5 m composites were used to derive these prior probability models for each area within the deposit. For example, for Lince, the 5 m composites tagged as volcanic breccias indicate that a node simulated as volcanic breccia has a 57 % probability of having less than 0.2 % TCu in grade. For Andesites, the same probability is 64 %, while for those simulated as Intrusives it is 71 %. Also, there is a 10 % probability that the nodes simulated as volcanic breccias have a grade greater than 3 % TCu, while this percentage is 3.2 % and 2.8 % for Andesites and Intrusives, respectively. This information is compiled for each threshold used, and used as soft information in the form of prior probability distributions.

The second step necessary to obtain the final simulated model is to use the prior grade distributions assigned to each node, as well as the 5 m composites themselves to simulate a Cu grade in each node. The SIS model used a search ellipsoid with a nominal 25 m search radius, with anisotropic ranges

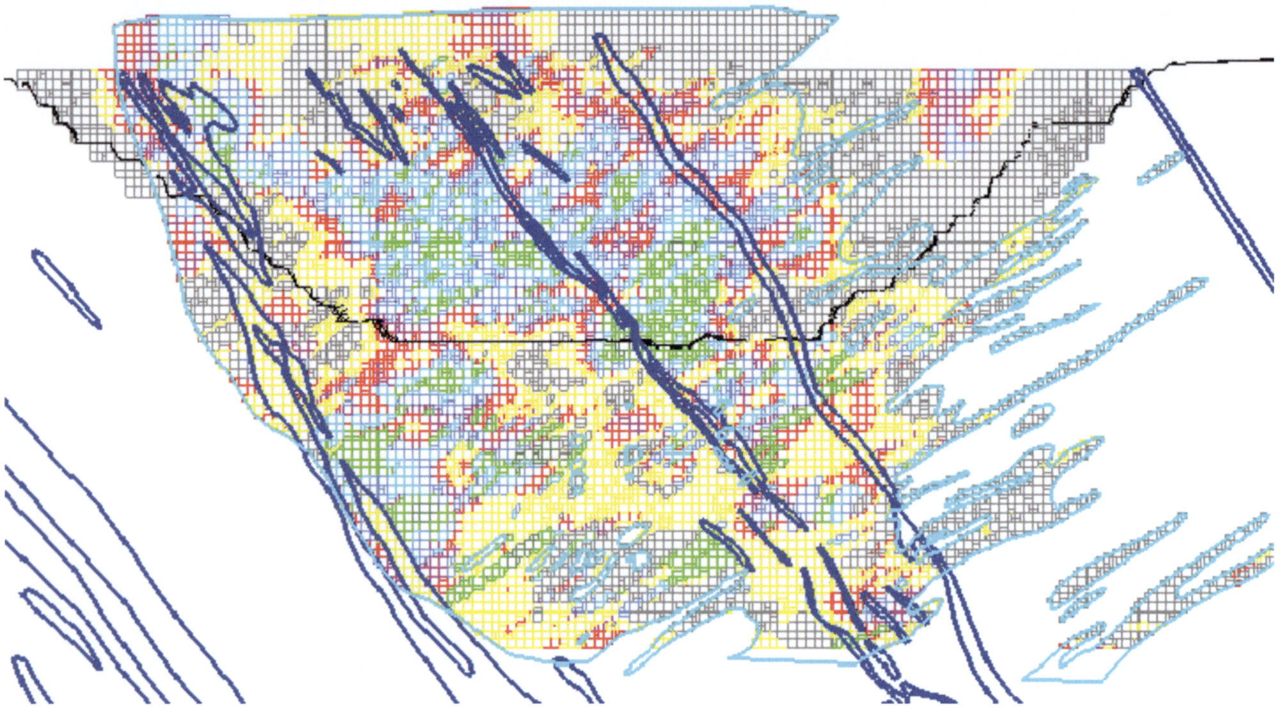

Fig. 14.52 Cross Section #8, simulation no. 1, Lince-Estefanía

corresponding to the same orientation of the median indicator variogram, and applying an octant search. A minimum of 2 composites were required to simulate a node, using a maximum of 10 composites and 10 previously simulated nodes.

These and other parameters are adapted to the characteristics of each area simulated within the deposit. Therefore, each simulation model reflects the different geologic characteristics of each subzone of the deposit, through the use of different indicator variogram models and simulation parameters.

After obtaining the simulation models, several checks are required in order to ensure that the simulated values obtained have the expected characteristics. It is important, for example, to verify that the distribution of the simulated values is similar to the distribution of the original 5 m composites, in average grade, variability, etc. Figure 14.51 shows as an example a quantile-quantile graph of the 5 m composites vs. the simulated nodes for Lince. Since the points approximately align on the 45° line, then the composites and the simulated values have very similar distributions. Other statistical checks include histograms, and reproduction of the variogram models used.

It is also important to check that the simulation model represents well the spatial distribution of grade, much like it is done for kriged resource models. Simple visual inspection of sections and plans of the output simulation model should show whether the composite grades and its patterns of grade distributions are reproduced as expected. Figure 14.52 shows a cross section looking NE (Cross Section #8), where current pit is shown in black. The geometry of the 0.1 % TCu envelope used as a hard limit that defines the simulation volume is clearly visible in both Figures, as is the definition of the cross sections used to interpret the envelope in the plan view. Simulated nodes can only occur inside the interpreted envelope, which defines the possible presence of mineralized mantos, and thus the simulation appears spotty, with significant areas without any TCu grades (grey in Fig. 14.52).

14.5.5 Probability Intervals by Area

After validating and checking the simulations, the models were reblocked and assigned to the same blocks of the resource models, in order to obtain the probability intervals for the corresponding resource estimates by areas. The block size for the open pit areas (Lince, D4, and Hilary) is $10 \times 10 \times 10$ m, while the block size for the underground mine (Estefanía) is $10 \times 10 \times 5$ m, these sizes corresponding to the nominal SMU size of each operation. Each block of the resource model has a classification code (measured, indicated, or inferred).

The model was evaluated for the 17 different subzones within Lince and Estefanía, corresponding to different production areas (current or planned for production within the next five years). For each of these subzones, an estimated grade was obtained from the block model average, as well as the average from the 10 simulated values for each block within the subzone. Note that the definition of these sub-

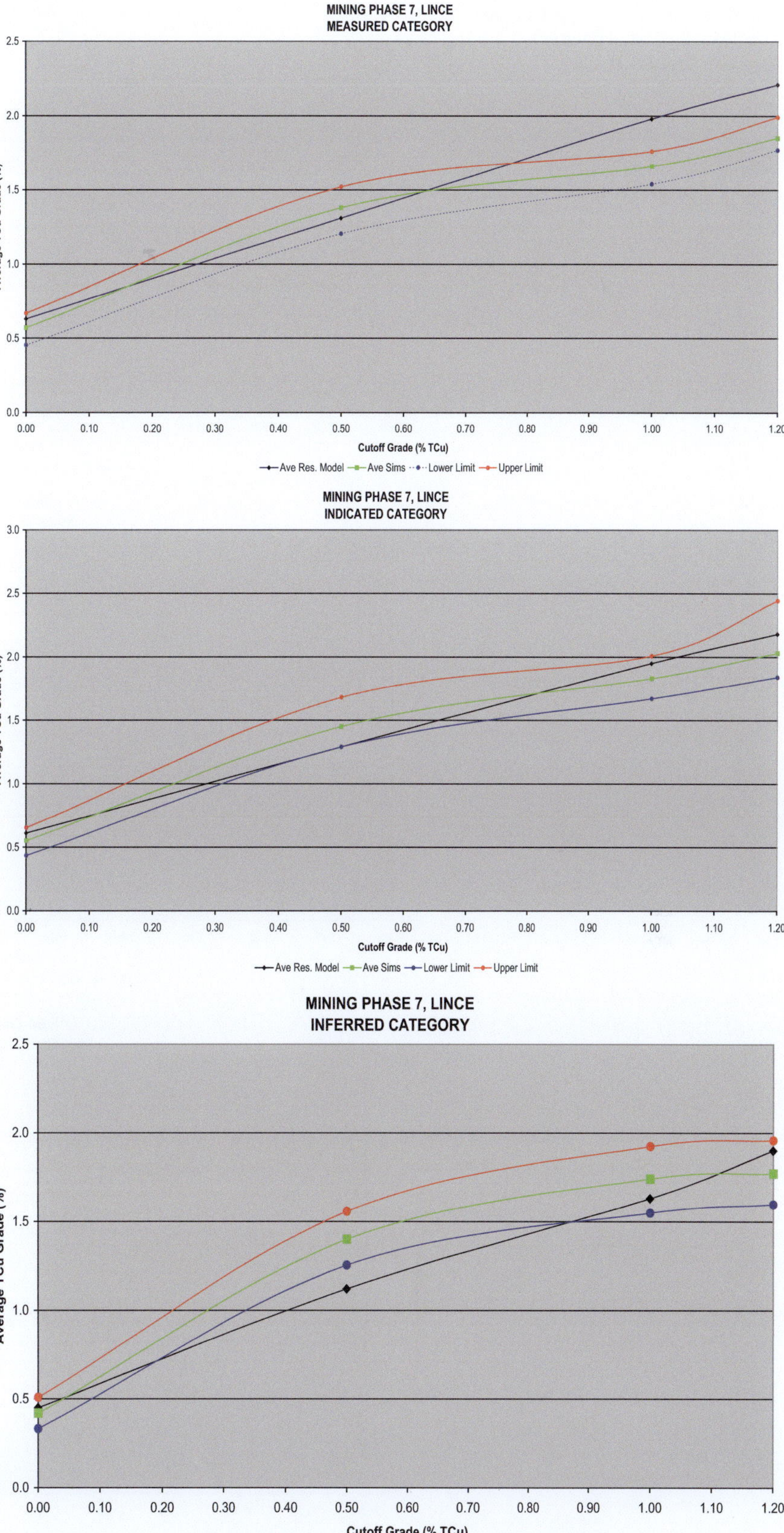

Fig. 14.53 Measured (top), indicated (middle) and inferred (bottom) resources, Area Phase 7 (Lince)

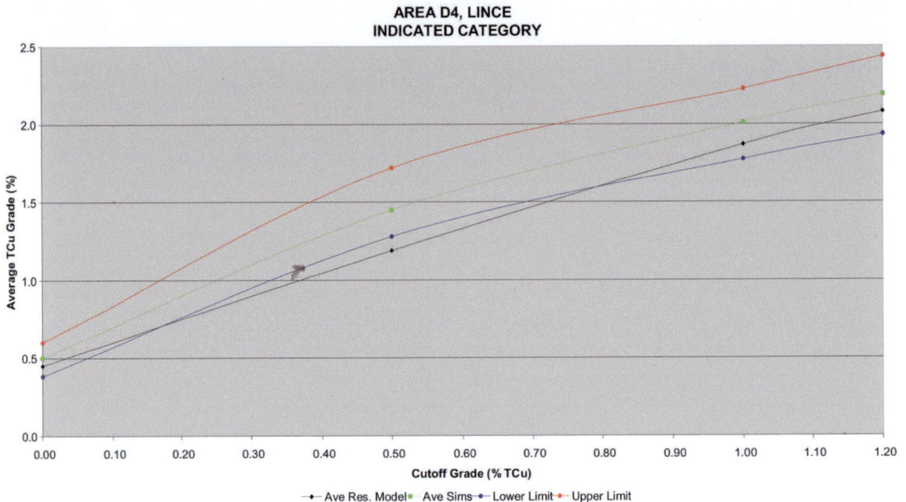

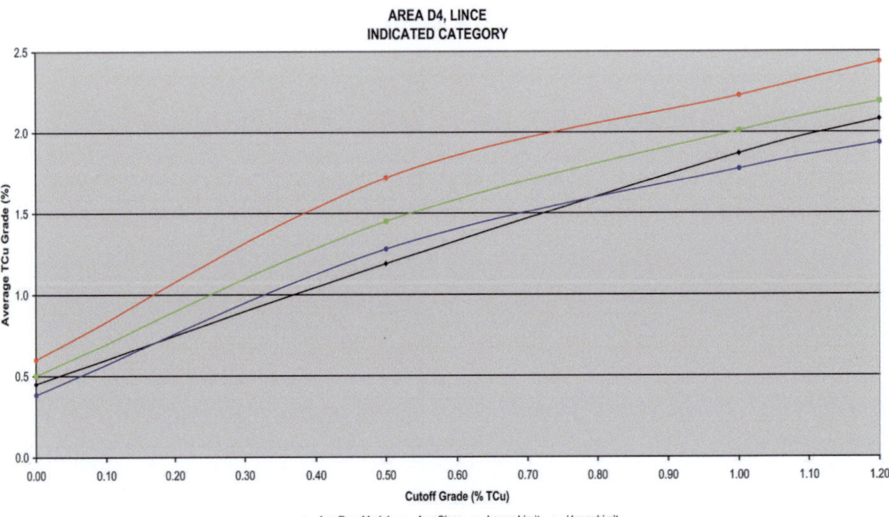

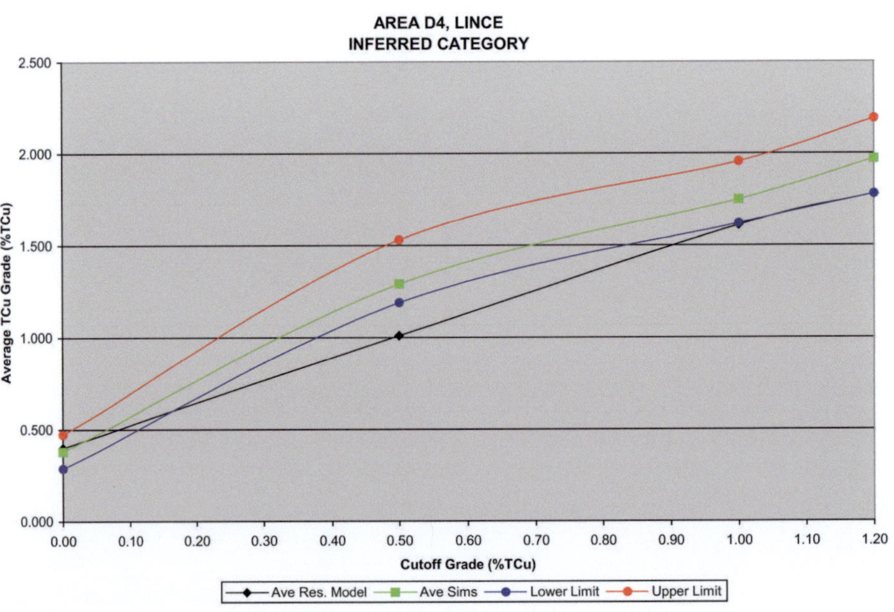

Fig. 14.54 Measured (top), indicated (middle) and inferred (bottom) resources, Area D4 (Lince)

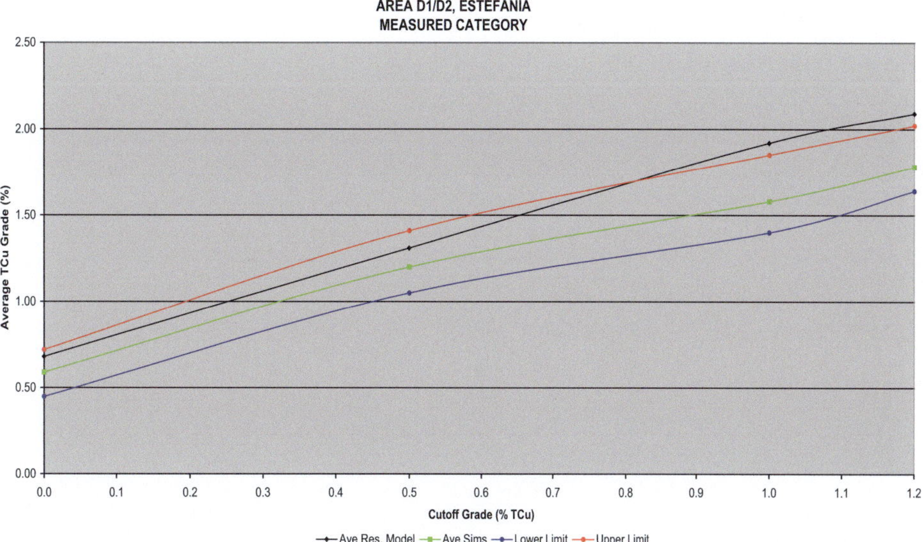

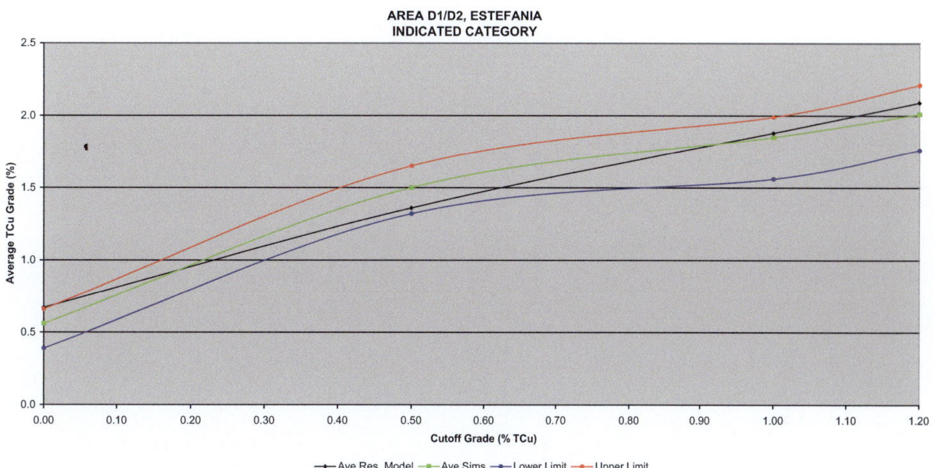

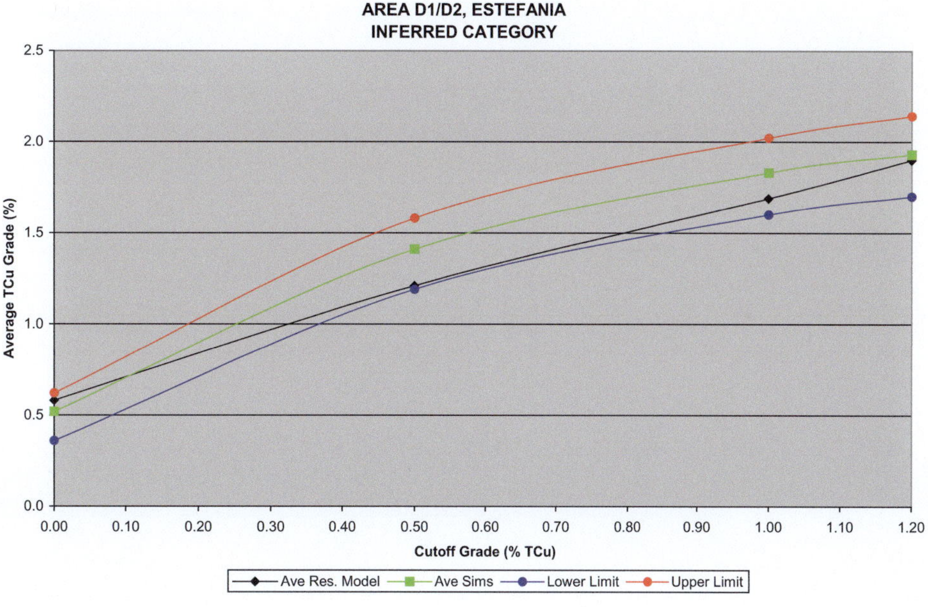

Fig. 14.55 Measured (top), indicated (middle) and inferred (bottom) resources, Area D1/D2 (Estefanía, Cut and Fill)

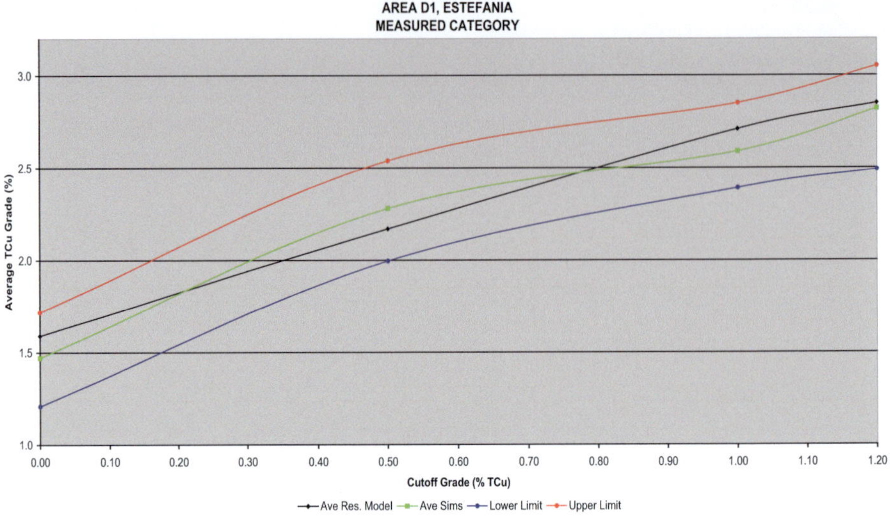

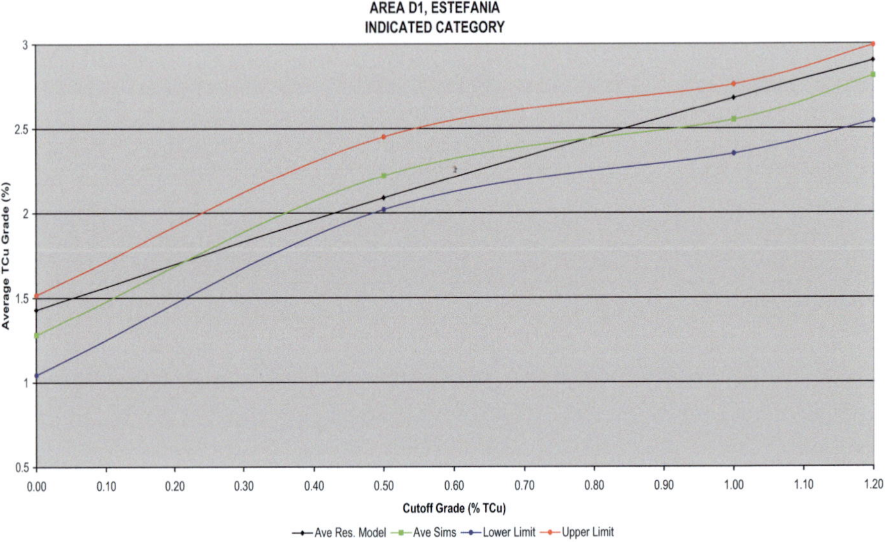

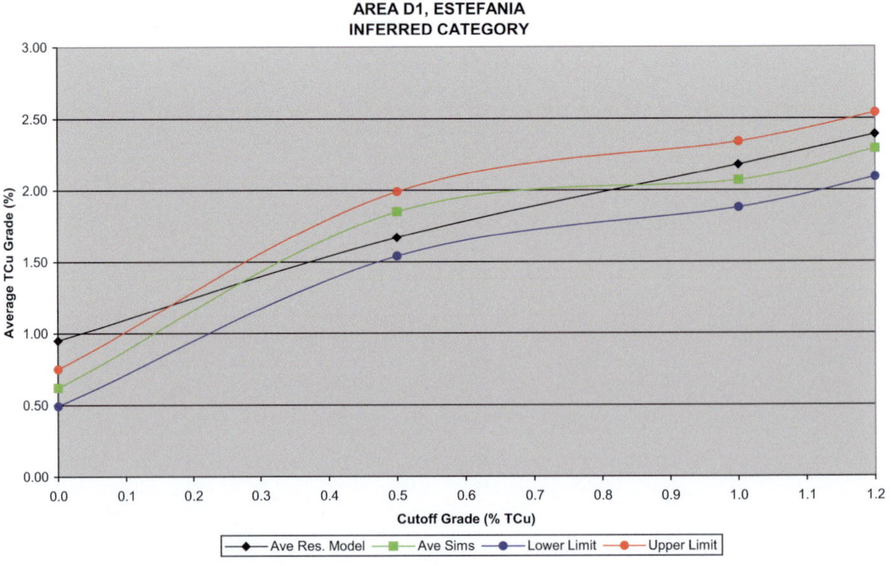

Fig. 14.56 Measured (top), indicated (middle) and inferred (bottom) resources, Area A1 (Estefanía, Cut and Fill)

zones within the deposit correspond to tonnages and grades planned to be extracted according to an existing mine schedule. In this sense, the set of 10 simulated values for each area is used to provide a risk assessment on the existing mine schedule and predicted cash flows, as well as comparing it with the more traditional resource classification scheme.

Only 4 of the 17 areas are discussed here, 2 open pit areas and 2 underground areas corresponding to a medium-term planning horizon. The open pit areas are the D4 and Mining Phase 7 within the Lince pit. The underground areas shown are the A1 and the D1/D2 combined area. For convenience, each of these areas was represented in the computer with a three-dimensional solid, such that these solids can be used to select the blocks of interest and calculate tonnages and grades.

There are several options available to both process the information and perform the mining risk analysis and to present the model of uncertainty. The following information is presented here:

- Average of the area according to the Resource Block Model ("Ave. Res. Model" on the graphs).
- Average of the area according to the simulated values ("Ave. Sims" on the graphs).
- Lower Probability Limit, defined as the 15th percentile of the distribution from the simulation model ("Lower Limit" on the graphs).
- Upper Probability Limit, defined as the 85th percentile of the distribution of possible values for each block ("Upper Limit" on the graphs).

The results are presented for four different cutoff grades: 0 % TCu (or global), 0.5 % TCu, 1.0 % TCu, and 1.2 % TCu. The results are expressed in % TCu, and the Upper and Lower Probability Limits are such that there is a 70 % probability that the true value is within those limits. In addition, for each area, the results are presented by Resource Classification categories (measured and indicated), in addition to total resources (which includes inferred as well as the previous two). Figures 14.53–14.56 present these results.

The results were obtained using volume weighing for each area; in other words, the metal content for each cutoff and classification category were first obtained, and from that the grades presented in Figs. 14.53–14.56 were derived.

14.5.6 Results

In addition to specific conclusions for each Area described below, the following are some of the more important general observations and conclusions:

1. The average grade of the simulations is somewhat different than the average grade from the resource model for most areas. This is a consequence of the differences in the internal dilution incorporated into each model; note how the averages at a 0 % cutoff are more similar. The issue of recoverable reserves is best addressed through conditional simulations, not a more traditional change of support model applied to the MIK model (Chap. 7, and also Journel and Kyriakidis 2004; Rossi and Alvarado 1998).
2. The simulation model results in probability intervals that are not symmetric with respect to the expected value. There is no reason why the probability of error on one side of the expected value has to be identical to the probability of error of the opposite side.
3. It is possible that the expected value (according to the Resource Model) falls outside the probability limits (P_{85}–P_{15}) defined; this can happen because the simulation model is obtained independently of the estimation model (even if applying the same random function), and is more likely when cutoff grades (conditional statistics) are considered.
4. The probability intervals are different for each cutoff grade analyzed. It is also different from measured to indicated to inferred resources. In general, higher cutoff grades result in wider probability intervals (higher uncertainty and risk, as expected), and the same can be said for the difference between measured, indicated, and inferred resources.
5. The measured, indicated, and inferred classes are not very useful when analyzing uncertainty and risk within local areas. For example, the measured category in one area will have a different uncertainty than the same measured category in a different area. The reason is that the resource classification scheme is usually developed on a global basis, and at best only appropriate for long-term risk assessments. The conditional simulation model shows that it is not appropriate to use the resource classification schemes for local risk assessment. That is, a certain block can be classified as measured within the long-term context, but it may not be even indicated when shorter term production periods are considered.
6. The resource classification categories for different areas show significant variability when described in terms of probabilities. For example, the measured resources for Phase 7 Expansion (for the 0.5 % TCu cutoff) show that the 70 % probability interval is within + 16 %/− 8 % (24 % total width) of the resource model grade, while the same "measured" resources for Area D1/D2 is within + 8 %/− 20 % of the resource model grade. What is called measured in one area with a given probability interval may have significantly different probability intervals in another area, but still be called "measured". This is to be expected, due to local geologic differences, additional complexities of the mineralization controls, and local differences in drill hole coverage.

Additional area-specific comments are:

- **Area Phase 7 (Lince, Fig.** 14.5**a–c):** the simulation model predicts that the block model is conservative (i.e., there is upside potential) for cutoffs below 0.7 % TCu, which is the cutoff of interest, while the opposite is true for 1.0 % TCu cutoff and above. This is true for each classification category considered, although more so for the inferred resources. At the 0.5 % TCu cutoff, blocks categorized as Measured are within +16 %/–8 % of the predicted Resource Model grade, while Indicated resources are within +30 %/–0 % of the expected grade. The Lower and Upper probability limits for the Inferred resources are within +12 %/+39 % of the expected resource model grade. The grade-tonnage curve for the simulated model was at the time a major concern, because it appeared as if the Resource Model was predicting much higher grade material at higher cutoffs. The issue was eventually resolved through infill drilling, and indeed the high grade distribution flattened as predicted by the conditional simulation model.
- **Area D4 (Open Pit, Fig.** 14.6**a–c):** the simulation model predicts for all cutoffs that the block model is conservative, although clearly not as much for the higher cutoffs. There is a clear difference in the width of the upper and lower limits between the measured and the indicated categories (Fig. 14.6a, b), and more so with respect to the inferred resources (Fig. 14.6c).
- **Area D1/D2 (Underground, Fig.** 14.7**a–c):** In the underground areas, the cutoff grade to be considered is the higher 1.0 % TCu. Therefore, the resource model appears optimistic, contrary to the open pit areas. At the time of this study, most of the resources in this area were classified as indicated, and prior to significant infill drilling already planned. The probability intervals for these indicated resources at the 1.0 % TCu cutoff is –17 %/+6 %, i.e., the actual grade can be up to 17 % lower and 6 % higher than the predicted grade, according to the simulation model.
- **Area A1 (Underground, Fig.** 14.8**a–c):** In this area the average of the simulation model predicts a lower grade for all classes (at a 1.0 % TCu). Note that the grades in this area are generally higher, compared to the other areas discussed, noting also that most of the resources in this area are indicated. However, the probability intervals are +3 %/–12 % of the predicted resource model grade for the indicated category, which implies a less variable grade distribution in this area compared to the other three areas.

The amount of information that can be derived from a simulation model is significantly larger than what has been presented here. As a consequence of a similar analysis to the one presented here, infill drilling campaigns, mine call factors, and other risk mitigation measures were taken to ensure that the predicted ore through the plant is achieved.

In addition, technical personnel and management have a tool to better understand the consequences of the resource classification schemes, and their significance. None of this detailed analysis is possible with traditional resource classifications, so it is reasonable to back up the traditional resource classification with a probabilistic analysis based on a conditional simulation model.

14.6 Grade Control at the San Cristóbal Mine

The most important task in daily life of an open-pit mining operation is to select ore and waste. This grade control process can be a simple ore/waste decision or a more complicated process because there may be different destinations or stockpiling and blending requirements.

Perfect selection, that is, making no mistakes in deciding the destination of every ton of material mined out, is impossible. Sampling errors, estimation errors, limited or bad information, and operational constraints and mistakes always result in ore loss and dilution, which in turn lead to economic losses. In extreme cases, these losses can be serious enough to compromise the profitability of the operation. Poor grade control may cause an operation to fail, such as the São Vicente mine in Brazil's Matto Grosso in the mid-1990s. Studies on mine failures and not realizing expectations have been completed be several researchers, for example Burmeister (1998) in Australia, and Knoll (1989) and Clow (1991) for Canadian operations. In many cases, the failures to meet expectations have been attributed to poor resource estimation and lack of grade control. Minimizing ore loss and dilution is critical to a successful operation, since every mistake made detracts from the maximum amount of ore that could, theoretically, be recovered from the pit. The San Cristóbal mine discussed here was on its way to join the unpleasant list of failed operations, until improvements in grade control turned around the mine.

In open-pit mines it is often difficult to define accurately the position of the dig boundaries prior to loading, particularly where there are few or no visual markers. Commonly used grade control methods include simple visual observations of the blast holes grades, some form of blast hole averaging into panels of arbitrary shape, or polygonal methods. In more recent years, several forms of kriging have gained acceptance, including ordinary and indicator kriging. Even more recently, grade control methods based on conditional simulations and economic optimization have gained some popularity.

Conditional simulation-based methods may be better than more traditional grade control methods, including kriging, when (a) ore and waste populations are intermixed, making it difficult to identify ore pods without leaving ore blast holes unrecovered; similarly, the recognizable ore pods may have significant amounts of waste within; (b) no visual markers are available; even if higher-grade controlling structures are identified, there is never assurance that they are mineralized; and (c) grade variability is significant.

The San Cristóbal case presented here illustrates some of the practical aspects of implementing a simulation-based grade control method (Isaaks 1990; Aguilar and Rossi 1996; Rossi 1999). The benefits of the method are evaluated based on production data, mine and mill reconciliation data, and cash flow analysis. San Cristóbal is a now-exhausted medium-sized open pit gold and silver mine, which processed about 10,800 metric tons daily of ore grading approximately 1 g/t Au and between 4 g/t and 6 g/t Ag. The mine operated on 5 m benches, and blast holes were used for breakage and to obtain grade samples from the pit, spaced at about 4.5 m and sampled over the entire 5 m bench height. Blast holes were drilled, sampled, and loaded with explosives on a daily basis, typically one blast of 300–400 holes per day. The ore, after being crushed in three stages, was heap-leached, and Au and Ag were recovered by passing the enriched solution through six activated carbon columns. Finally, an Au and Ag doré was produced with approximately 27–30 % Au. Up until the introduction of a new grade control method, the mine produced about 65,000 ounces of Au per year.

14.6.1 Geologic Setting

Gold mineralization is very erratic, with a highly skewed distribution that makes grade modeling and resource/reserve prediction difficult at any scale. Gold and silver minerals are associated to a sub-volcanic intrusion, mainly consisting of rhyolite, breccia, quartz-feldspar porphyries, and occasional dacitic dikes. Alteration is typical of porphyry intrusions, zoning from a central potassic alteration to an intermediate alteration zone characterized by actinolite and epidotes, and to an external halo of propyllitic alteration. Superimposed to these alterations a different sericite-quartz alteration has been identified, associated to veining and significant shearing.

Mineralization occurs within discontinuous structures, oriented North to NW-SE, within a dilational zone limited to the north and south by two shear zones. The structures hosting the mineralization are typically 0.1–1.0 m in width. Gold is present in a quartz-pyrite association. When the structures intercept more favorable lithologies, such as breccias, gold mineralization appears also disseminated into the host rock.

The geology is well understood but provides few markers with no visual indicators for the occurrence of gold and silver mineralization. The presence of veins does not ensure the occurrence of gold, and not all of the gold is strictly confined to the stockwork-like fractures. The short- and long-term production reconciliations obtained from 1991 (when the mine began operations) until mid-1994 were poor. Estimation methods for long-term mine planning were originally ordinary kriging estimates, controlled by the use of geologic and grade envelopes. A significant improvement was achieved by using Multiple Indicator Kriging instead of Ordinary kriging, still constrained by a low-grade envelope. But it was also determined that the grade control method used was losing significant quantities of gold and processing waste. This was because of the difficulty in drawing panels containing homogeneous ore zones. In an effort to remedy the situation, a conditional simulation method combined with economic optimization was designed, tested, and implemented.

14.6.2 Maximum Revenue (MR) Grade Control Method

Conditional simulations provide conditional probability distributions which, in conjunction with relevant economic parameters, can be used to minimize losses caused by imperfect selection. Imperfect pit selection and misclassification will always occur. The main objective of the method is to minimize economic losses. This optimization is achieved based on a set of economic parameters and the probability of occurrence of ore for every node within a blast. The maximum revenue grade control method (MR) uses loss functions as a basic tool for decision-making. The MR method requires two basic steps:

1. A set of conditional simulations is obtained, providing an uncertainty model about the grade at a specific point within the blast.
2. An economic optimization process using Loss Functions is implemented. It is designed to obtain the economically optimal ore/waste selection. The loss function quantifies the economic consequences of each possible decision, minimizing losses, see Chaps. 12 and 13 and Isaaks (1990).

The simulation model is based on blast holes, and the process is run on a daily basis. Quality of blast hole sampling, sample preparation and assaying had been addressed before the implementation of the MR method with a detailed sampling heterogeneity study (Pitard 1995). The procedures and protocols implemented were deemed to achieve a targeted 15 % fundamental sampling error variance for the blast holes.

The conditional simulations for the San Cristóbal open pit were built on a 1 m × 1 m × 5 m grid using the Sequential Indicator method. Shifts in the attitude of the ore controlling structures required the separation of the data into different populations. Evaluation of high-grade populations was required to control extreme grades in the simulation and correctly reproduce the observed variability (Parker 1991; Rossi and Parker 1993).

Indicator variograms were modeled from blast hole data, and a number of simulation parameters were optimized. This included minimum and maximum data value used; maximum simulated value allowed; maximum number of conditioning data to be used; and anisotropic search distances. The conditional simulation models were validated against original data. Figure 14.57 shows four conditional simulations obtained for Level 2,345 m.

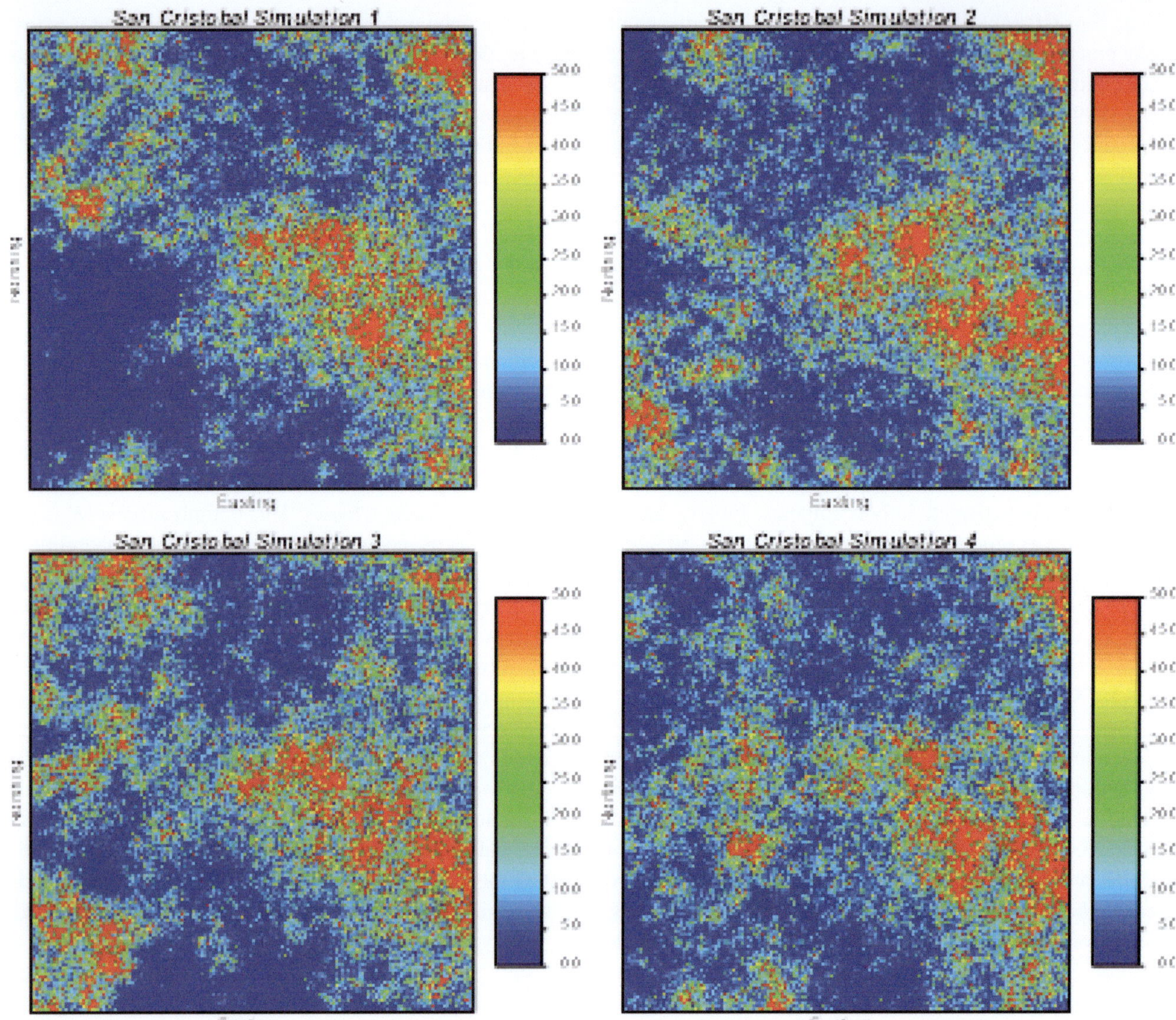

Fig. 14.57 Conditional simulations, San Cristóbal, Bench 2,345

In grade control, the selection decision (which material is ore and which is waste) has to be based on grade estimates, $z^*(\mathbf{u})$. Since the true grade value at each location is not known, an error can and will likely occur. The loss function $L(e)$ attaches an economical value to each possible error. The expected conditional loss can be found by applying a loss function to a set of equi-probable simulated grade values (conditional probability function) at each simulated node.

The minimum expected loss can then be found by simply calculating the conditional expected loss for all possible values for the grade estimates, and retaining the estimate that minimizes the expected loss. As described in Isaaks (1990), in grade control the expected conditional loss is a step function whose value depends on the operating costs, and the relative costs of miss-classification. This implies that the expected conditional loss depends only on the *classification* of the estimate $z^*(x)$, not on the estimated value itself. For example, the loss incurred when a block of leach ore is sent to the mill is a function of the difference in processing costs related to both leach and mill; it will, of course, also depend on the *true block grade*, but not on the *estimated block grade* value itself.

Minimizing the loss in such a way can be related to minimizing Type I (false positive) and Type II (false negative) errors (Fig. 14.58). In positively-skewed distributions, which are characteristic of minerals of high intrinsic value, such as precious and base metals, only a small proportion of the rock mass is economic. This implies that it is not the same to make a Type I (the material is thought to be ore, when it is in fact waste) or a Type II error (the material is thought to be waste, and in fact it is ore). The key difference in this method is that the process does not necessarily minimize grade estimation errors, but minimizes the economic consequences of such errors.

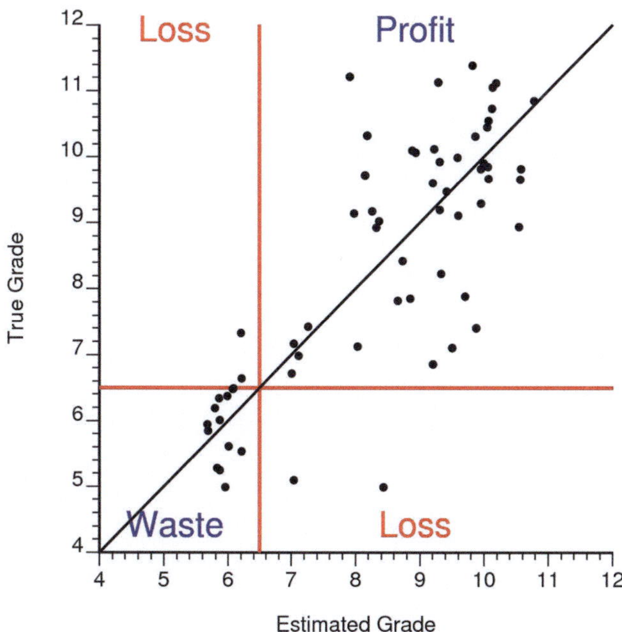

Fig. 14.58 Misclassification in grade control

14.6.3 Implementation of the MR Method

The work summarized here was initiated because the operation achieved poor grade and tonnage reconciliations for a number of years. The block model used for long-term mine planning was blamed for the problem. Several attempts were made to improve the resource model, and in the end the MIK model was believed to be as good a model as the drill hole data and the geologic knowledge would allow. Then, attention was turned to grade control, since the mineralization's variability at San Cristóbal was a significant challenge to the day-to-day operation, determining its operating profitability.

The existing grade control method was based on assignment of blast hole grades to panels. The blast holes were plotted on scaled maps; mine technicians would then draw polygons based on the observed grades at each blast hole; and then visually define areas of ore to be extracted. These maps with the polygons were passed on to the surveyors, who would then stake on the pit the ore and waste areas for the operators to load. The predicted average grade of the panel was calculated as a simple arithmetic average of all blast hole grades within the panel. A dilution band of 1 or 2 m. was added, estimating its grade by averaging surrounding blast holes. Then, the overall grade of the polygon was estimated as an area-weighted average of the ore zone and the dilution zone.

The case study described here details the changes introduced in the grade control method over a 13-month evaluation period from March 1995 to March 1996. All other aspects of grade control, including blast hole sampling, database creation, blasting practices, field demarcation, and operating practices were not affected by the change in grade control method.

The MR method is based on using blast hole data to obtain blast simulations. These simulations are processed using a loss function defined in terms of the operating profit as:

$$Loss = Actual - Potential$$

The revenue equation, usually expressed on per ton of ore basis, is given by:

$$\text{Profit} = (\text{Price}) \bullet (\text{Metallurgical Recovery}) \bullet (\text{Au grade}) - (\text{Treatment Costs} + \text{G\&A Costs})$$

The revenue function should include at least processing costs, and general and administration costs (G/A) assigned to the operation. Some mining companies assign to the mine other costs such as capital replacement and depreciation costs, or brownfield or greenfield exploration costs. This artificially increases the breakeven cutoff, which may be appropriate if the mill capacity is limited. In general, mining costs should not be included, since the blasted material will have to be moved regardless of its destination, that is, it is a sunk cost. Only if there are differences in transportation costs for different destinations (waste and ore) a differential mining cost term should appear.

A matrix can be built based on the alternative material destinations. In this case, only waste and leach ore have been considered; therefore, a 2×2 loss function matrix is obtained. Extending this matrix to multiple destinations, such as mill and stockpiles, is straightforward. The miss-classification matrix quantifies the cost in dollars of each possible mistake. Table 14.19 presents the simplest possible Loss Function corresponding to ore and waste destinations, which was used at San Cristóbal.

The diagonal cells in Table 14.19 are 0 because in those cases the right decision is made, i.e., there is no loss. Cell A_{21} represents the loss incurred when waste material is sent to the processing facility, in this case the leach pad. The potential benefit is negative (since the true grade is waste), the only cost that should have been incurred into is the G/A cost of operating the mine; a loss is added to the potential operating profit, stemming from the processing of waste material. Cell A_{12} presents the case where material that is ore is sent to the waste dump. The potential benefit is the revenue that would have been achieved if the right decision was made. The resulting loss in both cases is a negative dollar amount taking into account the actual and the opportunity costs.

The MR grade control method as implemented at San Cristóbal consists of the following steps:

1. Conditional simulations were obtained based on nearby blast holes. These conditional simulations provide the probability for each specific block to be either ore or waste.
2. Apply the loss function as defined in Table 14.19. The blocks were assigned a series of codes (waste or ore) representing the optimal selection in an economic sense.

Table 14.19 Loss function, in $/ton

	True grade is waste	True grade is leach ore
Estimated grade is waste	$A_{11}=0$	A_{12}=–G/A Costs -{Leach Revenues—Proc. Costs}
Estimated grade is leach ore	A_{21} = {Waste Revenues—Proc. Costs}—G/A Costs	$A_{22}=0$

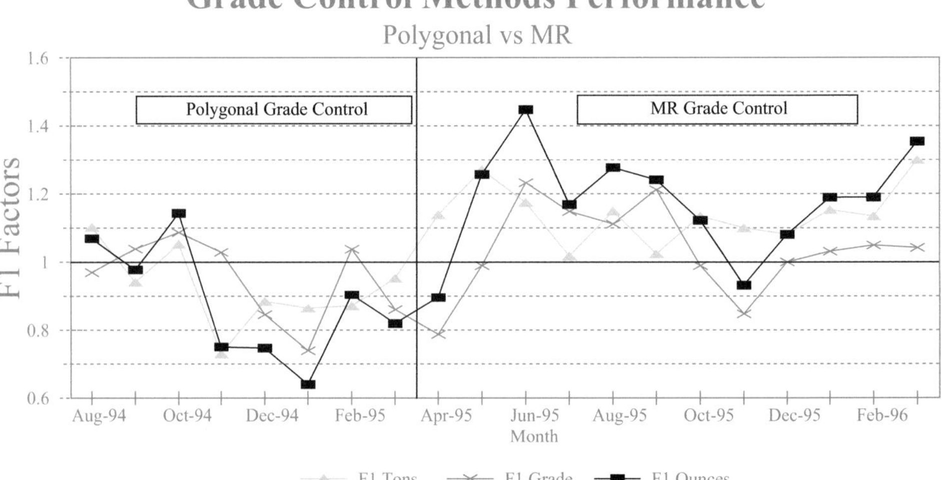

Fig. 14.59 Comparison of grade control to Resource Model (F_1 factors), San Cristóbal, August 1994–March 1996

3. The codes within each blast were visualized on screen and a polygon is drawn to define areas of ore and waste. This polygon was drawn manually by the grade control technician following operational constraints.
4. The MR grade control method was based on a geological model. This is necessary because the conditional simulations do not capture sharp mineralogical or structural transitions. The MR method is no different than any other grade control method in that it will only produce good results if properly controlled with in-pit mapping and a geologic model.
5. An estimate is generally required for the tons and grade to be recovered from each blast. At San Cristóbal the average grade of the simulations was used as the estimated grade for each panel.

Note that the decision of where to send each block or portion of the blast is made before any actual estimate of the grade is obtained. The decision only depends on the relative probabilities of each block of belonging to either the ore or waste category, and the potential cost of making a mistake.

14.6.4 Results

From March 1995 through March 1996 the operation implemented in parallel the existing polygonal and the new MR methods, allowing for a direct comparison based on production. The comparison demonstrated the remarkable improvement achieved by the MR method.

An F_1 factor is defined to compare block model results to grade control results. An F_2 factor is used to compare "loaded to heap" material to grade control predictions, as proposed by Parker (Rossi and Parker 1993). The F_3 factor ($F_3=F_1 * F_2$) is used to compare tons and grade predicted by the long-term block model (MIK) to tons and grade loaded to heaps.

Figure 14.59 shows the F_1 factors for the period, on a monthly basis; the introduction of the maximum revenue grade control method is evident. This was the only change introduced in the operation at the time. Table 14.20 shows the improvements achieved with the MR method. It compares over a 13-month period the tonnages, grades, and ounces predicted by the MIK block model to the tonnages, grades and ounces selected by the two grade control methods.

The F_2 factors averaged for the 13-month period in Table 14.20 reflect tonnages actually selected and loaded to the plant using the conditional simulation method. A 10 % unplanned dilution in ounces in an operation with an extremely erratic mineralization is quite reasonable. Polygonal-based method results in dig panels that make loading and selection in the pit much more difficult, which would make the resulting F_2 factors for the polygonal method even worse. The fine grid used by the MR method allows for an operational cut that incorporates less dilution. Another important conclusion is that in reality the Long-term block model (MIK) was conservative in tons and about unbiased in grade, and most of its perceived shortcomings were not in fact an issue.

Table 14.20 Comparison of the MIK long-term block model, MR grade control method, and polygon-based grade control method, March 1995–March 1996

	Tons of ore	Au grade	Ounces
F_1 *(polygonal grade control/MIK)*	0.91	0.94	0.86
F_2 *(plant/polygonal grade control)*	1.34	0.82	1.10
$F_3 = F_1 * F_2$ *(plant/MIK, polygonal GC)*	1.22	0.77	0.95
F_1 *(MR grade control/MIK)*	1.13	1.00	1.13
F_2 *(plant/MR grade control)*	1.01	0.89	0.90
$F_3 = F_1 * F_2$ *(plant/MIK, MR grade control)*	1.14	0.89	1.02

Table 14.20 can be translated into economic gains. For this particular operation, the MR method resulted in 34 % more available in-situ tons, a 10 % increment of available in-situ grade, for a net increment of 48 % in-situ metal. These results had a major impact in the company's cash flows, operating revenues, and costs. Overall in-situ revenues increased by US$ 11.2 million in 13 months, and the net increase in cash flow was US$ 4.8 million, or a monthly average of US$ 370,000, without including gains in diminished stripping cost. Recall that the pit shape was unchanged. Overall gold production jumped from approximately 65,000 ounces annually to about 80,000 ounces. As a reference, during the 13-month period, gold prices fluctuated between US$ 370 and US$ 415 per ounce.

Aided by an improved knowledge of the geology at San Cristóbal, the shortcomings of previously used methods for grade estimation were better understood. The implementation of the MR method resulted in production records and significant economic improvements to the operation. Not only higher recovery of ore and a better ore/waste selection overall was achieved, but there were also other operational improvements, allowing for a reduction of the unplanned dilution. The MR grade control method was implemented such that technicians with little geostatistical background can operate and control the system. The method was in use at San Cristóbal from February 1995 until the operation closed down in the late 1990s.

14.7 Geometallurgical Modeling at Olympic DAM, South Australia[4]

Conventional resource estimation is focused on one or a few metals that will be sold at a profit. Increasingly, however, it is becoming important to understand many other characteristics of the ore that affect processing performance and recovery. The detailed spatial distribution of these variables permits a more holistic optimization of the mining operation. This case study relates to BHP Billiton's Olympic Dam project in South Australia. Two important topics are addressed with the wealth of measurements taken at Olympic Dam: (1) recovery and other performance variables are related to measured rock properties through a multivariate regression model; and (2) geostatistical models of the key rock properties are constructed by simulation.

14.7.1 Part I: Hierarchical Multivariate Regression for Mineral Recovery and Performance Prediction

Mineral recovery and expected plant performance are difficult to predict because they are influenced by a large number of variables such as mineralogy, grade, grain size, plant operation parameters, etc. Often constant recovery factors and plant efficiencies are assumed for a given mine based on past experience and empirical rules. Such methods are acceptable during the feasibility stages of mineral exploration; however, when results of pilot plant trials are available, statistical methods can be utilized to better predict recovery and plant performance. In this case study, 841 bulk samples from flotation and leach tests are used for the calibration of a predictive model. The result is a model that can be used to predict recovery and plant performance based on available geometallurgical data.

Over 200 variables are available to develop a regression model. A danger with this many variables is that the relationship to recovery and plant performance variables would be over fit. Steps must be taken to avoid over fitting. Redundant and unimportant variables are identified and removed from the modeling process, reducing the number of variables to 112. Through a sequence of hierarchical variable amalgamation steps the variables are condensed into four major subcategories. A linear model based on these four amalgamated variables provides a predictive model that is used to estimate potential mineral recovery and plant performance over the entire deposit. Minerals of interest in this mine include copper, uranium, gold, and silver.

Plant performance is dependent on a large number of variables, such as (1) plant feed (2) operational parameters (3) equipment efficiencies (4) and equipment repairs. The purpose of this case study is to relate available geometallurgical data to plant performance. This is done by correlating the available data to pilot plant trials. A total of 841 pilot runs are available with associated plant feed mineralogy, head assays and mineral association data; the data is described in Table 14.21. Important plant performance indicators include recovery of Cu and U_3O_8, acid consumption (used in the leaching process), net recovery, drop weight index (DWi) and bond mill work index (BMWi); using the

[4] BHP Billiton—Uranium is gratefully acknowledged for allowing the publication of this case study. This case study is based on the paper "Geometallurgical Modeling at Olympic Dam", by Boisvert, J., Rossi, M. Ehrig, K., and Deutsch, C., accepted for publication in Mathematical Geosciences, 2013.

Table 14.21 Data available

Data type	Description	Notes
Head assays	This data contains the % content of various important elements, including: Co, As, Mo, Ni, Pb, Zn, Zr, Sr, Bi, Cd, Cs, Ga, In, Sb, Se, Te, Th, Tl	This data is compositional in nature
Mineralogy	A total of 10 identified minerals and minestals set make up the bulk of the deposit. These include: brannerite, coffinite, uraninite, pyrite, chalcopyrite, bornite, chalcocite, other sulphides, acid soluble gangue and acid insoluble gangue	This data is also compositional in nature
Association data	A number of thin sections are available. These have been analyzed and the complete matrix of associations between minerals is available. This describes the contact area between two adjacent minerals within a single grain of crushed material	This data is also compositional in nature

Table 14.22 Description of predictive models generated

Model	Input variables	Output	Comments
Full model	Head assays (i.e. %cu, %U …) 10 Mineralogy 10 × 11 matrix of associations Specific gravity	Cu, U, Au, Ag recoveries Acid consumption Net recovery (U) BMWi and DWi	This model represents the maximum data available
Typical model	Head assays (i.e. %cu, %U …) 10 Mineralogy Specific gravity	Cu, U, Au, Ag recoveries Acid consumption Net recovery (U) BMWi and DWi	This is the base case model. Field data will most likely contain all these variables
Limited model	Limited head assays 7 Mineralogy variables Specific gravity	Cu, U, Au, Ag recoveries Acid consumption Net recovery (U) BMWi and DWi	Only head assays that have many samples in the available database are considered

data in Table 14.21 as input to a regression model, these six plant performance variables can be predicted at all locations in the deposit.

14.7.2 Methodology

A linear regression model is used to predict the plant performance variables. One drawback with a linear regression model is that all input variables are required for prediction. Thus, if a single input variable is missing from a sample, the regression model cannot be applied. For this reason, three regression models are generated (Table 14.22). Each model represents a decreasing number of input parameters. For example, for locations in the deposit where association data is not known the "full model" cannot be applied and the "typical model" would be appropriate.

The regression models are based on a large set of input variables. The variables are merged into super secondary variables based on the correlations between variables. This is done because there are too few sample data available to accurately determine regression coefficients for the 204 input variables available. The final model is a linear regression on four super secondary variables. The methodology consists of six steps:

1. Normal score the input variables.
2. Merge the variables (level 1). This step reduces the 112 input variables to 23 merged variables.
3. Merge the variables (level 2). This step reduces the 23 merged variables to 4.
4. Regression on the four variables.
5. Back transform the estimated variables (DWi, BMWi, Cu recovery, U_3O_8 recovery, acid consumption and net recovery).
6. Determine uncertainty in the model.

Step 1: Normal Score Data First, the number of variables must be reduced. Variables are removed from the analysis because (1) they have a low correlation to the six output variables or (2) they are highly redundant with one of the other input variables. A variable was considered to have a low correlation if the maximum correlation to any of the output variables was less than 0.13. A variable was considered redundant with another input variable if it had a correlation greater than 0.94. This reduces the number of input variables to 112.

There are a total of 841 samples available for modeling; however, not all samples contain all 112 variables used in the calibration of this model. Due to the nature of a regression model, it is necessary that all 112 variables be present for a sample to be used for calibration. Of the 841 samples, 328 samples were retained for modeling. More data are available if the mineral associations are ignored (for example, if using the "typical" model).

All 118 variables (112 input + 6 output) are independently normal score transformed. A visual assessment of the

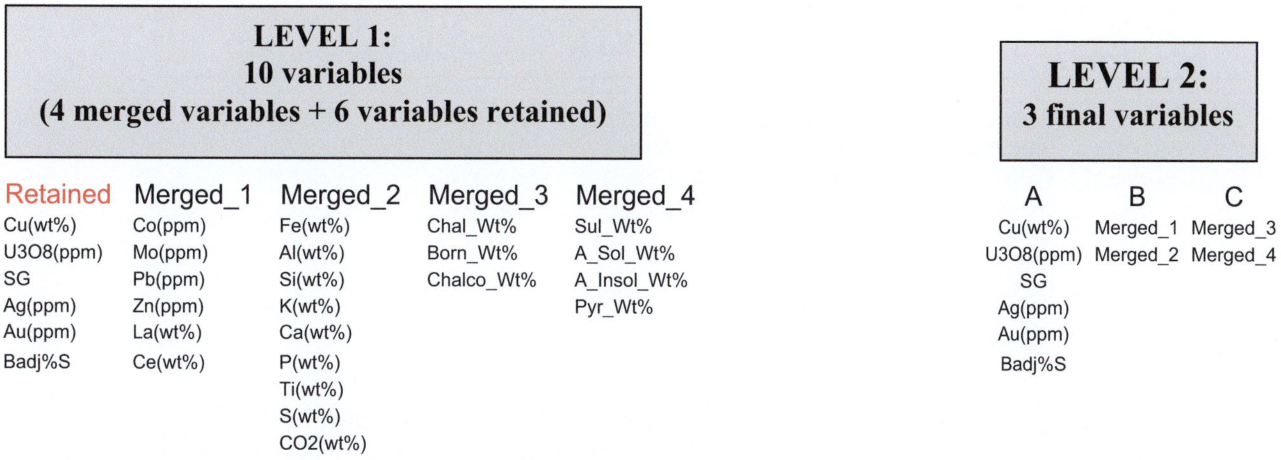

LEVEL 1:
10 variables
(4 merged variables + 6 variables retained)

Retained	Merged_1	Merged_2	Merged_3	Merged_4
Cu(wt%)	Co(ppm)	Fe(wt%)	Chal_Wt%	Sul_Wt%
U3O8(ppm)	Mo(ppm)	Al(wt%)	Born_Wt%	A_Sol_Wt%
SG	Pb(ppm)	Si(wt%)	Chalco_Wt%	A_Insol_Wt%
Ag(ppm)	Zn(ppm)	K(wt%)		Pyr_Wt%
Au(ppm)	La(wt%)	Ca(wt%)		
Badj%S	Ce(wt%)	P(wt%)		
		Ti(wt%)		
		S(wt%)		
		CO2(wt%)		

LEVEL 2:
3 final variables

A	B	C
Cu(wt%)	Merged_1	Merged_3
U3O8(ppm)	Merged_2	Merged_4
SG		
Ag(ppm)		
Au(ppm)		
Badj%S		

Fig. 14.60 Variables used in the limited model. A total of 28 input variables are considered

bivariate relationships between the input data indicated very few non-linear relationships; therefore, stepwise conditional transformations are not considered.

Step 2: Merge variables—reduce 112 input variables to 23 merged super secondary variables There is a danger of over fitting the available calibration data if a regression model is constructed on all 112 input variables. Therefore, subsets of the input data were amalgamated to construct super secondary merged variables. These merged variables are linear combinations of a subset of variables and significantly reduces the dimensionality of the problem while also reducing the risk of over fitting. The selection of subsets is based on the nature of the measurements; similar rock measurements are kept together.

The merged "super secondary" variables are generated by assigning weights to each variable:

$$M(v) = \sum_{i=1}^{n} \lambda_i v_i$$

where n is the number of variables to be merged based on the weights from a likelihood calculation. These weights are generated by solving the corresponding matrix for each merged variable and for each of the six output variables:

$$\begin{bmatrix} \rho_{1,1} & \rho_{2,1} & \cdots & \rho_{n,1} \\ \rho_{1,2} & \rho_{2,2} & \cdots & \rho_{n,2} \\ \vdots & \vdots & \ddots & \vdots \\ \rho_{1,n} & \rho_{2,n} & \cdots & \rho_{n,n} \end{bmatrix} \begin{bmatrix} \lambda_1 \\ \lambda_2 \\ \vdots \\ \lambda_n \end{bmatrix} = \begin{bmatrix} \rho_{0,1} \\ \rho_{0,2} \\ \vdots \\ \rho_{0,n} \end{bmatrix}$$

The right-hand side of this equation contains the correlation between one of the variables of interest and the n input variables to be merged, while the left hand side is the correlation between all n variables to be merged.

These correlation matrices may be poorly conditioned with few data. Poorly conditioned matrices are the cause of extreme weights (λ_i) and introduce unwarranted noise in the predictions. To prevent this, the correlation matrices are *fixed* to improve their stability. This correction is accomplished by decreasing the values of the off diagonal elements of the matrix, which increases the value of the smallest eigenvalue for the matrix and increases stability. The minimum eigenvalue for the correlation matrices was set to 0.05. Twenty-four of the correlation matrices for the full model required a correction, 18 of the correlation matrices for the typical model required a correction and 12 of the correlation matrices for the limited model required a correction.

The merged variables are a linear combination of $N(0,1)$ variables. Thus, the mean of the merged variables will be 0 but the variance will not be 1. The merged variables are standardized by the standard deviation determined from the following classical relationship:

$$\sigma^2(M(v)) = \sum_{i=1}^{n}\sum_{j=1}^{n} \lambda_i \lambda_j Cov(v_i, v_j)$$

Thus, the final merged variable becomes:

$$M(v) = \frac{\sum_{i=1}^{n} \lambda_i v_i}{\sum_{i=1}^{n}\sum_{j=1}^{n} \lambda_i \lambda_j Cov(v_i, v_j)}$$

Step 3: Merge variables—reduce 23 input variables to 4 merged variables for regression There are two levels of variable amalgamation. The first level grouped related variables into 16 merged variables and retains 7 additional variables for a total of 23 variables. Figure 14.60 shows the variables used in the limited model, while Fig. 14.61 shows the variables used in the regression models.

LEVEL 1: 19 variables (15 merged variables + 4 variables retained)

Retained	Merged_1	Merged_2	Merged_3	Merged_4	Merged_5
Cu(wt%)	Ba(wt%)	La(wt%)	Uran_Wt%	Chal_Wt%	Sul_Wt%
U3O8(ppm)	Fe(wt%)	Mg(wt%)	Cof_Wt%	Born_Wt%	A_Sol_Wt%
SG	Al(wt%)	Mn(wt%)	Bran_Wt%	Chal_Wt%	A_Insol_Wt%
K:Al	Si(wt%)	Na(wt%)			Pyr_Wt%
	K(wt%)	P(wt%)			
	Ca(wt%)	Ti(wt%)			
	S(wt%)				
	CO2(wt%)				
	F(wt%)				

Merged_6	Merged_7	Merged_8	Merged_9
Bran_Pyr_assoc	Cof_Bran_assoc	Uran_Cof_assoc	Pyr_Cof_assoc
Bran_Chalcopy_assoc	Cof_Uran_assoc	Uran_Chalcopy_assoc	Pyr_Uran_assoc
Bran_Bornite_assoc	Cof_Pyr_assoc	Uran_Bornite_assoc	Pyr_Chalcopy_assoc
Bran_Chalcocite_assoc	Cof_Chalcopy_assoc	Uran_A_Sol_assoc	Pyr_Sulphides_assoc
Bran_A_Sol_assoc	Cof_Chalcocite_assoc	Uran_A_Insol_assoc	Pyr_A_Sol_assoc
Bran_A_Insol_assoc	Cof_Sulphides_assoc		Pyr_Free_Surf_assoc
Bran_Free_Surf_assoc	Cof_A_Sol_assoc		
	Cof_A_Insol_assoc		
	Cof_Free_Surf_assoc		

Merged_10	Merged_11	Merged_12
Chalcopy_Bran_assoc	Bornite_Cof_assoc	Chalcocite_Chalcopy_assoc
Chalcopy_Cof_assoc	Bornite_Pyr_assoc	Chalcocite_ Bornite_assoc
Chalcopy_Uran_assoc	Bornite_Chalcopy_assoc	Chalcocite_Sulphides_assoc
Chalcopy_Pyr_assoc	Bornite_Chalcocite_assoc	Chalcocite_A_Sol_assoc
Chalcopy_Bornite_assoc	Bornite_Sulphides_assoc	Chalcocite_A_Insol_assoc
Chalcopy_Chalcocite_assoc	Bornite_A_Sol_assoc	Chalcocite_Free_Surf_assoc
Chalcopy_Sulphides_assoc	Bornite_A_Insol_assoc	
Chalcopy_A_Sol_assoc	Bornite_Free_Surf_assoc	
Chalcopy_A_Insol_assoc		
Chalcopy_Free_Surf_assoc		

Merged_13	Merged_14	Merged_15
Sulphides_Uran_assoc	A_Sol_Bran_assoc	A_Insol_Bran_assoc
Sulphides_Pyr_assoc	A_Sol_Cof_assoc	A_Insol_Cof_assoc
Sulphides_Chalcopy_assoc	A_Sol_Uran_assoc	A_Insol_Uran_assoc
Sulphides_Bornite_assoc	A_Sol_Pyr_assoc	A_Insol_Chalcopy_assoc
Sulphides_A_Sol_assoc	A_Sol_Chalcopy_assoc	A_Insol_Bornite_assoc
Sulphides_A_Insol_assoc	A_Sol_Bornite_assoc	A_Insol_Sulphides_assoc
	A_Sol_Chalcocite_assoc	A_Insol_A_Sol_assoc
	A_Sol_Sulphides_assoc	A_Insol_Free_Surf_assoc
	A_Sol_A_Insol_assoc	
	A_Sol_Free_Surf_assoc	

LEVEL 2: 4 Final Variables

A	B	C	D
Cu(wt%)	Merged_1	Merged_4	Merged_7
U3O8(ppm)	Merged_2	Merged_5	Merged_8
SG	Merged_3	Merged_6	Merged_9
K:Al			Merged_10
Ag(ppm)			Merged_11
Au(ppm)			Merged_12
Badj%S			Merged_13
			Merged_14
			Merged_15
			Merged_16

Variable A contains individual variables retained.
Variable B contains the remainder of the head assays.
Variable C contains all mineralogy variables.
Variable D contains all association variables.

Fig. 14.61 Variables used in the regression models

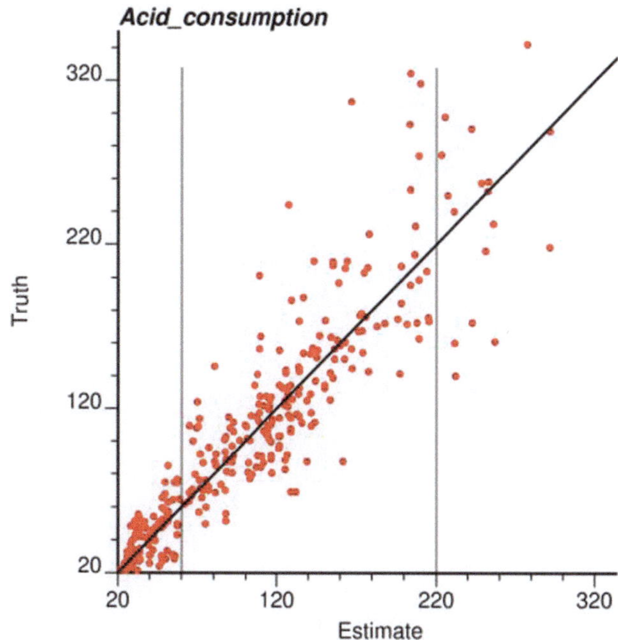

Fig. 14.62 The uncertainty in an estimate of 60 vs. 220 kg/ton. There is more uncertainty at 220 kg/ton

The second level groups the variables into the final 4 super-secondary variables used for regression:

A. Retained variables
B. Head assays
C. Mineralogy
D. Associations

Step 4: Regression The typical and limited models are generated by regression on variables A, B, and C while the full model considers variables A–D. Regression is performed with both linear and quadratic terms. However, through cross validation it was found that increasing the number of terms beyond the linear coefficients resulted in little consistent gain and the linear model is sufficient. Thus, the final model becomes:

$$Prediction = av_1 + bv_2 + cv_3 + dv_4$$

Step 5: Back Transformation Once the predictions are made in normal units for each of the six output variables, they must be transformed back into original units using the original transformation tables.

Step 6: Determine Uncertainty in the Model When a prediction is made, the uncertainty in that prediction is also determined. The uncertainty is obtained by examining the distribution of true values for a given estimate. Consider the difference in making an acid consumption prediction of 60 vs. 220 kg/ton (Fig. 14.62). There is more uncertainty in the estimate of 220 kg/ton. The measure of uncertainty used is the spread of the true values around the estimate. In this case, the p90-p10 range was chosen.

14.7.3 Analysis

All samples were used to generate the regression models with the above methodology. High correlation between the estimate and the truth is desirable. Rather than show the 768 coefficients for variable merging and the 24 regression coefficients, a tornado chart (Fig. 14.63) is used to illustrate the influence of each of the 112 variables on the overall model. The lower limit is determined by selecting the p_{10} value for the input variable of interest and setting all remaining 111 variables to their p_{50} value. An estimate is made for each of the six output variables, giving the lower limit on the tornado chart. Similarly, the p_{90} value is selected for the variable of interest to generate the upper limit on the tornado chart. A short horizontal line to the left of the variable indicates that the variable is negatively correlated with the output variable. Bars are shaded based on the origin of the variable: White—head assays; Gray—associations; Red—mineralogy; Black—specific gravity.

Figure 14.64 shows the models built on all possible data points available for the different models. Some interesting relationships were discovered in the cross plots and the tornado charts:

- Na is a significant contributor for DWi/BMWi—indicates different mineralogy.
- SG is important for DWi but not BMWi—this is expected as it matters whether the rock is brittle or not, and this is related to the ratio of iron/silica content in the rock matrix.
- BMWi is heavily influenced by the head assays (top 6 variables contributing to BMWi are from head assays).
- Individual mineralogy variables have little significance (Cu recovery is the exception).
- Presence of Chalcopyrite and acid insoluble gangue are critical to Cu recovery.
- Cu wt% has a large effect on U_3O_8 recovery but little effect on Cu recovery. This is because Cu recovery is approximately constant for the high Cu grades found in the deposit.
- Based on the tornado charts, associations are important for DWI, Cu recovery, acid consumption and net recovery. This is also seen in the comparison of the typical and full models (Fig. 14.64) as the BMWi and U_3O_8 recovery predictions are not significantly altered by removing the association data.
- Recoveries are the most difficult variables to predict (lowest correlation on Fig. 14.64). This is expected, as recovery is dependent on a large number of complex interactions.

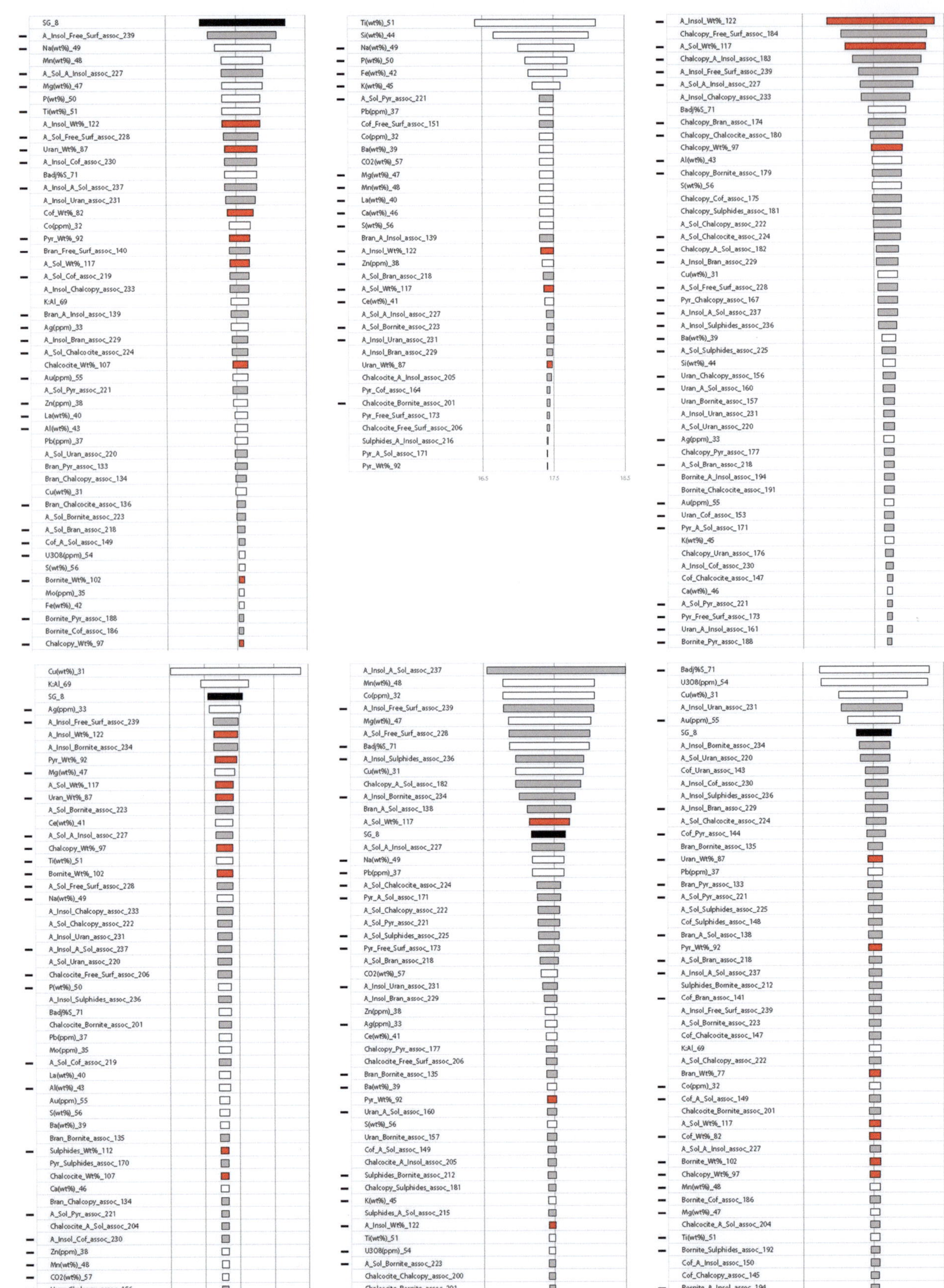

Fig. 14.63 Full model Tornado charts for DWi, BMWi and Cu recovery (*top*) as well as U_3O_8 recovery, acid consumption and net recovery (*bottom*). *White*: head assays; *Gray*: associations; *Red*: mineralogy; *Black*: SG

FULL MODEL

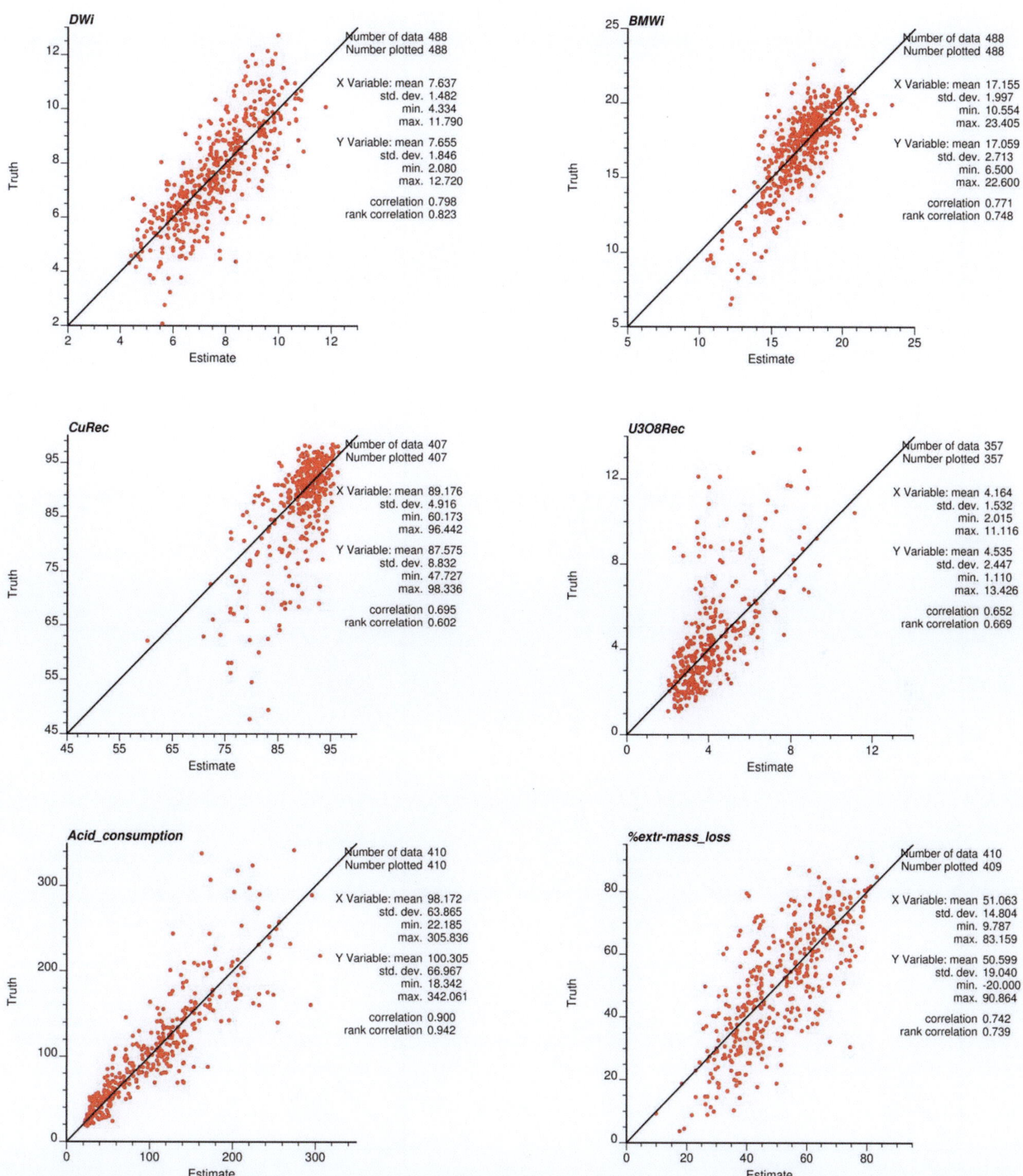

Fig. 14.64 Cross plots of the truth/estimated values based on the full model (top) and the typical model (middle) and the limited model (bottom)

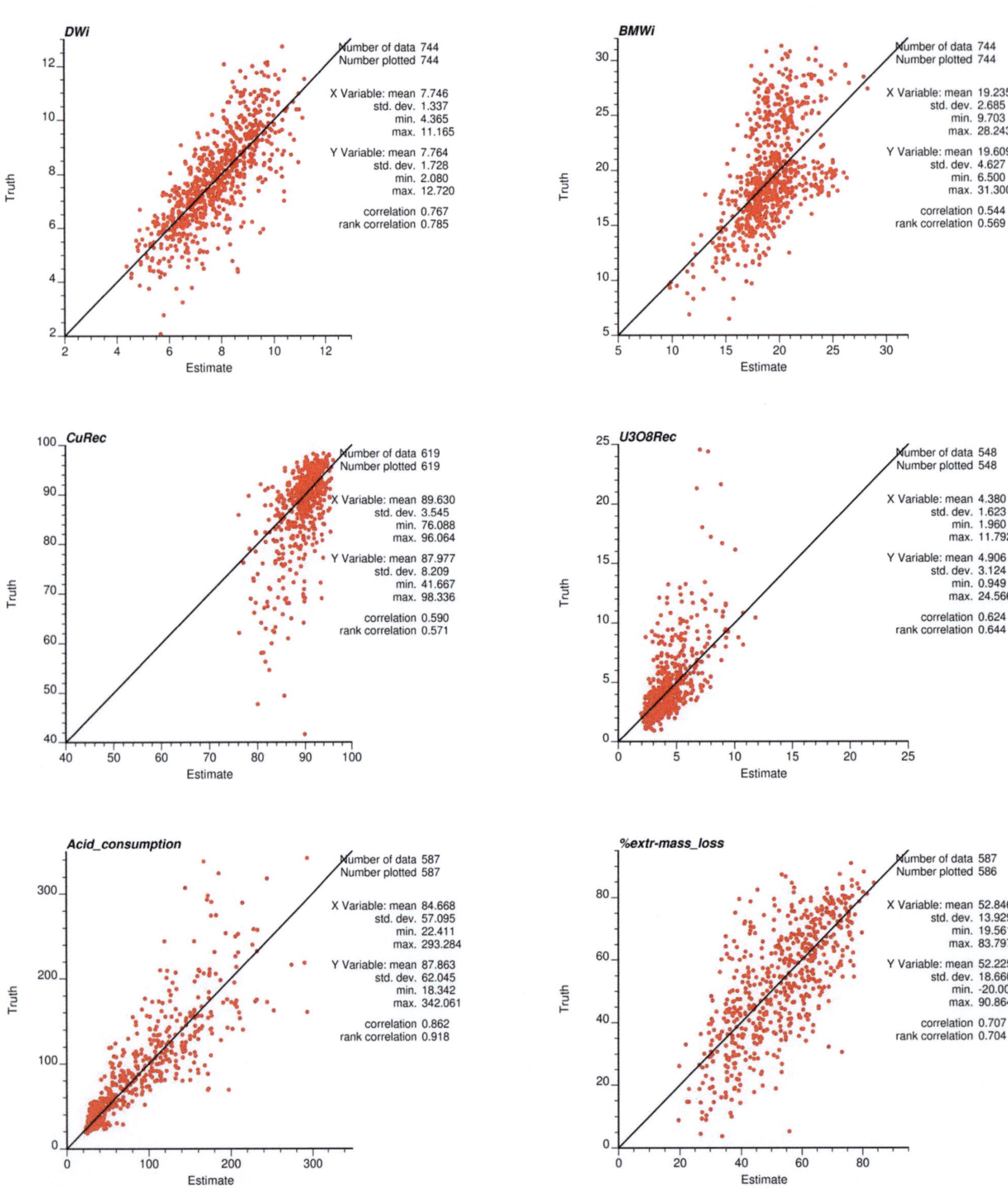

Fig. 14.64 (continued)

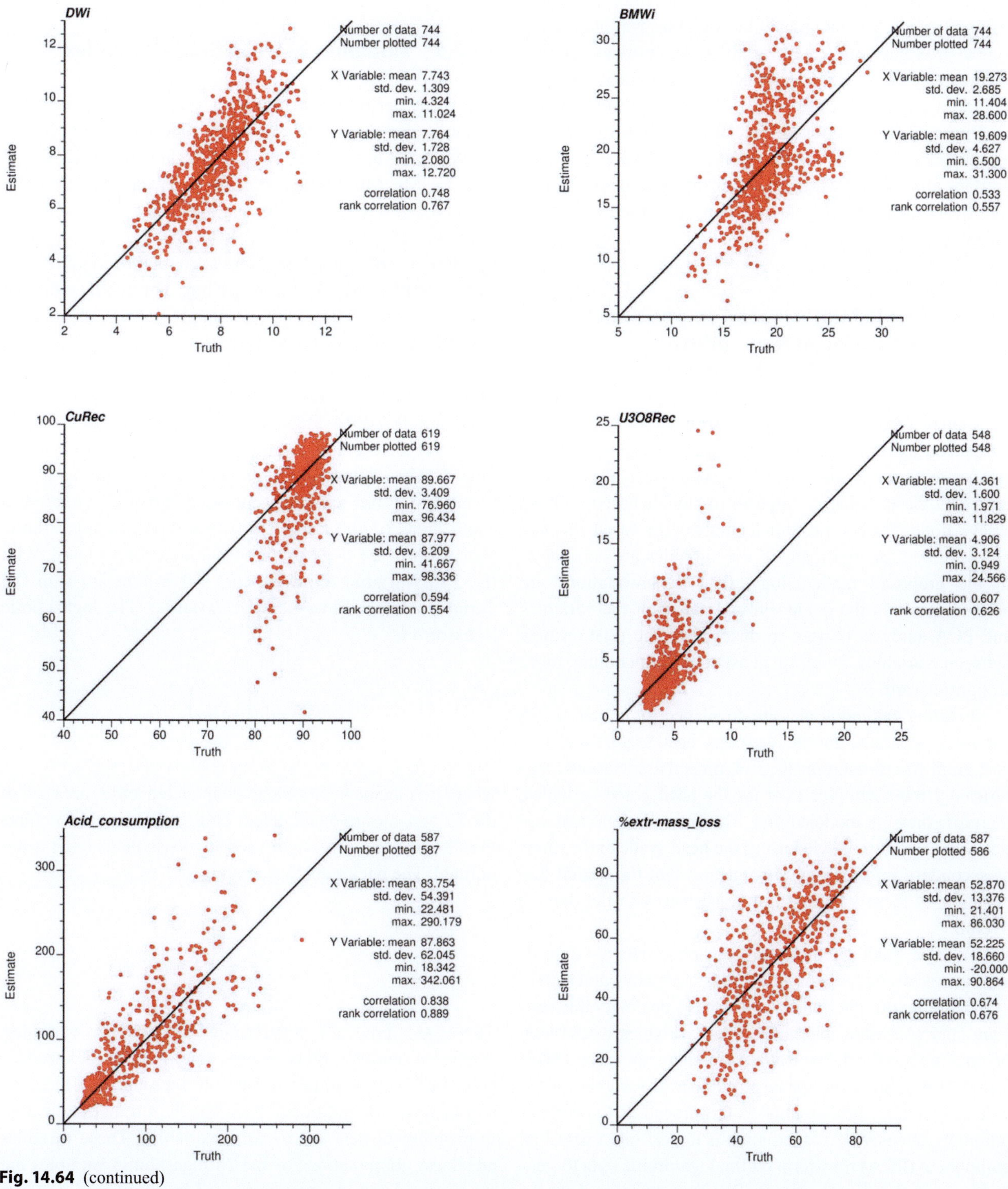

Fig. 14.64 (continued)

There are a number of opportunities for potential improvement on the modeling methodology presented in this case study: (1) optimize the merging of the variables at the two different levels. The merging of the variables was done using logical groupings of the 112 variables. An optimization procedure could be developed to select ideal subsets of variables to increase the predictive power of the regression model. (2) Improve the selection of the set of variables to use for each variable predicted. In this work, all 112 variables were used for all 6 output variables. Eliminating some of the less significant variables may reduce noise and increase model accuracy.

14.7.4 Part II: Multivariate Compositional Simulation of Non-additive Geometallurgical Variables

As shown, recovery and plant performance outcomes are influenced by a large number of variables, including head assays, mineralogy and mineral associations. Models that utilize all these variables outperform models based on head assays alone. The compositional nature of the variables must be accounted for, and many of the variables are correlated. In the proposed methodology, data transformations are used to maintain the compositional nature of the variables and PCA analysis is used to decorrelate the relationships between variables to make geostatistical modeling more straightforward.

Modeling methodologies are developed for a total of 135 variables, separated into three groups: head grade assay values; grain size measurements; and mineral associations. Significantly more samples exist for the head grade variables, therefore they are modeled first. The grain size and association variables are modeled using the head grade realizations as secondary information. This ensures that the spatial distribution of these variables are consistent with the deposit overall.

The head grade and mineral association data are considered compositional, that is, they are non-negative and sum to 100 %. A logarithmic transform is used to deal with this constant sum constraint. Normally, these variables would then be co-simulated with sequential Gaussian simulation (SGS; Isaaks 1990); however, the large number of variables available and the large grid size makes this procedure too computationally intensive. An alternative is to perform a principal component (PCA) transform on the logarithmic data to generate uncorrelated variables. SGS is then preformed on the uncorrelated PCA values. The values are back-transformed into original units to generate the realizations. This procedure is used to model the head grade and mineral association data. The grain size data, which are not compositional, are modeled using sequential Gaussian co-simulation for the p_{20}, p_{50} and p_{80} values of each mineral.

14.7.5 Modeling 23 Head Grade Variables

The plant performance modeling requires 23 head grade variables for input into the linear regression models: Cu, U3O8, Ag, Au, Co, Mo, Pb, Zn, Ba, Fe, Al, Si, K, Ca, S, Co2, La, Mg, Mn, Na, P, Ti, Ce. These 23 variables are simulated on a grid with the following dimensions: $x_{min}=56{,}105$; $y_{min}=30{,}515$; $z_{min}=-1932.5$; $x_{siz}=10$; $y_{siz}=10$; $z_{siz}=15$; $n_x=360$; $n_y=624$; $n_z=119$. There are a total of 111,572 head assay samples used in the modeling. The K:AL ratio and $B_{adj}S$ are also required, but are simply calculated from the realizations of K, Al, Ba, and S.

The head grade variables are considered compositional because all chemical and mineral rock components must sum to 100 %. Because not all elements in a sample are assayed, the sum of the head grades is always less than 100 %. However, in geostatistical modeling if this constraint is not explicitly imposed it can be violated in some areas of the model. For this reason a logarithmic transform of 24 head grade variables is considered, with the 24th variable imposing the 100 % constant sum (23 variables listed above + 1 filler variable). The logarithmic transform is:

$$y_i = \ln\left(\frac{x_i}{x_{filler}}\right)$$

where y_i is the new variable to be modeled and x_i are each of the 23 variables to be modeled. This transformation requires that there are no zero values for any variable as ln(0) is undefined. The back transformation is

$$x_i = \frac{e^{y_i}}{\sum_{i=1}^{24} e^{y_i} + 1}$$

There are now 23 logarithmic transformed variables. There are complex relationships among these 23 variables (Fig. 14.65). It would be difficult to reproduce all these relationships with traditional SGS. The PCA transform is used to generate 23 new *uncorrelated* variables. These variables are linear combinations of the 23 logarithmic variables but are uncorrelated. An assumption of independence between the 23 variables is then made and all 23 PCA variables are modeled independently with SGS. This ensures good reproduction of the correlation between the 23 variables in the final realizations.

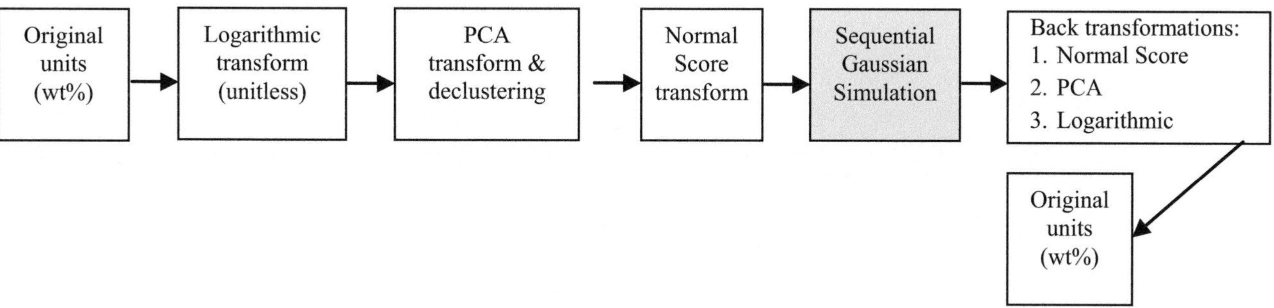

Table 14.23 Variograms for the normal score of the PCA head grade variables. A nugget (C0) and two spherical structures (C1 and C2) were used with no plunge angle

Variable Name	C0	C1	C2	Azimuth 1	Dip 1	Range 1			Azimuth 2	Dip 2	Range 2		
						Major	Minor	Vertical			Major	Minor	Vertical
NS:PCA 1	0.11	0.345	0.544	104	-75	118	79	65	100	-86	1141	1556	548
NS:PCA 2	0.035	0.608	0.357	186	83	67	54	63	158	-56	1417	606	482
NS:PCA 3	0.219	0.348	0.432	360	-80	282	110	197	360	-80	294	1193	945
NS:PCA 4	0.212	0.283	0.505	38	-76	314	79	108	349	-82	530	1627	1488
NS:PCA 5	0.292	0.378	0.33	290	-40	166	166	209	290	-40	670	1449	1303
NS:PCA 6	0.081	0.716	0.202	106	-89	59	54	48	113	-68	535	350	192
NS:PCA 7	0.107	0.302	0.59	50	-76	85	44	55	38	-61	716	1571	947
NS:PCA 8	0.168	0.415	0.417	88	-89	101	60	53	106	-79	471	606	247
NS:PCA 9	0.19	0.455	0.356	89	90	80	64	54	109	-69	496	454	237
NS:PCA 10	0.19	0.545	0.266	311	-12	54	62	73	354	-31	398	210	1020
NS:PCA 11	0.216	0.442	0.342	130	-80	96	68	72	130	-80	550	442	284
NS:PCA 12	0.188	0.426	0.386	281	-16	53	57	81	353	-39	296	247	672
NS:PCA 13	0.239	0.376	0.385	21	83	76	50	55	101	-42	446	713	311
NS:PCA 14	0.201	0.544	0.254	214	-2	49	42	61	224	-45	272	169	290
NS:PCA 15	0.451	0.463	0.085	292	-15	104	141	263	283	24	3791	943	25404
NS:PCA 16	0.234	0.561	0.205	23	-83	68	46	55	44	-58	280	280	784
NS:PCA 17	0.465	0.45	0.085	307	-7	99	122	203	283	-81	43720	1311	35267
NS:PCA 18	0.29	0.424	0.286	198	-5	52	52	67	194	-34	999	374	487
NS:PCA 19	0.211	0.559	0.23	100	-70	55	55	47	145	-73	839	220	148
NS:PCA 20	0.195	0.564	0.241	326	-5	53	57	65	5	-16	684	480	1160
NS:PCA 21	0.332	0.627	0.042	280	-20	51	57	70	280	-20	25464	535	8428
NS:PCA 22	0.305	0.25	0.445	294	-30	81	106	157	281	-61	683	683	365
NS:PCA 23	0.598	0.19	0.212	232	70	142	106	132	231	-53	2037	1022	786

An overall summary of the transformations used is shown below:

This methodology assumes that the normal score values of the principal components are independent. The PCA transform ensures that the components are uncorrelated, but they may not be independent. Poor histogram reproduction is seen in original units due to this lack of independence. There are a large number of head assay samples which makes the input histogram reliable; they should be reproduced in the simulation. To obtain reasonable histogram reproduction the final simulations are post processed to better match the declustered input histogram (Fig. 14.66). This has little effect on the correlations between variables and individual variable variograms, but improves histogram reproduction.

14.7.6 Details of the Sequential Gaussian Simulation

Implementation of SGS requires the use of variograms for each PCA variable as well as a number of other important parameters. For all variables considered in this case study, simulation was performed with 50 nearby data (25 data and 25 previously simulated nodes). Parameters for each variogram can be found in Table 14.23. Because of the large number of variables, variogram fitting software was used with a visual assessment to locate any major inconsistencies with data.

Declustering was used on the 23 PCA variables to obtain global histograms. A locally varying mean was used in the

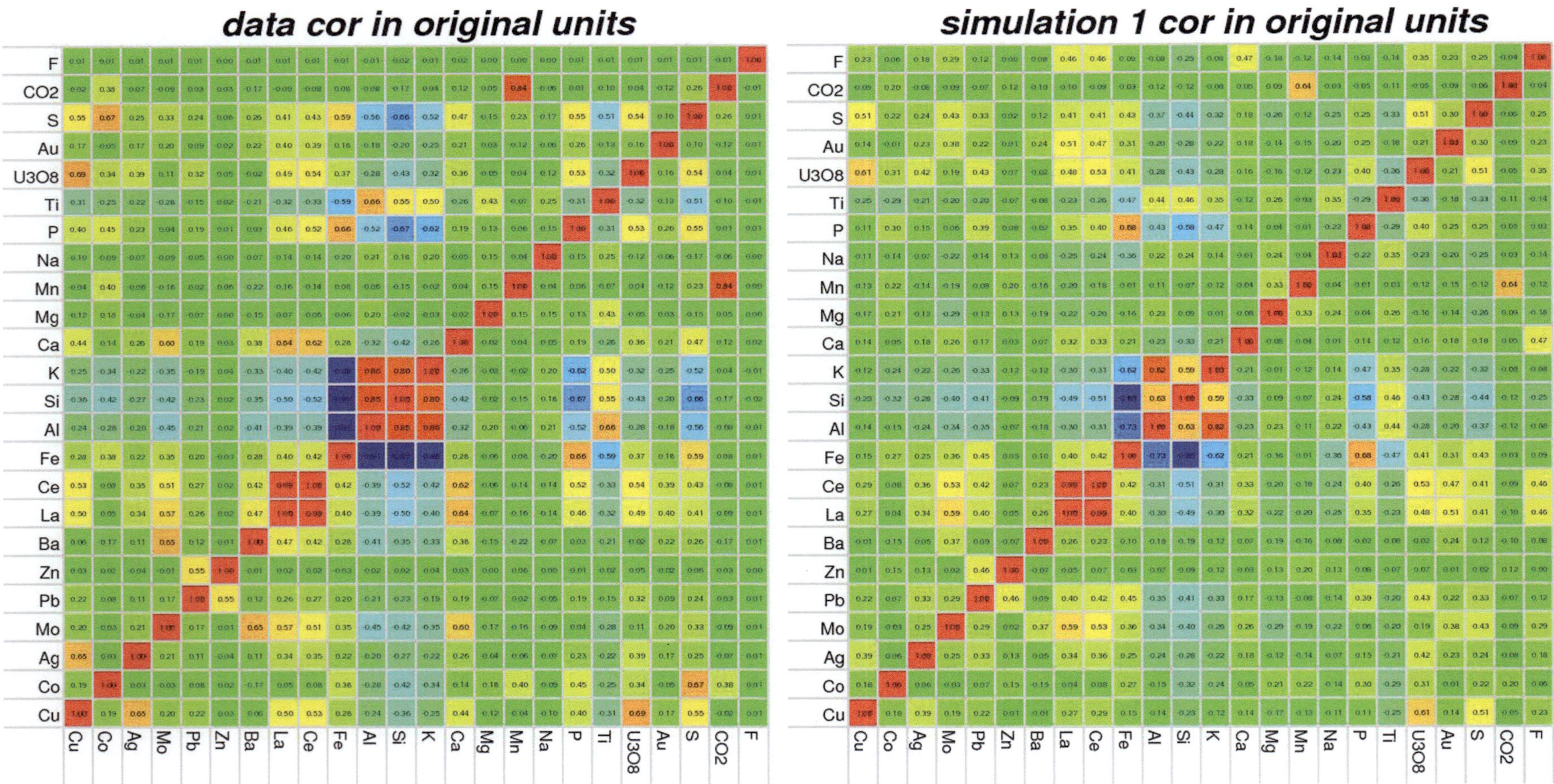

Fig. 14.65 Correlation between the head grade variables (*left*) and correlation in one simulation (*right*). Correlations calculated in original units

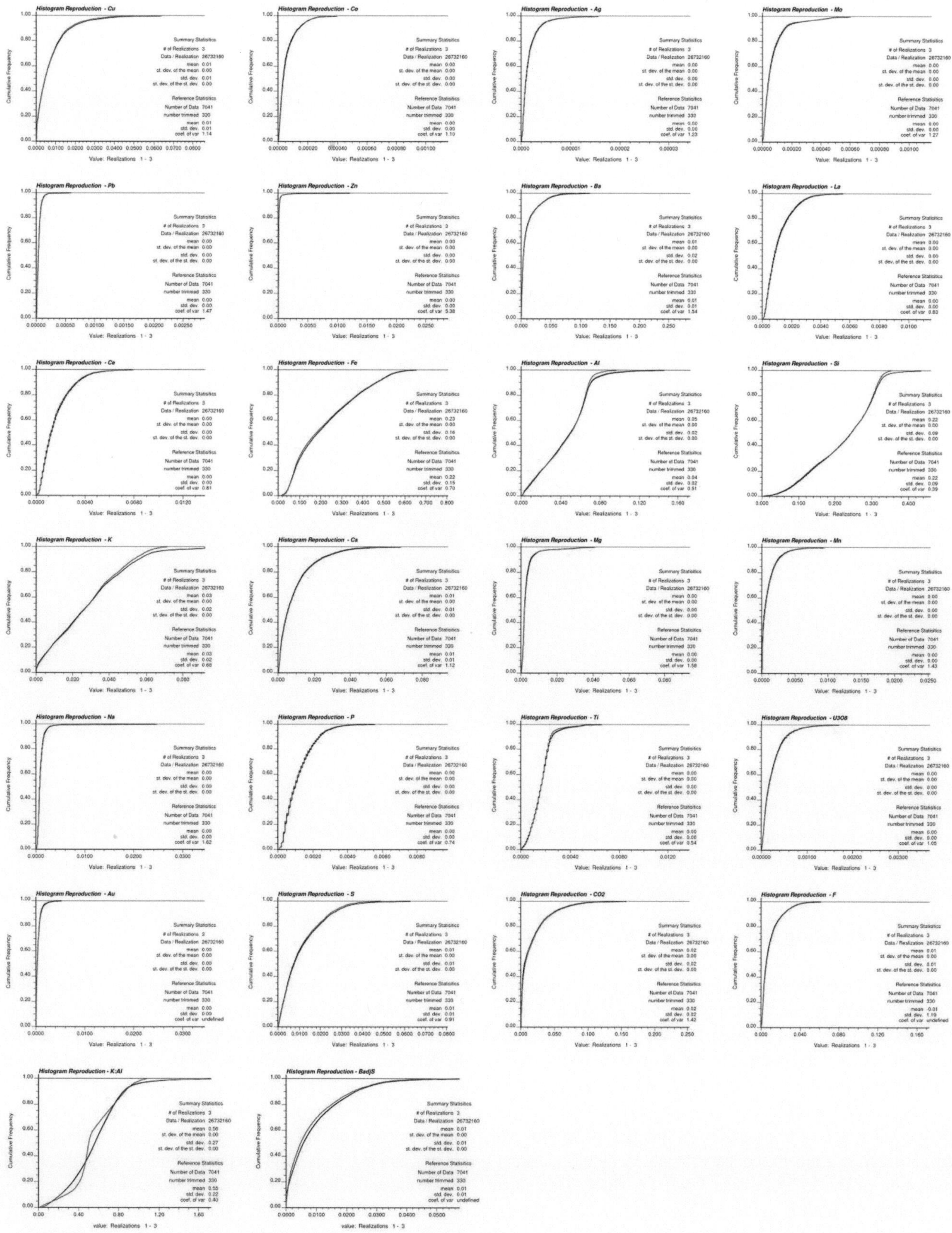

Fig. 14.66 Histogram reproduction for 25 head grade variables after post processing. *Black*: 3 realizations; *Red*: Input histogram with 7038 data

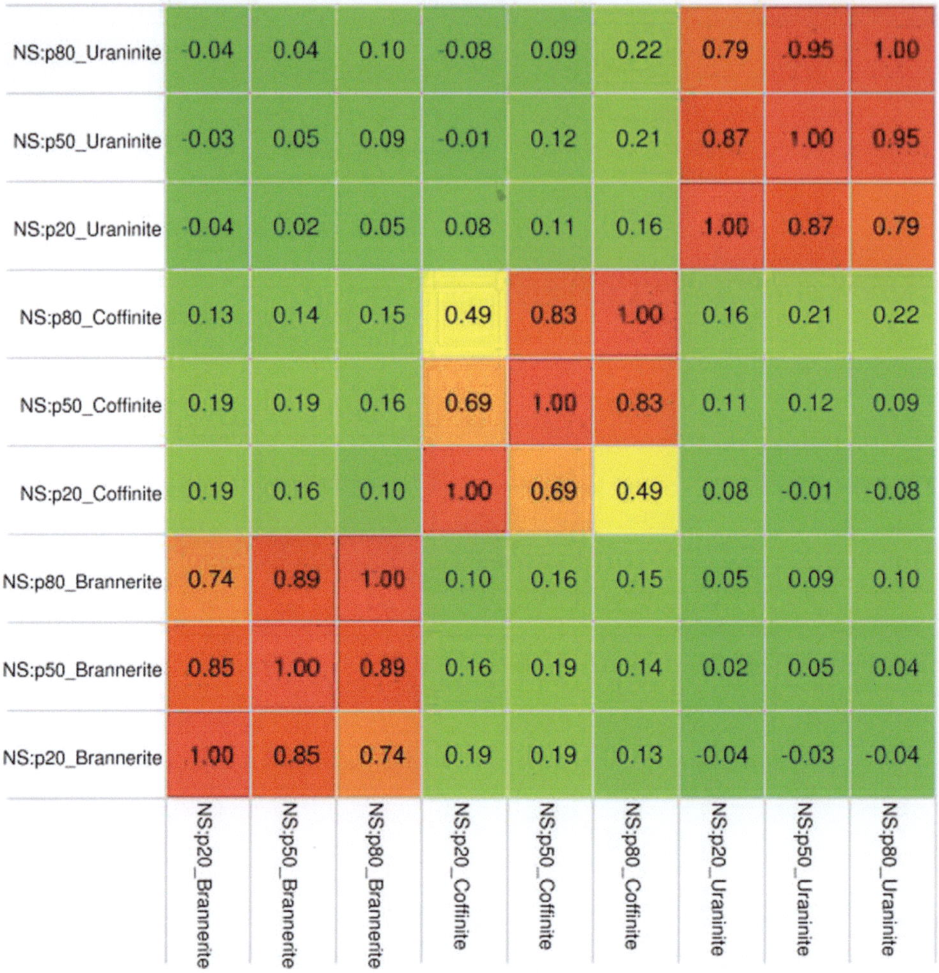

	NS:p20_Brannerite	NS:p50_Brannerite	NS:p80_Brannerite	NS:p20_Coffinite	NS:p50_Coffinite	NS:p80_Coffinite	NS:p20_Uraninite	NS:p50_Uraninite	NS:p80_Uraninite
NS:p80_Uraninite	-0.04	0.04	0.10	-0.08	0.09	0.22	0.79	0.95	1.00
NS:p50_Uraninite	-0.03	0.05	0.09	-0.01	0.12	0.21	0.87	1.00	0.95
NS:p20_Uraninite	-0.04	0.02	0.05	0.08	0.11	0.16	1.00	0.87	0.79
NS:p80_Coffinite	0.13	0.14	0.15	0.49	0.83	1.00	0.16	0.21	0.22
NS:p50_Coffinite	0.19	0.19	0.16	0.69	1.00	0.83	0.11	0.12	0.09
NS:p20_Coffinite	0.19	0.16	0.10	1.00	0.69	0.49	0.08	-0.01	-0.08
NS:p80_Brannerite	0.74	0.89	1.00	0.10	0.16	0.15	0.05	0.09	0.10
NS:p50_Brannerite	0.85	1.00	0.89	0.16	0.19	0.14	0.02	0.05	0.04
NS:p20_Brannerite	1.00	0.85	0.74	0.19	0.19	0.13	-0.04	-0.03	-0.04

Fig. 14.67 Correlation between the grain size variables. Minerals are simulated independently because of the small correlation between minerals

simulation to consider the non-stationary present throughout the deposit. The mean for each PCA was determined using a moving window average with a radius of 400 m in the horizontal direction and 50 % anisotropy in the vertical direction.

14.7.7 Modeling Nine Grain Size Variables

There are three Uranium minerals of interest: Brannerite, Coffinite and Uraninite. The p_{20}, p_{50} and p_{80} grain size for each mineral has been measured at 497 locations. The correlation between the percentiles of each grain size is reproduced by co-simulating the three percentiles, see Fig. 14.67.

The densely sampled 23 head grade values is used to supplement the lack of information for the grain size variables by considering a super secondary variable which is the amalgamation of the 23 PCA head grade variables. This super secondary variable is created differently for each mineral because the correlations between the mineral grain sizes and the PCA head grade variables differ. To generate this super secondary variable, a linear combination of the PCA head grades is determined from the following equations:

$$\begin{bmatrix} \rho_{1,1} & \rho_{2,1} & \cdots & \rho_{n,1} \\ \rho_{1,2} & \rho_{2,2} & \cdots & \rho_{n,2} \\ \vdots & \vdots & \ddots & \vdots \\ \rho_{1,n} & \rho_{2,n} & \cdots & \rho_{n,n} \end{bmatrix} \begin{bmatrix} \lambda_1 \\ \lambda_2 \\ \vdots \\ \lambda_n \end{bmatrix} = \begin{bmatrix} \rho_{0,1} \\ \rho_{0,2} \\ \vdots \\ \rho_{0,n} \end{bmatrix}$$

The right-hand side of this equation contains the correlation between one of the grain size variables and the 23 input head grade variables to be merged. The left hand side is the correlation between all 23 PCA head grade variables; note that the left hand side contains 1.0 on the diagonal and 0.0 for all off diagonal terms because the PCA values are uncorrelated. This is done for the p_{50} value for each mineral and the same super secondary variable is used for modeling the p_{20}, p_{50} and p_{80}. This single super secondary variable allows for the cosimulation of the three percentiles and only one exhaustive secondary variable. Without merging all secondary variables into a super secondary, the grain size simulations would have to consider 23 separate secondary variables in order to use all the available information from the head grade variables.

Brannerite

sup_sec6	0.37	0.47	0.49	1.00
P80	0.74	0.89	1.00	0.49
P50	0.85	1.00	0.89	0.47
P20	1.00	0.85	0.74	0.37
	P20	P50	P80	sup_sec6

Coffinite

sup_sec9	0.37	0.41	0.27	1.00
P80	0.49	0.83	1.00	0.27
P50	0.69	1.00	0.83	0.41
P20	1.00	0.69	0.49	0.37
	P20	P50	P80	sup_sec9

Uraninite

sup_sec12	0.28	0.44	0.56	1.00
P80	0.79	0.95	1.00	0.56
P50	0.87	1.00	0.95	0.44
P20	1.00	0.87	0.79	0.28
	P20	P50	P80	sup_sec12

sup_sec Uraninite	0.10	0.10	0.13	-0.05	-0.08	-0.06	0.42	0.51	0.57	0.17	-0.03	1.00
sup_sec Coffinite	0.18	0.28	0.33	0.23	0.29	0.32	0.03	0.02	0.01	0.66	1.00	-0.03
sup_sec Brannerite	0.32	0.48	0.57	0.16	0.19	0.21	0.10	0.12	0.13	1.00	0.66	0.17
p80_Uraninite	0.17	0.18	0.22	-0.01	-0.04	-0.03	0.80	0.95	1.00	0.13	0.01	0.57
p50_Uraninite	0.14	0.17	0.19	-0.01	-0.04	-0.01	0.89	1.00	0.95	0.12	0.02	0.51
p20_Uraninite	0.11	0.12	0.14	-0.03	-0.02	0.00	1.00	0.89	0.80	0.10	0.03	0.42
p80_Coffinite	0.07	0.19	0.21	0.44	0.75	1.00	0.00	-0.01	-0.03	0.21	0.32	-0.06
p50_Coffinite	0.10	0.16	0.19	0.64	1.00	0.75	-0.02	-0.04	-0.04	0.19	0.29	-0.08
p20_Coffinite	0.09	0.16	0.19	1.00	0.64	0.44	-0.03	-0.01	-0.01	0.16	0.23	-0.05
p80_Brannerite	0.71	0.88	1.00	0.19	0.19	0.21	0.14	0.19	0.22	0.57	0.33	0.13
p50_Brannerite	0.79	1.00	0.88	0.16	0.16	0.19	0.12	0.17	0.18	0.48	0.28	0.10
p20_Brannerite	1.00	0.79	0.71	0.09	0.10	0.07	0.11	0.14	0.17	0.32	0.18	0.10
	p20_Brannerite	p50_Brannerite	p80_Brannerite	p20_Coffinite	p50_Coffinite	p80_Coffinite	p20_Uraninite	p50_Uraninite	p80_Uraninite	sup_sec Brannerite	sup_sec Coffinite	sup_sec Uraninite

Fig. 14.68 Correlation between the grain size variables. Above: correlations from 497 data to the super secondary variables; below: correlations from one grain size simulation

Table 14.24 Variograms for the grain size data. A nugget (C0) and two spherical structures (C1 and C2) were used with no plunge/dip angle and no horizontal anisotropy

Variable Name	C0	C1	C2	Range 1		Range 2	
				Horizontal	Vertical	Horizontal	Vertical
Brannerite	0.4	0.2	0.4	200	20	200	150
Coffinite	0.4	0.2	0.4	400	20	400	300
Uraninite	0.4	0.2	0.4	200	20	200	350

The super secondary variable is used as a collocated secondary variable for each of the grain size models. Note that for the grain size variables neither a logarithmic nor a PCA transformation is considered because there are only three variables (p_{20}, p_{50} and p_{80}) for each mineral. Cosimulation of three variables can be accomplished in a reasonable amount of CPU time. This procedure is repeated for brannerite, coffinite and uraninite. This includes building a new super secondary variable for each mineral. Figure 14.68 shows the correlations between grain size variables for each uranium mineral.

Very few data exist for the grain size variables and the variograms are unstable, and so the same variograms are used for the p_{20}, p_{50} and p_{80} of each mineral. The spatial structure for the p_{20}, p_{50} and p_{80} are similar, with the small differences likely due to lack of data. Parameters for the variograms used are shown in Table 14.24.

14.7.8 Modeling 100 Association Matrix Variables

Modeling the association matrix utilizes a combination of the techniques previously discussed. The matrix is a 10×11 matrix where each row sums to 1.0 (or 100 %). Consider one particular sample:

A variogram is required for each of the 90 principal components (10 sets/rows with 9 principal components in each). As with the head grade variables these variograms were fit with automatic variogram fitting software and visually inspected for inconsistencies.

	Brannerite	Coffinite	Uraninite	Pyrite	Chalcopyrite	Bornite	Chalcocite	Other Sulphides	Acid Soluble Gangue	Acid Insoluble Gangue	Free Surface
Brannerite		8.02								88.18	3.80
Coffinite	1.71		1.64			0.25	0.24		3.50	90.67	2.00
Uraninite		23.51								76.49	
Pyrite											
Chalcopyrite						2.83			2.59	88.43	6.15
Bornite		0.18			0.93				15.50	75.89	7.49
Chalcocite		0.30							0.87	97.91	0.92
Other Sulphides										100.00	
Acid Soluble Gangue		0.05			0.02	0.32	0.01			91.16	8.44
Acid Insoluble Gangue	0.04	0.19	0.01		0.08	0.22	0.16	0.02	12.82		86.45

Each element in the matrix represents the % surface area of interaction between minerals determined from mineral liberation analysis. Each row sums to 1.0; however, each column does not sum to a constant value as the values are standardized by the proportions. There are a total of 100 elements in the matrix to be modeled, ignoring the diagonals. An assumption that the rows are independent is made to reduce the problem to simulating 10 independent sets (rows) of 10 dependent variables (columns). To maintain the constant sum constraint the logarithmic transformation is applied to each row resulting in the need to model 9 logarithmic variables. The PCA transformation is applied to reproduce the correlation between variables in each row. The principal components of each row are normal score transformed and then simulated with SGS. There are a total of 490 data available for simulation of association variables.

As with the grain size variables, the head grade simulations provide a super secondary variable to use in collocated SGS. There are a total of 23 (normal score PCA) head grade simulations to be combined into a single super secondary variable for each of the 100 elements in the association matrix. The PCA transform is done in such a way that the amount of data explained by each principal component can be measured. Some components 'contain' more information than others. In this case the first five components of the head grade realizations contain over 75 % of the information in the original head grades. Only the first 5 principal components generated in the head grade modeling are combined into the super secondary variable to reduce the computational requirements of the methodology. Moreover, the super secondary variable is only used for the first 4 of the 9 principal components of the association variables. Because there are 100 association variables to model, CPU time becomes an issue.

14.7.9 Special Considerations for the Association Data

Missing or "null" values always pose a problem in compositional data modeling. In this instance there are some entries that are missing because a particular mineral does not appear in a given sample. For rows that have some missing values but still sum to 1.0, the missing values are reset to 0.0001 or 0.01 %. In some cases there are entire rows that are missing. This is because the mineral does not appear at that location; however, in these cases all values cannot be set to a small value as they would not sum to 1.0. The solution undertaken in this study was to remove the samples where the entire row was missing.

When performing SGS at this location the values in that particular row are simulated as if the data did not exist (in fact this data does exist and has a value of zero). The mis match between the missing values at this location and the simulated values given the surrounding data can be fixed by assigning a 0.0 proportion to the missing minerals, and the mismatched association values can be ignored.

14.7.10 Histogram/Variogram Reproduction

There are 135 variables modeled in total. The histograms and variograms reproduction for the first 3 realizations have been analyzed. The following discussion compares the input histograms and variograms to the realization outputs.

14.7.10.1 Head Grade Variables

The head grade variables reproduce the histogram quite well (Fig. 14.69) because of post-processing. Variogram reproduction is checked in normal score units of the principal components.

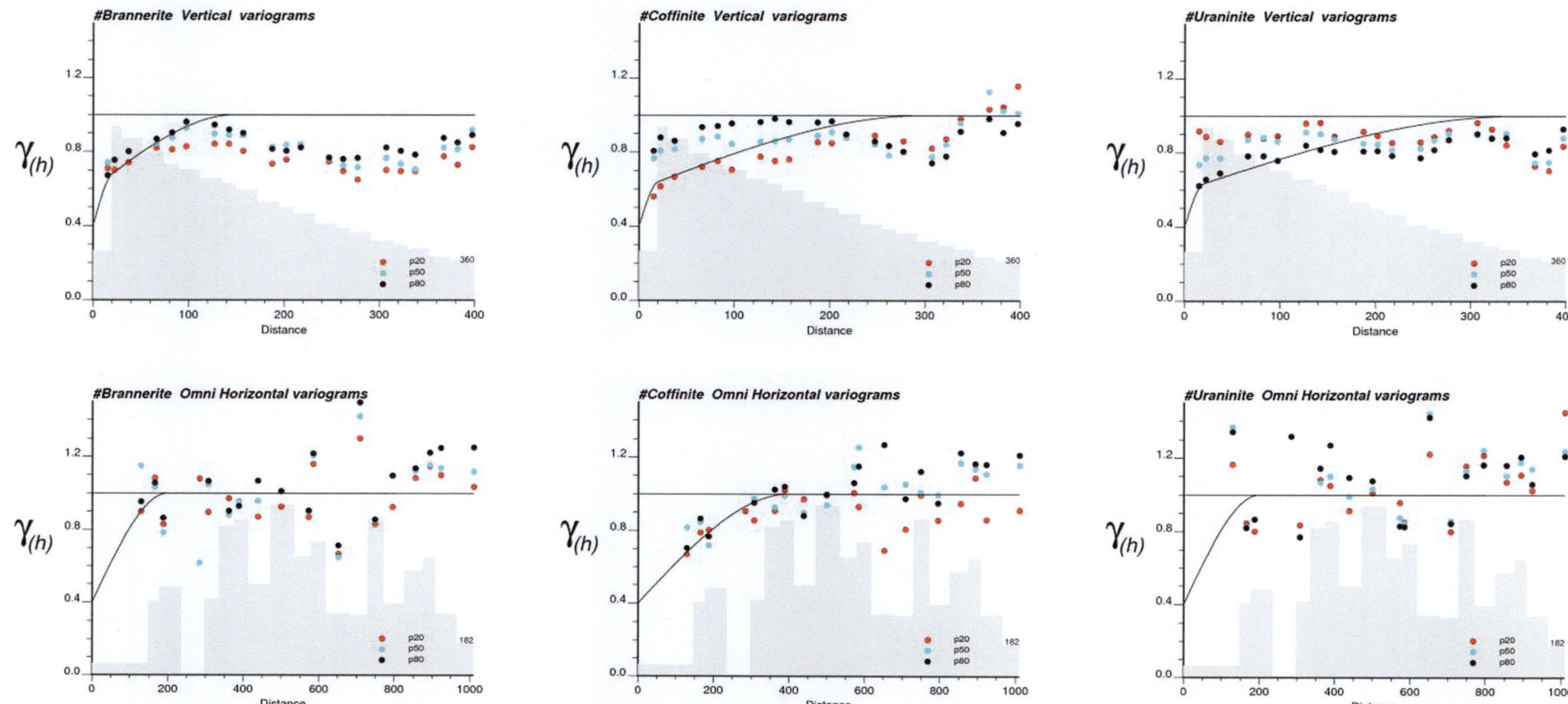

Fig. 14.69 Modeled variograms for the 9 grain size variables. The same variogram was used for the percentiles of each mineral

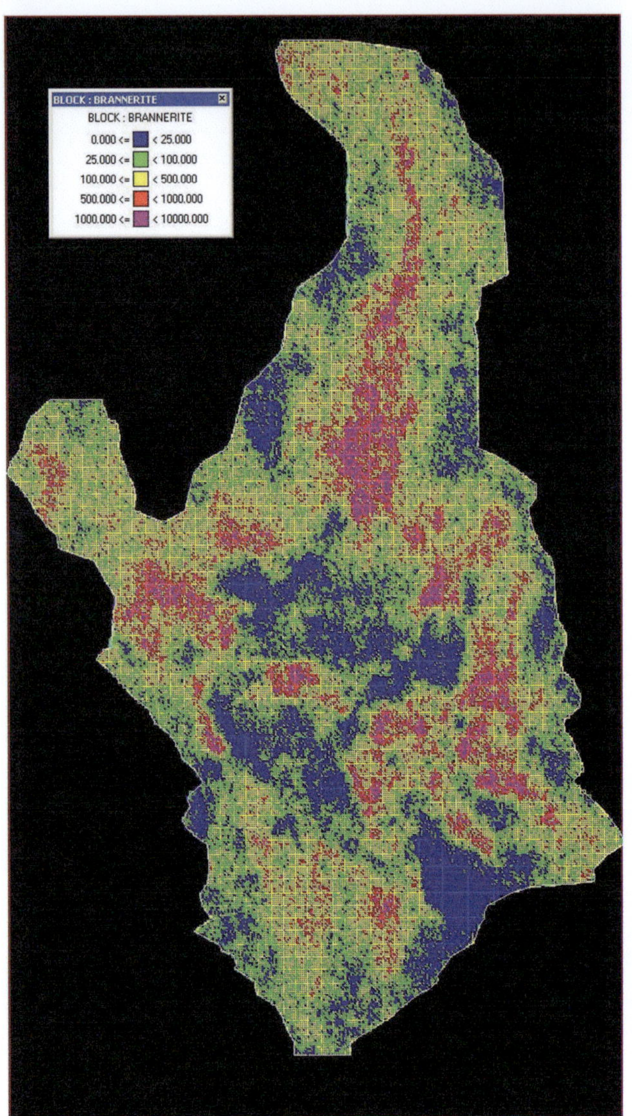

Fig. 14.70 Brannerite, −450 m elevation

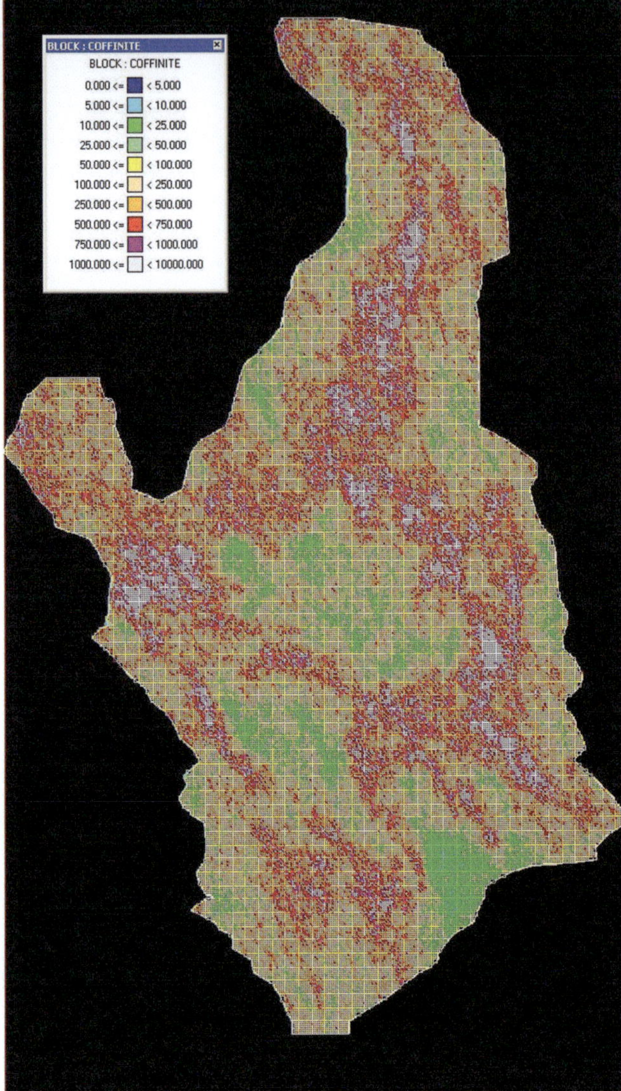

Fig. 14.71 Coffinite, −450 m elevation

14.7.10.2 Grain Size Variables

Histogram and variogram reproduction is heavily influenced by the secondary variables as there was very little grain size data. For this reason the histogram and variogram reproduction for the grain size variables does not exactly match the input. Moreover, the grain size variables are sparsely sampled suggesting that the input histogram and variogram may be unreliable. Some deviation from the input parameters due to the secondary information is warranted.

14.7.10.3 Association Matrix Variables

There are a total of 100 association variables. Histogram reproduction is not perfect. The resulting histograms and variograms deviate from the input because of (a) lack of independence of the principal components, and (b) influence of the super secondary attributes on the models.

14.8 Conclusions

Three linear regression models for the prediction of plant performance from head assay, mineralogy and association variables. This case study presented a methodology for the spatial modeling of these variables. The intention is to use the regression model with the spatial model to predict plant performance. The cost of obtaining samples of plant performance (i.e. pilot plant runs) is very high. Building models based on the sparse sampling of mineral recovery,

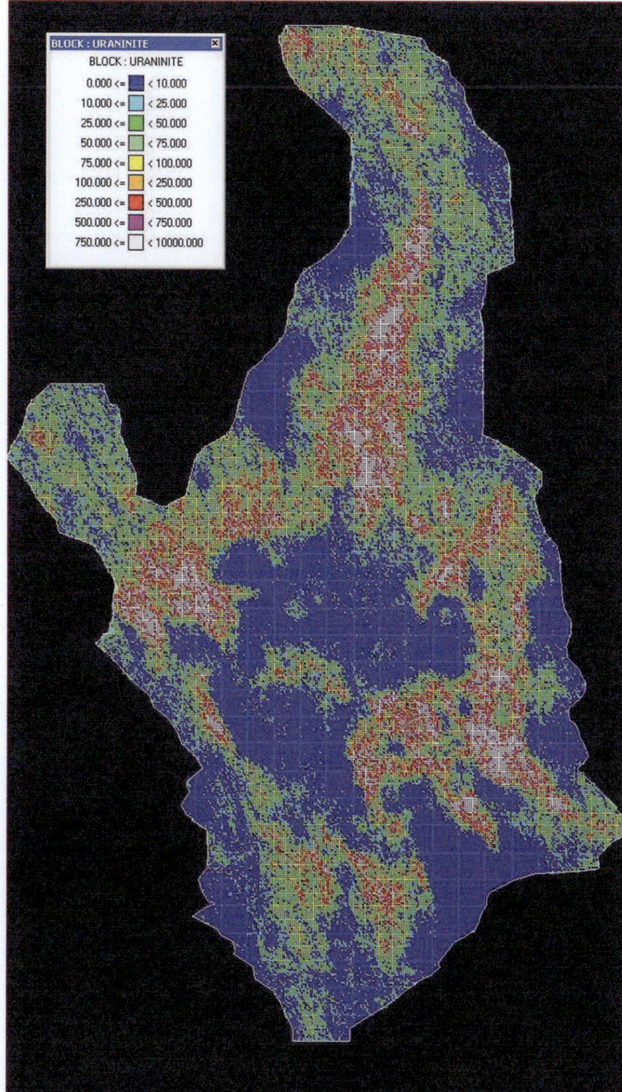

Fig. 14.72 Uraninite, −450 m elevation

acid consumption and work indexes allows for the mapping of these variables for all locations in the deposit. This provides a prediction of complex process-based variables that rarely have sufficient data density to generate appropriate variograms and prove difficult to effectively model. Figures 14.70, 14.71, and 14.72 show plan views at −450 m elevation of the three main Uranium minerals (brannerite, coffinite, and uraninite, respectively), while Figs. 14.73 and 14.74 show the predicted overall copper and uranium recoveries, respectively.

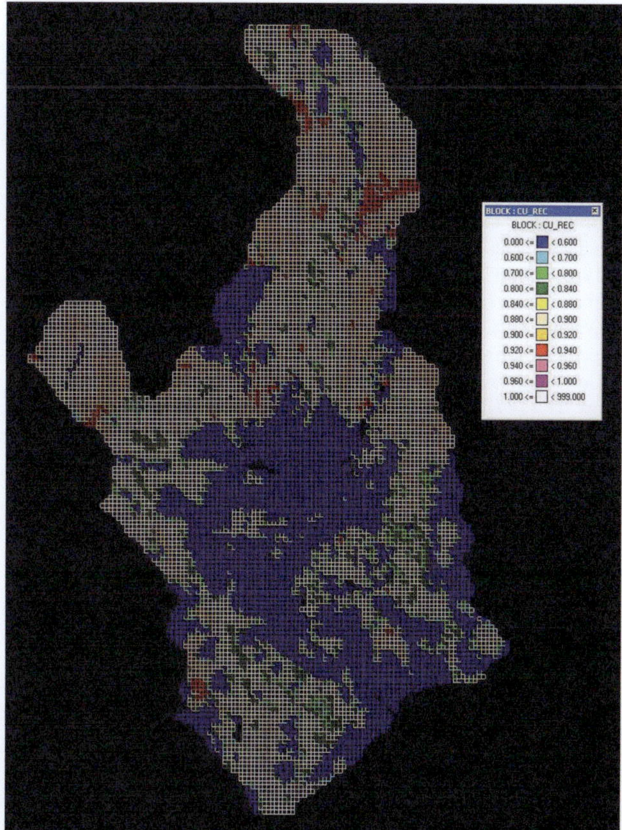

Fig. 14.73 Cu recovery, −450 m elevation

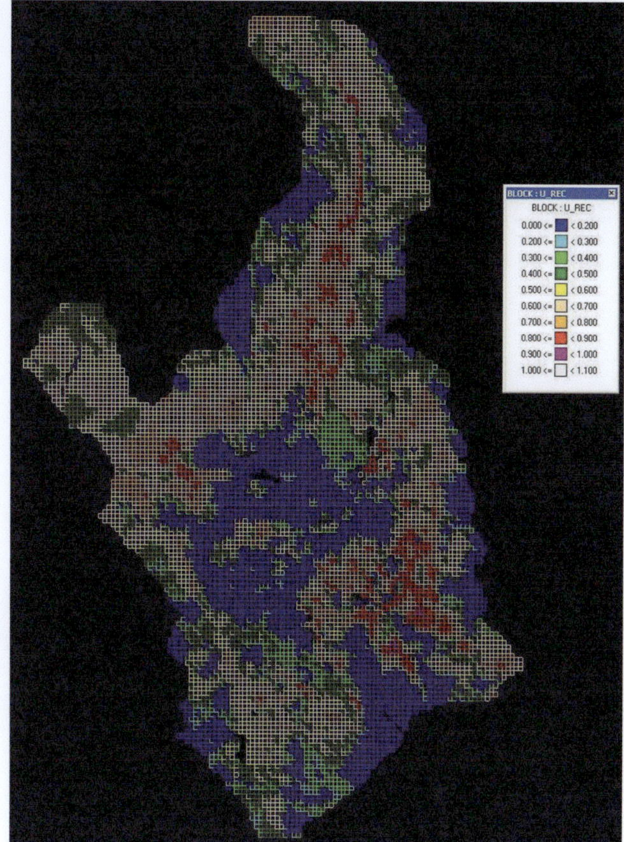

Fig. 14.74 Uranium recovery, −450 m elevation

References

Aguilar CA, Rossi ME (January 1996) Método para Maximizar Ganancias en San Cristóbal, Minería Chilena, Santiago, Chile, Ed. Antártica, No. 175, pp 63–69

Alabert FG (1987) Stochastic imaging of spatial distributions using hard and soft information. MSc Thesis, Stanford University, p 197

Burmeister B (1998) From resource to reality: a critical review of the achievements of new Australian gold mining projects during 1983–1987. MSc Thesis, Macquarie University

Campbell EA (1994) Geology and results of Stage I and II drilling, West Zone. Cerro Colorado: Internal RioChilex report

Cepeda A, Ditson GM, Mato G (1982) Geological assessment of the Cerro Colorado Project. internal RioChilex report, Chile

Clow G (1991) Why gold mines fail. Northern Miner Magazine 6(2):31–34

Davis BM (1987) Uses and abuses of cross-validation in geostatistics. Math Geol 17:563–586

Deustch CV (1989) DECLUS: Aa FORTRAN 77 program for determining optimum spatial declustering weights. Comput Geosci 15(3):325–332

Ferraris F, Di Biase F (1978) Hoja Antofagasta. Instituto de Investigaciones Geológicas de Chile, Carta Geológica de Chile, No. 30, p 48

Guilbert JM, Park CF Jr (1985) The geology of ore deposits. Freeman, New York, p 985

Isaaks EH (1990) The application of Monte Carlo methods to the analysis of spatially correlated data. PhD. Thesis, Stanford University, p 213

Isaaks EH (1999) SAGE2001 User's manual, software license and documentation. www.isaaks.com

Journel AG, Huijbregts ChJ (1978) Mining geostatistics. Academic, New York

Journel AG, Kyriakidis P (2004) Evaluation of mineral reserves, a simulation approach. Oxford University Press, New York

Knoll K (1989) And now the bad news. Northern Miner Magazine 4(6):48–52

Leuangthong O (2003) Stepwise conditional transformation for multivariate geostatistical simulation. Ph.D. Thesis, University of Alberta, Edmonton

Leuangthong O, Deutsch CV (2003) Stepwise conditional transformation for simulation of multiple variables. Math Geol 35(2):155–173

Leuangthong O, Hodson T, Rolley P, Deutsch CV (2006) Multivariate geostatistical simulation at Red Dog mine, Alaska, USA. CIM Inst Min Metall Petroleum May, pp 1-26

Luster GR (1985) Raw materials for portland cement: applications of conditional simulation of coregionalization. Ph.D. Thesis, Stanford University, Stanford, p 531

Moore D, Young L, Modene J, Plahuta J (1986) Geologic setting and genesis of the Red Dog zinc–lead–silver deposit, western Brooks Range, Alaska. Econ Geol 81:1696–1727

Parker HM (1991) Statistical treatment of outlier data in epithermal gold deposit reserve estimation. Math Geol 23:125–199

Pitard FF (1995) Sampling, ore grade control and statistical process at the San Cristóbal mine. Internal report, Inversiones Mineras Del Inca

Rosenblatt M (1952) Remarks on a multivariate transformation. Ann Math Stat 23(3):470–472

Rossi ME (1999) Optimizing grade control: a detailed case study. In: Proceedings of the 101st annual meeting of the Canadian Institute of Mining, Metallurgy, and Petroleum (CIM), Calgary, 2–5 May 1999

Rossi ME, Alvarado CSB (1998) Conditional simulations applied to recoverable reserves. In: Proceedings, 27th international symposium on computer applications in the minerals industries (APCOM), London, 19–23 April 1998, peer-reviewed

Rossi ME, Parker HM (1993) Estimating recoverable reserves: is it hopeless? Forum 'Geostatistics for the next century', Montreal, Quebec, 3–5 June 1993

Solow AR (1990) Geostatistical cross-validation: a cautionary note. Math Geol 22:637–639

Srivastava RM, Parker HM (1988) Robust measures of spatial continuity. In: Amstrong M (ed) Geostatistics. Reidel, Dordrecht, pp 295–308

Conclusions

15

Abstract

Significant decisions are made on the basis of mineral resource estimates. There is significant uncertainty associated with mineral resources because we sample a relatively small amount of the deposit. The framework, techniques and numerical/statistical tools that have evolved to address resource estimation in presence of sparse data and significant uncertainty are summarized.

15.1 Building a Mineral Resource Model

This section includes a summary of the steps involved in building a block model, as discussed throughout the book. A typical work flow is summarized. The summary also includes a review of the applications described, emphasizing the practical usefulness of the tools described, and the (potential) benefits obtained by the owner/company.

More time is spent on getting ready to do resource modeling than on actually applying specific geostatistical tools. It takes significant time to understand the geological setting, the data, the study objectives and ensure that the modeling workflow is designed to meet those objectives. Cleaning the data takes a great deal of time. Often, the data are not *dirty* or incorrect, but the format is different and inconsistent, there is missing data, there are different vintages of data, different companies involved and so on. Preparing the site specific data takes significant time. Understanding the geological context of the data is essential to supplement sparse data and to make good choices of model setup and modeling workflow.

Sufficient time must be allocated to sort out the study objectives, site specific data, analogue data and a conceptual understanding of the site. Of course, time must be left to perform the geostatistical study and meet the study objectives. Often, some data must be left out, some risk of error in the database must be accepted and an incomplete understanding of the geological context must also be accepted. Careful documentation must be assembled of the data inventory and the limitations that exist in the database and conceptual understanding. There must be a balance between satisfying prerequisites and getting on with the resource estimation to meet the study objectives.

Most geostatistical studies are repeated as more data become available or the objectives change. It is rare that a particular geostatistical study is the first analysis of completely new data for a site that has never been modeled before. It is important to assemble and review all relevant prior work such as reports, maps, models, and data files. Those that have studied the site in the past should be contacted to avoid making preventable mistakes and to address improvements that previous studies never had the time, data or resources to address.

A generic workflow for geostatistics could be summarized by eight steps. (1) Specify the goals of the study and take inventory of the available measurements and conceptual data. (2) Divide the area/volume of interest into subsets that are relevant for the specific situation. (3) Choose how the mean of each variable depends on location within each chosen subset. (4) Infer all required statistical parameters for creating spatial models of each variable within each subset. (5) Estimate the value of each variable at each unsampled location. (6) Thoroughly validate the estimated model, ensuring that the geologic and grade models are consistent with the assumptions, data, domaining geology, and methodology used in the estimation. (7) Simulate multiple realizations to assess joint uncertainty at different scales. Finally, (8) Post process the statistics, estimated models and simulated realizations to provide decision support information. The detailed implementation of these steps will depend on the purpose of the study.

M. E. Rossi, C. V. Deutsch, *Mineral Resource Estimation,* DOI 10.1007/978-1-4020-5717-5_15,

First: the goals of the study must be specified to determine the work effort required for the study, the variables to be predicted, the scale relevant for evaluation and the specific estimation, simulation and post processing steps. A data inventory must be taken to review all available measured data from drillholes, other sampling and historical production data. The numerical models should reproduce all of these measured data within the scale and accuracy of the data. Conceptual data must also be assembled including a geological understanding of the spatial distribution and analogue data. The conceptual model expressed in this first step may include schematic pictures and illustrations of the features that should be contained in the final model.

Second: the entire volume being modeled will not be combined and modeled with one technique and set of parameters. There are logical subsets based on geological zones and rock types. The estimation domains will define rock volumes that genetically belong together. The domains must be large enough to contain sufficient data for reliable statistics, yet small enough to isolate geological features for local accuracy. A hierarchical system may be used where large scale zones are modeled first, then smaller scale geological units, then continuous grades within the final domain definition.

Third: the mean value of each variable may depend on location within the chosen estimation domains. There are often significant trends in the distribution of rock types. These trends are understood even with few data. Continuous grade variables may also have local variations that are important. The results of the second step (subsets of the volume for geostatistical analysis) and third step (modeling the location dependence of the mean) are collectively known as the decision of stationarity.

Fourth: infer all required statistical parameters. The required statistical parameters will depend on the chosen technique that, in turn, depends on the conceptual model chosen for each stationary subset of the domain. Almost always, there will be a need to infer the univariate proportions and histograms of each grade variable within each domain. These univariate distributions are computed from the data and calculated to be representative of the entire subset. Some measures of spatial variability must also be inferred. In traditional Matheronian geostatistics (Matheron 1971), variograms are the measures that quantify the spatial variability of each category and rock property. In presence of sparse data, these statistical parameters are considered uncertain and a number of scenarios are documented.

Fifth: calculate an estimate of each variable at each unsampled location. These estimates are based on the data and do not involve any random numbers. The estimation is commonly a form of kriging considering indicators, data transformation, cokriging, and/or locally varying means as required. Whenever possible, the uncertainty is estimated directly with indicators for categorical variables and normal scores in a multivariate Gaussian context for continuous variables. This provides a single best estimate at each unsampled location together with a measure of uncertainty. This is based entirely on the data and decisions taken in the first four steps. The results are useful for resource assessment and checking.

Sixth: thorough validation of the estimated values is necessary. Often, the effort involved in validating and even calibrating a model is under-appreciated. Model calibration implies running multiple iterations of the estimation process, varying some specific parameters, in order to reproduce a certain reference (for example, a blast hole model). Validations are done to ensure the internal consistency of the model; that is, the estimated values are consistent with all assumptions, data, and geologic model used to build it. Comparisons with previous models are made also, and if available, against a production or reference model deemed to be a reasonably accurate representation of true grades and tonnages in the deposit.

Seventh: multiple realizations of all surfaces, rock types and grades could be obtained to quantify joint uncertainty and to provide a model of variability suitable for the assessment of dilution and recoverable reserves. The simulation techniques are often closely linked to the estimation techniques. The estimation results are used for checking the realizations and for a first estimate of the resource/reserve. Uncertainty over a large volume depends on the simultaneous uncertainty at many locations; simulating multiple realizations is the only practical approach to quantify such large-scale uncertainty. Also, the details of the geological heterogeneity may have a large influence on recovery and reserve calculations.

Eighth: post-process all of the model results. Sometimes, the statistical parameters from Steps 3 and 4 are useful in themselves; variogram ranges may be used to understand data spacing and expected length scales of geological features. The estimated model provides expected results at unsampled locations and measures of local uncertainty that are useful for data collection and management. Models of different grades must be combined and important resource variables calculated. The simulated models provide large scale uncertainty and input to subsequent engineering design.

These eight steps provide an overview of the workflow of resource estimation and geostatistics to address specific study objectives. Some of the details have been explained in preceding chapters; other details are inevitably learned by tradecraft, other textbooks, technical papers and software user's guides. Invariably, many assumptions are made during the course of a resource estimation study. The consequences of these assumptions and the limitations of the resulting models must be understood by the modeler and those using the resulting estimates.

15.2 Assumptions and Limitations of the Models Used

Healthy skepticism is encouraged. One illogical extreme viewpoint would be to accept resource models at face value since best practices were followed and significant cost was incurred for professionals and software. Another illogical extreme viewpoint would be to dismiss the resource models because of the large number of assumptions required. The famous statistician George E.P. Box wrote that "essentially all models are wrong, but some are useful." Healthy skepticism must be maintained while constructing the best resource models possible, and then using them for engineering design as required.

An important assumption relates to the reasonableness and correctness of the available data. There will be a variety of QA/QC procedures in place, but there are many possible sources of bias and error that may not be fully considered or accounted for. Moreover, we would assume that there is some geological continuity of those data values. We also assume that our geologic models are reasonable representations of the in-situ geology. And that based on the geologic model, the (stationary) domains defined are adequate for grade estimation. The geologic continuity is related to grade continuity, which we assume it is adequately captured with our variogram models. Most of our estimation/kriging techniques smooth the data and create models with relatively large areas of high and low grades. We assume that the data used to predict the degree of variability in the mineralization is adequate, and realistically represents for each domain the local and global variances. The engineers will assume that these models reasonably represent the reality and plan the details of a mining method; we assume that the final achieved ore/waste limits are similar in character to those predicted by early resource models. Of course, the local details are not as important as the overall assessment of dilution, lost ore and continuity.

Also, more detailed descriptions of the degree of checking should be made. Also, mining engineers and management should pre-define expected uncertainty (errors) in the predictions. For example, it is appropriate to define an acceptable error margin for the ore resource model for specified volumes, for example yearly or quarterly. All validation, checking, and reconciliation of the model can then refer back to the expected error for those volumes.

15.3 Documentation and Audit Trail Required

A major typical shortcoming in most resource models is lack of, or poor, documentation. It is significant because resource modeling can be a long and complex process, with many subjective decisions made along the way. The reserve model is in fact the most important asset on which the mining company bases its value. By extension, the resource model from which the reserves are obtained is the single most important asset a mining company has. As such, they are always subject to scrutiny.

Auditors are not the only ones that benefit from good documentation. The project or mine owner does as well, since the lessons learned from one modeling exercise can be better applied in future resource modeling iterations. But above all, ethics and transparency, as required by the current Reporting Standards, necessitate that all relevant aspects of the work be laid down in a comprehensive resource model report. The report must discuss and document what was done, the limitations of the work, and the degree of detail achieved.

Any auditor will strictly follow basic steps in checking a resource model. The resources estimator is well advised to be aware of them, and anticipate the documentation that will be required at a later date. The auditor's duty is to find flaws, errors, and inadequacies, which is the reason for an often intense level of scrutiny. The checks may include from the simplest graphical checks to verification of the database against original data collection/compilation documents, running an alternative check model, and bench-marking the software used in modeling against other software, often the auditor's own programs.

To cover (and pass) all possible checks, no assumptions can be made. H.M. Parker's basic auditing axioms simply and eloquently state the concept: (1) Trust no one; (2) Assume nothing; (3) Check everything.

A characteristic of resource estimation work is that it can be organized in modular form, since it a serial process. Therefore, documentation can also be arranged in such a way, and developed as work progresses. Assuming that the database has already been validated in its original repository, the process may begin with (1) loading the data into the modeling software; then, (2) data checks to ensure that the loading process was correct; then (3) geological interpretation and modeling takes place; (4) exploratory data analysis follows; (5) domain definition; (6) construction of the block model; (7) variography; (8) grade estimation; (9) resource classification; (10) resource validation; and finally (11) resource reporting.

A good practice is to lay down the procedure on a flowchart, which specifies the inputs and outputs of each step, and also the documentation required. The data files manipulation should be detailed, including the fields, and the associated scripts, run files, and programs that are used in each procedure. In those instances where the runs are for checking or validation purpose, both runs and run specifications (parameters and other input files) need to be preserved and archived.

All relevant variables need to be stored. If the kriging variance is used for resource classification, or the kriging pass, or the number of composites used, and so on. All scripts, legends, and data should be available for easy plotting to compare original data with estimated block grades. It is still important to have, on a large scale map, key sections and plans, properly signed by the operator, the internal quality control person, and the auditor.

For each general task in the resource modeling process, the following suggested audit trail and documentation can be considered adequate to satisfy an auditor in its request for data and documentation.

Database: Description of data fields and tables; description of prior audits and reviews; description of procedures, checks, validations (maintenance) that jointly ensure that the database is kept clean, and that new data is incorporated with the same standards.

Loading Data Into Modeling Software: Scripts or run files used to upload; is it an OBDC connection? Document the connection setup, the fields and files manipulation; and how the data is selected from the database.

Checking the Loaded Data: Document what checks have been made; why are they sufficient to ensure correct loading?

Geologic Modeling: Document and describe rationale for methodology chosen. Is it appropriate for this type of deposit? What criteria have been used to interpret geology? What checks have been made to ensure the quality of the output models?

Exploratory Data Analysis: Easy to access, organized archive of parameter files, run files, and plots. Use backup binders, not just electronic files. Document the assumptions and conclusions reached.

Estimation Domain Definition: Justification for logic; description of the process; supporting geology and statistical evidence; description of the sensitivities, if possible. Checks performed to confirm the decision.

Block Model: Documentation and description of model limits; block size(s), with partial or whole blocks; is it rotated? Was it built using a coordinate rotations and transformation? Project and world coordinates. How was the geology and estimation domains assigned?

Variography: Documentation of the estimators used; data transformations? Document parameters for obtaining the directional variograms, and the criteria used in modeling them. Document any data selection. By domain, or have domains been combined for practical reasons? Document all other assumptions made.

Grade Estimation: Document the method, with the parameters files used, and validations performed. Maintain all relevant files and ensure that the block model has all the necessary variables for checking.

Resource Classification: Document methodology. What criteria were used? How was it implemented? Document the checks performed.

Resource Validation: Document in detail, summarizing all the checks performed, and explaining why the resource model is deemed adequate for its objective. Include statements regarding the main assumptions, limitations, and risk areas perceived. What are the recommended risk mitigation procedures?

Resource Reporting: Document in detail the checks performed that the reported tonnages and grades correspond to the estimated model. Use appropriate number of significant digits. Include comparisons with previous models, and with reference (production models), and explain the reasons for the observed differences.

In summary, the audit trail has to demonstrate to third parties that each step of the process was completed with appropriate methodology, which was implemented correctly, and was thoroughly validated.

15.4 Future Trends

Trained professionals for resource estimation are hard to come by. The cyclic nature of the mining industry means that a relative lack of staff today likely means a glut tomorrow; however, it is still a general trend that there are relatively few highly trained people for resource estimation. In the last few years, during the recent and unprecedented non-renewable resources boom, this has become a major hurdle in project development. The larger mining companies are likely to outsource more work to consulting companies, but consulting companies also have challenges finding and retaining professionals. Moreover, there are few undergraduate programs that teach geostatistics or resource estimation; new professionals are not always willing to make the sacrifices that a solid professional development would require, and if so, much of the training is done on the job through postgraduate courses and mentoring.

There will be a trend toward increasing automation of many steps in resource estimation. This has many advantages including less professional time required, repeatability, transparency and easy updating with additional drilling. This also has many disadvantages. It will be easier to make mistakes, create models that are not geologically realistic and portray a false sense of confidence in the models because they appear realistic and appear to have used the latest available methodology and software. Senior professionals and gatekeepers in various organizations will guard against this and promote better work practices.

Regarding the data for resource estimation, it is likely that there will be improved use of secondary data coming from geophysical measurements of less expensive drilling and sampling techniques. Cokriging and other techniques to simultaneously use high quality data together with lower quality data will be developed and implemented in the widely used commercial software.

The use of multiple point statistics and advanced multivariate methods will be increased to permit improved realism in the spatial distribution of the models and capture the complex mineralogical relationships present in most of our deposits. Challenges posed by incomplete sampling of all the data and local variations in data quality will be met with improved techniques for data imputation and dealing with non-stationarity.

The longstanding challenge of non-stationarity will not go away, but improved techniques will become available to help subset the data appropriately, model the chosen domains, understand and quantify the nature of contacts and consider trends in the important variables.

We also feel that professionals (and the available software tools) will become better equipped to manage the high degree of uncertainty associated with resource estimation. Probabilistic models, scenarios, risk assessment and simulation techniques will become more commonplace and less in the hands of a few specialists.

References

Matheron G (1971) The theory of regionalized variables and its applications. In: Fasc 5, Paris School of Mines, p 212

Index

M. E. Rossi, C. V. Deutsch, *Mineral Resource Estimation,* DOI 10.1007/978-1-4020-5717-5,

MIX
Papier aus verantwortungsvollen Quellen
Paper from responsible sources
FSC® C105338

If you have any concerns about our products,
you can contact us on
ProductSafety@springernature.com

In case Publisher is established outside the EU,
the EU authorized representative is:
Springer Nature Customer Service Center GmbH
Europaplatz 3, 69115 Heidelberg, Germany

Printed by Libri Plureos GmbH
in Hamburg, Germany